Plant Biotechnology: Techniques and Applications

Plant Biotechnology: Techniques and Applications

Edited by Zoe Eastwood

SYRAWOOD
PUBLISHING HOUSE
New York

Published by Syrawood Publishing House,
750 Third Avenue, 9th Floor,
New York, NY 10017, USA
www.syrawoodpublishinghouse.com

Plant Biotechnology: Techniques and Applications
Edited by Zoe Eastwood

International Standard Book Number: 978-1-68286-715-0 (Hardback)

Cataloging-in-Publication Data

Plant biotechnology : techniques and applications / edited by Zoe Eastwood.
    p. cm.
Includes bibliographical references and index.
ISBN 978-1-68286-715-0
1. Plant biotechnology. 2. Agricultural biotechnology. 3. Plant genetic engineering.
I. Eastwood, Zoe.
SB106.B56 P53 2019
631.523 3--dc23

TABLE OF CONTENTS

PREFACE

This book was inspired by the evolution of our times; to answer the curiosity of inquisitive minds. Many developments have occurred across the globe in the recent past which has transformed the progress in the field.

Plant biotechnology is concerned with the application of the techniques of biotechnology to manipulate the characteristics of economically important plants for use in agriculture. Some of these characteristics are improved yield, disease and pest resistance, abiotic stress tolerance, etc. In recent years, various innovative methods of plant breeding have come into practice, like marker assisted selection, genetic modification, etc. Such techniques facilitate the introduction of new traits as well as control over a plant's genetic structure. This book is a valuable compilation of topics, ranging from the basic to the most complex advancements in the field of plant biotechnology. It brings forth some of the most innovative concepts and elucidates the unexplored aspects of this field. This book aims to serve as a reference to a broad spectrum of readers.

This book was developed from a mere concept to drafts to chapters and finally compiled together as a complete text to benefit the readers across all nations. To ensure the quality of the content we instilled two significant steps in our procedure. The first was to appoint an editorial team that would verify the data and statistics provided in the book and also select the most appropriate and valuable contributions from the plentiful contributions we received from authors worldwide. The next step was to appoint an expert of the topic as the Editor-in-Chief, who would head the project and finally make the necessary amendments and modifications to make the text reader-friendly. I was then commissioned to examine all the material to present the topics in the most comprehensible and productive format.

I would like to take this opportunity to thank all the contributing authors who were supportive enough to contribute their time and knowledge to this project. I also wish to convey my regards to my family who have been extremely supportive during the entire project.

Editor

H+-pyrophosphatase IbVP1 promotes efficient iron use in sweet potato [*Ipomoea batatas* (L.) Lam.]

Weijuan Fan[1], Hongxia Wang[1], Yinliang Wu[1], Nan Yang[1], Jun Yang[2] and Peng Zhang[1,*]

[1]*National Key Laboratory of Plant Molecular Genetics, CAS Center for Excellence in Molecular Plant Sciences, Institute of Plant Physiology and Ecology, Shanghai Institutes for Biological Sciences, Chinese Academy of Sciences, Shanghai, China*
[2]*Shanghai Key Laboratory of Plant Functional Genomics and Resources, Shanghai Chenshan Plant Science Research Center, Chinese Academy of Sciences, Shanghai Chenshan Botanical Garden, Shanghai, China*

*Correspondence
email zhangpeng@sibs.ac.cn*

Summary

Iron (Fe) deficiency is one of the most common micronutrient deficiencies limiting crop production globally, especially in arid regions because of decreased availability of iron in alkaline soils. Sweet potato [*Ipomoea batatas* (L.) Lam.] grows well in arid regions and is tolerant to Fe deficiency. Here, we report that the transcription of type I H+-pyrophosphatase (H+-PPase) gene *IbVP1* in sweet potato plants was strongly induced by Fe deficiency and auxin in hydroponics, improving Fe acquisition via increased rhizosphere acidification and auxin regulation. When overexpressed, transgenic plants show higher pyrophosphate hydrolysis and plasma membrane H+-ATPase activity compared with the wild type, leading to increased rhizosphere acidification. The *IbVP1*-overexpressing plants showed better growth, including enlarged root systems, under Fe-sufficient or Fe-deficient conditions. Increased ferric precipitation and ferric chelate reductase activity in the roots of transgenic lines indicate improved iron uptake, which is also confirmed by increased Fe content and up-regulation of Fe uptake genes, e.g. *FRO2*, *IRT1* and *FIT*. Carbohydrate metabolism is significantly affected in the transgenic lines, showing increased sugar and starch content associated with the increased expression of *AGPase* and *SUT1* genes and the decrease in β-amylase gene expression. Improved antioxidant capacities were also detected in the transgenic plants, which showed reduced H_2O_2 accumulation associated with up-regulated ROS-scavenging activity. Therefore, H+-PPase plays a key role in the response to Fe deficiency by sweet potato and effectively improves the Fe acquisition by overexpressing *IbVP1* in crops cultivated in micronutrient-deficient soils.

Keywords: sweet potato, iron acquisition, H+-pyrophosphatase, rhizosphere acidification, auxin regulation, carbohydrate metabolism.

Introduction

Iron (Fe) deficiency inhibits plant growth and reduces crop yields and quality, especially in alkaline and calcareous soils. Approximately 30% of the world's cultivated soil is susceptible to Fe deficiency due to severely limited bioavailability of Fe (Guerinot and Yi, 1994). In plants grown under Fe deficiency, the ferric chelate reductase (FCR) and H+-ATPase (PM-ATPase) activity are up-regulated in the plasma membrane. In addition, the expression of other Fe uptake genes is increased in response to altered levels of hormones (e.g. auxin; Schmidt et al., 2000; Santi and Schmidt, 2008, 2009; Bacaicoa et al., 2011; Kabir et al., 2013), carbohydrates (e.g. sucrose; Zargar et al., 2015; Lin et al., 2016) and reactive oxygen species (ROS; Molassiotis et al., 2006; Donnini et al., 2011), explaining the diverse effects of Fe deficiency on plant development and growth (Briat et al., 2015; Kobayashi and Nishizawa, 2012).

Sweet potato [*Ipomoea batatas* (L.) Lam.] is an important root crop ranking seventh in the annual production worldwide. It plays a key role in food security and nutritional intervention, especially in Africa and Asia (Bovell-Benjamin, 2007; Low, 2011). Sweet potato thrives in both fertile- and nutrient-deficient soils (Bovell-Benjamin, 2007; Woolfe, 1992). Most lands used for its cultivation are relatively infertile and lack bioavailable Fe (White and

Zasoski, 1999; Yan et al., 2006; Zuo and Zhang, 2011). Therefore, genetic improvement in sweet potato is essential to enhance tolerance to iron deficiency and increase iron content in storage roots (White and Broadley, 2009). Plants absorb iron from the soil via increased plasmalemma FCR activity in the roots. Fe absorption is also enhanced by PM-ATPase-mediated proton extrusion that acidifies the root apoplast and rhizosphere, as well as activation of high-affinity transport systems (Grusak and Pezeshgi, 1996; Jin et al., 2007, 2008, 2009). Unfortunately, no such study was conducted in sweet potato. Adamski et al. (2011, 2012) recently reported the effect of different iron concentrations on the morphological, anatomical, enzymatic, physiological and photosynthetic characteristics of sweet potato. It is necessary to study the response of sweet potato to Fe deficiency and develop useful approaches to increase Fe acquisition from the soil.

The H+-translocating inorganic pyrophosphatase (H+-PPase) plays a key role in plant energy metabolism by hydrolysing pyrophosphate (PPi) into sodium and/or proton gradients and transporting these ions across the membrane (Luoto et al., 2013). The hydrolytic activity balances the cytoplasmic pyrophosphate generated in various anabolic reactions such as DNA, RNA and protein synthesis in which PPi released as a by-product of ATP hydrolysis, and inhibits gluconeogenesis and cellulose synthesis (Baykov et al., 1999). Up-regulation of H+-PPase expression and activity has been reported in plants grown under abiotic stress conditions such as chilling, anoxia, mineral deficiency and salt stress (Maeshima, 2000). Up-regulated H$^+$-PPases induce plant tolerance to various stresses, such as drought, high salinity, and N and Pi deprivation (Arif et al., 2013; Paez-Valencia et al., 2013; Yang et al., 2007). It also affects auxin and sugar transport and accumulation during root and shoot growth (Gaxiola et al., 2012; Khadilkar et al., 2016; Li et al., 2005; Pasapula et al., 2011; Pizzio et al., 2015).

The H$^+$-PPase overexpression enhances apoplastic acidification by increasing the abundance and activity of H$^+$-ATPase at the plasma membrane (Gaxiola et al., 2016; Li et al., 2005; Undurraga et al., 2012). It increases root and shoot biomass, photosynthetic capacity and nutrient uptake under normal or nutrient-limited conditions, such as low NO_3^- and Pi (Khadilkar et al., 2016; Li et al., 2005; Lv et al., 2015; Paez-Valencia et al., 2013; Pizzio et al., 2015; Yang et al., 2007, 2014). Rhizosphere acidification is a principal mechanism in plant mineral nutrition, because it contributes to nutrient solubility and proton-motive force in the plasma membrane (Palmgren, 2001). Acidification also accelerates auxin transport to increase lateral root branching via up-regulated H$^+$-ATPase and FCR activity (Cho and Hong, 1995; Frías et al., 1996; Li et al., 2000; Zheng et al., 2003), and secretion of organic acids (Ruan et al., 2013; Yang et al., 2007), suggesting increased resilience to Fe deficiency. Nevertheless, the role of H$^+$-PPase in enhancing Fe absorption in plants is largely unknown.

In this study, we reported the functional characterization of a single-copy sweet potato H$^+$-PPase gene IbVP1 to improve Fe uptake and utilization via increased rhizosphere acidification. Overexpression of IbVP1 in sweet potato increases auxin transport, sugar content and ROS scavenging reflecting the intrinsic mechanisms of Fe acquisition in this important hexaploid species.

Results

IbVP1 is single-copied type I H$^+$-PPase gene in hexaploid sweet potato

The full-length cDNA of IbVP1 gene consists of 2644 nucleotides and encodes a predicted polypeptide of 737 amino acids with a molecular weight of 77.7 kDa and pI 5.84. Amino acid sequence alignment reveals that the IbVP1 shares high level of identity with H$^+$-PPases in other plant species, such as the dicot Arabidopsis thaliana (85.5%) and the monocot Oryza sativa (85.3%), which contain the highly conserved domains reported by Drozdowicz and Rea (2001) (Figure 1a). Further, phylogenetic analysis suggests that IbVP1 belongs to vacuolar H$^+$-pyrophosphatases and is clustered to type I (K$^+$-sensitive) H$^+$-PPases (Figure 1b). IbVP1 was shown to contain 13 conserved membrane-spanning domains using TMpred program for transmembrane prediction (Figure 1c). Southern blot of genomic DNA digested with EcoRV, XbaI, XhoI and HindIII revealed a single band when probed with the 5′-UTR of IbVP1, indicating that IbVP1 is a single-copied gene in the hexaploid species (Figure S1).

IbVP1 is ubiquitously expressed and induced by Fe deficiency and auxin in sweet potato

To determine the expression pattern in various tissues of greenhouse-grown sweet potato plant, qRT-PCR and semi-quantitative RT-PCR analyses were performed in the leaf, petiole, stem, fibrous root and storage root tissues at three different developmental stages: fibrous root (diameter <0.2 cm),

developmental root (0.2 cm < diameter <0.5 cm) and young storage root (diameter 0.5–1.0 cm), denoted by FR, DR and YSR (Figure 2a). IbVP1 transcription was detected in all tissues, with the highest expression in young storage root tissues (Figures 2a and S2a), indicating its role in storage root development.

In a 48-h Fe deficiency regimen using hydroponic growth system, the transcripts of IbVP1 in both roots and leaves were significantly induced by Fe deficiency, reaching a maximum at 12 h followed by a gradual decrease (Figures 2b and S2b). A fourfold increase in IbVP1 transcription in root was observed at 12 h, and twofold higher in leaf at 12 h, when compared with that at 0 h. Exogenous auxin treatment of sweet potato in a solution containing various concentrations of 3-indole acetic acid (IAA; 0, 50, 100, 200, 500 and 1000 μM) for 3 h also dramatically up-regulated the IbVP1 transcription (Figures 2c and S2c). A maximum transcript level (~17-fold that of control) was detected with 200 μM IAA (Figure 2c). A time-course study of IbVP1 expression (0, 1, 3, 6, 12, 24 and 48 h) in fibrous roots cultured in 200 μM IAA solution revealed that the maximum IbVP1 transcription level was detected at 6 h after treatment (Figures 2d and S2d), which was about 15-fold higher than the initial levels. These results suggest that the transcription of IbVP1 is extensively regulated in response to exogenous IAA. Taken together, the strong IbVP1 response in Fe deficiency and auxin suggests its critical role in plant response to Fe deficiency and plant growth.

IbVP1-overexpressed sweet potato shows enhanced V-PPase and PM-ATPase activities

At least nine independent transgenic lines of Taizhong 6 cultivar were generated using the Agrobacterium-mediated transformation and confirmed by Southern blot for T-DNA integration (Figure S3a,b) and by qRT-PCR for IbVP1 expression (Figure S3c). Three single-integrated transgenic lines IA4, IA7 and IA8 with the highest levels of expression showed normal phenotype and growth and were used in the following study.

Consistent with IbVP1 expression, the three transgenic plants overexpressing IbVP1 also showed higher V-H$^+$-PPase activities in the tonoplasts of the root apex than in the WT under normal conditions (Figure 3a). Further, the PM-H$^+$-ATPase hydrolytic activities in root microsomal fractions were also increased, at least 35% higher, in IA transgenic plants than in WT (Figure 3b), indicating the coupling effect between H$^+$-PPase and PM-H$^+$-ATPase activities (Undurraga et al., 2012).

Overexpression of IbVP1 promotes root development

Under normal conditions, the transgenic plants exhibited improved plant morphology compared with the WT plants (Figure 4a), noticeably in root. The IA transgenic plants produced more lateral roots (65% in average, Figure 4b) and longer lateral root length (95% in average, Figure 4c) compared with WT. To validate the increased root system following auxin flux in the transgenic plants, the expression of several key genes involved in auxin polar transport was analysed by qRT-PCR. Higher expression of PIN1a and PIN1b, the two genes encoding auxin efflux carriers, and AUX1, encoding an auxin influx carrier, were detected in IA transgenic plants compared with WT (Figure 4d). Accordingly, IAA content was significantly increased in the roots of IA transgenic plants (Figure 4e). No significant difference in leaf IAA content was found between IA plants and WT, indicating that the enhanced root growth was due to increased auxin transport from shoot into root in transgenic plants, which is consistent with

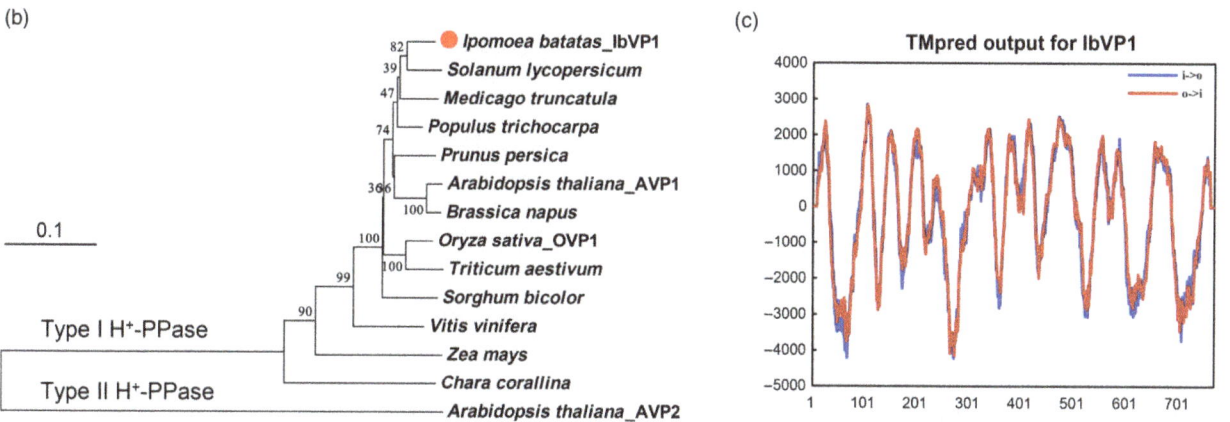

Figure 1 Amino acid sequence alignment and phylogenetic analysis of IbVP1 compared with other species. (a) Multiple alignment of the deduced amino acid sequences of H+-PPase proteins from *Ipomoea batatas* (IbVP1, AFQ00710.1), *Arabidopsis thaliana* (AVP1, NP_173021.1) and *Oryza sativa* (OVP1, BAA08232.1). Residues are highlighted in black and grey according to the level of conservation. The highly conserved motifs reported by Drozdowicz and Rea (2001) for H+-PPase proteins are boxed. (b) Phylogenetic tree of typical vacuolar H+-PPase proteins derived from various species, including *Medicago truncatula* (XP_003609463), *Brassica napus* (NP_001302829.1), *Populus trichocarpa* (XP_006381091.1), *Prunus persica* (AF367446_1), *Solanum lycopersicum* (BAM65603.1), *Sorghum bicolor* (BAM65603), *Triticum aestivum* (ABX10014.1), *Vitis vinifera* (NP_001268072.1), *Zea mays* (NP_001152459.1) and *Arabidopsis thaliana* (AVP2, ABX10014.1). (c) The predicted transmembrane domains of IbVP1 protein using TMpred program.

the previous observations of *AVP1*-overexpressing Arabidopsis plants (Li *et al.*, 2005).

Overexpression of *IbVP1* increases tolerance of transgenic plants to Fe deficiency

Under Fe deficiency, the IA lines showed better plant growth in solid medium with more greenish leaves and lateral roots, unlike the WT with obvious chlorotic leaves and weak roots. Even under Fe sufficiency, the differences in plant phenotype were obvious, with more lateral roots in the IA plants (Figure 5a). Increase in the

leaf and root biomass was detected in IA plants under both conditions (Figure 5b,c) and more significantly in roots under the Fe deficiency condition. Under adequate Fe levels, the leaf biomass increase in transgenic lines ranged from 10.5% to 21.1% in fresh weight (FW) and from 11.9% to 28.8% in dry weight (DW) compared with that of WT. In roots, the biomass increase in transgenic lines ranged from 35.4% to 93.2% in FW and from 14.8% to 18.5% in DW. Under Fe deficiency, the leaf biomass increase in transgenic lines ranged from 41.6% to 50.6% in FW and from 23.8% to 42.9% in DW compared with

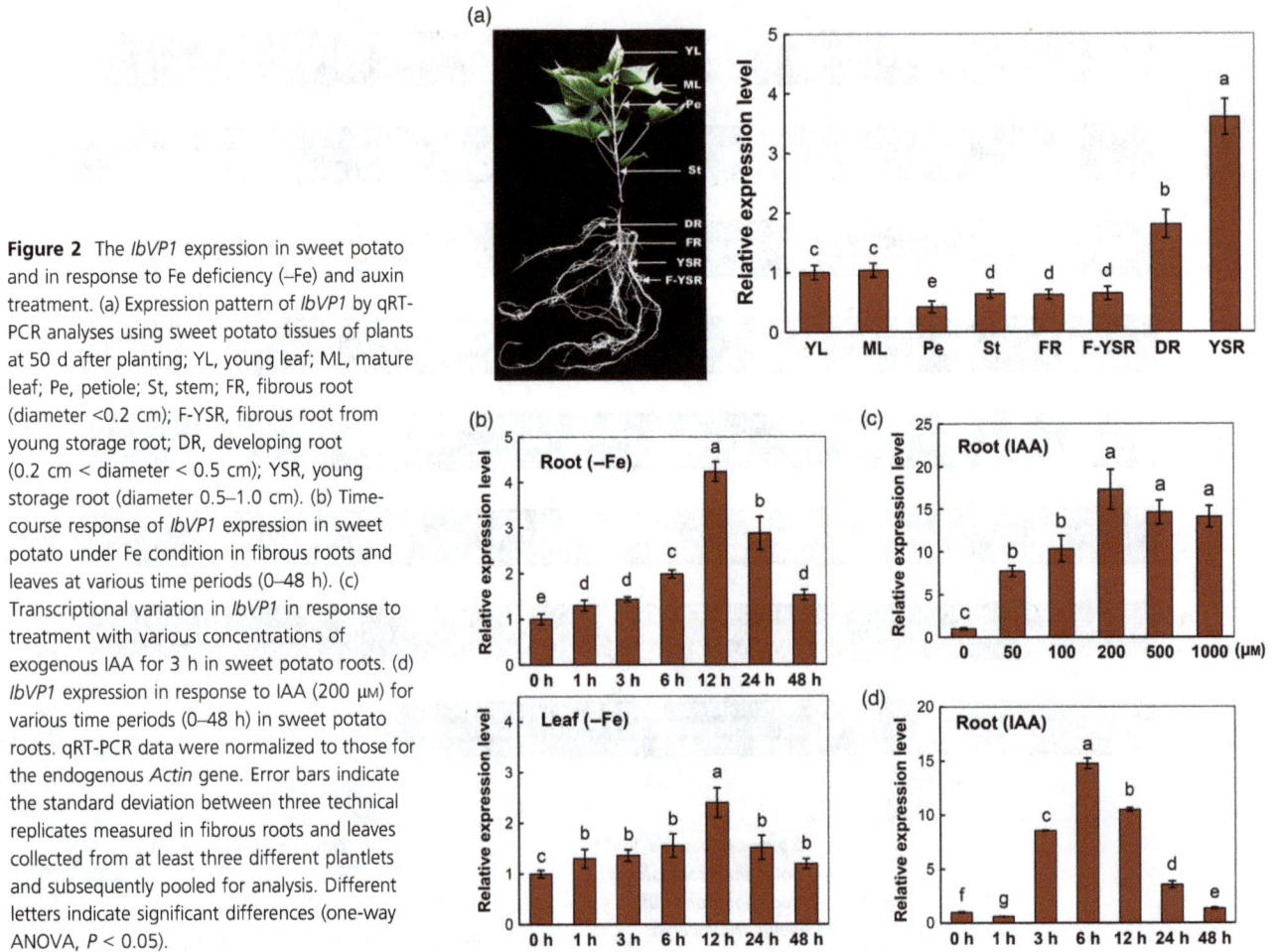

Figure 2 The *IbVP1* expression in sweet potato and in response to Fe deficiency (–Fe) and auxin treatment. (a) Expression pattern of *IbVP1* by qRT-PCR analyses using sweet potato tissues of plants at 50 d after planting; YL, young leaf; ML, mature leaf; Pe, petiole; St, stem; FR, fibrous root (diameter <0.2 cm); F-YSR, fibrous root from young storage root; DR, developing root (0.2 cm < diameter < 0.5 cm); YSR, young storage root (diameter 0.5–1.0 cm). (b) Time-course response of *IbVP1* expression in sweet potato under Fe condition in fibrous roots and leaves at various time periods (0–48 h). (c) Transcriptional variation in *IbVP1* in response to treatment with various concentrations of exogenous IAA for 3 h in sweet potato roots. (d) *IbVP1* expression in response to IAA (200 μM) for various time periods (0–48 h) in sweet potato roots. qRT-PCR data were normalized to those for the endogenous *Actin* gene. Error bars indicate the standard deviation between three technical replicates measured in fibrous roots and leaves collected from at least three different plantlets and subsequently pooled for analysis. Different letters indicate significant differences (one-way ANOVA, $P < 0.05$).

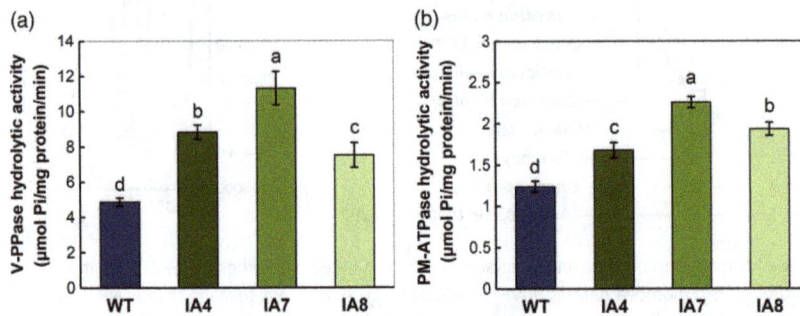

Figure 3 Hydrolytic activities of vacuolar H^+-PPase and PM-H^+-ATPase in sweet potato plants. (a) Vacuolar H^+-PPase activities of root tonoplast vesicles monitored by the release of Pi from PPi; (b) PM-H^+-ATPase activities in root microsomal fractions monitored by phosphate release. IA, *IbVP1* transgenic line; WT, wild-type plant. Data represent means ± SD of three independent assays. Different letters indicate significant differences (one-way ANOVA, $P < 0.05$).

that of WT. In roots, the biomass increase in transgenic lines ranged from 30.3% to 54.6% in FW and from 90.2% to 123% in DW (Figure 5b,c).

When the IA transgenic and WT plants were grown in hydroponic Fe sufficiency condition for 2 weeks and transferred to Fe-sufficient or Fe-deficient hydroponic solution for another 2 weeks, leaf chlorosis was observed in newly emerging WT leaves (Figure 6a). Approximately 2.3-fold-reduced SPAD reading was found in chlorotic leaves. The transgenic plants were only reduced about 37% compared with plants grown under

Fe-sufficient conditions (Figure 6b), which is also reflected by the significant decrease in total chlorophyll content (Figure 6b, bottom panel), including chlorophyll a and b (Figure 6c).

Overexpression of *IbVP1* promotes rhizosphere acidification and Fe acquisition

As indicated above, IA transgenic lines show enhanced V-PPase and PM-ATPase hydrolytic activities (Figure 3). Theoretically, rhizosphere acidification is increased in IA lines. Indeed, compared with WT, IA plants showed greater acidification zone indicated by

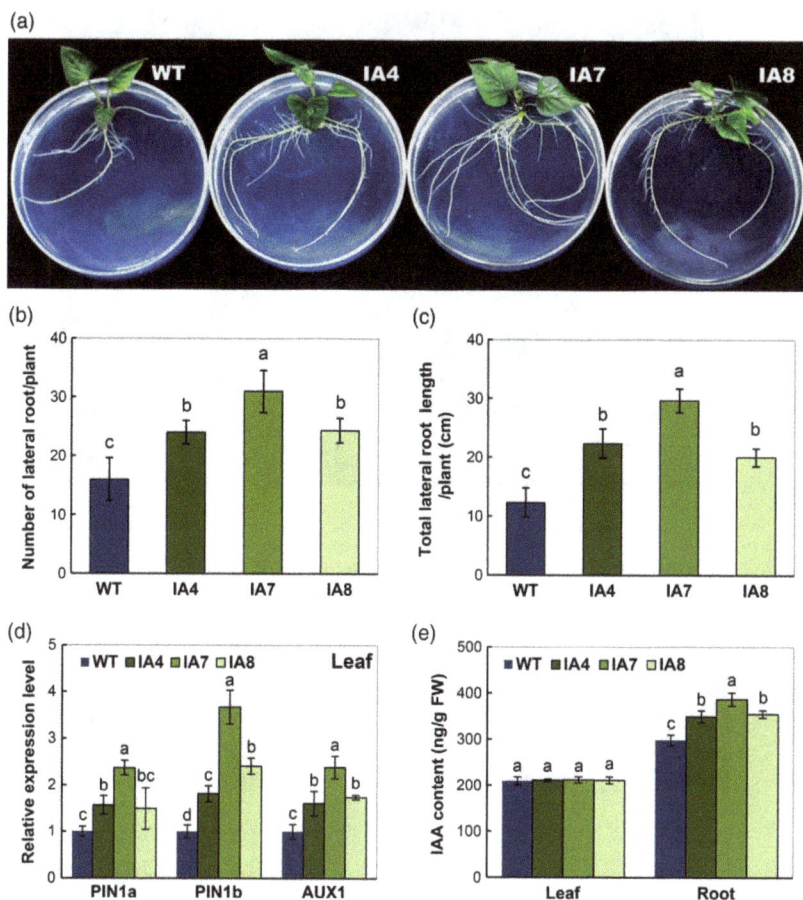

Figure 4 Growth status and auxin response in sweet potato plants. (a) Phenotypes in normal MS solid medium; (b, c) lateral root characters; (d) expression of auxin transport-associated genes in leaves; (e) indole-3-acetic acid (IAA) content of roots and leaves. IA, *IbVP1* transgenic line; WT, wild-type plant. Values represent means ± SD using three plants per line of three independent experiments. Different letters indicate significant differences (one-way ANOVA, P < 0.05).

the yellow colour in agar plates containing the pH indicator bromocresol purple for 4 h, regardless of iron levels (Figure 7a). Further, Perls' Prussian blue staining analysis of root Fe^{3+} precipitation also confirmed increased Fe^{3+} intensity in the IA plants (Figure 7b). The root FCR activity was increased in the IA lines under both Fe-sufficient and Fe-deficient conditions (Figure 7c).

Fe contents in the young leaves and roots of IA lines were significantly higher than in the WT (Figure 8a), under Fe-sufficient or Fe-deficient conditions. The average increase in Fe content of IA lines was 40.9% in leaves and 60.3% in roots compared with that of WT under Fe-deficient conditions. Notably, a ten-fold difference in Fe content was found between leaf and root. The highest Fe content was detected in the roots of IA7, reaching 2709 and 2135 µg/g (DW) under Fe sufficiency or Fe deficiency, respectively. Further, the expression of Fe acquisition genes that are involved in Fe uptake and proton release was also affected. The qRT-PCR analysis of *FRO2*, *IRT1*, *FIT* and H^+-ATPase gene *AHA2* revealed a dramatic increase in the expression of IA transgenic plants, grown under Fe deficiency (Figure 8b). All these findings suggested up-regulation of Fe uptake in the IA plants. These results indicate that *IbVP1* promoted Fe acquisition in sweet potato.

IbVP1-overexpression enhances antioxidant activity under Fe deficiency

As Fe is a cofactor of many antioxidant enzymes and generates reactive oxygen species (ROS) via Fenton reaction (Halliwell and Gutteridge, 1984), its deficiency leads to oxidative stress. Fe deficiency enhances H_2O_2 concentration and lipid peroxidation in roots of both IA transgenic and WT plants, particularly in the WT (Figure 9a). Compared with WT, the IA plants show less H_2O_2 concentration and lipid peroxidation. In parallel, the antioxidant activities of SOD, CAT and APX in IA lines are higher than in WT (Figure 9b), suggesting that overproduction of *IbVP1* leads to redox homoeostasis by up-regulating ROS scavenging, especially under Fe deficiency. Nevertheless, all plants showed increased SOD activity and reduced APX and CAT activity, indicating differential response of ROS-scavenging enzymes under Fe deficiency in sweet potato.

Altered carbohydrate metabolism in *IbVP1* overexpressed sweet potato

To enhance growth in the IA plants as stated previously (Figures 5–7), the glucose, fructose and sucrose levels in both leaves and roots, and starch content in leaves were measured. All the tissues were harvested at 12:00 pm. Irrespective of Fe levels, all transgenic plants contain relatively higher levels of all soluble sugars in leaves and roots compared with those in WT (Figure 10).

For example, under Fe-deficient conditions, the fructose content was 32.3–35.9 mg/g DW in the leaves of transgenic plants, 25%–39% higher than in WT (25.8 mg/g DW). The glucose levels increased greatly under treatment for Fe deficiency: from 33.2 to 36.4 mg/g DW in the IA lines, 14%–25% higher than in WT. Higher sucrose levels in leaves and roots of transgenic plants were also observed. The sucrose levels increased significantly under Fe-deficient conditions (Figure 10a,b). In leaves, relatively more starch was detected in the IA lines compared with WT (Figure 10c).

(a)

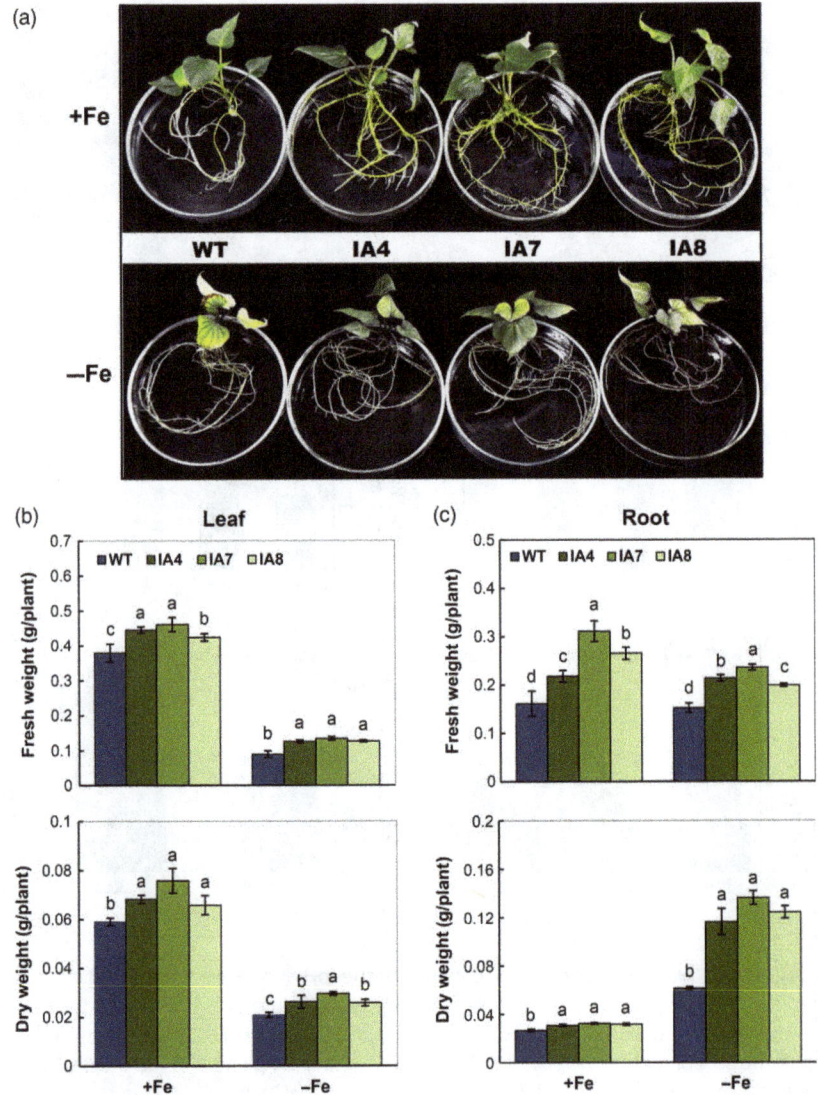

Figure 5 Growth response to Fe deficiency in sweet potato plants cultured *in vitro*. (a) Growth phenotype under exposure to Fe-sufficient (+Fe) or Fe-deficient (−Fe) medium after 15 days of culture; (b, c) fresh weights and dry weights of leaves (b) and roots (c). IA, *IbVP1* transgenic line; WT, wild-type plant. Values represent means ± SD using three plants per line of three independent experiments. Different letters indicate significant differences (one-way ANOVA, $P < 0.05$).

Further expression analysis of genes related to sugar and starch metabolism in roots showed significant up-regulation of *AGPase* and down-regulation of β-*amylase* in IA lines (Figure 11). Increased expression of *SUT1*, the sucrose transporter critical for cell-to-cell sucrose movement and sucrose phloem loading (Srivastava *et al.*, 2008; Wang *et al.*, 2015), was also demonstrated in the roots using qRT-PCR (Figure 11). Other genes, such as α-*amylase*, *INV1*, *INV2*, *SUS* and *SUT2*, were not affected in IA lines under either condition. These results indicate that carbon availability and allocation are affected by *IbVP1* overexpression in sweet potato.

Discussion

Since the discovery of membrane-bound H+-PPases in plants (Karlsson, 1975), intensive studies of biological function in stress tolerance and plant development have been conducted. Under stressful conditions, up-regulation of enzyme expression enables plant cells to use PPi as an energy source and maintain membrane integrity and intracellular transport via hydrolysis-mediated ion transport (Gaxiola *et al.*, 2016; Greenway and Gibbs, 2003; Stitt, 1998). Increased abiotic stress tolerance and enhanced growth performance in Pi or nitrogen deficiency were observed during H+-

PPase overexpression in Arabidopsis (Gaxiola *et al.*, 2001), tomato (Dong *et al.*, 2011; Park *et al.*, 2005; Yang *et al.*, 2007), tobacco (Khoudi *et al.*, 2012), rice (Zhang *et al.*, 2011; Zhao *et al.*, 2006), maize (Li *et al.*, 2008), lettuce (Paez-Valencia *et al.*, 2013) and cotton (Lv *et al.*, 2009; Pasapula *et al.*, 2011). Nevertheless, the role of H+-PPase-induced Fe uptake is unknown. In this study, we reported the role of *IbVP1* in improving Fe acquisition via increased rhizosphere acidification, carbohydrate metabolism and auxin regulation under Fe deficiency. Overexpression of *IbVP1* in sweet potato facilitates fortification of crop plants against Fe deficiency.

Sweet potato is mainly cultivated in semi-arid and arid lands that lack bioavailable Fe, especially in alkaline calcareous soils. For example, in China, sweet potato is mainly grown in the mountainous and infertile sandy lands, which are prone to Fe deficiency (Zuo and Zhang, 2011). Iron is essential for plant growth, especially in the early and middle stages, which are key phases of aerial growth. A level of 33 mg Fe/kg DW in the young leaves of sweet potato represents the threshold for iron deficiency (O'Sullivan *et al.*, 1997). Nevertheless, severe Fe deficiency in sweet potato is uncommon, possibly due to up-regulated H+-PPase expression, resulting in increased rhizosphere acidification and enhanced Fe acquisition, as shown in our study. Even under

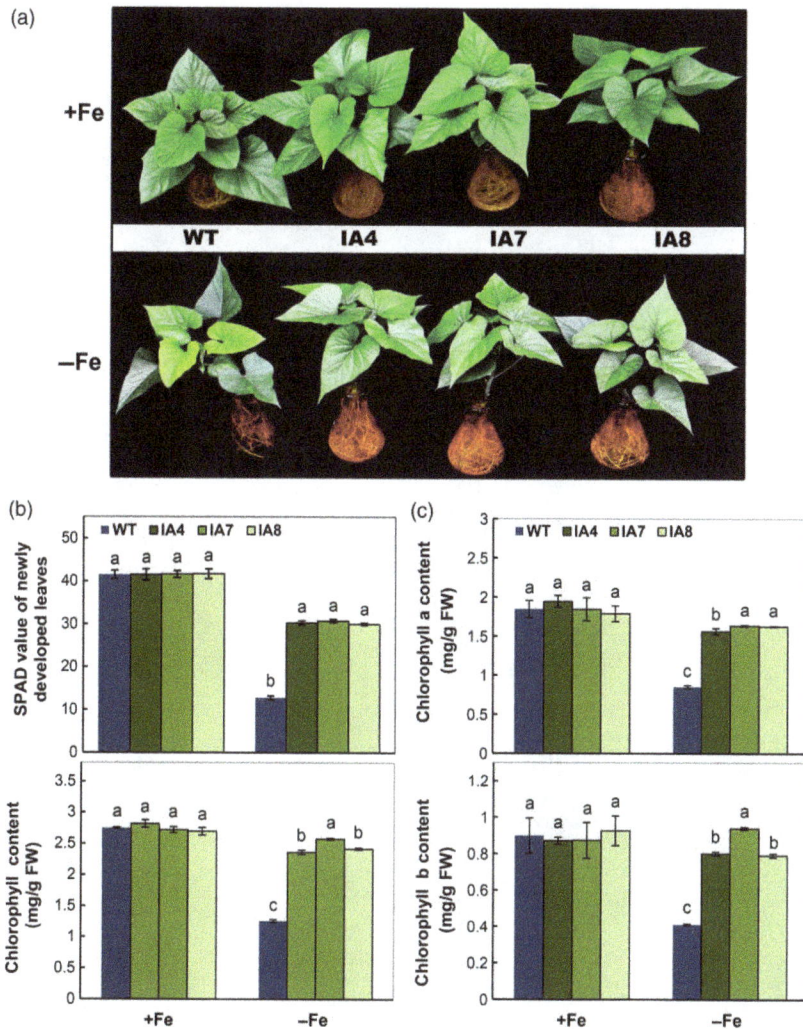

Figure 6 Growth and photosynthetic response to Fe deficiency in sweet potato plants in hydroponic culture. (a) Growth status under Fe-sufficient (+Fe) or Fe-deficient (−Fe) conditions; (b) SPAD index and total chlorophyll content of the newly developed leaves; (c) contents of chlorophyll a and chlorophyll b. IA, *IbVP1* transgenic line; WT, wild-type plant. Values represent means ± SD using three plants per line of three independent experiments. Different letters indicate significant differences (one-way ANOVA, *P* < 0.05).

Fe deficiency, the Fe content in WT roots is approximately 1300 mg Fe/kg DW, which is 10 times higher than in WT leaves. Overexpression of *IbVP1* in sweet potato grown under Fe deficiency restored the Fe levels to normal values found in WT grown under Fe sufficiency. Therefore, IbVP1 is one of the key contributors to Fe acquisition in the species.

H$^+$-PPase has been considered as a yield-enhancing factor (Gonzalez *et al.*, 2010; Khadilkar *et al.*, 2016). Increased biomass, especially root system, has been reported in plants overexpressing type I H$^+$-PPases, under deficiencies of phosphate (Gaxiola *et al.*, 2012; Pei *et al.*, 2012; Yang *et al.*, 2007, 2014) and nitrogen (Lv *et al.*, 2015; Paez-Valencia *et al.*, 2013). The type I H$^+$-PPase AVP1 triggers auxin efflux by up-regulating the expression of Pinformed 1 auxin efflux facilitator and P-adenosine triphosphatase, in Arabidopsis (Li *et al.*, 2005) and *AUX1*, *PIN1a* and *PIN1b* in maize (Pei *et al.*, 2012). In this study, the expression of *IbVP1* was also induced by auxin (IAA) treatment (Figure 2c,d). As auxin and its signalling pathways play a key function in plant response to Fe deficiency (Chen *et al.*, 2010), application of exogenous auxin increased the expression of Fe acquisition genes, such as *FCR* in several plant species (Jin *et al.*, 2011; Li *et al.*, 2000; Zheng *et al.*, 2003). The overexpression of *IbVP1* in sweet potato enlarged the root systems associated with increased IAA content in roots and significantly elevated the expression of auxin transport genes (i.e.

PIN1a, *PIN1b* and *AUX1*) (Figure 4). As a result, auxin polar transport was increased in these IA lines. These results establish a relationship between Fe deficiency, *IbVP1* expression, auxin flux and root development in *IbVP1*-overexpressing plants. Indeed, the expression of *IbVP1* was up-regulated by IAA treatment (Figure 2) explaining the high yield of sweet potato grown in Fe-deficient soils.

Enhanced rhizosphere acidification by increased PM-H$^+$-ATPase activity is strongly associated with up-regulated H$^+$-PPase activity, as demonstrated by the *IbVP1* overexpressing sweet potato under normal and Fe-deficient conditions (Figures 3 and 7) consistent with other reports (Paez-Valencia *et al.*, 2013; Yang *et al.*, 2007). Changes in the rhizosphere represent a central mechanism in plant mineral nutrition, contributing to nutrient solubility and proton-motive force (Li *et al.*, 2015; Palmgren, 2001; Palmgren, 1998; Santi and Schmidt, 2009; Zhu *et al.*, 2009). The increase in root iron precipitation, Fe content and FCR activity in the IA lines demonstrated regulation of Fe uptake, which is associated with the up-regulation of key genes such as *FRO2*, *IRT* and *FIT* involved in Fe acquisition (Figure 8).

Pyrophosphatases generate energy and release free Pi, which affects plant carbon metabolism (Gaxiola *et al.*, 2012; Geigenberger *et al.*, 1998). The altered partitioning of photoassimilate in tobacco and potato expressing an *Escherichia coli* cytosolic inorganic pyrophosphatase gene (Sonnewald, 1992) and the

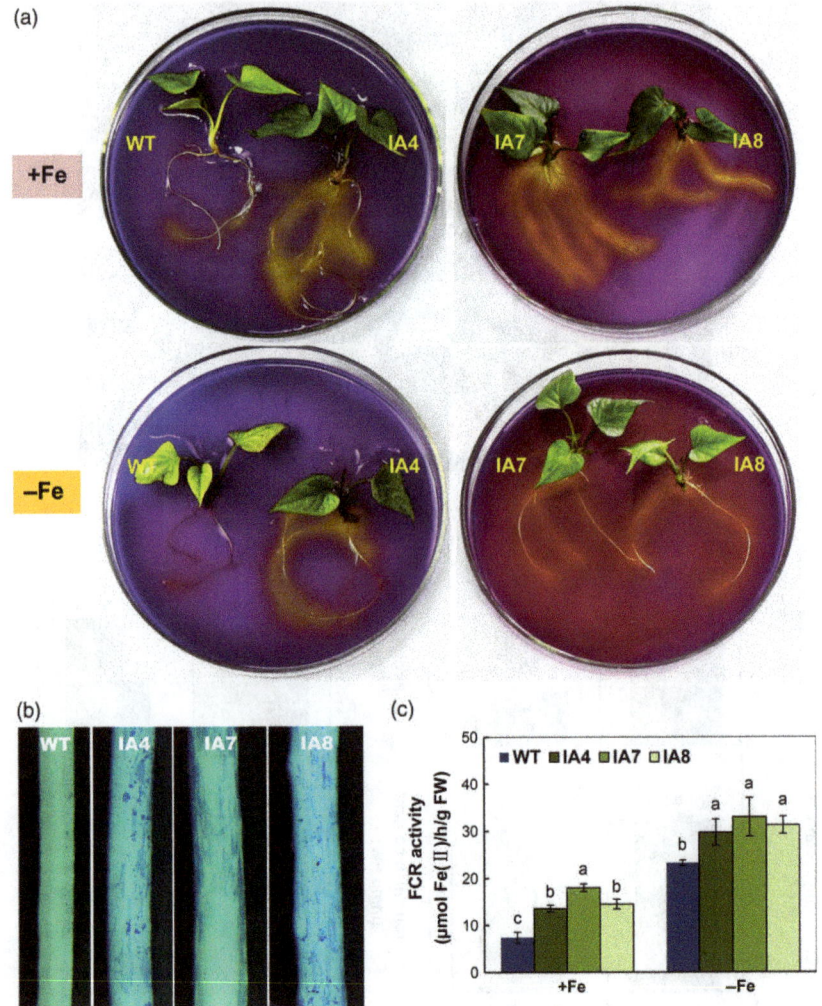

Figure 7 Fe acquisition analysis of sweet potato plants in response to Fe deficiency. (a) Rhizosphere acidification assay using agar plates containing bromocresol purple for 4 h; (b) root ferric precipitation assay of IA plants grown on Fe-sufficient (+Fe) medium with Perls' Prussian blue staining; (c) ferric chelate reductase (FCR) activity in the roots of plants grown under Fe-sufficient (+Fe) or Fe-deficient (−Fe) conditions. IA, *IbVP1* transgenic line; WT, wild-type plant. Values represent means ± SD using three plants per line of three independent experiments. Different letters indicate significant differences (one-way ANOVA, $P < 0.05$).

suppression of sucrose synthesis in the cotyledons of *avp* Arabidopsis mutant (Ferjani *et al.*, 2011) support the significant role of PPases in sucrose metabolism. The concentrations of glucose, fructose and sucrose were significantly increased in leaves of IA transgenic lines grown in Fe-sufficient or Fe-deficient conditions (Figure 10). Accordingly, sugar transporter gene *SUT1*, which plays a critical role in sucrose phloem loading and transport as well as cellular sugar partitioning, was up-regulated in the transgenic sweet potato (Figure 11). Further, the high expression of *IbVP1* in young storage root indicates an important function related to storage root development (Figure 2). These results further confirmed that up-regulation of H⁺-PPase promoted translocation of photosynthates from leaf to root, as hypothesized previously (Gaxiola *et al.*, 2012; Hermans *et al.*, 2006; Khadilkar *et al.*, 2016; Paez-Valencia *et al.*, 2011). A recent study also demonstrated that increased sucrose accumulation was required for the regulation of auxin-mediated Fe deficiency in plants (Lin *et al.*, 2016). We hypothesize that the root phenotype of *IbVP1*-overexpressing sweet potato is improved by increased auxin transport and enhanced sugar partitioning.

Fe deficiency in plants disrupts normal electron transfer in mitochondria and chloroplasts, resulting in overproduction of ROS and subsequent oxidative damage (Jelali *et al.*, 2014; Ranieri *et al.*, 2003; Tewari *et al.*, 2005; Vigani *et al.*, 2013). As an important cofactor, Fe is a constituent of antioxidant enzymes such as Fe-SOD, catalase (CAT), ascorbate peroxidase (APX) and peroxidases (POD), and acts as a pro-oxidant via the Fenton reaction (Ravet and Pilon, 2013). The activity of CAT or APX, both containing a heme group, was decreased under iron deficiency in plant species such as maize, sunflower, *Pyrus communis* and *Pisum sativum* (Donnini *et al.*, 2011; Jelali *et al.*, 2014; Ranieri *et al.*, 2003; Sun *et al.*, 2007). Therefore, enzyme activity is essential to maintain redox homeostasis in plants under stressful condition. Indeed, under Fe-deficient conditions, decreased APX and CAT activities and increased SOD activity were detected in sweet potato, leading to accumulation of H_2O_2. In the IA lines, their activities were significantly higher than in WT, especially APX and CAT, showing improved ROS-scavenging capacity (Figure 9), which is consistent with the previous reports (Jelali *et al.*, 2014). The results confirmed that a high Fe availability facilitates detoxification of cellular enzyme systems.

In summary, our study suggests that *IbVP1* regulates Fe acquisition in sweet potato, and its overexpression promotes tolerance of transgenic plants to Fe deficiency. It induces plant growth via altered carbohydrate metabolism, improved auxin polar transport and increased rhizosphere acidification. This study also provides a rationale for induction of tolerance to iron deficiency in sweet potato grown in arid soils and provides a useful strategy for Fe acquisition in crops.

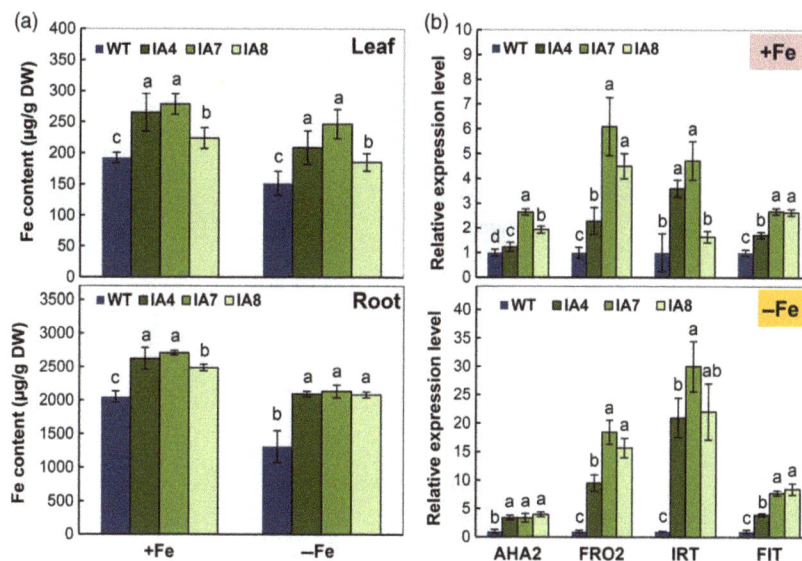

Figure 8 Iron accumulation and expression profile of Fe uptake genes in sweet potato grown under Fe-sufficient (+Fe) and Fe-deficient (−Fe) conditions. (a) Fe content in leaves and roots; values represent means ± SD using three plants per line in three independent experiments. Different letters indicate significant differences (one-way ANOVA, $P < 0.05$); (b) qRT-PCR analysis of altered expression of H+-ATPase gene *AHA2*, FERRIC REDUCTION OXIDASE2 (*FRO2*), IRON-REGULATED TRANSPORTER1 (*IRT1*) and FER-LIKE FE DEFICIENCY-INDUCED TRANSCRIPTION FACTOR (*FIT*). IA, *IbVP1* transgenic line; WT, wild-type plant. Values were normalized to those of the endogenous *Actin* gene. Error bars indicate standard deviation between three technical replicates measured in fibrous roots collected from at least three different sweet potato plantlets and subsequently pooled for analysis.

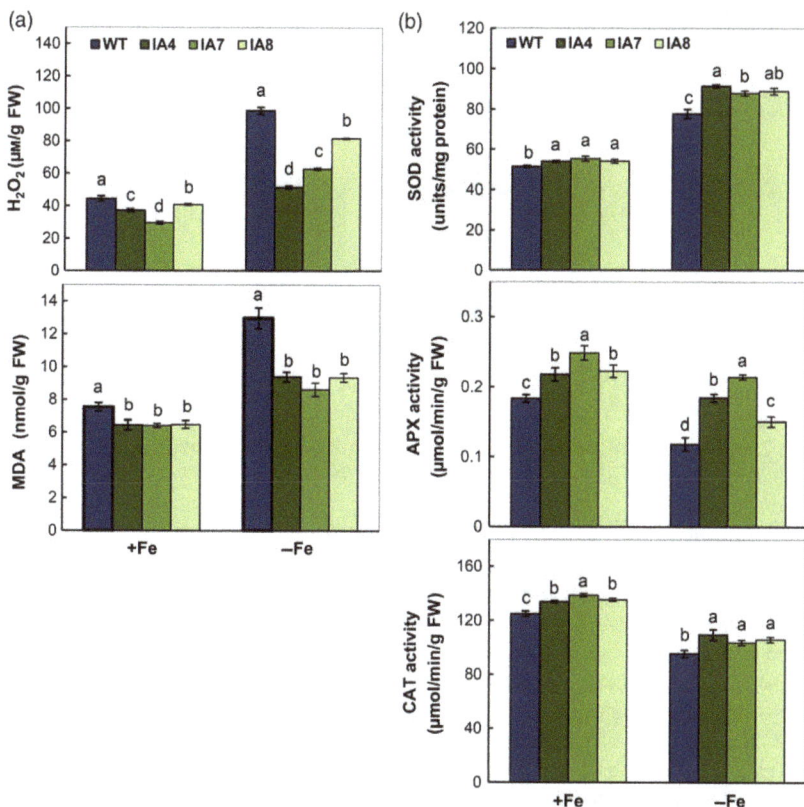

Figure 9 Antioxidant activity analysis in sweet potato plants under Fe-sufficient (+Fe) and Fe-deficient (−Fe) conditions. (a) H_2O_2 and malondialdehyde (MDA) content in the roots; (b) enzymatic activity of SOD, APX and CAT. IA, *IbVP1* transgenic line; WT, wild-type plant. Values indicate means ± SD using three plants per line of three independent experiments. Different letters indicate significant differences (one-way ANOVA, $P < 0.05$).

Experimental procedures

Plant materials, growth conditions and hydroponic treatments

The genome of sweet potato [*Ipomoea batatas* (L.) Lam.] Taizhong 6, a new farmer-preferred cultivar released at our centre, was recently sequenced and released into public domain (http://public-genomes-ngs.molgen.mpg.de/SweetPotato, Yang et al., 2016). The cultivar was used in the experiment. Sweet potato plants were propagated by sprouts or vine cutting in field. Apical stems bearing 2–3 leaves were planted in plastic pots containing well-mixed soil (soil : peat : perlite, 1 : 1 : 1) and

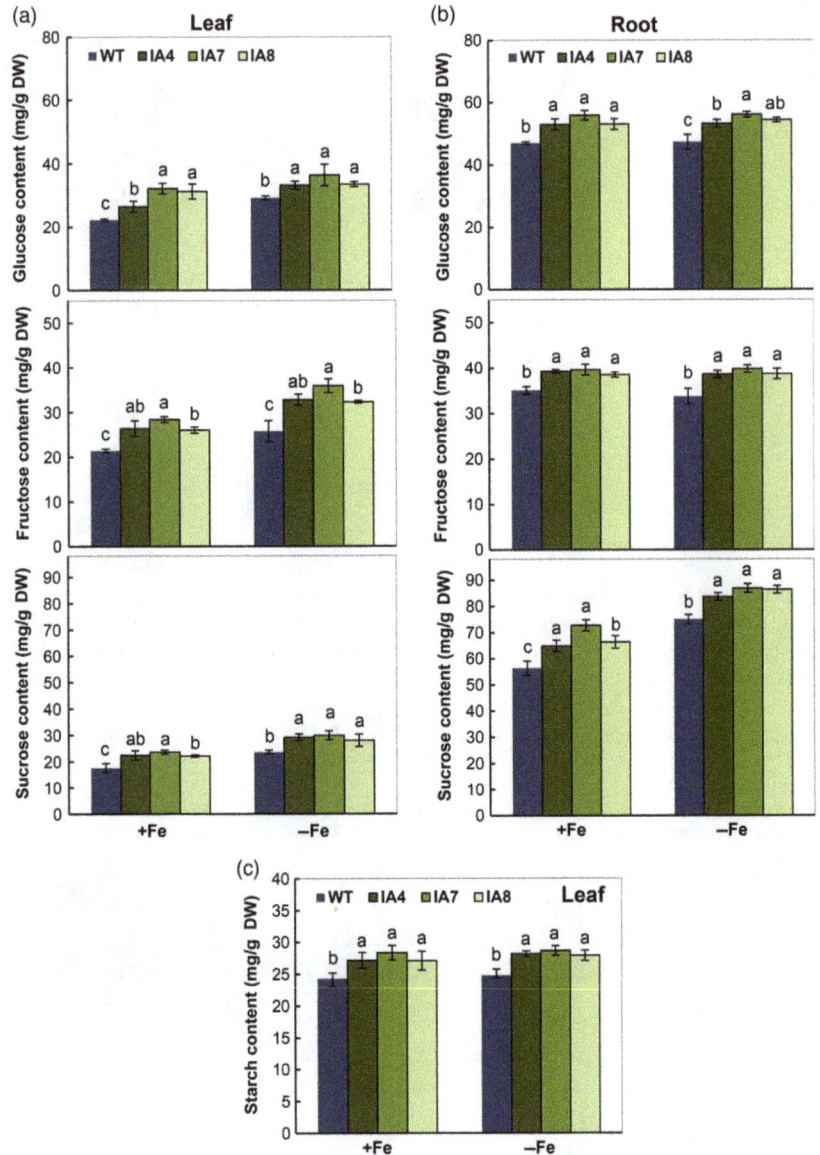

Figure 10 Changes in carbohydrate metabolism of sweet potato plants under Fe-sufficient (+Fe) and Fe-deficient (−Fe) conditions. (a, b) Glucose, fructose and sucrose levels in leaves (a) and roots (b); (c) starch content in leaves. IA, *IbVP1* transgenic line; WT, wild-type plant. Values represent means ± SD using three plants per line of three independent experiments. Different letters indicate significant differences (one-way ANOVA, $P < 0.05$).

grown in the greenhouse (16 h/8 h light/dark cycle, 25 °C day/night). *In vitro* shoot cultures were subcultured on SBM medium (MS salts including vitamins + 0.3 mg/L VB1 + 30 g/L sucrose, pH5.8) in plant growth chambers under a 16 h photoperiod provided by cool-white fluorescent tubes ($\sim$50 µmol/m^2/s), at 25 °C and 50% relative humidity (RH).

Apical stems bearing two leaves and a petiole (two-leaf plantlets) were harvested from sweet potato plants and incubated in brown flasks containing Fe-EDTA nutrient solutions (+Fe, pH 6.5) for 2 weeks. The Fe-EDTA nutrient solution had the following composition (in µM): Ca(NO$_3$)$_2$ (300), K$_2$SO$_4$ (50), MgSO$_4$ (50), NaH$_2$PO$_4$ (30), H$_3$BO$_3$ (3), MnCl$_2$ (0.5), ZnSO$_4$ (0.4), CuSO$_4$ (0.2), (NH$_4$)$_6$Mo$_7$O$_{24}$ (1) and Fe-EDTA (100). The solution was refreshed every 2 days. After the development of the fibrous roots from the distal end of the petiole, the plantlets were transferred to nutrient solution that contained 0 µM Fe-EDTA (−Fe, pH6.5) or hormone solutions for treatment. The plants were incubated with various concentrations (0, 50, 100, 200, 500 and 1000 µM) of 3-indole acetic acid (IAA, Sigma, St Louis, MO, USA; 13750-25G-A) at 25 °C in the dark for 3 h. The

plantlets were incubated in a 200 µM IAA solution at 25 °C in the dark for different durations (0–48 h).

IbVP1 cloning and sequence analyses

A sweet potato cDNA library was used to screen the H$^+$-PPase genes using *Arabidopsis thaliana* AVP1 cDNA probe (Genbank accession No. NM_101437). The full-length cDNA was acquired by re-sequencing the corresponding positive cDNA clones and submitted to NCBI under the Genbank accession No. JN688962.1. The deduced amino acid sequence of IbVP1 was aligned with H$^+$-PPases from other species. The molecular weight and isoelectric point (pI) of IbVP1 were predicted using the DNASTAR program (DNASTAR, Madison, WI). Sequence alignments were performed with ClustalW (http://www.ebi.ac.uk/clustalw, Chenna *et al.*, 2003). The phylogenetic relationship of the sequences was analysed using the neighbour-joining method with a bootstrap value of 500 replications under the Mega 6.0 program (Tamura *et al.*, 2013). Transmembrane domains were predicted by the TMpred program (www.ch.embnet.org/software/TMPRED_form.html).

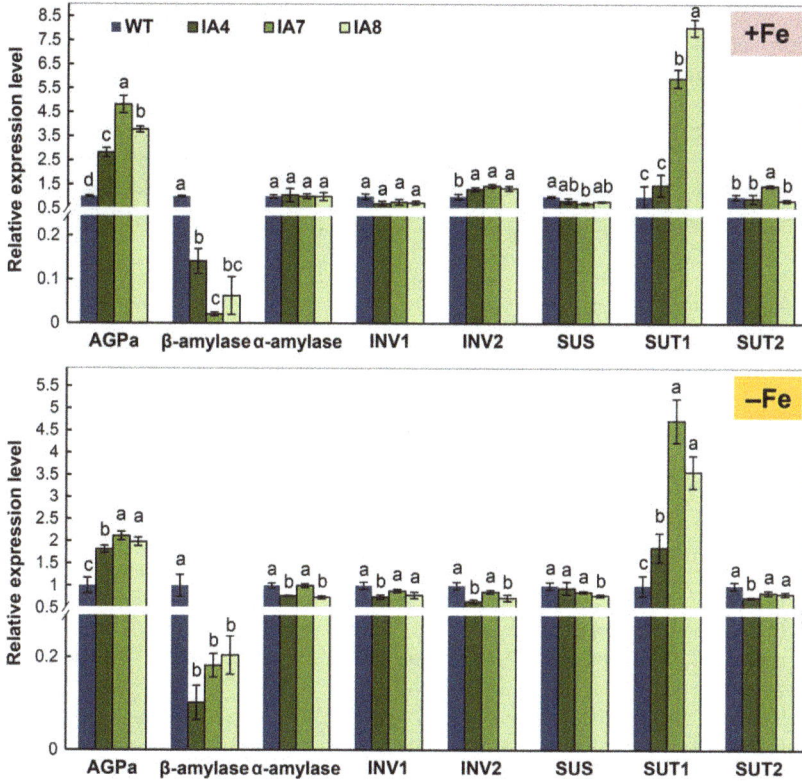

Figure 11 Expression patterns of sugar and starch metabolic genes in the roots of sweet potato plants under Fe-sufficient (+Fe) and Fe-deficient (−Fe) conditions. IA, *IbVP1* transgenic line; WT, wild-type plant. AGPa, ADP-glucose pyrophosphorylase; INV1, invertase 1; INV2, invertase 2; SUS, sucrose synthesis; SUT1, sucrose H^+-symporter; SUT2, sucrose transporter 2. qRT-PCR data were normalized to those of the endogenous *Actin* gene. Error bars indicate the standard deviation between three technical replicates measured in fibrous roots collected from at least three different sweet potato plantlets and subsequently pooled for analysis. Different letters indicate significant differences (one-way ANOVA, $P < 0.05$).

Production of *IbVP1*-overexpressed sweet potato plants

The PCR product of the full-size *IbVP1* cDNA targeting *Kpn*I and *Sal*I restriction sites at the ends of the forward and reverse primers, respectively, was subcloned into the pMD-18T vector (TakaRa Dalian Co. Ltd., Dalian, China). After digestion with *Kpn*I and *Sal*I, the *IbVP1* cDNA segment was inserted into the pCAMBIA1301-based plant expression vector to generate the binary vector pC1301-IbVP1. The *Agrobacterium tumefaciens* strain LBA4404 harbouring the pC1301-IbVP1 was used for sweet potato transformation. Shoot meristems from apical and axillary buds of sweet potato cultivar Taizhong 6 were used as material for embryogenic calli induction. Propagation, inoculation with *A. tumefaciens* strain and plant regeneration of embryonic calli were performed as described by Yang *et al.* (2011). Putative transgenic plants were monitored for rooting in the SBM medium supplemented with 10 mg/L hygromycin.

Southern blot and qRT-PCR analyses

Genomic DNA was isolated from plants cultured *in vitro* using the method described by Kim and Hamada (2005). Genomic DNA (20 μg) from wild type (WT) and transgenic lines was digested with corresponding restriction enzymes. After purification, the digested DNA samples were fractionated by 0.8% agarose gel and subsequently capillary transferred onto an Amersham Hybond N^+ nylon membrane (GE Healthcare Life Sciences, Indianapolis, IN). The probes of *IbVP1* and hygromycin phosphor-transferase gene (*HPT*) were labelled using a PCR DIG Probe Synthesis Kit (Roche Applied Science, Mannheim, Germany) with the primer pairs of *IbVP1* (VP1-F and VP1-R) and *HPT* (HPT-F and HPT-R) (Table S1). The hybridization with the DIG-labelled oligonucleotide probes and the immunological detection were performed according to the manufacturer's instructions using the

DIG-High Prime DNA Labeling and Detection Starter Kit II (Roche Applied Science, Mannheim, Germany).

Extracted total RNA from the sweet potato leaves and roots using the RNAprep Pure Plant Kit (Tiangen, Beijing, China) was treated with DNase and reverse transcribed with M-MLV Reverse Transcriptase RNaseH (Toyobo, Osaka, Japan). For the quantitative real-time reverse transcription polymerase chain reaction (qRT-PCR), specific primers of sweet potato genes (Table S1) were designed for analysing their expression levels using the SYBR Green PCR master mix (Toyobo, Osaka, Japan) in a Bio-Rad CFX96 thermocycler. Amplification conditions were as follows: 95 °C for 1 min, followed by 40 cycles of 95 °C for 15 s and 60 °C for 30 s. To calibrate the expression level, the sweet potato *Actin* gene was chosen as the internal control.

Semi-quantitative RT-PCR was performed under the following amplification conditions for the *IbVP1* fragment (using similar primers as in the qRT-PCR): initial denaturation at 94 °C for 5 min, followed by 30 cycles of 94 °C for 30 s, 60 °C for 30 s, 72 °C for 30 s and then a final extension at 72 °C for 10 min. Amplification conditions for the actin fragment were as follows: initial denaturation at 94 °C for 4 min, followed by 26 cycles of 94 °C for 30 s, 60 °C for 30 s, 72 °C for 30 s and followed by a final extension at 72 °C for 10 min. The PCR products were separated by 1.2% (w/v) agarose gel electrophoresis.

Vacuole isolation and vacuolar H^+-PPase activity measurement

For measurement of vacuolar H^+-PPase activity, sweet potato root tonoplast vesicles were isolated by sucrose density gradient ultracentrifugation according to the method of De Michelis and Spanswick (1986). Root tip segments (~20 g) were harvested and rinsed in cold distilled water, and then homogenized in 80 mL of ice-cold buffer containing 50 mM HEPES/Tris (pH 7.8),

10% glycerol, 2 mM EGTA, 0.25 M sucrose and 0.5% BSA (w/v). After filtration, precentrifugation of the homogenate was performed at 4200 g for 10 min to eliminate cell debris. The supernatant was then centrifuged at 50228 g for 2 h in a BECKMAN XL-70 (Beckman, Palo Alto, CA, USA). After gently resuspension, the microsomal suspension was layered directly onto discontinuous sucrose gradients and centrifuged at 100 000 g for 2 h; the collected pellets were resuspended. After centrifugation at 146 000 g for 1 h, the vesicle pellets were collected and resuspended in storage buffer containing 5 mM HEPES/Tris (pH 7.4), 2 mM DTT and 2 mM $MgSO_4$. V-H$^+$-PPase activities were determined by the release of Pi from PPi after incubation for 30 min at 25 °C. Each of the activity measurements used approximately 20 μg of protein. Pi determination was based on the method of Lin and Morales (1977). V-H$^+$-PPase hydrolytic activity was calculated as the difference in values measured in the presence and absence of 50 mM KCl (K$^+$-stimulated PPase activity).

Plasma membrane vesicle isolation and PM-ATPase assay

Plasma membrane vesicles were isolated from sweet potato roots using the phase partitioning method reported by Klobus and Buczek (1995). Sweet potato roots (10 g) were used for the isolation. PM-H$^+$-ATPase hydrolytic activity was measured as described by Janicka-Russak et al. (2008). Fifty-microlitre membrane vesicle preparations were added to 0.5 mL reaction solution containing ATP in a tube and were incubated at 37 °C for 20 min. Absorption at 660 nm (A660) was measured with Beckman Coulter DU730 (Fullerton, CA, USA) spectrophotometer. The net PM-ATPase A660 was calculated by subtracting the A660 with Na$_3$VO$_3$ from A660 without Na$_3$VO$_3$. The enzyme activity was calculated according to the protein and inorganic phosphorus content obtained and the reaction time (20 min).

Root ferric reductase activity, acidification capacity and Perls' Prussian blue staining

Ferric reductase activity was determined using the method of Grusak and Pezeshgi (1996). Briefly, 0.1 g of the whole excised root was placed in a tube filled with 10 mL of assay solution, which consisted of 0.5 mM CaSO$_4$, 0.1 mM BPDS, 0.1 mM MES and 100 μM Fe-EDTA at pH 5.5. During storage in a dark room at 25 °C for 2 h, the tubes were periodically hand-swirled at 20-min intervals. The absorbance of the assay solutions was measured at 535 nm, and the concentration of Fe (II) [BPDS]$_3$ was quantified using a standard curve.

Acidification capacity was detected using the method of Yi and Guerinot (1996). Briefly, one-week-old subculture plantlets grown on agar plastic jars were transferred to Fe-deficient medium for 3 d, and the plantlets were transferred to a 1% agar plate containing 0.006% bromocresol purple and 0.2 mM CaSO$_4$ (pH 6.5) for 4 h.

To localize Fe^{3+}, 2-week-old plantlets grown on Fe-sufficient medium were vacuum-infiltrated with fresh Perls' Prussian blue staining solution (equal volumes of 4% (w/v) potassium ferro-cyanide and 4% (v/v) HCl) for 30 min. After rinsing with water, plantlets were observed and photographed.

Chlorophyll content measurement

The chlorophyll content of young leaves was detected with the Minolta SPAD-502 leaf chlorophyll meter using the SPAD reading. To compensate for leaf variables, the transmittance at 940 nm

served as a reference, as chlorophyll concentration influences the absorption at 650 nm (Azia and Stewart, 2001). For spectrophotometric determination of chlorophyll content, the leaf samples (200 mg) collected from sweet potato plants were homogenized with 95% ethanol (v/v) and kept in dark for 2 days. After filtering through a filter paper, the homogenate was used for the total chlorophyll content determination as described by Arnon (1949).

Sugar and starch content analysis

Materials harvested at 12:00 pm from WT and transgenic plants were ground and baked at 80 °C for two days until a constant dry weight (DW). The dried sample 30 mg was dissolved in 0.7 mL of 80% ethanol, thoroughly vortexed and incubated at 70 °C for 2 h. Aliquots of 0.7 mL HPLC-grade water and 0.7 mL chloroform were added to the sample. After shaking several times, the mixtures were centrifuged at 12 000 g for 10 min. The pellet was analysed for starch content while the supernatant was composed of soluble sugar. The starch pellet was washed three times with 80% ethanol, and the total starch content was analysed using total starch kit (Megazyme International Ireland Limited, Wicklow, Ireland). The aqueous supernatant of 0.7 mL was transferred into 1.5-mL tube and mixed with 0.7 mL chloroform. After centrifugation at 12 000 g for 10 min, 0.5 mL of the supernatant was transferred to a glass tube for HPLC analysis of sugars. The sugar separation was carried out using the Agilent HPLC column (ZORBAX Carbohydrate column; 4.6 × 150 mm, 5 μm) with a differential refraction detector. The mobile phase consisted of 75% acetonitrile with a flow rate of 0.8 mL/min, and the column was kept at 35 °C. The sugar types were characterized according to the retention time of the standard references. The concentration in samples was calculated from the external standard curve.

IAA measurement

3-indole acetic acid was extracted as described by Pan et al. (2010) with modifications. Using a mortar and pestle, 0.5 g new fresh tissue from sweet potato plants was ground in fine powder in liquid nitrogen. After incubation with 5 mL extraction solvent (β-propanol:H$_2$O:concentrated HCl, 2 : 1 : 0.002, v/v/v) for 30 min at 4 °C under continuous shaking in the dark, 5 mL dichloromethane was added to each sample and vortexed for 30 min in a cold room at 4 °C. The samples were transferred to a refrigerated microcentrifuge at 4 °C and centrifuged at 13 000 g for 5 min. About 4.5 mL of the solvent from the lower phase was transferred into a screwcap vial using a Pasteur pipette, and the solvent mixture was concentrated (not completely dry) using N$_2$ gas. The samples were re-dissolved in 0.1 mL methanol. The IAA ELISA kit was used for IAA content measurement (IAA ELISA kit, Sigma, St Louis, MO, USA). The samples were measured four times, and the standard error was calculated.

Metal content assay

Sweet potato seedlings grown in hydroponics were washed as described (Gong et al., 2003) before sampling root and leaf tissues. The control plant tissues and plant tissues treated for Fe deficiency were digested in concentrated nitric acid for 5–7 days at room temperature, and the samples were boiled and completely digested. After dilution with Millipore-filtered deionized water, samples were briefly centrifuged. The diluted samples were measured using ICP-Mass spectrometer (ELAN DRC-e, Perkin Elmer, Toronto, ON, Canada).

Measurement of lipid peroxidation and H_2O_2 content

Lipid peroxidation in the leaf tissues (200 mg) from the sweet potato plants was assayed for malondialdehyde (MDA) as essentially described by Dhindsa and Matowe (1981). The H_2O_2 content was assessed using 1 g leaf tissues according to the method of Sairam and Srivastava (2002). All analyses were carried out in triplicate.

ROS-scavenging enzyme activity assays

Superoxide dismutase activity was assessed using the method of Beauchamp and Fridovich (1971), by measuring the photochemical inhibition of NBT at 560 nm. CAT activity was measured according to Xu et al. (2013) by monitoring the consumption of H_2O_2 at 240 nm for 4 min. The APX activity was determined according to Xu et al. (2014). Ascorbate oxidation was measured spectrophotometrically based on the decrease in absorption at 290 nm, using the absorption coefficient of 2.8/mM/cm. All measurements were conducted in triplicate.

Statistical analysis

All data were represented as mean ± SD from at least three independent experiments with three replicates each. Significant differences between treatments were analysed with one-way analysis of variance (ANOVA) using the program SigmaPlot 10.0 (Systat Software, San Jose, CA). A value of $P < 0.05$ was considered statistically significant difference.

Acknowledgements

We thank Ms Hongying Yi from SIPPE for ICP-MS analysis. This work was supported by grants from the National Natural Science Foundation of China (31501356), the International Science & Technology Cooperation Program of China (2015DFG32370) and the National Key Technology Research and Development Program of China (2015BAD15B01). The authors declare no conflict of interest.

References

Adamski, J.M., Peters, J.A., Danieloski, R. and Bacarin, M.A. (2011) Excess iron-induced changes in the photosynthetic characteristics of sweet potato. J. Plant Physiol. **168**, 2056–2062.

Adamski, J.M., Danieloski, R., Deuner, S., Braga, E.J.B., de Castro, L.A.S. and Peters, J.A. (2012) Responses to excess iron in sweet potato: impacts on growth, enzyme activities, mineral concentrations, and anatomy. Acta Physiol. Plant. **34**, 1827–1836.

Arif, A., Zafar, Y., Arif, M. and Blumwald, E. (2013) Improved growth, drought tolerance, and ultrastructural evidence of increased turgidity in tobacco plants overexpressing Arabidopsis vacuolar pyrophosphatase (AVP1). Mol. Biotechnol. **54**, 379–392.

Arnon, D.I. (1949) Copper enzymes in isolated chloroplasts – polyphenoloxidase in beta-vulgaris. Plant Physiol. **24**, 1–15.

Azia, F. and Stewart, K.A. (2001) Relationships between extractable chlorophyll and SPAD values in muskmelon leaves. J. Plant Nutr. **24**, 961–966.

Bacaicoa, E., Mora, V., Maria Zamarreno, A., Fuentes, M., Casanova, E. and Maria Garcia-Mina, J. (2011) Auxin: A major player in the shoot-to-root regulation of root Fe-stress physiological responses to Fe deficiency in cucumber plants. Plant Physiol. Biochem. **49**, 545–556.

Baykov, A.A., Cooperman, B.S., Goldman, A. and Lahti, R. (1999) Cytoplasmic inorganic pyrophosphatase. In Inorganic Polyphosphates (23) (Schröder, H.C., ed.), pp. 127–150. Berlin: Springer Verlag.

Beauchamp, C. and Fridovich, I. (1971) Superoxide dismutase – improved assays and an assay applicable to acrylamide gels. Anal. Biochem. **44**, 276–287.

Bovell-Benjamin, A.C. (2007) Sweet potato: a review of its past, present, and future role in human nutrition. In Advances in Food and Nutrition Research, Vol. 52 (Taylor, S.L. ed.), pp. 1–59.

Briat, J.-F., Dubos, C. and Gaymard, F. (2015) Iron nutrition, biomass production, and plant product quality. Trends Plant Sci. **20**, 33–40.

Chen, W.W., Yang, J.L., Qin, C., Jin, C.W., Mo, J.H., Ye, T. and Zheng, S.J. (2010) Nitric oxide acts downstream of auxin to trigger root ferric-chelate reductase activity in response to iron deficiency in Arabidopsis. Plant Physiol. **154**, 810–819.

Chenna, R., Sugawara, H., Koike, T., Lopez, R., Gibson, T.J., Higgins, D.G. and Thompson, J.D. (2003) Multiple sequence alignment with the Clustal series of programs. Nucleic Acids Res. **31**, 3497–3500.

Cho, H.T. and Hong, Y.N. (1995) Effect of IAA on synthesis and activity of the plasma-membrane H⁺-ATPase of sunflower hypocotyls, in relation to IAA-induced cell elongation and H⁺ excretion. J. Plant Physiol. **145**, 717–725.

De Michelis, M.I. and Spanswick, R.M. (1986) H⁺-pumping driven by vanadate sensitive ATPase in membrane vesicles from corns roots. Plant Physiol. **81**, 542–547.

Dhindsa, R.S. and Matowe, W. (1981) Drought tolerance in 2 mosses – correlated with enzymatic defense against lipid-peroxidation. J. Exp. Bot. **32**, 79–91.

Dong, Q.L., Liu, D.D., An, X.H., Hu, D.G., Yao, Y.X. and Hao, Y.J. (2011) MdVHP1 encodes an apple vacuolar H⁺-PPase and enhances stress tolerance in transgenic apple callus and tomato. J. Plant Physiol. **168**, 2124–2133.

Donnini, S., Dell'Orto, M. and Zocchi, G. (2011) Oxidative stress responses and root lignification induced by Fe deficiency conditions in pear and quince genotypes. Tree Physiol., **31**, 102–113.

Drozdowicz, Y.M. and Rea, P.A. (2001) Vacuolar H⁺ pyrophosphatases: from the evolutionary backwaters into the mainstream. Trends Plant Sci. **6**, 206–211.

Ferjani, A., Segami, S., Horiguchi, G., Muto, Y., Maeshima, M. and Tsukaya, H. (2011) Keep an eye on PPi: The vacuolar-type H⁺-pyrophosphatase regulates postgerminative development in Arabidopsis. Plant Cell, **23**, 2895–2908.

Frías, I., Caldeira, M.T., Pérez-Castiñeira, J.R., Navarro-Aviñó, J.P., Culiañez-Maciá, F.A., Kuppinger, O., Stransky, H. et al. (1996) A major isoform of the maize plasma membrane H⁺-ATPase: characterization and induction by auxin in coleoptiles. Plant Cell, **8**, 1533–1544.

Gaxiola, R.A., Li, J., Undurraga, S., Dang, L.M., Allen, G.J., Alper, S.L. and Fink, G.R. (2001) Drought -and salt-tolerant plants result from overexpression of the AVP1 H⁺-pump. Proc. Natl Acad. Sci. USA, **98**, 11444–11449.

Gaxiola, R.A., Sanchez, C.A., Paez-Valencia, J., Ayre, B.G. and Elser, J.J. (2012) Genetic manipulation of a "vacuolar" H⁺-PPase: from salt tolerance to yield enhancement under phosphorus-deficient soils. Plant Physiol. **159**, 3–11.

Gaxiola, R.A., Regmi, K., Paez-Valencia, J., Pizzio, G. and Zhang, S. (2016) Plant H⁺-PPases: reversible enzymes with contrasting functions dependent on membrane environment. Molecular Plant, **9**(3), 317–319.

Geigenberger, P., Hajirezaei, M., Geiger, M., Deiting, U., Sonnewald, U. and Stitt, M. (1998) Overexpression of pyrophosphatase leads to increased sucrose degradation and starch synthesis, increased activities of enzymes for sucrose-starch interconversions, and increased levels of nucleotides in growing potato tubers. Planta, **205**, 428–437.

Gong, J.M., Lee, D.A. and Schroeder, J.I. (2003) Long-distance root-to-shoot transport of phytochelatins and cadmium in Arabidopsis. Proc. Natl Acad. Sci. USA, **100**, 10118–10123.

Gonzalez, N., De Bodt, S., Sulpice, R., Jikumaru, Y., Chae, E., Dhondt, S., Van Daele, T. et al. (2010) Increased leaf size: different means to an end. Plant Physiol. **153**, 1261–1279.

Greenway, H. and Gibbs, J. (2003) Mechanism of anoxia tolerance in plants. II. Energy requirements for maintenance and energy distribution to essential processes. Funct. Plant Biol. **30**, 999–1036.

Grusak, M.A. and Pezeshgi, S. (1996) Shoot-to-root signal transmission regulates root Fe(III) reductase activity in the dgl mutant of pea. Plant Physiol. **110**, 329–334.

Guerinot, M.L. and Yi, Y. (1994) Iron – nutritious, noxious, and not readily available. Plant Physiol. **104**, 815–820.

Halliwell, B. and Gutteridge, J.M.C. (1984) Oxygen-toxicity, oxygen radicals, transition-metals and disease. Biochem. J. **219**, 1–14.

Hermans, C., Hammond, J.P., White, P.J. and Verbruggen, N. (2006) How do plants respond to nutrient shortage by biomass allocation? *Trends Plant Sci.* **11**, 610–617.

Janicka-Russak, M., Kabala, K., Burzynski, M. and Klobus, G. (2008) Response of plasma membrane H+-ATPase to heavy metal stress in *Cucumis sativus* roots. *J. Exp. Bot.* **59**, 3721–3728.

Jelali, N., Donnini, S., Dell'Orto, M., Abdelly, C., Gharsalli, M. and Zocchi, G. (2014) Root antioxidant responses of two *Pisum sativum* cultivars to direct and induced Fe deficiency. *Plant Biol.*, **16**, 607–614.

Jin, C.W., You, G.Y., He, Y.F., Tang, C.X., Wu, P. and Zheng, S.J. (2007) Iron deficiency-induced secretion of phenolics facilitates the reutilization of root apoplastic iron in red clover. *Plant Physiol.* **144**, 278–285.

Jin, C.W., Chen, W.W., Meng, Z.B. and Zheng, S.J. (2008) Iron deficiency-induced increase of root branching contributes to the enhanced root ferric chelate reductase activity. *J. Integr. Plant Biol.* **50**, 1557–1562.

Jin, C.W., Du, S.T., Chen, W.W., Li, G.X., Zhang, Y.S. and Zheng, S.J. (2009) Elevated carbon dioxide improves plant iron nutrition through enhancing the iron-deficiency-induced responses under iron-limited conditions in tomato. *Plant Physiol.* **150**, 272–280.

Jin, C.W., Du, S.T., Shamsi, I.H., Luo, B.F. and Lin, X.Y. (2011) NO synthase-generated NO acts downstream of auxin in regulating Fe-deficiency-induced root branching that enhances Fe-deficiency tolerance in tomato plants. *J. Exp. Bot.* **62**, 3875–3884.

Kabir, A.H., Paltridge, N.G., Roessner, U. and Stangoulis, J.C.R. (2013) Mechanisms associated with Fe-deficiency tolerance and signaling in shoots of *Pisum sativum*. *Physiol. Plant.* **147**, 381–395.

Karlsson, J. (1975) Membrane-bound potassium and magnesium ion-stimulated inorganic pyrophosphatase from roots and cotyledons of sugar beet. *Biochim. Biophys. Acta*, **399**, 356–363.

Khadilkar, A.S., Yadav, U.P., Salazar, C., Shulaev, V., Paez-Valencia, J., Pizzio, G.A., Gaxiola, R.A. et al. (2016) Constitutive and companion cell-specific overexpression of AVP1, encoding a proton-pumping pyrophosphatase, enhances biomass accumulation, phloem loading, and long-distance transport. *Plant Physiol.* **170**, 401–414.

Khoudi, H., Maatar, Y., Gouiaa, S. and Masmoudi, K. (2012) Transgenic tobacco plants expressing ectopically wheat H+-pyrophosphatase (H+-PPase) gene TaVP1 show enhanced accumulation and tolerance to cadmium. *J. Plant Physiol.* **169**, 98–103.

Kim, S.H. and Hamada, T. (2005) Rapid and reliable method of extracting DNA and RNA from sweet potato, *Ipomoea batatas* (L). Lam. *Biotechnol. Lett.* **27**, 1841–1845.

Klobus, G. and Buczek, J. (1995) The role of plasma-membrane oxidoreductase activity in proton transport. *J. Plant Physiol.* **146**, 103–107.

Kobayashi, T. and Nishizawa, N.K. (2012) Iron uptake, translocation, and regulation in higher plants. *Ann. Rev. Plant Biol.* **63**, 131–152.

Li, C.J., Zhu, X.P. and Zhang, F.S. (2000) Role of shoot in regulation of iron deficiency responses in cucumber and bean plants. *J. Plant Nutr.* **23**, 1809–1818.

Li, J.S., Yang, H.B., Peer, W.A., Richter, G., Blakeslee, J., Bandyopadhyay, A., Titapiwantakun, B. et al. (2005) Arabidopsis H+-PPase AVP1 regulates auxin-mediated organ development. *Science*, **310**, 121–125.

Li, B., Wei, A., Song, C., Li, N. and Zhang, J. (2008) Heterologous expression of the TsVP gene improves the drought resistance of maize. *Plant Biotechnol. J.* **6**, 146–159.

Li, S., Pan, X.-X., Berry, J.O., Wang, Y., Naren Ma, S., Tan, S. et al. (2015) OsSEC24, a functional SEC24-like protein in rice, improves tolerance to iron deficiency and high pH by enhancing H+ secretion mediated by PM-H+-ATPase. *Plant Sci.*, **233**, 61–71.

Lin, T.I. and Morales, M.F. (1977) Application of a one-step procedure for measuring inorganic-phosphate in presence of proteins – actomyosin atpase system. *Anal. Biochem.* **77**, 10–17.

Lin, X.Y., Ye, Y.Q., Fan, S.K., Jin, C.W. and Zheng, S.J. (2016) Increased sucrose accumulation regulates iron-deficiency responses by promoting auxin signaling in Arabidopsis plants. *Plant Physiol.* **170**, 907–920.

Low, J.W. (2011) Unleashing the potential of sweetpotato to combat poverty and malnutrition in sub-Saharan Africa through a comprehensive initiative. *Acta Horticult.* **921**, 171–179.

Luoto, H.H., Baykov, A.A., Lahti, R. and Malinen, A.M. (2013) Membrane-integral pyrophosphatase subfamily capable of translocating both Na+ and H+. *Proc. Natl Acad. Sci. USA*, **110**(4), 1255–1260.

Lv, S.L., Lian, L.J., Tao, P.L., Li, Z.X., Zhang, K.W. and Zhang, J.R. (2009) Overexpression of *Thellungiella halophila* H+-PPase (TsVP) in cotton enhances drought stress resistance of plants. *Planta*, **229**, 899–910.

Lv, S.L., Jiang, P., Nie, L.L., Chen, X.Y., Tai, F., Wang, D.L., Fan, P.X. et al. (2015) H+-pyrophosphatase from *Salicornia europaea* confers tolerance to simultaneously occurring salt stress and nitrogen deficiency in Arabidopsis and wheat. *Plant Cell Environ.* **38**, 2433–2449.

Maeshima, M. (2000) Vacuolar H+-pyrophosphatase. *Biochim. Biophys. Acta*, **1465**, 37–51.

Molassiotis, A., Tanou, G., Diamantidis, G., Patakas, A. and Therios, I. (2006) Effects of 4-month Fe deficiency exposure on Fe reduction mechanism, photosynthetic gas exchange, chlorophyll fluorescence and antioxidant defense in two peach rootstocks differing in Fe deficiency tolerance. *J. Plant Physiol.* **163**, 176–185.

O'Sullivan, J.N., Asher, C.J. and Blamey, F.P.C. (1997) *Nutrient Disorders of Sweet Potato*, Vol. 48. Canberra: Australian Centre for International Agricultural Research.

Paez-Valencia, J., Patron-Soberano, A., Rodriguez-Leviz, A., Sanchez-Lares, J., Sanchez-Gomez, C., Valencia-Mayoral, P., Diaz-Rosas, G. et al. (2011) Plasma membrane localization of the type I H+-PPase AVP1 in sieve element-companion cell complexes from *Arabidopsis thaliana*. *Plant Sci.* **181**, 23–30.

Paez-Valencia, J., Sanchez-Lares, J., Marsh, E., Dorneles, L.T., Santos, M.P., Sanchez, D., Winter, A. et al. (2013) Enhanced proton translocating pyrophosphatase activity improves nitrogen use efficiency in romaine lettuce. *Plant Physiol.* **161**, 1557–1569.

Palmgren, M.G. (2001) Plant plasma membrane H+-ATPases: powerhouses for nutrient uptake. *Ann. Rev. Plant Physiol. Plant Mol. Biol.* **52**, 817–845.

Palmgren, M.G. (1998) Proton gradients and plant growth: role of the plasma membrane H+-ATPase. *Adv. Bot. Res.* **28**, 1–70.

Pan, X., Welti, R. and Wang, X. (2010) Quantitative analysis of major plant hormones in crude plant extracts by high-performance liquid chromatography-mass spectrometry. *Nat. Protoc.* **5**, 986–992.

Park, S., Li, J.S., Pittman, J.K., Berkowitz, G.A., Yang, H.B., Undurraga, S., Morris, J. et al. (2005) Up-regulation of a H+-pyrophosphatase (H+-PPase) as a strategy to engineer drought-resistant crop plants. *Proc. Natl Acad. Sci. USA*, **102**, 18830–18835.

Pasapula, V., Shen, G.X., Kuppu, S., Paez-Valencia, J., Mendoza, M., Hou, P., Chen, J.A. et al. (2011) Expression of an Arabidopsis vacuolar H+-pyrophosphatase gene (AVP1) in cotton improves drought- and salt tolerance and increases fibre yield in the field conditions. *Plant Biotechnol. J.* **9**, 88–99.

Pei, L., Wang, J., Li, K., Li, Y., Li, B., Gao, F. and Yang, A. (2012) Overexpression of *Thellungiella halophila* H+-pyrophosphatase gene improves low phosphate tolerance in maize. *PLoS ONE*, **7**(8), e43501.

Pizzio, G.A., Paez-Valencia, J., Khadilkar, A.S., Regmi, K., Patron-Soberano, A., Zhang, S., Sanchez-Lares, J. et al. (2015) Arabidopsis type I proton-pumping pyrophosphatase expresses strongly in phloem, where it is required for pyrophosphate metabolism and photosynthate partitioning. *Plant Physiol.* **167**, 1541–1553.

Ranieri, A., Castagna, A., Baldan, B., Sebastiani, L. and Soldatini, G.F. (2003) H2O2 accumulation in sunflower leaves as a consequence of iron deprivation. *J. Plant Nutr.* **26**, 2187–2196.

Ravet, K. and Pilon, M. (2013) Copper and iron homeostasis in plants: the challenges of oxidative stress. *Antioxid. Redox Signal.* **19**, 919–932.

Ruan, L., Zhang, J., Xin, X., Miller, A.J. and Tong, Y. (2013) Elymus dahuricus H+-PPase EdVP1 enhances potassium uptake and utilization of wheat through the development of root system. *J. Soil Sci. Plant Nutrit.* **13**(3), 716–729.

Sairam, R.K. and Srivastava, G.C. (2002) Changes in antioxidant activity in sub-cellular fractions of tolerant and susceptible wheat genotypes in response to long term salt stress. *Plant Sci.* **162**, 897–904.

Santi, S. and Schmidt, W. (2008) Laser microdissection-assisted analysis of the functional fate of iron deficiency-induced root hairs in cucumber. *J. Exp. Bot.* **59**, 697–704.

Santi, S. and Schmidt, W. (2009) Dissecting iron deficiency-induced proton extrusion in Arabidopsis roots. *New Phytol.* **183**, 1072–1084.

Schmidt, W., Tittel, J. and Schikora, A. (2000) Role of hormones in the induction of iron deficiency responses in Arabidopsis roots. *Plant Physiol.* **122**, 1109–1118.

Sonnewald, U. (1992) Expression of *Escherichia-coli* inorganic pyrophosphatase in transgenic plants alters photoassimilate partitioning. *Plant J.* **2**, 571–581.

Srivastava, A.C., Ganesan, S., Ismail, I.O. and Ayre, B.G. (2008) Functional characterization of the Arabidopsis AtSUC2 sucrose/H$^+$ symporter by tissue-specific complementation reveals an essential role in phloem loading but not in long-distance transport. *Plant Physiol.* **148**, 200–211.

Stitt, M. (1998) Pyrophosphate as an energy donor in the cytosol of plant cells: an enigmatic alternative to ATP. *Bot. Acta*, **111**, 167–175.

Sun, B., Jing, Y., Chen, K., Song, L., Chen, F. and Zhang, L. (2007) Protective effect of nitric oxide on iron deficiency-induced oxidative stress in maize (*Zea mays*). *J. Plant Physiol.* **164**, 536–543.

Tamura, K., Stecher, G., Peterson, D., Filipski, A. and Kumar, S. (2013) MEGA6: molecular evolutionary genetics analysis version 6.0. *Mol. Biol. Evolut.* **30**, 2725–2729.

Tewari, R.K., Kumar, P., Neetu and Sharma, P.N. (2005) Signs of oxidative stress in the chlorotic leaves of iron starved plants. *Plant Sci.* **169**, 1037–1045.

Undurraga, S.F., Santos, M.P., Paez-Valencia, J., Yang, H.B., Hepler, P.K., Facanha, A.R., Hirschi, K.D. *et al.* (2012) Arabidopsis sodium dependent and independent phenotypes triggered by H$^+$-PPase up-regulation are SOS1 dependent. *Plant Sci.* **183**, 96–105.

Vigani, G., Zocchi, G., Bashir, K., Philippar, K. and Briat, J.F. (2013) Signals from chloroplasts and mitochondria for iron homeostasis regulation. *Trends Plant Sci.* **18**, 305–311.

Wang, L., Lu, Q., Wen, X. and Lu, C. (2015) Enhanced sucrose loading improves rice yield by increasing grain size. *Plant Physiol.* **169**, 2848–2862.

White, P.J. and Broadley, M.R. (2009) Biofortification of crops with seven mineral elements often lacking in human diets - iron, zinc, copper, calcium, magnesium, selenium and iodine. *New Phytol.* **182**, 49–84.

White, J.G. and Zasoski, R.J. (1999) Mapping soil micronutrients. *Field Crop. Res.* **60**, 11–26.

Woolfe, J.A. (1992) *Sweet Potato: An Untapped Food Resource.* New York: Cambridge University Press.

Xu, J., Duan, X.G., Yang, J., Beeching, J.R. and Zhang, P. (2013) Enhanced reactive oxygen species scavenging by overproduction of superoxide dismutase and catalase delays postharvest physiological deterioration of cassava storage roots. *Plant Physiol.* **161**, 1517–1528.

Xu, J., Yang, J., Duan, X.G., Jiang, Y.M. and Zhang, P. (2014) Increased expression of native cytosolic Cu/Zn superoxide dismutase and ascorbate peroxidase improves tolerance to oxidative and chilling stresses in cassava (*Manihot esculenta* Crantz). *BMC Plant Biol.* **14**(1), 208.

Yan, X., Wu, P., Ling, H., Xu, G., Xu, F. and Zhang, Q. (2006) Plant nutriomics in China: an overview. *Ann. Bot.* **98**, 473–482.

Yang, H., Knapp, J., Koirala, P., Rajagopal, D., Peer, W.A., Silbart, L.K., Murphy, A. *et al.* (2007) Enhanced phosphorus nutrition in monocots and dicots over-expressing a phosphorus-responsive type I H$^+$-pyrophosphatase. *Plant Biotechnol. J.* **5**, 735–745.

Yang, J., Bi, H.-P., Fan, W.-J., Zhang, M., Wang, H.-X. and Zhang, P. (2011) Efficient embryogenic suspension culturing and rapid transformation of a range of elite genotypes of sweet potato (*Ipomoea batatas* L. Lam.). *Plant Sci.* **181**, 701–711.

Yang, H., Zhang, X., Gaxiola, R.A., Xu, G., Peer, W.A. and Murphy, A.S. (2014) Over-expression of the Arabidopsis proton-pyrophosphatase AVP1 enhances transplant survival, root mass, and fruit development under limiting phosphorus conditions. *J. Exp. Bot.* **65**, 3045–3053.

Yang, J., Moeinzadeh, M-H., Kuhl, H., Helmuth, J., Xiao, P., Liu, G., Zheng, J. *et al.* (2016) The haplotype-resolved genome sequence of hexaploid *Ipomoea batatas* reveals its evolutionary history. *bioRxiv*, 064428, doi: http://dx.doi.org/10.1101/064428.

Yi, Y. and Guerinot, M.L. (1996) Genetic evidence that induction of root Fe(III) chelate reductase activity is necessary for iron uptake under iron deficiency. *Plant J.* **10**, 835–844.

Zargar, S.M., Kurata, R., Inaba, S., Oikawa, A., Fukui, R., Ogata, Y., Agrawal, G.K. *et al.* (2015) Quantitative proteomics of Arabidopsis shoot microsomal proteins reveals a cross-talk between excess zinc and iron deficiency. *Proteomics*, **15**, 1196–1201.

Zhang, J., Li, J., Wang, X. and Chen, J. (2011) OVP1, a vacuolar H$^+$-translocating inorganic pyrophosphatase (V-PPase), overexpression improved rice cold tolerance. *Plant Physiol. Biochem.* **49**, 33–38.

Zhao, F.Y., Zhang, X.J., Li, P.H., Zhao, Y.X. and Zhang, Y. (2006) Co-expression of the *Sueda salsa* SsNHX1 and Arabidopsis AVP1 confer greater salt tolerance to transgenic rice than the single SsNHX1. *Mol. Breed.* **17**, 341–353.

Zheng, S.J., Tang, C.X., Arakawa, Y. and Masaoka, Y. (2003) The responses of red clover (*Trifolium pratense* L.) to iron deficiency: a root Fe(III) chelate reductase. *Plant Sci.* **164**, 679–687.

Zhu, Y., Di, T., Xu, G., Chen, X., Zeng, H., Yan, F. and Shen, Q. (2009) Adaptation of plasma membrane H$^+$-ATPase of rice roots to low pH as related to ammonium nutrition. *Plant Cell Environ.* **32**, 1428–1440.

Zuo, Y. and Zhang, F. (2011) Soil and crop management strategies to prevent iron deficiency in crops. *Plant Soil*, **339**, 83–95.

Tomato facultative parthenocarpy results from Sl*AGAMOUS-LIKE 6* loss of function

Chen Klap[1,†], Ester Yeshayahou[1,†], Anthony M. Bolger[2], Tzahi Arazi[1], Suresh K. Gupta[1], Sara Shabtai[1], Björn Usadel[2,3], Yehiam Salts[1] and Rivka Barg[1,*]

[1]*The Institute of Plant Sciences, The Volcani Center, Agricultural Research Organization, Rishon LeZion, Israel*
[2]*Institut für Biologie I, RWTH Aachen, Aachen, Germany*
[3]*Institut für Bio-und Geowissenschaften 2 (IBG-2) Plant Sciences, Forschungszentrum Jülich, Jülich, Germany*

*Correspondence
email rivkabarg@gmail.com or
rivkab@volcani.agri.gov.il
†These authors contributed equally to this
work.

Keywords: ovary arrest, fruit set,
SlAGL6, tomato fruit size, CRISPR/
Cas9, Solyc01g093960.

Summary

The extreme sensitivity of the microsporogenesis process to moderately high or low temperatures is a major hindrance for tomato (*Solanum lycopersicum*) sexual reproduction and hence year-round cropping. Consequently, breeding for parthenocarpy, namely, fertilization-independent fruit set, is considered a valuable goal especially for maintaining sustainable agriculture in the face of global warming. A mutant capable of setting high-quality seedless (parthenocarpic) fruit was found following a screen of EMS-mutagenized tomato population for yielding under heat stress. Next-generation sequencing followed by marker-assisted mapping and CRISPR/Cas9 gene knockout confirmed that a mutation in Sl*AGAMOUS-LIKE 6* (SlAGL6) was responsible for the parthenocarpic phenotype. The mutant is capable of fruit production under heat stress conditions that severely hamper fertilization-dependent fruit set. Different from other tomato recessive monogenic mutants for parthenocarpy, Sl*agl6* mutations impose no homeotic changes, the seedless fruits are of normal weight and shape, pollen viability is unaffected, and sexual reproduction capacity is maintained, thus making Sl*agl6* an attractive gene for facultative parthenocarpy. The characteristics of the analysed mutant combined with the gene's mode of expression imply Sl*AGL6* as a key regulator of the transition between the state of 'ovary arrest' imposed towards anthesis and the fertilization-triggered fruit set.

Introduction

Fruit development following fertilization is critical for the completion of the plant life cycle. In tomato, the ovary, which develops in concert with the rest of the flower organs (growth phase I, according to Gillaspy *et al.*, 1993), ceases to undergo cell divisions shortly (1–2 days) before anthesis and enters an 'ovary arrest' state. Only if fertilization is successfully completed, a signal believed to be produced by the young embryo provokes the ovary to resume growth. This growth involves initially a phase of rapid cell division and expansion (designated phase II or 'fruit set') for 5–10 days (Bohner and Bangerth, 1988; Varga and Bruinsma, 1986), and subsequently (during phase III) growth is driven mainly by cell enlargement concomitant with nuclear polyploidization (Chevalier *et al.*, 2014, and references therein). Once reaching full size, ripening processes initiate.

The default programme of fertilization-dependent fruit development ensures that resources are not wasted sustaining purposeless fruit development, whereas parthenocarpy, that is fertilization-independent seedless fruit development, is a counterproductive trait in all the sexually reproducing plant species. Sexual reproduction entails that the hypersensitivity of the microsporogenesis process, and the mature male gametes to moderately high or low temperatures, and to extreme humidity or light intensity (El Ahmadi and Stevens, 1979; Mesihovic *et al.*, 2016; Picken, 1984; Sato *et al.*, 2006), is a major hindrance for year-round fertilization-dependent tomato yielding. Consequently, breeding for parthenocarpy is considered a valuable

goal especially in the context of maintaining sustainable agriculture in the face of global warming (Ariizumi *et al.*, 2013; Gorguet *et al.*, 2005; Ruan *et al.*, 2012). Other advantages of parthenocarpy relate to consumers' preference of seedless over seeded fruits, improved fruit quality due to elevated content of total soluble solids (TSS) (Carmi *et al.*, 2003; Casas Diaz *et al.*, 1987; Falavigna *et al.*, 1978; Ficcadenti *et al.*, 1999) and saving of energy invested in separating the seeds from processed products.

Since tomato and other vegetables that could benefit from parthenocarpy are commonly propagated from seeds, hence only genetic sources for facultative parthenocarpy, where seeded fruits can develop following successful fertilization (Varoquaux *et al.*, 2000), are of practical value. Presently, the most extensively characterized nontransgenic sources for facultative parthenocarpy in tomato are as follows: the three monogenic sources, *pat* (Beraldi *et al.*, 2004; presumably a mutated Solyc03g120910, Selleri, 2011; Soressi and Salamini, 1975), *procera* (a mutated Sl*DELLA*, Bassel *et al.*, 2008) and *entire* (mutated Sl*AUX/IAA9*, Mazzucato *et al.*, 2015; Saito *et al.*, 2011), all of which manifest undesired pleiotropic effects; and the three digenic sources, *pat-2* (Hazra and Dutta, 2010; Vardy *et al.*, 1989), IL5-1 and IVT-line 1 (Gorguet *et al.*, 2008), all manifesting acceptable parthenocarpic phenotype; and the inferior oligogenic source *pat-3/pat-4* (Nuez *et al.*, 1986; Philouze and Maisonneuve, 1978). Despite the importance of this trait, exploitation of these mutants in breeding programmes is still rather limited. Some of them are associated with mild or severely undesirable pleiotropic effects (e.g. Ariizumi *et al.*, 2013; Carrera *et al.*, 2012; Lin *et al.*, 1984; Mazzucato et al., 1998; Philouze, 1989). And introgression of a digenic source is much more laborious, especially since the identity of the genes underlying any of these three sources was not reported so far. Besides their applicative importance, parthenocarpic mutants are indispensable in the study of the mechanism underlying 'ovary arrest' at pre-anthesis and its fertilization-triggered release leading to fruit set.

In the present study, we demonstrate that mutated alleles of the MADS-box gene SlAGL6 enable tomato yielding under heat stress. The mutations confer facultative parthenocarpy manifested in the development of seedless fruits comparable in both weight and shape to wild-type (WT) seeded fruits and that without pleiotropic effects, thus making it a novel valuable source for parthenocarpy in tomato. The pivotal role of SlAGL6 in controlling the transition from the state of 'ovary arrest' to fertilization-triggered fruit set is also discussed.

Results

Line 2012 is a new monogenic recessive mutant for parthenocarpy

A chemically, EMS-mutagenized M_2 population generated in the M82 cultivar (generated by J. Hirschenhoren and Y. Kapulnik, The Volcani Center, ARO) was screened for mutants yielding under extremely high temperatures (Figure S3a), which prevented fertilization-dependent fruit set, as described in Data S1. Family No. 2012 included two plants that set high-quality parthenocarpic fruits with good jelly fill under these conditions, whereas the parental line set only tiny, hollow fruits, commonly dubbed 'nuts'.

One of these two plants served to pollinate emasculated flowers of M82 plant, which set seeded fruits. These BC_1 plants were not parthenocarpic. However, 7 out of 40 BC_1F_2 progenies set seedless fruits under the extremely hot conditions prevailing in the late summer (Figure S3b) when the parental line M82 managed to set tiny 'nuts' fruitlets only (Figure 1a vs. b). This indicated that the trait is governed by a single recessive mutation. Further, out of 30 plants from the same BC_1F_2 population grown in the winter (Figure S3c), in a nonheated glass house, nine set seedless fruits, whereas the rest of the siblings set no normal fruits, indicating that the mutation enables parthenocarpic fruit development also under temperatures too low to allow fertilization-dependent fruit set.

Mapping the 2012 mutation

To map the mutation, we chose to adopt the bulk segregation approach (Michelmore et al., 1991) and perform Illumina sequencing of two genomic libraries one coming from a bulk of 2012 BC_1F_2 parthenocarpic plants and the other from their nonparthenocarpic siblings. For this purpose, BC_1F_2 plants were grown again in the late summer when the day temperatures were high enough (Figure S3d) to seriously damage microsporogenesis and hence prevented fertilization-dependent fruit development. Under these conditions, at the date of harvest, the parthenocarpic plants produced high-quality red parthenocarpic fruits with good jelly fill, whereas the nonparthenocarpic (NP) siblings produced only small distorted, green puffy 'nuts' fruits, which weight was significantly smaller than that of the red fruits harvested from their parthenocarpic siblings (Figure 1c).

The genomic library coming from a pool of 20 BC_1F_2 plants characterized as 'clearly parthenocarpic' was designated '2012 library', and the other coming from a pool of 23 of their siblings

characterized as 'clearly nonparthenocarpic' was designated 'NP (nonparthenocarpic) library'. Bioinformatics analysis of the sequenced libraries was performed as detailed in Data S1. This analysis pointed to a segment of 3.85 million nucleotides in chromosome 1, spanning between SL2.5ch01:85115654 and 88965277, as the likely location of the mutation (Table 1). The analysis revealed nine homozygous mutated single nucleotide polymorphisms (SNPs) in this region in the '2012 library' that were heterozygous in the 'NP library', that is only 19–37% of the reads in this library showed the mutated alleles, while the others included the expected wild-type (WT) allele.

To restrict the location of the mutation underlying the 2012 mutant, we tested, in two segregating populations, for cosegregation of SNPs dispersed along the chromosomal interval suggested as the mutation location (Table 1, column 2), with the parthenocarpic phenotype. Cosegregation with six SNPs was examined in a testcross (TC) population (designed to segregate in a 1 : 1 ratio of parthenocarpic and nonparthenocarpic plants) which was allowed to set fruit under heat stress (Figure S3e). As shown in Table 2a, this analysis eliminated the candidacy of the mutations represented by SNP Nos. 1, 5 and 6 as they are not closely linked with parthenocarpy. To further zoom in on the location of the mutation, a 2012 BC_2F_2 population was similarly analysed for cosegregation of the parthenocarpic phenotype with the mutated version of SNP Nos. 2 and 3. This analysis, which is summarized in Table 2b, eliminated the candidacy of SNP No. 2, since six nonparthenocarpic plants were homozygous for its mutated version. Because of the incomplete linkage between the parthenocarpic phenotype and the mutated SNP No. 3, plants homozygous for mutated SNP No. 3 were genotyped also for SNP No. 4. Yet SNP No. 4 remained a less likely candidate because the three nonparthenocarpic plants were homozygous also for its mutated version (Table 2b). Furthermore, the progenies (BC_2F_3) of these three plants manifested clear parthenocarpy when allowed to set fruit in the winter, under suboptimal temperatures (Figure S3g), suggesting that the nonparthenocarpic phenotype of their parents reflects the facultative nature of the mutation.

Together, these analyses strongly suggested the mutated SlAGL6 (represented by SNP No. 3, see Table 1), as the gene underlying the parthenocarpic mutation 2012. SlAGL6 encodes for a MADS-box protein belonging to the type II lineage $MIKC^C$, subfamily AGL6, of the MADS-box transcription factor family (Smaczniak et al., 2012).

CRISPR/Cas9-mutated SlAGL6 confers parthenocarpy

To confirm that the mutated SlAGL6 is the gene underlying the 2012 parthenocarpy, CRISPR/Cas9 technology was exploited to knockout the SlAGL6 gene (Solyc01g093960). Synthetic gRNA was designed to target the second exon of SlAGL6 (Figure 2a). It was incorporated into a Cas9 expressing binary vector and transformed into tomato line MP-1.

Progenies of three R_0 plants designated sg1, sg4 and sg5 differing in the nature of their mutation (Data S1), but all leading to premature stop codon (Figure 2b), were chosen for analysis of their ability to set parthenocarpic fruits. R_1 plants genotyped as heterozygous (+/m) or homozygous (m/m) for mutated SlAGL6 were grown in a greenhouse, side by side with the parental line MP-1. Red fruits were picked, weighed and analysed for seeds bearing. As demonstrated in Figure 2c, heterozygous progenies produced seeded fruits only, whereas the progenies homozygous or bi-allelic for mutated versions of SlAGL6 set mostly parthenocarpic fruits, as well as a few underseeded fruits (bearing up to 10

(a)   (b)

(c)

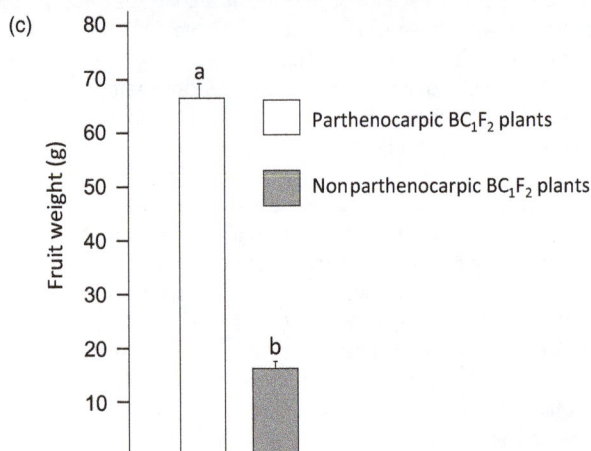

Figure 1 The 2012 mutation enables parthenocarpic tomato fruit development under heat stress. In BC_1F_2 population segregating for the 2012 mutation, when yielding in the late summer: (a) parthenocarpic siblings set high-quality seedless fruits, whereas (b) M82 plants set small distorted fruitlets. (c) Within the segregating BC_1F_2 population, the weight of the fruit harvested from 23 parthenocarpic plants was significantly higher than that from 24 nonparthenocarpic siblings (t-test, $P < 0.001$).

seeds), and fruits containing more than 10 seeds (defined as seeded fruits). Thus similar to the 2012 mutant, they are manifesting facultative parthenocarpy. Besides being seedless, the parthenocarpic fruits are similar in shape and jelly fill to those of the parental line fruits or their heterozygous seeded siblings (Figure S1).

The conclusive proof for allelism of the 2012 mutation and the CRISPR/Cas9-generated SlAGL6 mutants came from the parthenocarpic phenotype of most of the fruits produced on F_1 hybrid between sg1 plant homozygous for mutated SlAGL6 and 2012 plant homozygous for mutated SNP No. 3, whereas F_1 hybrid between MP-1 (+/+) and the same 2012 plant produced seeded fruits only (Figure 2d).

If the observed parthenocarpy is true ('vegetative'), namely requires no external trigger, it will set fruit under pollination restrictive conditions; if the mutant is 'stenospermocarpic', that is requires an external stimulus (provided, e.g., by damaged pollen

or the young embryos before their abortion), it implies that this parthenocarpy still relies on pollination, even though an unsuccessful one (Varoquaux et al., 2000). To distinguish between the two, flowers were emasculated at pre-anthesis and tested for fruit set. As shown in Figure 3a, all the emasculated flowers of the R_1 heterozygous progenies and MP-1 aborted, whereas over 60% of the emasculated flowers of mutated homozygous progenies set fruits, testifying to the true vegetative nature of the Slagl6-induced parthenocarpy. In agreement, the enlarged ovules collected from fruits developed from emasculated flowers are similar in size and appearance to those collected from nonemasculated parthenocarpic fruits, both of which are substantially smaller than normal seeds (Figure 3b).

Mutated SlAGL6 maintains fruit weight

In many cases, parthenocarpy was claimed to reduce fruit size compared to that of seeded fruits. As a first step towards

Table 1 Description of the SNPs along the predicted location of the 2012 mutation in chromosome 1. Bioinformatics analysis of two sequenced libraries, one representing nonparthenocarpic (NP) and the other parthenocarpic (2012) siblings from the BC_1F_2 population, provided the presented numbers of WT and mutated nucleotide reads for each of the nine SNPs in each of the two libraries. ORF – open reading frame

| SNP | SNP No.* | Position on Ch 1 (M82) SL2.50 | No. WT reads/No. mutated reads | | Number and annotation of the mutated gene (in parenthesis: position of the SNP) |
			NP library	2012 library	
I	1	85 115 654	52G/28A	0G/60A	Solyc01g091480; Armadillo repeat kinesin 2 (first intron)
II	2	85 400 236	59A/29C	0A/67C	Solyc01g091860, SET domain protein, possibly involved in peptidyl-lysine monomethylation (10th intron)
III	3	85 536 662	32G/25A	0G/53A	Solyc01g093960, Agamous-like MADS-box 6 (ORF premature stop)
IV		85 785 954	32G/25A	0G/53A	Solyc01g094230, Protein phosphatase-2C (ORF, silent)
V	4	86 070 972	55T/16C	0T/71C	Intergenic
VI		86 587 877	25G/6A	0G/55A	Solyc01g095250, Chitinase, Glycoside hydrolase X2 (ORF, mis-sense)
VII		87 367 821	22T/13A	0T/45A	Intergenic
VIII	5	88 007 968	61C/24T	0C/99T	Solyc01g097030, MUSTANG transposase Zn fingers (first intron)
IX	6	88 965 277	49A/26G	0A/65G	Intergenic
Total Distance (bp)		3 849 623			

*The corresponding no. of the SNP when genotyped by DYN R&D.

Table 2 Analysis of cosegregation of the 2012 parthenocarpic mutation with candidate SNPs. (a) The testcross (TC) population was genotyped for the six SNPs specified in Table 1 (column 2), in the few cases where the genotyping was inconclusive less than 96 results are presented. (b) The BC_2F_2 population was genotyped for SNP Nos. 2 and 3. The 126 plants homozygous for mutated version of SNP No. 3 were also genotyped for SNP No. 4. The three nonparthenocarpic plants homozygous for mutated SNP No. 3 are also homozygous for mutated SNP No. 4. The analysed populations are detailed in Data S1. m: mutated and +: WT versions of a SNP. N.T. – not tested

| SNP Site # | Distance between consecutive SNPs | Phenotype | No. of plants carrying the various SNP genotypes in: | | | | | |
| | | | (a) The TC population ($n = 96$) | | | (b) The BC_2F_2 population ($n = 498$) | | |
			m/m	+/m	+/+	m/m	+/m	+/+
1	0	Parthenocarpic	0	0	50	N.T	N.T	N.T
		Nonparthenocarpic.	0	0	46	N.T	N.T	N.T
2	284 582	Parthenocarpic	44	4	0	123	1	0
		Nonparthenocarpic	1	44	0	6	254	114
3	136 426	Parthenocarpic.	47	3	0	124	0	0
		Nonparthenocarpic	1	45	0	3	257	114
4	534 310	Parthenocarpic	46	3	0	120	3	0
		Nonparthenocarpic.	3	42	0	3	0	0
5	1 936 996	Parthenocarpic	44	5	0	N.T	N.T	N.T
		Nonparthenocarpic	5	41	0	N.T	N.T	N.T
6	957 309	Parthenocarpic	29	17	3	N.T	N.T	N.T
		Nonparthenocarpic	3	24	18	N.T	N.T	N.T
Total Distance (bp)			3 849 623			670 736		

assessing a possible penalty of the *SlAGL6* mutation on yielding parameters, its effect on fruit weight was examined in three different populations/genetic backgrounds: (i) the segregating 2012 BC_2F_2 population, in the background of the determinate cultivar M82, (ii) the segregating F_2 population derived from a cross between the big fruit semi-determinate cultivar Marmande and parthenocarpic 2012 plant, and (iii) R_1 progenies of the CRISPR/Cas9-derived line sg1 in the indeterminate MP-1 line background.

The effect of the mutation on fruit weight and yielding potential was assessed on the same 2012 BC_2F_2 plants used

to map the mutation (Table 2). This population grew under near-optimal temperatures for fertilization-dependent fruit setting (detailed in Data S1 and Figure S3f). Three and a half months after planting in the net house, all the red, breaker and mature green fruits were harvested from 11 to 14 plants carrying the following three alternative genotypes of SlAGL6: homozygous WT (+/+), heterozygous (+/m) or homozygous mutated (m/m) allele, as well as from eight M82 plants. As shown in Figure 4a, the yielding potential, expressed as the yield including mature green, breaker and red fruits, harvested from the plants homozygous for the mutated allele

(a)

(b) **Hypothetical translated products of the four Sl*AGL6* mutations tested**

AGL6 MGRGRVELKRIENKINRQVTFSKRRNGLLKKAYELSVLCEAEVALIIFSSRGKLYEFGSAGITKTLERYQRCCLNPQDNCGERETQSWYQEV...

2012 MGRGRVELKRIENKINRQVTFSKRRNGLLKKAYELSVLCEAEVALIIFSSRGKLYEFGSAGITKTLERYQRCCLNPQDNCGERETQSWY*

sg1   MGRGRVELKRIENKINRQVTFSKRRNGLLKKAYELSVLCEAEVALIIFSSRGKLYEFGSAELVPRGL*

sg4   MGRGRVELKRIENKINRQVTFSKRRNGLLKKAYELSVLCEAEVALIIFSSRGKLYEFGSAGITKTLERYQRLP*

sg5   MGRGRVELKRIENKINRQVTFSKRRNGLLKKAYELSVLCEAEVALIIFSSRGKLYEFGSAGITKTLERYQRVALILKTIVVKEKHRAGTKRSLN*

(c)

(d)

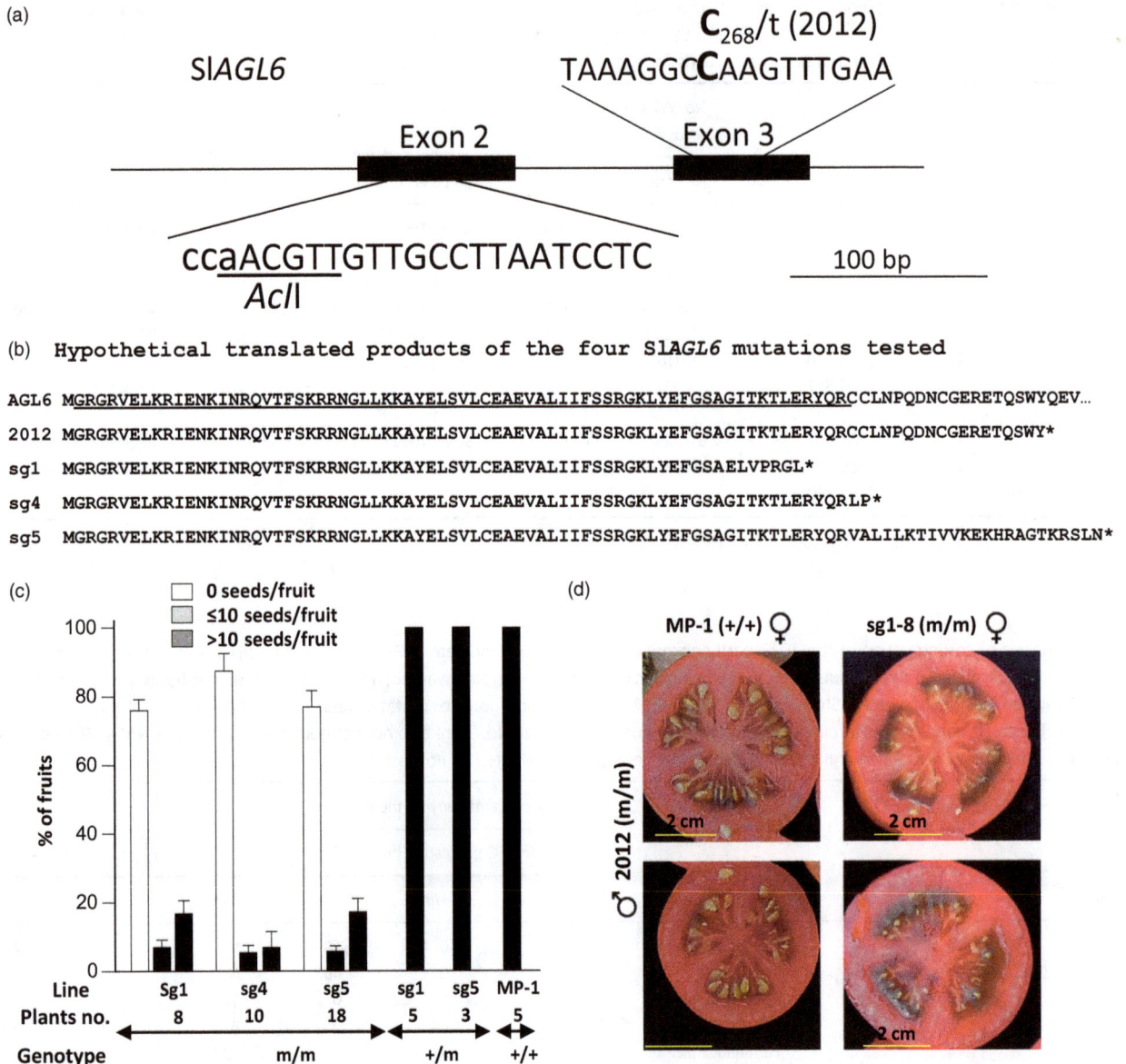

Figure 2 The CRISPR/Cas9-generated Sl*AGL6* mutations are parthenocarpic. (a) The chosen target guiding sequence for CRISPR/Cas9 modification in exon 2 of SlAGL6 is presented, the PAM (its reverse complement) is depicted in lower case letters, and the AclI restriction site expected to be destroyed by Cas9-induced mutations is underlined. The EMS-induced mutation 2012 in exon 3 is also presented. (b) The hypothetical translated products of the four SlAGL6 mutations tested: the EMS-induced 2012 and the CRISPR/Cas9-generated mutations sg1, sg4 and sg5. For the WT SlAGL6 protein, only the first 92 amino acids (AA) are presented, and the MADS-box domain is underlined. Grey coloured AAs differ from those of the WT protein. The asterisk (*) denotes premature stop codon. (c) The CRISPR/Cas9-derived SlAGL6 mutations sg1, sg4 and sg5 all lead to facultative parthenocarpy. Presented is the mean rate (%) $\pm$ SEM of seedless, underseeded and seeded fruits in plants homozygous, or heterozygous for the mutated allele, and in the parental line MP-1. (No heterozygous sg4 progenies were found among the screened ones). d) F_1 hybrid between plant homozygous for the 2012 mutated allele of SlAGL6 and plant homozygous for the sg1 mutated allele produced seedless fruits, whereas a hybrid between the same 2012 plant and MP-1 produced seeded fruits only, testifying to allelism of 2012 and sg1 mutated version of Sl*AGL6*.

does not differ from that of line M82, or the siblings either homozygous (+/+) or heterozygous (+/m) for the WT Sl*AGL6* allele. Further, the plants homozygous for the mutated allele were characterized by a profoundly earlier and more concentrated yielding, manifested in a significantly higher yield of red fruits at the date of harvest (Figure 4b, and d vs. e). The average weight of the red fruits developed on the (m/m) plants is significantly higher than that of the seeded red fruits harvested from the parental line M82, or its (+/+) and (+/m) siblings (Figure 4c).

In order to start estimating the potential of the 2012 mutation to support development of parthenocarpic fruits also when introduced into large fruit background, 2012 plant was crossed with the medium–large, multilocular fruit, semi-determinate open variety Marmande (www.rareseeds.com/marmande-tomato/), and the effect of the mutation on parthenocarpic fruit weight was estimated in the segregating F_2 population (described in Data S1). Fruit weight is governed by several QTLs (Tanksley, 2004), and in segregating F_2 population, the average fruit weight is always similar to that of the small fruit parent, with only small

(a)

Line	SlAGL6 genotype	No. emasc.	No. set fruits	No. aborted
MP-1	+/+	10	0	10
sg1, sg5	+/m	25	0	25
Sg1, sg4, sg5	m/m	22	14	8

(b)

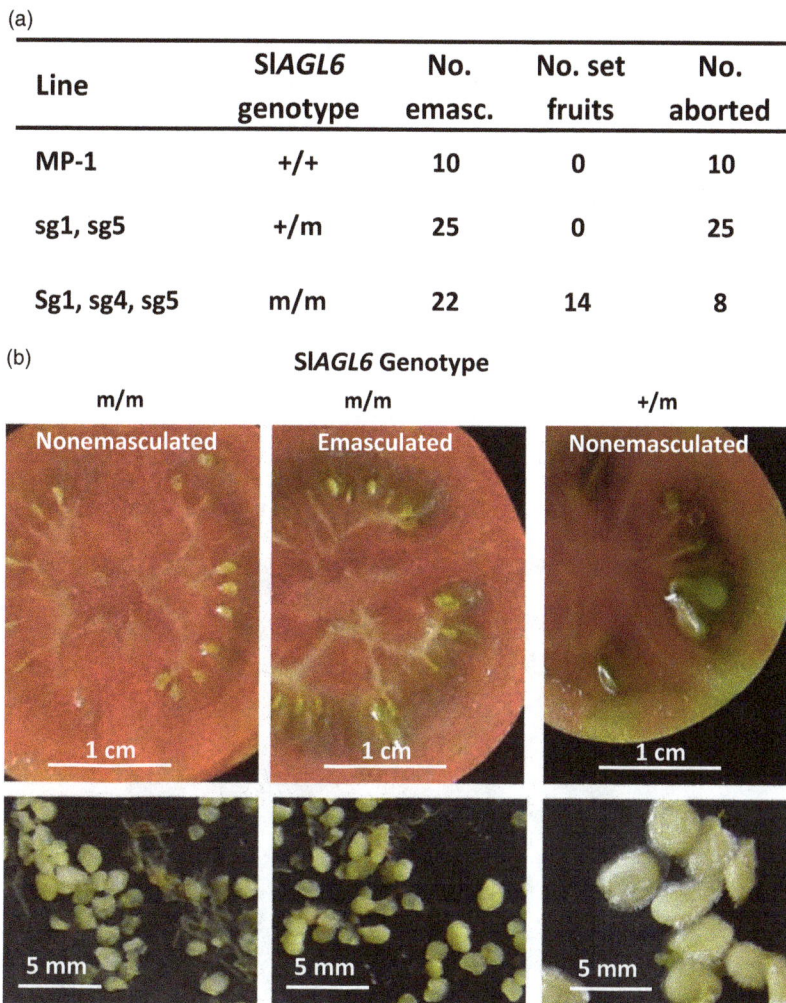

Figure 3 The CRISPR/Cas9-derived SlAGL6 mutations induce vegetative parthenocarpy. (a) Fruit set from flowers emasculated at pre-anthesis from plants of the indicated SlAGL6 genotype. (b) The enlarged ovules collected from red fruits developed from nonemasculated or emasculated flowers developed on a plant homozygous for CRISPR/Cas9-mutated SlAGL6 allele are comparable in size and appearance. Both are similarly smaller than true seeds collected from fruits of a heterozygous (+/m) sibling. The three presented fruits were harvested on the same date.

portion of the progenies close in weight to that of the big fruit parent (Lippman and Tanksley, 2001; Perry, 1915). Thus, we examined whether, in this F_2 population, the mutation affects fruit weight and the tendency to generate parthenocarpic fruits close in size to the big fruit parent. Analysis performed on a small population including 48 F_2 plants grown in 5-L pots in a greenhouse indicated that among the 11 progenies homozygous for the mutated SlAGL6 (SNP No. 3), two produced seedless fruits of weight higher than that of the seeded fruits of the big fruit parent, and the average fruit weight of the mutated progenies was somewhat higher than in their nonmutated siblings (Figure S2a). Plants from the same F_2 population were also grown in the soil, in a net house, next to the BC_2F_2 population described above. Fruits were picked and analysed from the 26 homozygous mutated plants. As shown in Figure S2b, c, two out the five plants with the highest average fruit weight showed a very strong parthenocarpic phenotype, as nearly all of their fruits were seedless. These analyses indicate that the mutation does not prevent the parthenocarpic F_2 progenies from reaching weight similar to that of the big fruit parent.

Surprisingly, 12 out of the 13 tested R_1 progenies of plant sg1 were found to be devoid of the transgenic cassette, while all the tested progenies of plants sg4 and sg5 contained it. Absence of the transgene enabled to assess in sg1 R_1 progenies the effect of the SlAGL6 mutation *per se*, on fruit weight. As shown in Figure 4f, the weight of seedless fruits harvested from sg1 R_1 progenies homozygous for the mutation was comparable to that of seeded fruits developed on MP-1 plants grown side by side under the same ambient conditions.

Slagl6 improves yielding under heat stress

Yielding under natural heat stress conditions was examined comparing MP-1 and sg1 line homozygous for Slagl6 and devoid of the Cas9 cassette (at R_2). Plants were planted in a net house on 20 April 2016 and the first harvest was performed 67 days later (as detailed in Data S1). During the months of May and June, the day temperatures were very high including a 3 days long spell (between 14 and 16 of May 2016) of extremely high temperatures (maximum day temperature 38 °C and above, Figure S3h). These naturally occurring heat stress conditions fall under the definition of 'chronic mild heat stress' known to hamper microsporogenesis and hence fertilization-dependent fruit set (see Mesihovic et al., 2016; and references therein). As demonstrated in Figure 5a, under these climatic conditions, the red fruit yield of the parental line MP-1 was significantly lower than that of line sg1 (ca. 85% lower). This difference reflects mainly a dramatic difference in the number of fruits produced (Figure 5b, f, g), which was 83% lower in line MP-1, and also a significant lower fruit weight in the latter (Figure 5c), although by 13% only. Similar to other parthenocarpic mutants (e.g. Carmi et al., 2003;

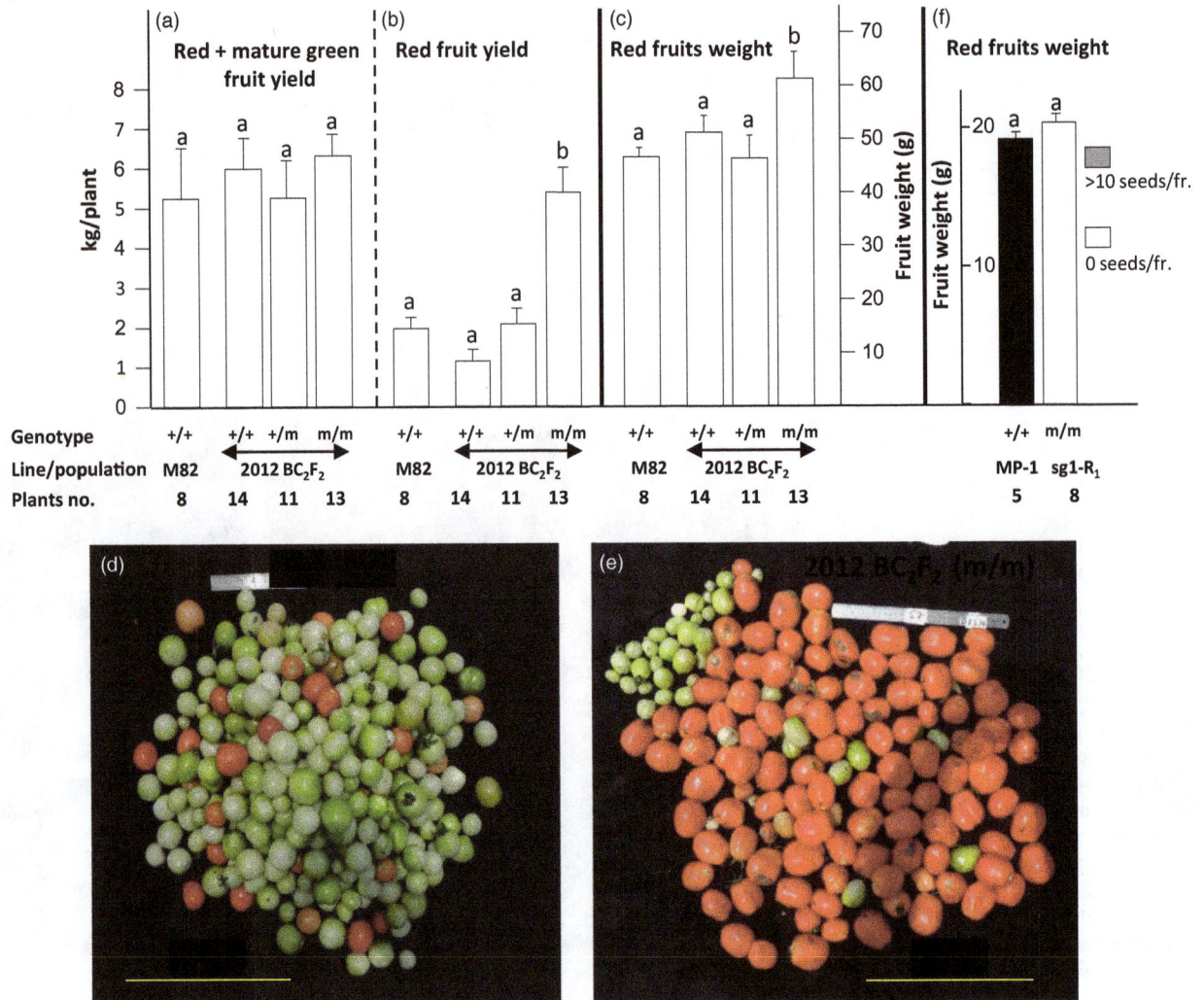

Figure 4 Effect of Sl*AGL6* genotype on yield components in the M82 and MP-1 backgrounds, under ambient temperatures. (a–c) Analysis of 2012 BC_2F_2 plants with different genotypes of SNP No. 3 (Sl*AGL6*) compared to line M82. (a) Yield potential, defined as the yield (kg/plant) of all fruits reaching at least the mature green stage, is not affected by the mutation (one-way analysis of variance, $P = 0.75$); (b) compared to M82, only siblings homozygous for the mutation produce significantly (*t*-test, $P < 0.001$) higher yield of marketable red fruit; (c) the weight of their red fruits is significantly (*t*-test, $P < 0.05$) higher than that of M82 At the date of harvest; (d) in M82, most of the harvested fruits were still green, while (e) in (m/m), plant most of them were red ripe; and (f) parthenocarpic fruits of line sg1 plants homozygous for the mutation (at R_1 generation) are comparable in weight to seeded MP-1 fruits (*t*-test, $P = 0.946$). In panels a, b, c and f, columns accompanied by different lowercase letters differ significantly.

Casas Diaz *et al.*, 1987), the TSS content, expressed as Brix, of red ripe seedless fruits was significantly higher than that of seeded fruits of MP-1 (Figure 5d), while the acidity (pH) of the fruits remained similar (Figure 5e).

Sl*AGL6* mode of expression

To gain additional insight into the function of Sl*AGL6*, its expression was queried in publically available data and complemented by quantitative RT-PCR of developing ovaries. According to the Expression Atlas of Tomato Tissues (http://tomatolab.cshl.edu/~lippmanlab2/allexp_query.html, Park *et al.*, 2012; Tomato Genome Consortium 2012), Sl*AGL6* is highly expressed only in the flower meristem, flower bud and open flower, yet it sharply declines in the developing fruit (Figure S4a). This explains why the Sl*agl6* mutation does not affect vegetative development, transition to reproductive stage,

pollen viability, or fruit size and shape (Figure 6a–c, e–n). However, despite the reported expression in the flower meristem, the flower bud and the flower (Figure S4a), the only subtle difference noticed in the flowers is that the petals are paler and somewhat narrower and longer than in the WT (Figure 6d).

In agreement with the Expression Atlas of Tomato Tissues, four transcriptomic analyses concerning tomato fruit set reported that relative to pre-anthesis (−2 days postanthesis, DPA) (Tang *et al.*, 2015; Wang *et al.*, 2009) or anthesis (Pattison *et al.*, 2015; Zhang *et al.*, 2016), Sl*AGL6* expression sharply declines in fertilized fruit at 4-5DPA (Figure S4b,c,d). However, in unpollinated emasculated ovaries, it scarcely declined (by 1.4-fold only, Figure S4b), suggesting that its decline following fertilization is inherent to fruit set.

To examine Sl*AGL6* mode of expression during growth phase I, its expression was quantified in developing ovaries of line M82. As demonstrated in Figure 7, Sl*AGL6* expression elevates in the

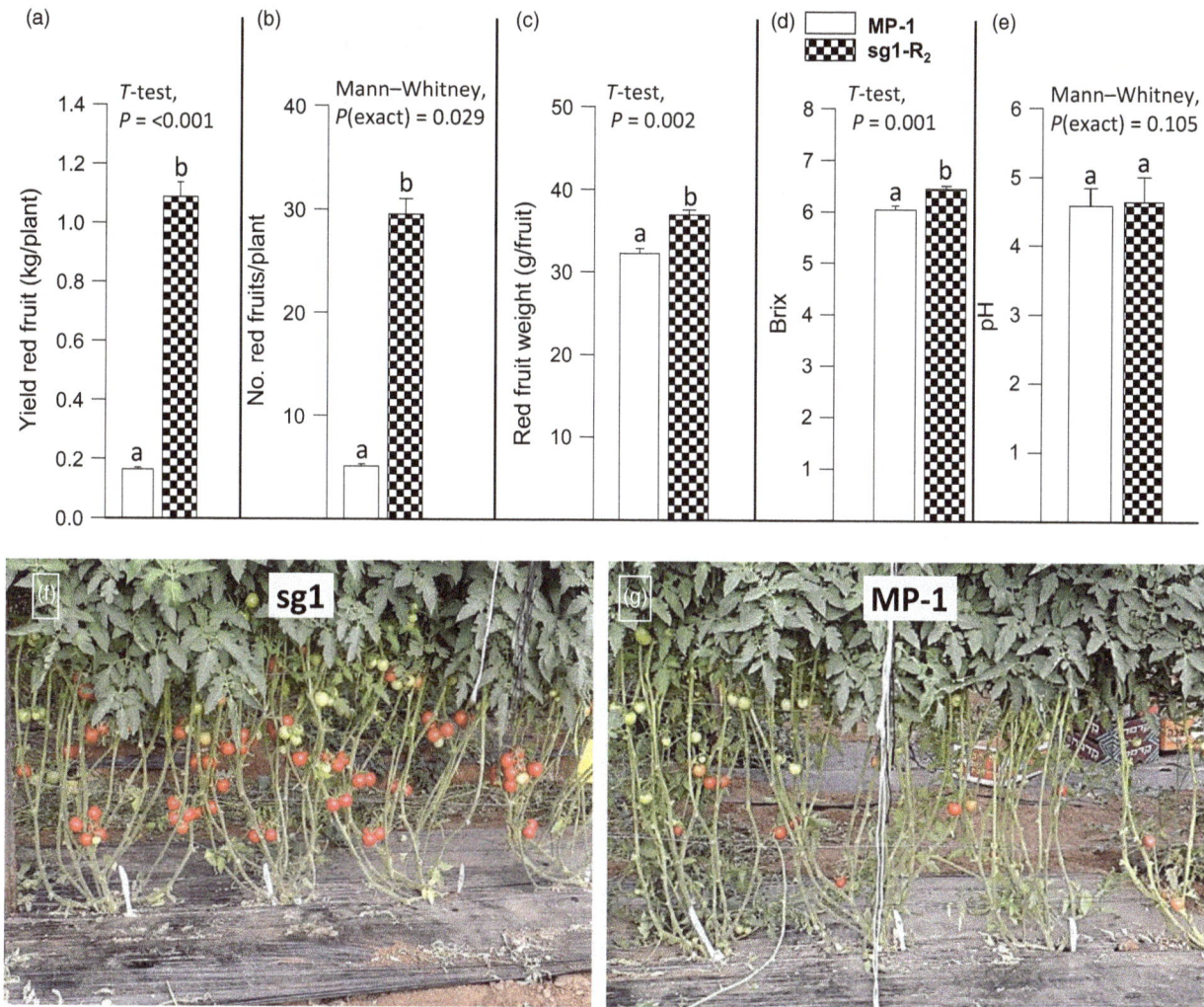

Figure 5 Slagl6 parthenocarpy improves yield under heat stress. Compared to line MP-1, line sg1 manifests significantly higher: (a) red fruit yield, (b) number of red fruits, (c) red fruit weight and (d) Brix, while (e) the fruit pH remains unchanged. Data presented in a-e are derived from experiment performed on four replicates, as detailed in Data S1. (f, g) Difference in fruit load between MP-1 (f) and sg1 (g) plants partly defoliated and photographed before harvest. Growth and climatic conditions are detailed in the results section. Line sg1 (at R_2) is homozygous for the sg1 mutation depicted in Figure 2b and devoid of the Cas9 cassette.

developing ovaries, and it peaks towards the stage of 'ovary arrest' (10-mm-long buds correspond to pre-anthesis) and remains high at anthesis. Yet 5DPA it sharply declines to the level found in ovaries of the young 4-mm-long flower buds, hence associating 'ovary arrest' with elevated SlAGL6 expression and fruit set with its decline. Genetic variation between M82 and Micro Tom or *Solanum pimpinellifolium* might explain the higher fold decline between anthesis and 4DPA reported before (Pattison *et al.*, 2015; Tang *et al.*, 2015; Wang *et al.*, 2009), whereas smaller variation between M82 and Moneymaker could account for the observed similarity in fold decline (Figure 7 *vs.* Figure S4d, queried from Zhang *et al.*, 2016).

Evidence for preferential sublocalization of the SlAGL6 transcript within the ovules of the mature arrested ovary was provided by Pattison *et al.* (2015). Following transcriptomic analysis of laser-captured tissues, they show that SlAGL6 level in the ovules is at least fourfold higher than in the other tissues comprising the ovary. However, at 4DPA, its expression in the embryo is already 15-fold lower than in the ovule (Fig. S4c). This finding was further corroborated by Zhang *et al.* (2016) (Fig. S4d).

Discussion

Loss of SlAGL6 function results in facultative parthenocarpy that ensures fruit production under high temperatures

After isolating the EMS-induced 2012 parthenocarpic mutant (Figure 1a *vs.* b), next-generation sequencing (Table 1) followed by marker-assisted mapping (Table 2) and CRISPR/Cas9 gene knockout (Figure 2) confirmed that mutated SlAGAMOUS-LIKE 6 confers facultative parthenocarpy in tomato.

To fully determine whether mutated SlAGL6 imposes any unacceptable penalty on yielding potential or fruit characteristics, it waits to be introduced as a single mutation into elite tomato cultivars and tested when grown under ambient environmental conditions and established horticultural practices. Nonetheless, in three different genetic backgrounds, the weight of Slagl6 seedless fruits was comparable or even higher than that of the seeded WT ones (Figures 4c, f, 5c, Figures S1 and S2), and yielding potential was comparable to that of the parental cultivar M82 (Figure 4a).

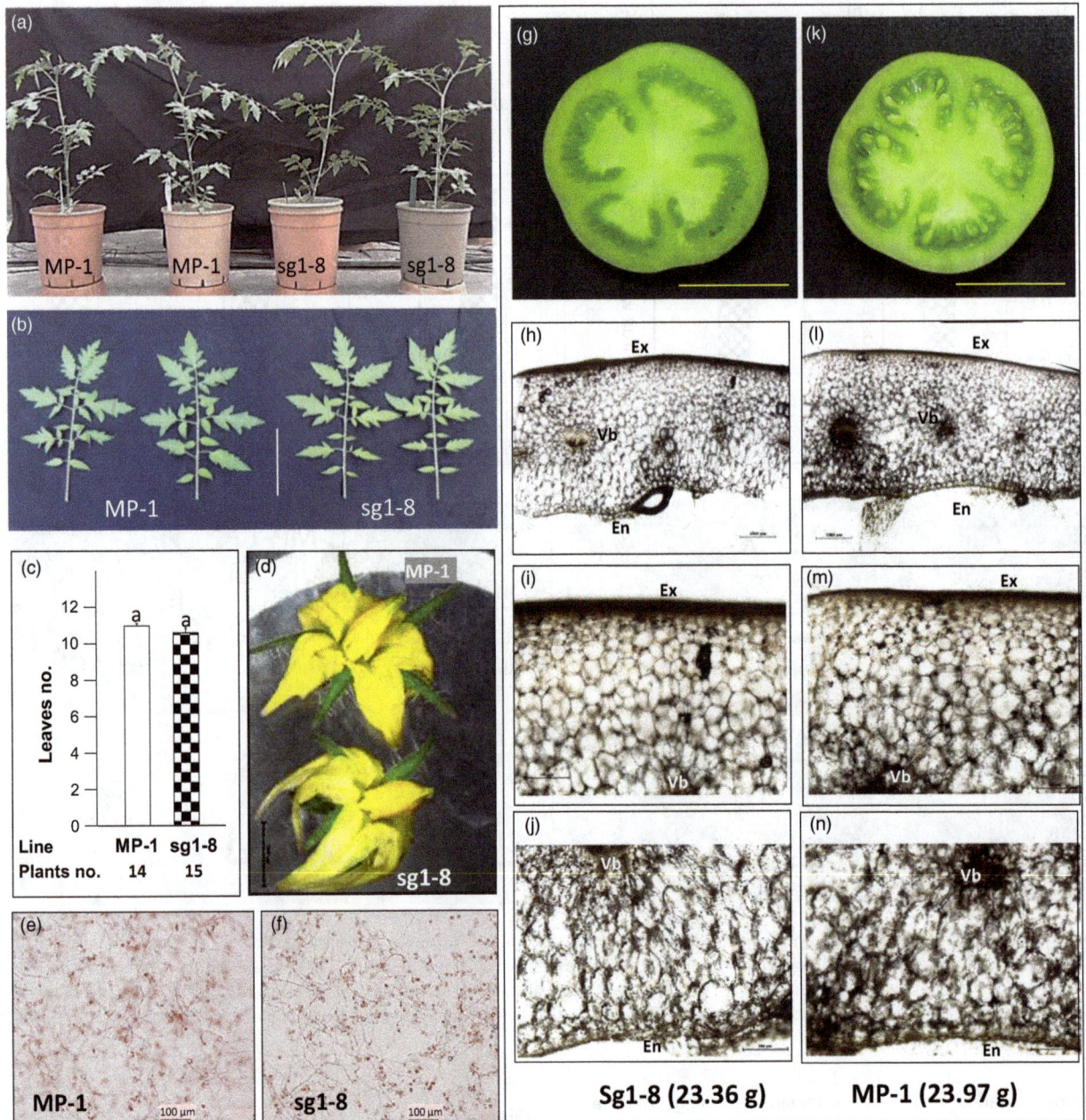

Figure 6 Phenotypic similarity between MP-1 and SlAGL6-mutated line sg1-8. (a) The plants do not differ in growth habit; (b) the shape of the leaves is similar, scale = 8 cm; and (c) the first inflorescence appeared after a similar number of true leaves (Mann–Whitney rank sum test, $P = 0.371$). (d) the flowers are similar except the petals of sg1-8 being paler and somewhat narrower and longer than those of MP-1, scale = 5 mm. (e, f) Pollen fertility is similar. Presented are in vitro germinated pollen grains with similarly elongated pollen tubes, photographed after 18-h incubation, scale = 100 μm. (g–n) The pericarp of parthenocarpic fruit of line sg1-8 (g–j) is similar to that of seeded fruit of MP-1 (k–n) of similar weight (ca. 23.6 g) and size (g, k scale = 2 cm). (h, l) The pericarp is of similar width and shape, scale = 1000 μm. The cells in the layers between the vascular bundle rim and the exodermis (the exocarp) (i, m) and those between the vascular bundle rim and the endodermis (endocarp) (j, n) are of similar appearance. In i, j, m, n, scale = 500 μm. Photographs (h–j) and (l–n) are of thin freehand transverse sections taken from the middle (equator) of the nearly mature green fruits presented in g and k, and photographed under light microscope. Line sg1-8 is described in legend to Figure 5. Ex, exodermis; En, endodermis; Vb, vascular bundle.

The exact conditions favouring seed setting in the mutated plants under fertilization permissive conditions remain to be elucidated. Vigorous inflorescence vibration resulted in many seeded fruits from plants that otherwise set mainly seedless ones (data not shown). In the absence of intentional vibration, enhanced tendency was clearly associated with two parameters: first, the small fruits developed on old plants, frequently

bear seeds. This is an unusual phenomenon, as it was shown that seeded tomato fruits are larger than underseeded ones (Carmi et al., 2003; Imanshi and Hiura, 1975; Varga and Bruinsma, 1976). Second, fruits that set at temperatures mildly lower than optimal were frequently found to contain seeds. In both cases, seed production presumably reflects conditions slowing the rate of ovary expansion into fruit, thus allowing the

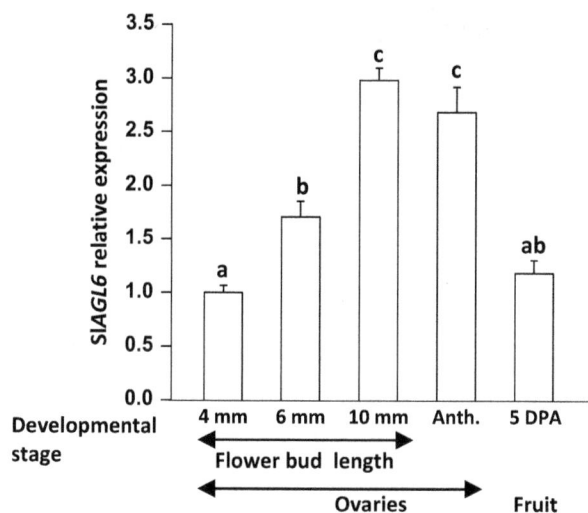

Figure 7 Relative expression of SlAGL6 during ovary development in M82 cultivar. qRT-PCR analysis of SlAGL6 in developing ovaries (developmental stage defined by flower bud length), ovaries at anthesis (Anth.) and young fruits harvested 5DPA. Relative expression levels were normalized to SlTIP41 (Solyc10g049850) as the reference gene and calculated by the comparative delta delta Ct ($\Delta\Delta$Ct) method. The analysis was performed on three biological replicates. Columns accompanies by different lowercase letters differ significantly (Tukey–Kramer HSD test, $P \leq 0.01$).

pollen grains to complete germination, elongation and fertilization of ovules before the style is detached from the otherwise rapidly expanding ovary/fruit. The genetic background apparently affects the facultative manifestation as well: while in 50% (13/26) of the Slagl6/Slagl6 F_2 progenies of the 2012 × Marmande hybrid, grown under ambient conditions, 8/8 fruits tested per plants were seeded (Figure S2b), only 2.4% (3/126) of the Slagl6/Slagl6 plants of the 2012 BC_2F_2 population grown side by side were completely seeded (see Table 2b). Although not tested during the relevant flowering period, genetic differences in anther dehiscence under the given humidity conditions, the rate of ovary enlargement pre- and postanthesis and/or the duration of stigma receptivity could be among the factors underlying the observed difference between these genetic backgrounds.

The ability of the Slagl6-induced parthenocarpy to solve the problem of yielding under fertilization restrictive conditions such as imposed by chronic mild heat stress (Mesihovic et al., 2016) is its most important agronomic attribute. When challenged by continuous mild heat stress, which was worsen by a 3-day spell of acute stress (Figure S3f), the mutated line yielded over sixfold higher than the parental line (Figure 5a) and that mainly because of the profoundly higher number of flowers that set (seedless) fruits under these fertilization restrictive conditions (Figure 5b, f, g). The demonstrated capability of Slagl6 to yield under microsporogenesis restrictive conditions (Sato et al., 2006), together with the ripening uniformity along consecutive trusses exhibited in the M82 determinate background (Figure 4b, d-e), makes Slagl6 particularly suitable for breeding of processing tomato cultivars. This is because under fluctuating climatic conditions, it maximizes the marketable yield that can be obtained in a single mechanical harvest.

Slagl6 is an attractive gene for parthenocarpy

Three digenic sources for parthenocarpy which manifest only mild or no adverse pleiotropic effects and hence of practical value are as follows: pat-2 (Hazra and Dutta, 2010; Vardy et al., 1989), IL5-1 and IVT-line 1 (Gorguet et al., 2008) (as reviewed by Ariizumi et al., 2013). A reliable comparative assessment of the horticultural/parthenocarpic performance of these three sources with that of the Slagl6 mutation waits the identification of the mutated genes underlying these sources and the introducing of all of them to a common genetic background. Nonetheless, different from them, Slagl6 is a single recessive source for facultative parthenocarpy, which is not allelic to any of these three, since none of them was suggested to map to chromosome 1 (Gorguet et al., 2008; Nunome et al., 2014). Taken together, the simple mode of inheritance, the lack of pleiotropic effects (Figure 6) or adverse effects on fruit weight or shape (Figures 4c, f, 5b, 6g-n, and Figures S1 and S2), the true vegetative nature of the induced parthenocarpy (Figure 3) and its facultative manifestation make Slagl6 an attractive single recessive gene for parthenocarpy. The CRISPR/Cas9 technology now enables expeditious integration of Slagl6 into any elite cultivar of interest. That can be done either by direct CRISPR/Cas9-mediated SlAGL6 knockout in the two parents of the elite hybrid cultivars, if they are amenable for transformation, or introgressed by backcrossing.

SlAGL6 functions before and after fertilization

Comprehensive transcriptomic studies of tomato fruit set found significant changes in the expression of hundreds of genes, including several down-regulated MADS-box genes (e.g. Pattison et al., 2015; Tang et al., 2015; Wang et al., 2009; Zhang et al., 2016). Among them, SlAG1 and SlAGL6 were suggested to play an important role in fruit set (Wang et al., 2009), and so was also the tomato MADS-box 29 (TM29) (Ruan et al., 2012). However, experimentally supported identification of any transcription factor actually involved in the regulation of 'ovary arrest' unless it is fertilized is still missing (Ruan et al., 2012). The severe homeotic malformations of the seedless fruit developed following silencing TM29 (Ampomah-Dwamena et al., 2002) or SlAG1 (Gimenez et al., 2016; Pan et al., 2010; Pnueli et al., 1994) indicate a role in stamen and carpel development rather than as activators of the normal fertilization-triggered fruit set, and experimental support for the SlAGL6 suggested function was not provided.

The studied SlAGL6 mutants together with relevant transcriptomic analyses identify SlAGL6 as a central player in the mechanism underpinning fertilization-dependent fruit development. The presented data strongly indicate that the principle, if not the sole, role of SlAGL6 is to serve as a key regulator of the transition between the 'ovary arrest' state reached at the end of growth phase I and the fertilization-triggered resumption of growth, that is fruit set. The information gathered in the current study led us to propose a model (Figure 8) according to which SlAGL6 accumulation at pre-anthesis, most likely in the maturing ovules, plays a pivotal role in the induction or at least retention of 'ovary arrest'. At anthesis, successful ovule fertilization signals for SlAGL6 down-regulation, and once its suppressive effect is alleviated, the ovary/fruit growth is resumed and continues to its full growth potential.

The reasoning behind the proposed model is as follows: first, although according to the Expression Atlas of Tomato Tissues (http://tomatolab.cshl.edu/~lippmanlab2/allexp_query.html; Park

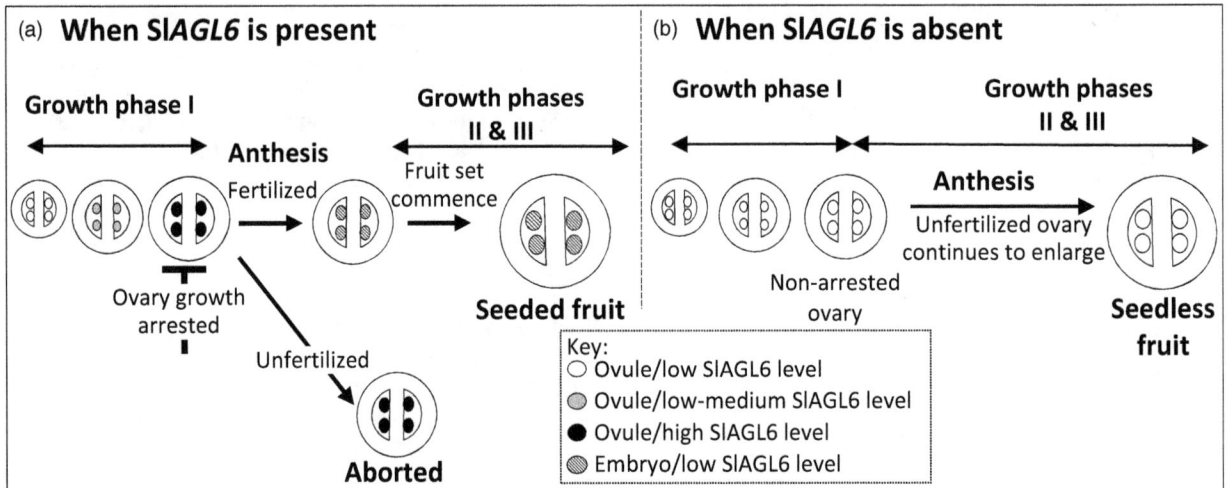

Figure 8 A model depicting the key role of SlAGL6 in the regulation of fertilization-dependent fruit set. (a) Low level of SlAGL6 in the ovules of the young ovary allows their growth until the high level, reached towards the end of growth phase I, inhibits further development of the ovary, resulting in 'ovary arrest'. Only the fertilization-induced down-regulation of SlAGL6 in the young embryos enables the resumption of growth and production of a fully developed seeded fruit. Alternatively, (b) loss of the 'ovary arrest' function in the Slagl6 mutant allows fertilization-independent seedless fruit development.

et al., 2012; Tomato Genome Consortium 2012), SlAGL6 is highly expressed in the flower meristem, the flower bud and the flower (Figure S4a); besides the subtle change in the petal hue and shape, the Slagl6 flowers are normal, and both male and female fertile (Figure 6d, e-f). This indicates that different from some other species where an E-function was attributed to the AGL6 clade (Dreni and Zhang, 2016), SlAGL6 is dispensable in determining the identity of any of the flower whorls according to the extended ABC(DE) and the quartet models (Coen and Meyerowitz, 1991; Theissen, 2001; Theissen and Saedler, 2001). Second, quantification of SlAGL6 transcripts in developing ovaries of line M82 demonstrates that it is not highly expressed throughout growth phase I (Figure 7), but rather gradually elevates and peaks at the stage when the ovary growth is arrested. However, 5DPA it already declines to the low level found in ovaries of the young 4-mm-long flower buds. Although our analysis was performed on whole ovaries, based on the demonstrated preferential expression of SlAGL6 in the ovules at anthesis (Figure S4c, d, queried from Pattison et al., 2015; Zhang et al., 2016), it is reasonable to assume that the observed peak at pre-anthesis represents mainly increased expression in the ovules of ovaries which growth is arrested. The lower level of the transcript in the younger growing ovaries, together with the Slagl6 parthenocarpy, strongly suggests that SlAGL6 accumulation in the mature ovules acts as a key suppressor of ovary growth beyond anthesis unless it is fertilized, thus preventing accidental development of unfertilized ovary into purposeless fruit. Lastly, the normal size and shape of the seedless fruits indicate that development of the fruit to its full growth potential relies predominantly on the removal of the SlAGL6 suppressive signal (either following fertilization or by mutation), rather than on promoting signals emitted by the developing embryos independent of SlAGL6 down-regulation. Continuously emitted SlAGL6 suppressive signal from unfertilized ovules within the developing WT fruit (as suggested by Tang et al., 2015; see Figure S4b) could explain the often observed restricted development of underfertilized fruits (Carmi et al., 2003; Imanshi and Hiura, 1975; Varga and Bruinsma, 1976). This assumption is

further supported by the significantly higher weight of the Slagl6 parthenocarpic fruits in the segregating 2012 BC$_2$F$_2$ population compared to that of the seeded fruits of the parental line M82 (Figure 4c), as well as that of the seedless sg1 fruits compared to the MP-1 seeded ones when developed under heat stress (Figures 5c). In both cases, not all of the ovules are necessarily fertilized in the seeded WT fruits.

Neofunctionalization of AGL6 in tomato

It is noteworthy that in apple (Malus domestica), where the fleshy fruit is a pome derived from the floral tube fused to the carpels, parthenocarpy was found to be governed by mutated MdPI, the homolog of Arabidopsis PISTILLATA (Yao et al., 2001), which belongs to the DEF/GLO rather than the AGL6 clade of MADS-box genes (Smaczniak et al., 2012). Alternatively, parthenocarpy was not reported for any of the mutated SlAGL6 homologs characterized so far. In rice (Oryza sativa), the homeotic changes manifested by mutated OsMADS6 testify to its role in determining floral organ and meristem identities (e.g. Duan et al., 2012; Li et al., 2011; Ohmori et al., 2009; Zhang et al., 2010). Knocking down its Nigella damascene homolog affected structure of sepals and petals indicating an A-function (Wang et al., 2015). The Arabidopsis thaliana homolog (At2g45650) affects flowering time and axillary bud formation (Huang et al., 2012; Koo et al., 2010; Yoo et al., 2011). Interestingly, similar to Slagl6, the mutated Petunia hybrida homolog, Phagl6, caused only subtle effect on petals colour and indentation. Yet, although similar to tomato and Arabidopsis (Schauer et al., 2009), PhAGL6 is highly expressed in the mature ovules, in this dry capsule fruit species, as in Arabidopsis, parthenocarpy was not reported (Rijpkema et al., 2009). Thus similar to other MADS-box genes that underwent neofunctionalization (Dreni and Kater, 2014; Scutt et al., 2006; Smaczniak et al., 2012; Zahn et al., 2006), in tomato, and presumably in other fleshy Solanaceae fruits, AGL6 acquired a new function, acting as the suppressor of ovary development beyond anthesis. Further identification of the signals and regulators involved in SlAGL6 down-regulation and its immediate targets could point to additional candidate genes for parthenocarpy.

Experimental procedures

Populations for mapping the 2012 mutant

Several populations were analysed, including 2012 BC_1F_2, a testcross (TC) population, 2012 BC_2F_2 and F_2 population derived from 2012 × Marmande cross. Generation of these populations, their growth conditions and parameters analysed in each of them are detailed in Data S1 and in Figure S3.

SNP genotyping of 2012 derived progenies

Genotyping was performed as a service by DYN R&D Ltd (Migdal Haemek, Israel), following the melting curve SNP method (Ye et al., 2002). Plants were genotyped in duplicate, on two separately sampled leaves.

Genomic DNA libraries, sequencing and bioinformatics analysis

From each of 20 plants derived from 2012 BC_1F_2 population, defined as 'strong parthenocarpic', one young leaf (ca. 150 mg FW) was picked, and from the pooled leaves, DNA was extract. Similarly, DNA was extracted from a pool of leaves sampled at the same date from 23 'nonparthenocarpic' plants. From the two DNA samples, in the Technion, (The Life Sciences and Engineering Infrastructure Center) Haifa, Israel, two sequencing libraries, one designated '2012 library' and the other 'NP (nonparthenocarpic) library', were prepared and sequenced using 100-bp paired end reads on an Illumina HiSeq 2000 platform. Bioinformatics analysis was performed following Bolger et al. (2014a,b), Li and Durbin (2009), Li et al. (2009) and Schneeberger et al. (2009), as detailed in Data S1.

Construction of a CRISPR/Cas9 knockout plasmid and tomato transformation

The CRISPR/Cas9 construct was designed to create a deletion/insertion after 212 bp of the Solyc01g093960 coding sequence (predicted exon 2, see Figure 2b). The 20-bp target sequence was chosen to be followed by protospacer adjacent motif (PAM), the requisite binding site for Cas9, TGG (depicted in Figure 2b). The selected sgRNA was amplified using the primers: SalI-gRNA-F: AGA*gtcgac*ATAGCGATT<u>GAGGATTAAGGCAACAACGTG</u>TTTTAG AGCTAGAAATAGCAAG and HindIII-gRNA-R: TAAGCT*aagctt*C GATCTAAAAAAAGCACCGACT (the added restriction sites are presented in italics lowercase letters, and the specific target sequence (cRNA) is underlined in the SalI-gRNA-F primer sequence). The PCR product was restricted and cloned into the pRCS binary vector SalI-HindIII sites as previously described (Chandrasekaran et al., 2016). The binary vector was transformed to *Agrobacterium tumefaciens* strain EHA105, which served to transform the indeterminate tomato line MP-1 as previously described (Barg et al., 1997).

Detection of CRISPR/Cas9-induced mutations

The Solyc01g093960 gRNA target site was designed to include AclI restriction enzyme site overlapping three bp upstream from the PAM (see Figure 2a), the predicted cut site of the Cas9 nuclease, so that DNA double-strand break repair could disrupt the restriction site. R_O plants were screened for the presence of chimeric section as detailed in Data S1, and the same procedure was used to genotype R_1 progenies. The presence of the Cas9/sgRNA cassette in the progenies was PCR-tested.

Quantitative RT-PCR analysis of SlAGL6

Analysis was performed on total RNA samples extracted from developing ovaries and young fruit collected from line M82 plants, as earlier described (Damodharan et al., 2016).

Pollen viability assay

Freshly harvested pollen grains were incubated in germination solution (Firon et al., 2012) for 18 h, at 24 °C, and examined under light microscope.

All primers used in this study are listed in Table S1.

Statistical analysis

ANOVA statistical analyses were performed with SIGMASTAT 2.0 program. (http://en.softonic.com/s/sigma-stat-version-2/).

Acknowledgements

We are indebted to Mr. Moshe Gabay for outstanding handling of the plants, and the students Sharon Israeli, Aviram Trachtenberg, Marina Gorodner and Shira Corem for excellent technical assistance. We thank Y. Kapulnik and J Hirschenhoren for the M_2-mutagenized seeds, Subha Damodharan for sharing RNA samples, the research groups of Y. Kapulnik, H. Kultai, N. Firon, E. Pressman and D. Granot from the ARO for participation in setting the mutagenized population screening experiment, A. Faigenboim and K. Machemer for bioinformatics assistance, M. Perel for meteorological data and A. Sherman and E. Grotewold for critically reviewing the manuscript. This work was supported by the Chief Scientist of the Israel Ministry of Agriculture and Rural Development grants no. 261-0925, 261-0916 and 256-0980. The authors declare no conflict of interests.

References

Ampomah-Dwamena, C., Morris, B.A., Sutherland, P., Veit, B. and Yao, J.-L. (2002) Down-regulation of TM29, a tomato SEPALLATA homolog, causes parthenocarpic fruit development and floral reversion. *Plant Physiol.* **130**, 605–617.

Ariizumi, T., Shinozaki, Y. and Ezura, H. (2013) Genes that influence yield in tomato. *Breed. Sci.* **63**, 3–13.

Barg, R., Pilowsky, M., Shabtai, S., Carmi, N., Szechtman, A.D., Dedicova, B. and Salts, Y. (1997) The TYLCV-tolerant tomato line MP-1 is characterized by superior transformation competence. *J. Exp. Bot.* **48**, 1919–1923.

Bassel, G.W., Mullen, R.T. and Bewley, J.D. (2008) *procera* is a putative DELLA mutant in tomato (*Solanum lycopersicum*): effects on the seed and vegetative plant. *J. Exp. Bot.* **59**, 585–593.

Beraldi, D., Picarella, M.E., Soressi, G.P. and Mazzucato, A. (2004) Fine mapping of the parthenocarpic fruit (*pat*) mutation in tomato. *Theor. Appl. Genet.* **108**, 209–216.

Bohner, J. and Bangerth, F. (1988) Cell number, cell size and hormone levels in semi-isogenic mutants of *Lycopersicon pimpinellifolium* differing in fruit size. *Physiologia Plant.* **72**, 316–320.

Bolger, A., Scossa, F., Bolger, M.E., Lanz, C., Maumus, F., Tohge, T., Quesneville, H. et al. (2014a) The genome of the stress-tolerant wild tomato species *Solanum pennellii*. *Nat. Genet.* **46**, 1034–1038.

Bolger, A.M., Lohse, M. and Usadel, B. (2014b) Trimmomatic: a flexible trimmer for Illumina sequence data. *Bioinformatics*, **30**, 2114–2120.

Carmi, N., Salts, Y., Dedicova, B., Shabtai, S. and Barg, R. (2003) Induction of parthenocarpy in tomato via specific expression of the *rolB* gene in the ovary. *Planta*, **217**, 726–735.

Carrera, E., Ruiz-Rivero, O., Peres, L/E., Atares, A. and Garcia-Martinez, J.L. (2012) Characterization of the procera tomato mutant shows novel functions of the Sl*DELLA* protein in the control of flower morphology, cell division and expansion, and the auxin-signaling pathway during fruit-set and development. *Plant Physiol.* **160**, 1581–1596.

Casas Diaz, A.V., Hewitt, J.D. and Lapushner, D. (1987) Effects of parthenocarpy on fruit quality in tomato. *J. Am. Soc. Hort. Sci.* **112**, 634–637.

Chandrasekaran, J., Brumin, M., Wolf, D., Leibman, D., Klap, C., Pearlsman, M., Sherman, A. *et al.* (2016) Development of broad virus resistance in non-transgenic cucumber using CRISPR/Cas9 technology. *Mol. Plant Pathol.* **17**, 1140–1153.

Chevalier, C., Bourdon, M., Pirrello, J., Cheniclet, C., Gévaudant, F. and Frangne, N. (2014) Endoreduplication and fruit growth in tomato: evidence in favour of the karyoplasmic ratio theory. *J. Exp. Bot.* **65**, 2731–2746.

Coen, E.S. and Meyerowitz, E.M. (1991) The war of the whorls - genetic interactions controlling flower development. *Nature*, **353**, 31–37.

Damodharan, S., Zhao, D. and Arazi, T. (2016) A common miRNA160-based mechanism regulates ovary patterning, floral organ abscission and lamina outgrowth in tomato. *Plant J.* **86**, 458–471.

Dreni, L. and Kater, M.M. (2014) MADS reloaded: evolution of the AGAMOUS subfamily genes. *New Phytol.* **201**, 717–732.

Dreni, L. and Zhang, D. (2016) Flower development: the evolutionary history and functions of the AGL6 subfamily MADS-box genes. *J. Exp. Bot.* **67**, 1625–1638.

Duan, Y., Xing, Z., Diao, Z., Xu, W., Li, S., Du, X., Wu, G. *et al.* (2012) Characterization of Osmads6-5, a null allele, reveals that OsMADS6 is a critical regulator for early flower development in rice (*Oryza sativa* L.). *Plant Mol. Biol.* **80**, 429–442.

El Ahmadi, A.B. and Stevens, M.A. (1979) Reproductive responses of heat-tolerant tomatoes to high temperatures. *J. Amer. Soc. Hort. Sci.* **104**, 686–691.

Falavigna, A., Badino, M. and Soressi, G.P. (1978) Potential of the monomendelian factor *pat* in the tomato breeding for industry. *Genetica Agraria*, **32**, 159–160.

Ficcadenti, N., Sestili, S., Pandolfini, T., Cirillo, C., Rotino, G.L. and Spena, A. (1999) Genetic engineering of parthenocarpic fruit development in tomato. *Mol. Breeding*, **5**, 463–470.

Firon, N., Pressman, E., Meir, S., Khoury, R. and Altahan, L. (2012) Ethylene is involved in maintaining tomato (*Solanum lycopersicum*) pollen quality under heat-stress conditions. *AoB. Plants*, **2012**, pls024.

Gillaspy, G., Ben-David, H. and Gruissem, W. (1993) Fruits: a developmental perspective. *Plant Cell*, **5**, 1439–1451.

Gimenez, E., Castañeda, L., Pineda, B., Pan, I.L., Moreno, V., Angosto, T. and Lozano, R. (2016) *TOMATO AGAMOUS1* and *ARLEQUIN/TOMATO AGAMOUS-LIKE1* MADS-box genes have redundant and divergent functions required for tomato reproductive development. *Plant Mol. Biol.* **91**, 513–531.

Gorguet, B., van Heusden, A.W. and Lindhout, P. (2005) Parthenocarpic fruit development in tomato. *Plant Biol.* **7**, 131–139.

Gorguet, B., Eggink, P.M., Ocana, J., Tiwari, A., Schipper, D., Finkers, R., Visser, R.G.F. *et al.* (2008) Mapping and characterization of novel parthenocarpy QTLs in tomato. *Theor. Appl. Genet.* **116**, 755–767.

Hazra, P. and Dutta, A.K. (2010) Inheritance of parthenocarpy in tomato (*Solanum lycopersicum*) and its association with two marker characters. *Int. J. Plant Sci.* **1**, 144–149.

Huang, X., Effgen, S., Meyer, R.C., Theres, K. and Koornneef, M. (2012) Epistatic natural allelic variation reveals a function of AGAMOUS-LIKE6 in axillary bud formation in *Arabidopsis. Plant Cell*, **24**, 2364–2379.

Imanshi, S. and Hiura, I. (1975) Relationship between fruit weight and seed content in the tomato. *J. Jpn. Soc. Hort. Sci.* **44**, 33–40.

Koo, S.C., Bracko, O., Park, M.S., Schwab, R., Chun, H.J., Park, K.M., Seo, J.S. *et al.* (2010) Control of lateral organ development and flowering time by the Arabidopsis thaliana MADS-box Gene *AGAMOUS-LIKE6. Plant J.* **62**, 807–816.

Li, H. and Durbin, R. (2009) Fast and accurate short read alignment with Burrows-Wheeler transform. *Bioinformatics*, **25**, 1754–1760.

Li, H., Handsaker, B., Wysoker, A., Fennell, T., Ruan, J., Homer, N., Marth, G. *et al.* and 1000 Genome Project Data Processing Subgroup. (2009) The sequence alignment/map format and SAMtools. *Bioinformatics*, **25**, 2078–2079.

Li, H., Liang, W., Hu, Y., Zhu, L., Yin, C., Xu, J., Dreni, L. *et al.* (2011) Rice MADS6 interacts with the floral homeotic genes SUPERWOMAN1, MADS3, MADS58, MADS13, and DROOPING LEAF in specifying floral organ identities and meristem fate. *Plant Cell*, **23**, 2536–2552.

Lin, S., George, W.L. and Splittstoesser, W.E. (1984) Expression and inheritance of parthenocarpy in 'Severianin' tomato. *J. Hered.* **75**, 62–66.

Lippman, Z. and Tanksley, S.D. (2001) Dissecting the genetic pathway to extreme fruit size in tomato using a cross between the small-fruited wild species *Lycopersicon pimpinellifolium* and *L. esculentum* var, Giant Heirloom. *Genetics*, **158**, 413–422.

Mazzucato, A., Taddei, A.R. and Soressi, G.P. (1998) The parthenocarpic fruit (*pat*) mutant of tomato (*Lycopersicon esculentum* Mill.) sets seedless fruits and has aberrant anther and ovule development. *Development*, **125**, 107–114.

Mazzucato, A., Cellini, F., Bouzayen, M., Zouine, M., Mila, I., Minoia, S., Petrozza, A. *et al.* (2015) A TILLING allele of the tomato Aux/IAA9 gene offers new insights into fruit set mechanisms and perspectives for breeding seedless tomatoes. *Mol. Breeding*, **35**, 22.

Mesihovic, A., Iannacone, R., Firon, N. and Fragkostefanakis, S. (2016) Heat stress regimes for the investigation of pollen thermotolerance in crop plants. *Plant Reprod.* **29**, 93–105.

Michelmore, R.W., Paran, I. and Kesseli, R.V. (1991) Identification of markers linked to disease-resistance genes by bulked segregant analysis: a rapid method to detect markers in specific genomic regions by using segregating populations. *Proc. Natl Acad. Sci. USA*, **88**, 9828–9832.

Nuez, F., Costa, J. and Cuartero, J. (1986) Genetics of the parthenocarpy for tomato varieties "Sub-Artic Plenty", "75/59" and "Severianin". *Z. Pflanzenzucht*, **96**, 200–206.

Nunome, T., Honda, I., Ohyama, A., Miyatake, K., Yamaguchi, H. and Fukuoka, H. (2014) *Map based cloning of tomato parthenocarpic fruit-2 (pat-2)* gene. Procs. XVIIIth EUCARPIA meeting, Avignon, France, 70 (abst.).

Ohmori, S., Kimizu, M., Sugita, M., Miyao, A., Hirochika, H., Uchida, E., Nagato, Y. *et al.* (2009) MOSAIC FLORAL ORGANS1, an AGL6-like MADS box gene, regulates floral organ identity and meristem fate in rice. *Plant Cell*, **21**, 3008–3025.

Pan, I.L., McQuinn, R., Giovannoni, J.J. and Irish, V.F. (2010) Functional diversification of AGAMOUS lineage genes in regulating tomato flower and fruit development. *J. Exp. Bot.* **61**, 1795–1806.

Park, S.J., Jiang, K., Schatz, M.C. and Lippman, Z.B. (2012) Rate of meristem maturation determines inflorescence architecture in tomato. *Proc. Natl Acad. Sci. USA*, **109**, 639–644.

Pattison, R.J., Csukasi, F., Zheng, Y., Fei, Z., van der Knaap, E. and Catalá, C. (2015) Comprehensive tissue-specific transcriptome analysis reveals distinct regulatory programs during early tomato fruit development. *Plant Physiol.* **168**, 1684–1701.

Perry, F.E. (1915) The inheritance of size in tomatoes. *Ohio Naturalist*, **15**, 473–497.

Philouze, J. (1989) Natural parthenocarpy in tomato. IV. A study of the polygenic control of parthenocarpy in line 75/59. *Agronomie*, **9**, 63–75.

Philouze, J. and Maisonneuve, B. (1978) Heredity of the natural ability to set parthenocarpic fruits in a German line. *Tomato Genet. Coop.* **28**, 12.

Picken, A.J.F. (1984) A review of pollination and fruit set in the tomato (*Lycopersicon esculentum* Mill.). *J. Hort. Sci.* **59**, 1–13.

Pnueli, L., Hareven, D., Rounsley, S.D., Yanofsky, M.F. and Lifschitz, E. (1994) Isolation of the tomato AGAMOUS gene TAG1 and analysis of its homeotic role in transgenic plants. *Plant Cell*, **6**, 163–173.

Rijpkema, A.S., Zethof, J., Gerats, T. and Vandenbussche, M. (2009) The petunia AGL6 gene has a SEPALLATA-like function in floral patterning. *Plant J.* **60**, 1–9.

Ruan, Y.L., Patrick, J.W., Bouzayen, M., Osorio, S. and Fernie, A.R. (2012) Molecular regulation of seed and fruit set. *Trends Plant Sci.* **17**, 656–665.

Saito, T., Ariizumi, T., Okabe, Y., Asamizu, E., Hiwasa-Tanase, K., Fukuda, N., Mizoguchi, T. *et al.* (2011) TOMATOMA: a novel tomato mutant database distributing Micro-Tom mutant collections. *Plant Cell Physiol.* **52**, 283–896.

Sato, S., Kamiyama, M., Iwata, T., Makita, N., Furukawa, H. and Ikeda, H. (2006) Moderate increase of mean daily temperature adversely affects fruit set of *Lycopersicon esculentum* by disrupting specific physiological processes in male reproductive development. *Ann. Bot. Lond.* **97**, 731–738.

Schauer, S.E., Schlüter, P.M., Baskar, R., Gheyselinck, J., Bolaños, A., Curtis, M.D. and Grossniklaus, U. (2009) Intronic regulatory elements determine the divergent expression patterns of AGAMOUS-LIKE6 subfamily members in Arabidopsis. *Plant J.* **59**, 987–1000.

Schneeberger, K., Ossowski, S., Lanz, C., Juul, T., Petersen, A.H., Nielsen, K.L., Jørgensen, J.E. *et al.* (2009) SHOREmap: simultaneous mapping and mutation identification by deep sequencing. *Nat. Methods*, **6**, 550–551.

Scutt, C.P., Vinauger-Douard, M., Fourquin, C., Finet, C. and Dumas, C. (2006) An evolutionary perspective on the regulation of carpel development. *J. Exp. Bot.* **57**, 2143–2152.

Selleri, L. (2011) *Ph. D. Thesis* http://hdl.handle.net/2067/2501.

Smaczniak, C., Immink, R.G., Angenent, G.C. and Kaufmann, K. (2012) Developmental and evolutionary diversity of plant MADS-domain factors: insights from recent studies. *Development*, **139**, 3081–3098.

Soressi, G.P. and Salamini, F. (1975) A monomendelian gene inducing parthenocarpic fruits. *Rep. Tomato Genet. Coop.* **25**, 22.

Tang, N., Deng, W., Hu, G., Hu, N. and Li, Z. (2015) Transcriptome profiling reveals the regulatory mechanism underlying pollination dependent and parthenocarpic fruit set mainly mediated by auxin and gibberellin. *PLoS ONE*, **10**, e0125355.

Tanksley, S.D. (2004) The genetic, developmental, and molecular bases of fruit size and shape variation in tomato. *Plant Cell*, **16**, 181–189.

Theissen, G. (2001) Development of floral organ identity: stories from the MADS house. *Curr. Opin. Plant Biol.* **4**, 75–85.

Theissen, G. and Saedler, H. (2001) Plant biology: floral quartets. *Nature*, **409**, 469–471.

Tomato Genome Consortium (2012) The tomato genome sequence provides insights into fleshy fruit evolution. *Nature*, **485**, 635–641.

Vardy, E., Lapushner, D., Genizi, A. and Hewitt, J. (1989) Genetics of parthenocarpy in tomato under a low temperature regime: II. Cultivar "Severianin". *Euphytica*, **41**, 9–15.

Varga, A. and Bruinsma, J. (1976) Roles of seeds and auxins in tomato fruit growth. *Z. Pflanzenphysiol.* **80**, 95–104.

Varga, A. and Bruinsma, J. (1986) Tomato. In *CRC Handbook of Fruit Set and Development* (Monselise, S.P., Ed.), pp. 461–481. CRC Press Inc, Boca Raton, FL.

Varoquaux, F., Blanvillain, R., Delseny, M. and Gallois, P. (2000) Less is better: new approaches for seedless fruit production. *Trends Biotechnol.* **18**, 233–242.

Wang, H., Schauer, N., Usadel, B., Frasse, P., Zouine, M., Hernould, M., Latché, A. *et al.* (2009) Regulatory features underlying pollination-dependent and -independent tomato fruit set revealed by transcript and primary metabolite profiling. *Plant Cell*, **21**, 1428–1452.

Wang, P., Liao, H., Zhang, W., Yu, X., Zhang, R., Shan, H., Duan, X. *et al.* (2015) Flexibility in the structure of spiral flowers and its underlying mechanisms. *Nat. Plants*, **2**, 15188.

Yao, J., Dong, Y. and Morris, B.A. (2001) Parthenocarpic apple fruit production conferred by transposon insertion mutations in a MADS-box transcription factor. *Proc. Natl Acad. Sci. USA*, **98**, 1306–1311.

Ye, J., Parra, E.J., Sosnoski, D.M., Hiester, K., Underhill, P.A. and Shriver, M.D. (2002) Melting curve SNP (McSNP) genotyping: a useful approach for diallelic genotyping in forensic science. *J. Forensic Sci.* **47**, 593–600.

Yoo, S.K., Wu, X., Lee, J.S. and Ahn, J.H. (2011) AGAMOUS-LIKE 6 is a floral promoter that negatively regulates the FLC/MAF clade genes and positively regulates FT in Arabidopsis. *Plant J.* **65**, 62–76.

Zahn, L.M., Leebens-Mack, J.H., Arrington, J.M., Hu, Y., Landherr, L.L., dePamphilis, C.W., Becker, A. *et al.* (2006) Conservation and divergence in the AGAMOUS subfamily of MADS-box genes: evidence of independent sub- and neofunctionalization events. *Evol. Dev.* **8**, 30–45.

Zhang, J., Nallamilli, B.R., Mujahid, H. and Peng, Z. (2010) OsMADS6 plays an essential role in endosperm nutrient accumulation and is subject to epigenetic regulation in rice (*Oryza sativa*). *Plant J.* **64**, 604–617.

Zhang, S., Xu, M., Qiu, Z., Wang, K., Du, Y., Gu, L. and Cui, X. (2016) Spatiotemporal transcriptome provides insights into early fruit development of tomato (*Solanum lycopersicum*). *Sci. Rep.* **6**, 23173.

3

Ectopic expression of a cyanobacterial flavodoxin in creeping bentgrass impacts plant development and confers broad abiotic stress tolerance

Zhigang Li[1,2,†], Shuangrong Yuan[2,†], Haiyan Jia[2,3], Fangyuan Gao[2,4], Man Zhou[2], Ning Yuan[2], Peipei Wu[2], Qian Hu[2], Dongfa Sun[1,*] and Hong Luo[2,*]

[1]College of Plant Science and Technology, Huazhong Agricultural University, Wuhan, Hubei, China
[2]Department of Genetics and Biochemistry, Clemson University, Clemson, SC, USA
[3]The Applied Plant Genomics Laboratory of Crop Genomics and Bioinformatics Centre, and National Key Laboratory of Crop Genetics and Germplasm Enhancement, Nanjing Agricultural University, Nanjing, Jiangsu, China
[4]Crop Research Institute, Sichuan Academy of Agricultural Sciences, Chengdu, Sichuan, China

*Correspondence
email hluo@clemson.edu;
sundongfa1@mail.hzau.edu.cn
[†]These authors contributed equally to this work.

Keywords: cyanobacterial flavodoxin, abiotic stress tolerance, drought tolerance, heat tolerance, methyl viologen resistance, nitrogen starvation, turfgrass, transgenics.

Summary

Flavodoxin (Fld) plays a pivotal role in photosynthetic microorganisms as an alternative electron carrier flavoprotein under adverse environmental conditions. Cyanobacterial Fld has been demonstrated to be able to substitute ferredoxin of higher plants in most electron transfer processes under stressful conditions. We have explored the potential of Fld for use in improving plant stress response in creeping bentgrass (*Agrostis stolonifera* L.). Overexpression of Fld altered plant growth and development. Most significantly, transgenic plants exhibited drastically enhanced performance under oxidative, drought and heat stress as well as nitrogen (N) starvation, which was associated with higher water retention and cell membrane integrity than wild-type controls, modified expression of heat-shock protein genes, production of more reduced thioredoxin, elevated N accumulation and total chlorophyll content as well as up-regulated expression of nitrite reductase and N transporter genes. Further analysis revealed that the expression of other stress-related genes was also impacted in Fld-expressing transgenics. Our data establish a key role of Fld in modulating plant growth and development and plant response to multiple sources of adverse environmental conditions in crop species. This demonstrates the feasibility of manipulating Fld in crop species for genetic engineering of plant stress tolerance.

Introduction

Abiotic stresses, such as drought, salinity and extreme temperatures, are the major factors impacting plant growth and crop productivity. Most of the adverse environmental conditions inflict damages on the stressed plants through the generation of oxidative and osmotic stresses. Under adverse conditions, plants have evolved numerous mechanisms adapting to diverse environmental challenges to avoid elimination by natural selection. Gene networks involved in plant response to various abiotic stresses including stress perception, signal transduction, production of stress-related proteins and enzymes, and synthesis of compatible osmolytes and antioxidant metabolites have been extensively studied. Information obtained has been used to develop strategies to genetically engineer crop species improving plant performance under various adverse environmental conditions (Apse and Blumwald, 2002; Flowers, 2004; Mittler, 2006; Seki et al., 2003; Vinocur and Altman, 2005; Wang et al., 2003; Zhang et al., 2004; Zhou and Luo, 2013). Besides molecular strategies manipulating the expression of individual genes for structural and regulatory proteins or noncoding RNA molecules to modify endogenous systems improving plant stress tolerance, alternative approaches based on novel mechanisms derived from microorganisms can also be explored to develop additional avenues of new biotechnology tools for use in plant genetic engineering achieving improved stress tolerance in crop species.

In photo-microorganisms, substitution of stress-sensitive enzymes and proteins by resistant isofunctional versions is a typical instance responding to unfriendly environmental conditions. Flavodoxins (Flds), found in some algae and cyanobacteria, are small soluble electron transfer flavoproteins, which participate in many oxido-reductive processes in prokaryotes and eukaryotes (Blanchard et al., 2007; Coba de la Peña et al., 2013; Lodeyro et al., 2012; Singh et al., 2004; Zurbriggen et al., 2007). As redox shuttles for essential oxido-reductive pathways in the stroma, Flds are largely equivalent to those of ferredoxins (Fds), which are ubiquitous small electron transfer proteins, and play a key role in photosynthetic organisms by transferring reducing equivalents produced in the photosynthetic electron transport chain (PETC) to key enzymes including Fd-NADP reductase (FNR), Fd-nitrite and sulphite reductases, thioredoxin (Trx)/Fd-Trx reductase (FTR) and other regulatory and metabolic enzymes involved in critical cellular pathways such as nitrogen (N) and sulphur assimilation, amino acid and fatty acid metabolism, the Calvin cycle, the malate valve and other relevant processes (Balmer *et al.*, 2003; Hanke *et al.*, 2004; Knaff, 1996; Sétif, 2001; Zurbriggen *et al.*, 2008). Moreover, Flds and Fds participate in different routes of cyclic electron flow to prevent stress-elicited excessive reducing power in the PETC and the stroma (Kramer *et al.*, 2004; Munekage *et al.*, 2004), maintaining a balanced Fd redox state, which also plays a role in the intracellular signalling pathway between chloroplast and nucleus (Knaff, 1996). Given their importance as the key proteins in major metabolic pathways crucial for cell function, impaired activities of Fds would negatively impact plant growth and productivity (Holtgrefe *et al.*, 2003). Unfortunately, Fds are extremely sensitive to environmental stress and their expression is down-regulated transcriptionally and post-transcriptionally by numerous adverse environmental conditions (Holtgrefe *et al.*, 2003; Petrack *et al.*, 1998; Tognetti *et al.*, 2006; Zurbriggen *et al.*, 2008).

In photosynthetic microorganisms, the Fd protein is the preferred electron carrier under normal conditions, whereas the *Fld* gene is typically induced under environmental or nutritional hardships (Singh *et al.*, 2004; Yousef *et al.*, 2003; Zheng *et al.*, 1999). Overexpression of Fld in *Escherichia coli* enhanced bacterial oxidative stress (Coba de la Pena *et al.*, 2013; Zheng *et al.*, 1999). In higher plants, Fd retention in the plant lineage is probably related to its higher efficiency as an electron carrier, compared with Fld, which was lost along with evolution (reviewed by Karlusich *et al.*, 2014). However, *in vitro* study has demonstrated that cyanobacterial Fld is still able to function as electron carrier efficiently interacting with Fd-dependent plant partner enzymes, such as chloroplast FNR (Nogués *et al.*, 2004). Transgenic (TG) studies in tobacco show that Fld is able to complement Fd deficiency in knocked-down TG plants (Blanco *et al.*, 2011). Further study exploring the potential use of bacterial Fld in higher plants for their protection from adverse environmental conditions revealed that TG tobacco expressing Fld exhibits significantly improved plant resistance to various adverse environmental conditions including oxidative stress, high light intensities, chilling, UV radiation, phytotoxicity, iron deficiency and water deficit (Ceccoli *et al.*, 2011; 2012; Tognetti *et al.*, 2006, 2007a,b). Fld-expressing TG *Medicago truncatula* plants also exhibited less-affected N fixation in nodules by salt stress than in wild-type (WT) control plants (Coba de la Peña *et al.*, 2010). This strongly suggests that introduction of Fld into plants could serve as a new strategy for genetic engineering of plant stress tolerance. So far, however, the feasibility of using Fld in agriculturally and economically important crop species, especially in perennial crops, for enhancing plant stress tolerance has not been extensively explored, and the molecular mechanisms underlying Fld-mediated plant stress response and plant development also remain elusive. In this study, we generated TG creeping bentgrass (*Agrostis stolonifera* L.) plants ectopically expressing Fld and conducted further analysis of the TG turfgrass to investigate the role Fld plays in controlling plant development and plant response to environmental stress. Using TG approach to study the impact of Fld in an important crop species, we attempt to address the following questions. Can Fld function in perennial crops to impact plant growth? Is Fld implicated in plant stress response that contributes to enhancing plant tolerance to various abiotic stresses in grasses? And if so, what is the molecular mechanism underlying Fld-mediated plant stress response in perennial grasses?

Results

Generation and molecular analysis of TG plants expressing the cyanobacterial *Fld* gene

To explore the effectiveness of a cyanobacterial Fld in perennial grasses for improving plant response to environmental stresses, we first synthesized a 669-bp DNA fragment containing the coding sequence of a pea FNR chloroplast-targeting transit signal peptide (Newman and Gray, 1988; Serra *et al.*, 1995) translationally fused to the cyanobacterial *Fld* gene (Fig. S1a). FNR transit peptide serves to target the fusion protein into the chloroplast. The *FNR-Fld* gene was then used to prepare a chimeric gene construct, pUbi:*FNR-Fld*/ p35S:*bar* (Fig. S1b). In this construct, the *FNR-Fld* gene under the control of the corn ubiquitin (Ubi) promoter was linked to the herbicide glufosinate (phosphinothricin) resistance gene, *bar*, which was driven by the cauliflower mosaic virus 35S (CaMV35S) promoter. The construct was introduced into the creeping bentgrass cultivar, Penn A-4, to produce a total of 28 independent TG lines. RT-PCR analysis and Northern hybridization using *Fld* gene as a probe demonstrated *Fld* expression in all TG lines (see examples in Fig. S1c). Real-time PCR analysis further confirmed the high-level expression of *Fld* in transgenics (see examples in Fig. S1d). When grown and evaluated in glasshouse, the TG lines were all similar to each other in morphology, development and response to various cultivation conditions. Seven TG lines, TG4, TG5, TG6, TG16, TG17, TG23 and TG24, were selected and clonally multiplied by vegetative propagation for further analysis.

Overexpression of Fld leads to modified plant growth and development

As shown in Figure 1a–c, Fld-expressing TG plants exhibited significant difference from WT controls. Overexpression of Fld caused dramatic change in plant morphology, resulting in retarded plant growth and development. When comparing plants maintained in Elite 1200 pots with pure sand for 22 weeks, the TG plants grew significantly slower than WT controls, with the total biomass being an average of 44.7% less in fresh weight and an average of 40.9% less in dry weight, respectively (Figure 1e). However, the TG plants exhibited significantly higher tillering rate than WT controls (Fig. S2a). Upon vernalization, both WT and TG plants flowered normally (Figure 1c). However, the TG plants produced smaller inflorescence with less branches and spikelets than WT controls (Figure 1d, Fig. S2b). Interestingly, the flag leaf of each inflorescence in TG plants was much more open than that of the WT controls, with a leaf–stem angle of more than 90° (Figure 1d). Taken together, cyanobacterial Fld impacted plant development in both vegetative and reproductive stages, causing changes in plant morphology and delayed plant growth in TG creeping bentgrass.

Overexpression of Fld improves plant oxidative stress tolerance

To investigate how Fld expression in TG plants would impact plant response to environmental stress, we first set to examine performance of the Fld-overexpressing TG lines grown under oxidative stress. Eight-week-old plants were treated with 30 μM of methyl viologen (MV, Sigma-Aldrich Co. LLC, MO), an artificial acceptor and donor which accepts electrons from photosystem I and transfers them to molecular oxygen to produce reactive oxygen species (ROS, Semenov *et al.*, 2003). As illustrated in Figure 2b for plants grown in cone-tainers and subjected to MV

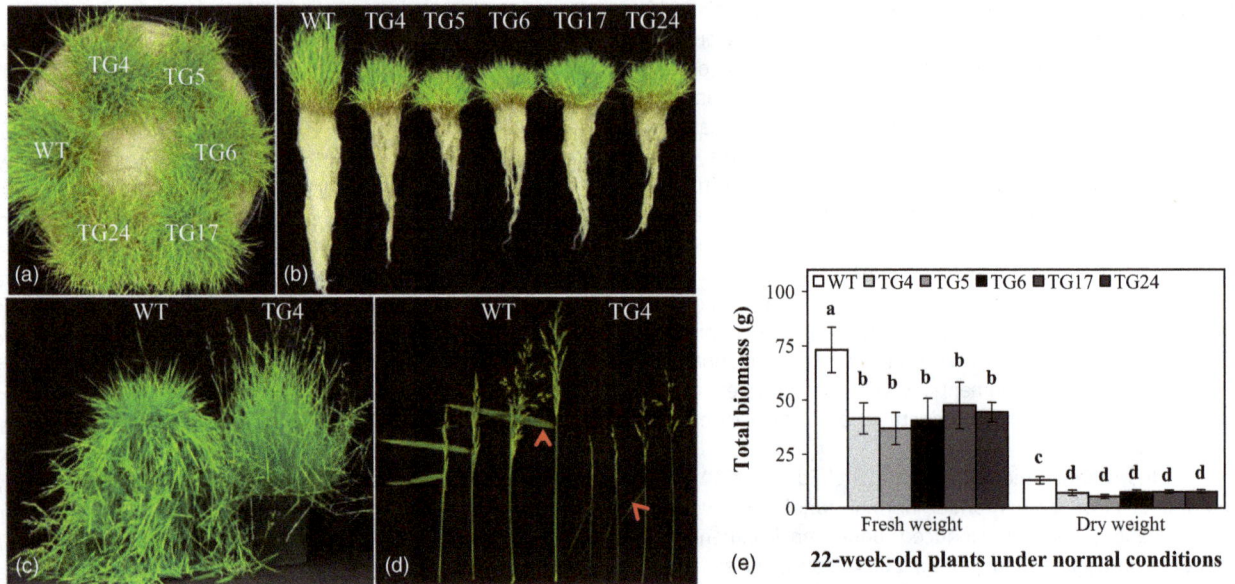

Figure 1 Overexpression of Fld leads to modified plant growth and development in transgenic (TG) creeping bentgrass. Wild-type (WT) and five TG plants developed in pure sand under normal conditions in a growth room for 22 weeks (a) were carefully removed from the Elite 1200 pots and washed briefly to display their root development (b). (c) The TG plants exhibited a different phenotype from WT controls at the reproductive stage. (d) The TG inflorescence was different from that of the WT controls in spike size and flag leaf angle. The leaf angle between the spikelet stem and the midrib of the ventral side of the flag leaf was indicated by arrows. (e) Total biomass (fresh and dry weights) of the 22-week-old TG and WT plants. The statistically significant difference between groups was determined by one-way ANOVA. Means not sharing the same letter are statistically significantly different (P < 0.05).

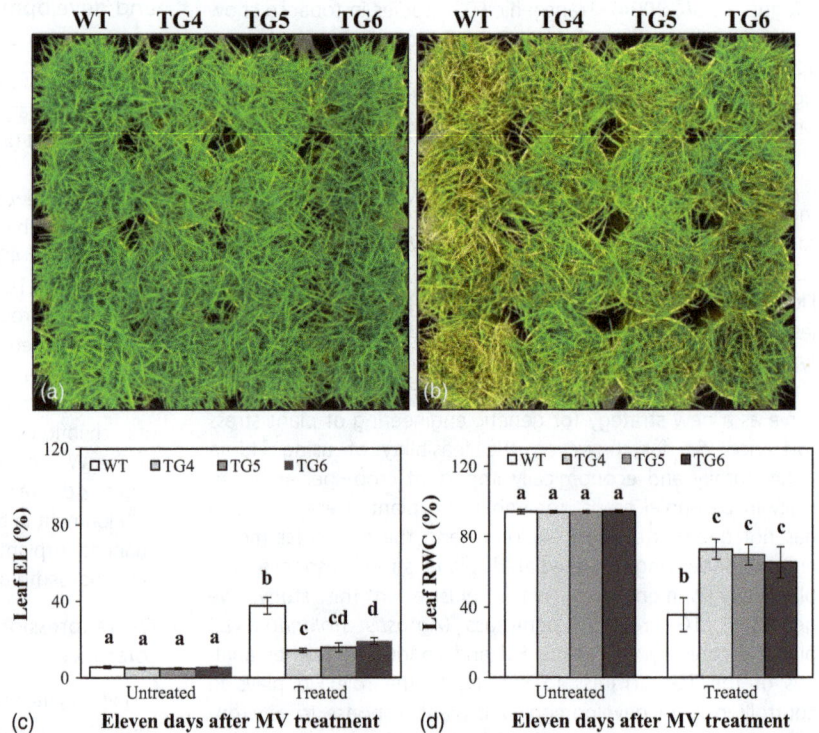

Figure 2 Transgenic (TG) turfgrass overexpressing flavodoxin (Fld) exhibits enhanced oxidative stress tolerance compared to wild-type (WT) controls. TG and WT plants were repotted in cone-tainers and grown for 10 weeks under normal maintenance (a). Fully developed plants were sprayed daily with redox-cycling herbicide and methyl viologen (MV, 30 μM with 0.02% Triton X-100) for 3 days. The TG plants exhibited enhanced resistance to MV (b) with significantly lower leaf electrolyte leakage (EL) (c), but higher relative water content (RWC) (d) than WT controls. Photographs were taken before (a) and 4 days after the 3-day MV treatments (b). The statistically significant difference in leaf EL and RWC between groups was determined by one-way ANOVA. Means not sharing the same letter are statistically significantly different (P < 0.05).

treatment for 3 days, Fld-expressing transgenics showed significantly less damage than the non-TG control plants without Fld. The difference in MV-elicited damage between TG and WT control plants became more pronounced with the increasing treatment times (Fig. S3a). Similar results were also observed for plants grown together in big pots (Fig. S3b).

Upon a 3-day MV treatment, the plants were allowed to recover by sufficient watering for 10 days, and the shoots were clipped for use in measuring EL and RWC. As shown in Figure 2c, d, although no differences were observed between the TG and WT control plants for both leaf EL and RWC under normal growth conditions, the leaf EL of the MV-treated TG plants was significantly lower than that of the MV-treated WT controls (Figure 2c), and the leaf water loss in the MV-treated Fld-expressing plants was also significantly less than that in the MV-treated control plants without Fld (Figure 2d), suggesting that under MV treatment, transgenics exhibited less cell membrane damage and enhanced water retention capacity.

Overexpression of Fld results in enhanced drought tolerance in TG plants

To examine how ectopic expression of the Fld in creeping bentgrass would impact plant response to water stress, we evaluated the performance of the Fld-overexpressing TG lines grown in sand under drought conditions. The results indicated that compared to WT controls, the TG lines tested all exhibited significantly enhanced drought tolerance. As exemplified in Figure 3a for line TG4, individual Fld-expressing TG plants (circled in red) and WT controls developed from five tillers were randomly planted together in a tray (50 × 35 × 10 cm) filled with sand. Forty days after development under normal growth conditions, drought stress was applied by water withholding for 12 days until plants were heavily suffered from water deficiency. Plants were then allowed to recover by sufficient watering for 14 days. As shown in Figure 3a–c, the TG plants were all recovered from the drought-elicited damage with more than 80% of the tillers alive, whereas almost all the WT control plants died. Moreover, the TG roots and shoots displayed significantly less growth inhibition than WT controls under drought stress conditions. As a result, TG plants produced more biomass than WT controls as reflected by significantly higher fresh weights of roots and shoots in TG plants than in WT controls (Figure 3d). Similar results of enhanced drought tolerance in Fld transgenics were also obtained in various TG lines tested in cone-tainers (Fig. S4).

Overexpression of Fld increases plant heat tolerance that is associated with modified expression of heat-shock protein (HSP) genes and the production of more reduced Trx

To investigate whether or not ectopic expression of the cyanobacterial Fld would impact plant thermotolerance, we evaluated the performance of the Fld-overexpressing TG plants under heat stress. As demonstrated in Figure 4, the TG plants grown in cone-tainers outperformed WT controls under heat stress conditions (Figure 4b–e). While the control plants without Fld became severely damaged (carbohydrate deprivation) 17 days after heat exposure, the Fld-expressing TG plants only exhibited minor heat-inflicted symptom (Figure 4b–e). All the TG plants tested recovered upon release from the stress, whereas most of the WT controls did not survive the treatment (Figure 4c, e). The enhanced heat tolerance observed in the TG plants was associated with a lower leaf cell EL than the WT controls under the heat stress (Figure 4f), suggesting an enhanced cell membrane integrity in Fld-expressing TG plants compared to the non-TG WT controls.

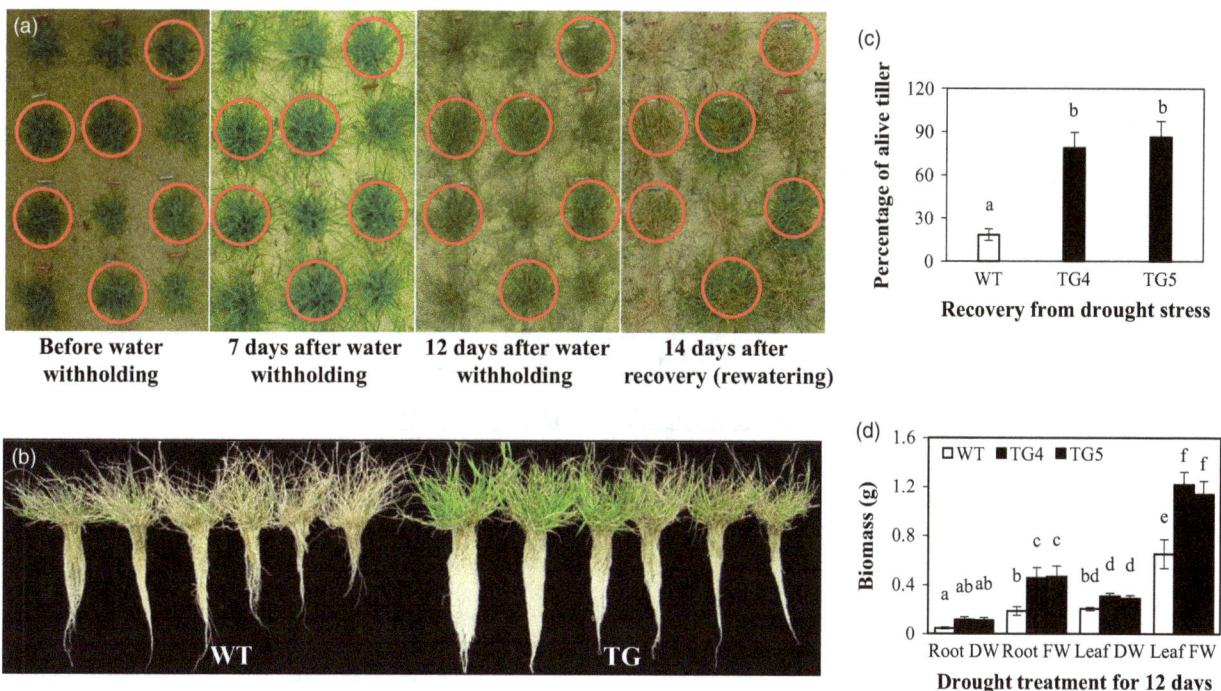

Figure 3 Transgenic (TG) turfgrass overexpressing flavodoxin (Fld) exhibits enhanced drought tolerance compared to wild-type (WT) controls. Individual TG and WT plants originated from five tillers were repotted randomly in a tray (50 × 35 × 10 cm) and developed for 40 days under normal growth conditions (a). The plants were then treated by water withholding for 12 days until plants were heavily suffered, followed by recovery with sufficient watering for 12 days (a). The TG plants exhibited much higher tolerance to drought stress under water deprivation conditions and recovered much faster after rewatering than WT controls (a, b). The percentage of the survived tillers in TG plants was significantly higher than that in WT controls (c). Similar results were also obtained for plant biomass. TG plants exhibited significantly higher biomass than WT controls (d). Photographs were taken before and during water withholding and after recovery. TG plants were circled in red. The figure shows results from two representative TG lines (TG4 and TG5). The statistically significant difference in percentage of the survived tillers and biomass between groups was determined by one-way ANOVA. Means not sharing the same letter are statistically significantly different ($P < 0.05$).

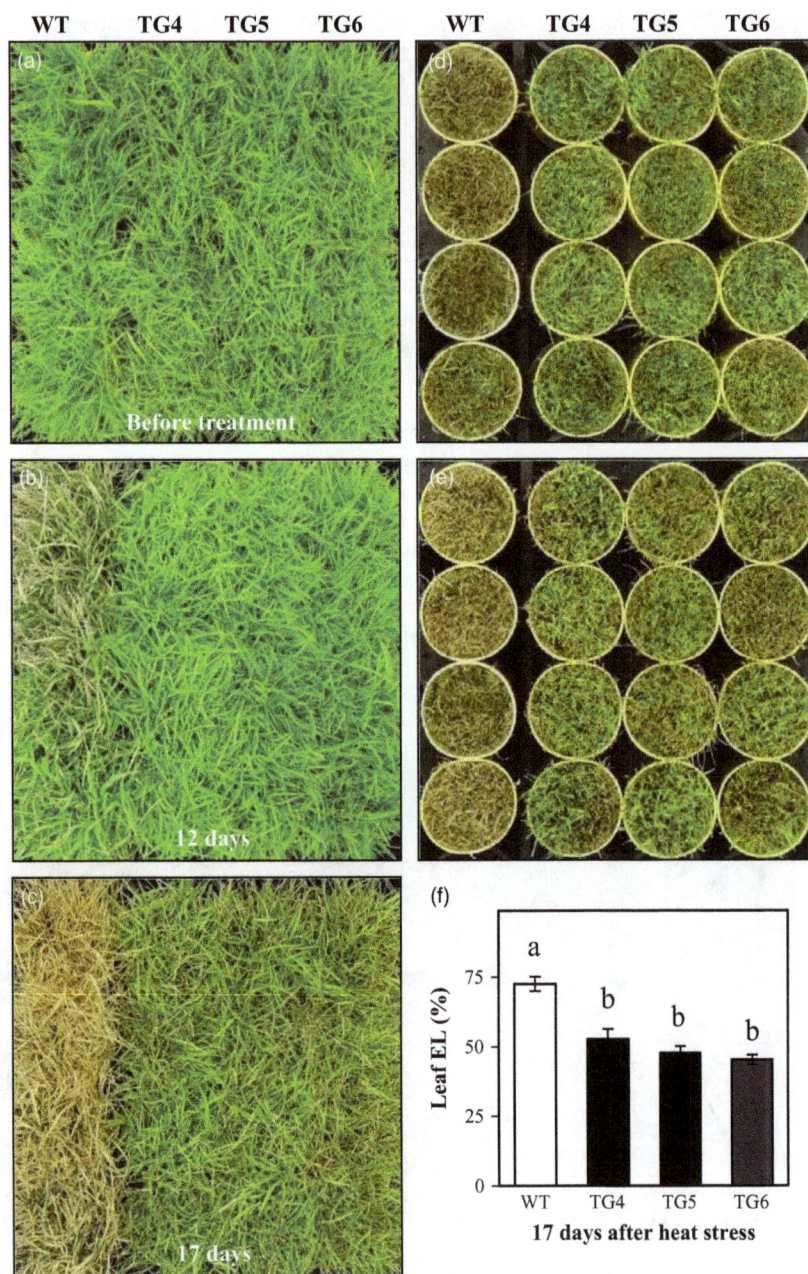

Figure 4 Overexpression of flavodoxin (Fld) enhances creeping bentgrass heat stress tolerance. Transgenic (TG) lines (TG4, 5 and 6) and wild-type (WT) control plants were repotted in cone-tainers and arranged in hexagon shape. After 10-week full development under normal maintenance (a), WT and TG plants were exposed to 35 °C/40 °C (night/daytime) treatment regime for 17 days (b, c). The TG lines showed significantly enhanced heat tolerance compared to WT controls (b, c). After heat stress, shoots were clipped (d) for electrolyte leakage (EL) measurement. The shoot recovery of TG plants was much better than that of the WT controls after clipping and recovery for 1 week (e). The EL of the WT plants was significantly higher than that of the three independent TG lines (f). Photographs were taken on indicated dates. The statistically significant difference in leaf EL between groups was determined by one-way ANOVA. Means not sharing the same letter are statistically significantly different ($P < 0.05$).

To better understand the molecular mechanism of Fld-mediated plant heat tolerance, we analysed the expression of the three creeping bentgrass small *HSP* (*sHSP*) genes, *AsHSP17* (Sun *et al.*, 2016), *AsHSP26.7* and *AsHSP26.8* (Wang and Luthe, 2003) with molecular weight of 17, 26.7 and 26.8 kD, respectively. The results showed that under normal growth conditions, the expression of the three *sHSP* genes was all extremely low in both the TG plants and WT controls (Fig. S5), and no significant difference was observed between the TG plants and WT controls for *AsHSP26.7* and *AsHSP26.8* (Figure 5c, e). However, the expression of the *AsHSP17* in the TG plants was significantly higher than that in WT controls (Figure 5a). Upon heat stress (40 °C for 4 h), the three *AsHSP* genes in both WT and the TG plants were all significantly induced. Compared to the normal growth conditions, the heat stress-induced expression of the *AsHSP17*, *AsHSP26.7* and *AsHSP26.8* increased 3.8×10^4, 2.8×10^5 and 350 times,

respectively, in WT controls, and 1300, 3.7×10^4 and 3500 times, respectively, in the Fld transgenics. The expression of the *AsHSP17.0* and *AsHSP26.7* was significantly lower (Figure 5b, d), whereas that of the *AsHSP26.8* was significantly higher (Figure 5f), in the TG plants than in WT controls.

To investigate what role Fld plays in plant redox shuttling under stress conditions, we conducted tests to compare reduced Trx contents between the Fld-expressing transgenics and WT controls under normal and heat stress conditions. As illustrated in Figure 6, although no significant difference was observed between the TG plants and WT controls in the expression of one of the *Trx* gene, *AsTrx h* (Buchanan, 2016), under both normal and heat stress conditions (Figure 6a), there was more reduced Trx produced in the TG plants than in WT controls under heat treatments, especially under prolonged heat stress (Figure 6b).

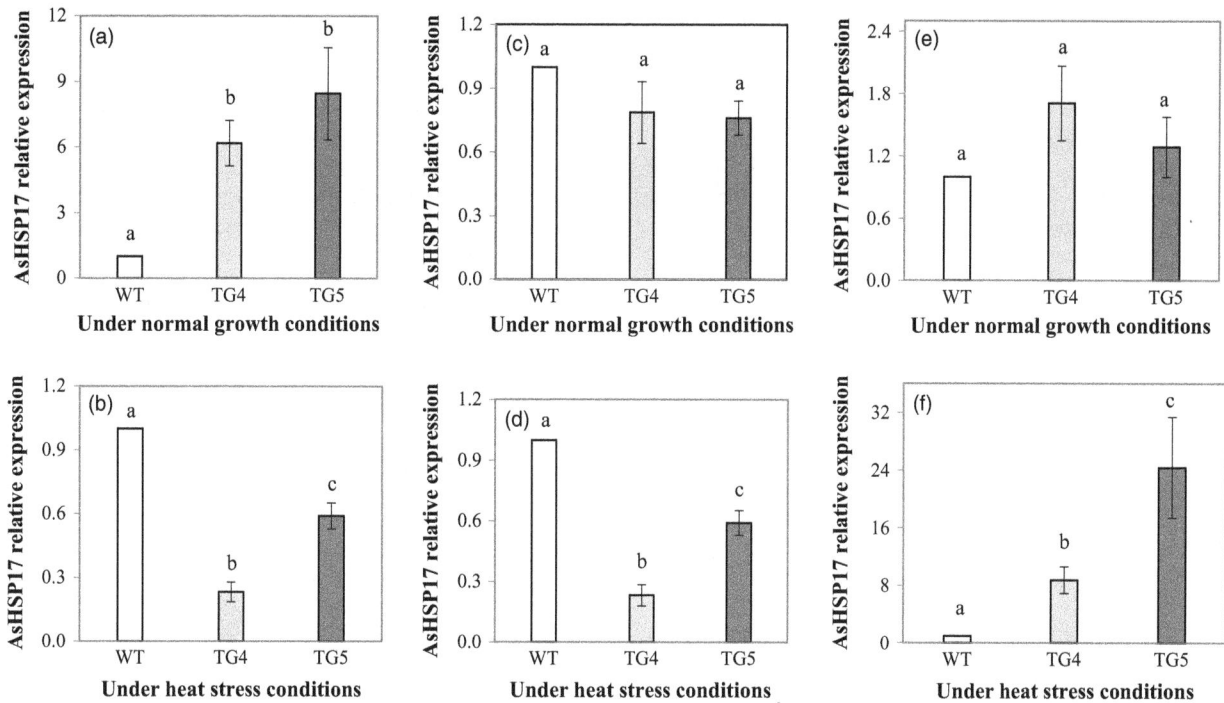

Figure 5 Expression profiles of three heat-shock protein (HSP) genes, *AsHSP17* (a, b), *AsHSP26.7* (c, d) and *AsHSP26.8* (e, f), in transgenic (TG) and wild-type (WT) control plants under normal growth and heat stress conditions. ΔΔCt method was used for real-time RT-PCR analysis. Two reference genes, *AsACT1* and *AsUBQ*, were used as endogenous controls and showed similar results. The data presented are those using *AsUBQ* as the endogenous controls (Zhou *et al.*, 2013). Three biological replicates and three technical replicates were used for statistic analysis. Error bars represent SE (n = 9). The statistically significant difference between groups was determined by one-way ANOVA. Means not sharing the same letter are statistically significantly different ($P < 0.05$).

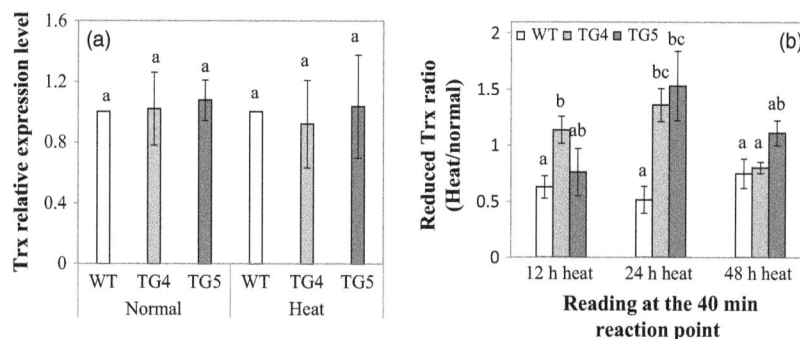

Figure 6 Thioredoxin (*Trx*) gene expression and reduced Trx content in transgenic (TG) and wild-type (WT) control plants. (a) Expression of the *Trx* gene, *Trxh* in TG and WT control plants under normal and heat stress conditions. ΔΔCt method was used for real-time RT-PCR analysis. Two reference genes, *AsACT1* and *AsUBQ*, were used as endogenous controls and showed similar results. The data presented are those using *AsUBQ* as the endogenous control (Zhou *et al.*, 2013). Three biological replicates and three technical replicates were used for statistic analysis. Error bars represent SE (n = 9). (b) Ratio of the reduced Trx contents under heat stress and normal conditions in TG and WT control plants. The statistically significant difference between groups was determined by one-way ANOVA. Means not sharing the same letter are statistically significantly different ($P < 0.05$).

Fld transgenics exhibit enhanced tolerance to N starvation associated with elevated N accumulation and total chlorophyll content as well as up-regulated expression of NiR and N transporter genes

N is an essential nutrient for plant growth and development. A number of plant regulatory and metabolic enzymes, such as Fd-NiR that catalyses the reduction of nitrite to ammonia, are involved in N assimilation and metabolism, and many of them use Fd or Fld as electron donor (Arizmendi and Serra, 1990; Zurbriggen *et al.*, 2008). This prompted us to investigate the potential role Fld may play in plant adaptation to N starvation and plant N assimilation. To this end, we compared plant growth under various N concentrations and measured relevant biochemical, physiological and molecular parameters in both TG and WT control plants. As demonstrated in Figure 7a for plants subjected to a 5-week-long treatment with different N supplies, the Fld-expressing TG plants displayed greener shoot colour than WT controls under N starvation conditions (0, 0.4 and 2 mM). Further analysis revealed that compared to N-sufficient (10 mM) or N-excessive plants (40 mM), plants grown under N deficiency conditions (0, 0.4 and 2 mM) had reduced total chlorophyll contents (Figure 7b). While the total chlorophyll contents were similar between the TG plants and WT controls under N-sufficient and N-excessive conditions, a significantly higher chlorophyll content was observed in the TG plants than in WT controls under N-starved conditions (Figure 7b), suggesting that the TG plants may be less prone to chlorophyll degradation and, therefore, likely maintain a higher capacity in photosynthesis than WT controls under N starvation conditions. Moreover, shoot and root dry weights of the TG plants were higher than those of the WT controls 5 weeks after 0.4 mM N treatment although this difference was statistically insignificant in shoot dry weight (Figure 7c, d). Interestingly, suboptimum N supplies inhibited shoot growth, but appeared to promote root growth in both the TG and WT control plants (Figure 7c, d, Fig. S6). It is also noteworthy that over-fertilization (40 mM N concentration) appeared to inhibit plant growth as reflected by the reduced shoot and root biomass compared to plants under optimum N supplies (Figure 7c, d, Fig. S6). This inhibition in WT controls was greater than that in the TG plants although the difference was statistically insignificant (Figure 7c, d). Taken together, these results indicated that overexpression of Fld

positively impacted plant response to N starvation and may have contributed to maintaining plant photosynthesis under N deficiency conditions.

To further understand what caused the differential growth between the Fld-expressing transgenics and WT control plants under N deprivation conditions, we measured shoot and root total N contents in the TG and WT control plants under various N supplies (0, 0.4, 2, 10 and 40 mM). The results indicated that the higher the concentration of the N solution applied, the more the total N amount that plants accumulated in shoots and roots (Figure 8a, b). This tendency in N accumulation was more pronounced in shoot than in root. TG shoots and roots accumulated more N than WT controls under N-starved conditions, and the differences in N accumulation between TG and WT control plants were mostly significant (0, 0.4 and 2 mM, Figure 8a, b), suggesting an enhanced N use efficiency (NUE) in TG plants. To investigate what caused the enhanced NUE in Fld transgenics, we examined the transcript levels of the genes encoding a high affinity nitrate transporter and the NiR, the key enzymes in N assimilation pathway. As shown in Figure 8c, the expression of the nitrite transporter (Kotur et al., 2013) gene, AsNRT, was significantly up-regulated in TG plants in comparison with WT controls. Further analysis revealed that there was no significant difference in AsNiR (KR911829) expression between TG and WT control plants (Figure 8d). However, the enzyme

Figure 7 Flavodoxin (Fld)-expressing transgenic (TG) plants exhibit enhanced tolerance to nitrogen (N) starvation. (a) Fully developed TG and wild-type (WT) plants in cone-tainers were maintained under normal growth conditions for 9 weeks and then treated with 1× MS medium containing 0, 0.4, 2, 10 or 40 mM nitrate (15 mL per day) for 5 weeks. Leaf total chlorophyll content was measured for both TG and WT control plants (b). Shoot and root tissues (n = 5) were harvested and processed for dry weight (c, d) measurement. TG4 was shown as a representative TG line for plant response to N starvation. The statistically significant difference in total chlorophyll, shoot and root dry weights between groups was determined by one-way ANOVA. Means not sharing the same letter are statistically significantly different (P < 0.05).

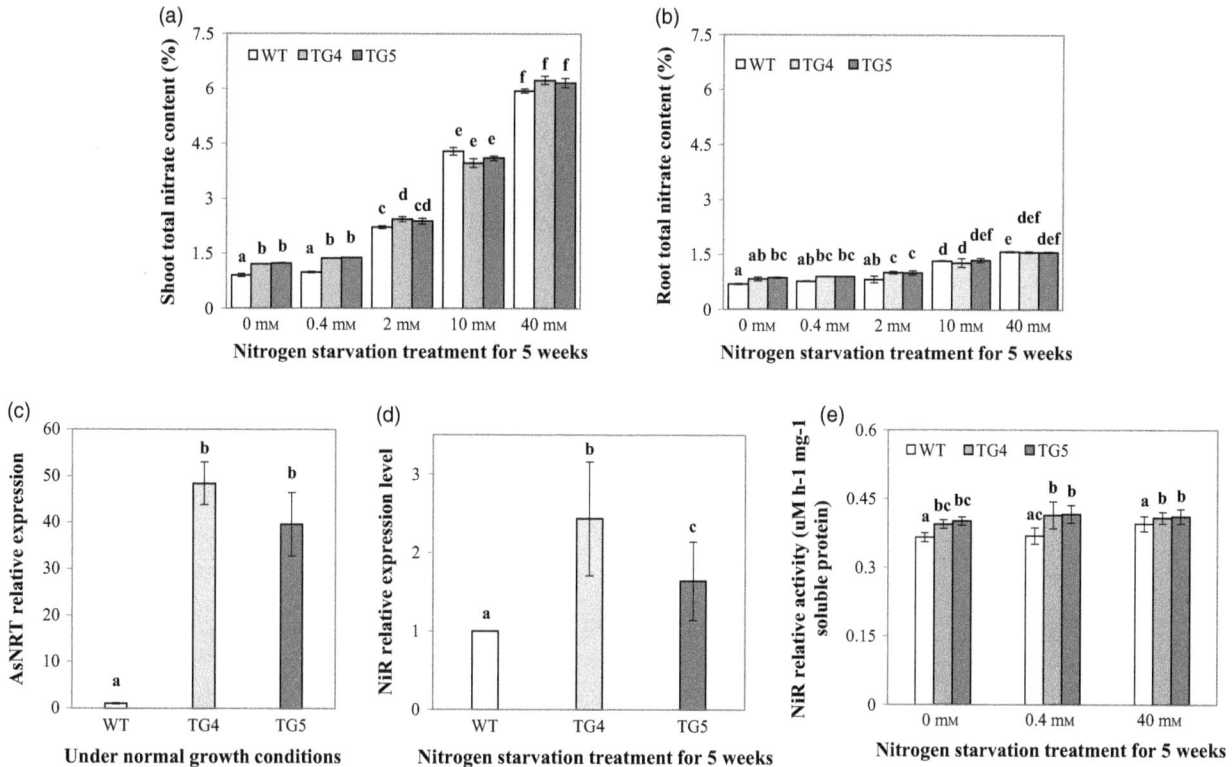

Figure 8 The impact of flavodoxin (Fld) on plant nitrogen (N) uptake and assimilation. (a) Shoot total N content (n = 4) of wild-type (WT) and transgenic (TG) plants measured 5 weeks after application of different N concentrations. (b) Root total N content (n = 4) of WT and TG plants measured 5 weeks after application of different N concentrations. (c) Expression of the N transporter *AsNRT* in TG and WT control plants under normal growth conditions. (d) Expression of the nitrate reductase (*NiR*) gene in TG and WT control plants 5 weeks after N starvation. (e) NiR activity in TG and WT control plants under normal and N starvation conditions. The statistically significant difference between groups was determined by one-way ANOVA. Means not sharing the same letter are statistically significantly different ($P < 0.05$).

activity of the NiR was significantly higher in the TG plants than in WT controls under N starvation conditions (Figure 8e).

The impact of Fld on the expression of other stress-related genes

To examine how Fld affects other stress-related genes in TG plants, we cloned partial sequences of six creeping bentgrass genes encoding AsDREB2A (dehydration response element binding protein, 2A), AsDREB2B (dehydration response element binding protein, 2B) (Matsukura *et al.*, 2010; Mizoi *et al.*, 2013), AsCP450 (cytochrome P450 94 family-like) (Aubert *et al.*, 2015), AsRAP (ethylene response transcription factor) (Hinz *et al.*, 2010), AsNAC (NAC domain protein) (Olsen *et al.*, 2005; Yao *et al.*, 2012; Zhou *et al.*, 2013) and AsPR1 (plant pathogenesis-related protein) (Sels *et al.*, 2008) that are highly homologous to their counterparts in *Brachypodium distachyon*. Gene expression analysis using real-time RT-PCR analyses demonstrated that under normal growth conditions, the expression of the *AsDREB2A* (Figure 9a), *AsDREB2B* (Figure 9b) and *AsCP450* (Figure 9c) all went down sharply in the Fld-expressing TG plants compared to WT controls. However, the expression of these genes in the Fld TG plants was significantly up-regulated upon heat stress, showing no significant difference from, or higher than that of the WT controls (Figure 9a–c). The expression of an ethylene response transcription factor (*ERF*) gene, *AsRAP*, was also down-regulated in the Fld TG plants under normal growth conditions, but significantly up-regulated upon heat stress compared to WT

controls (Figure 9d). Another stress-related transcription factor gene, *AsNAC*, of the Fld TG plants showed no difference in expression from that of the WT controls under normal growth conditions (Figure 9e, left panel). However, *AsNAC* expression was significantly induced in the majority of the TG plants upon heat stress and significantly higher than that of the WT controls (Figure 9e, right panel). Similarly, *AsPR1* expression in the Fld TG plants showed no difference from that in WT controls under normal conditions, but was markedly induced and significantly higher than that in WT controls upon heat stress (Figure 9f). Taken together, these data demonstrated that Fld overexpression resulted in enhanced expression of a number of important stress-related genes in TG creeping bentgrass plants under stressful conditions.

Discussion

Chloroplast Fd, a mobile electron carrier to distribute reducing equivalents generated in the PETC during photosynthesis to various essential metabolic, regulatory and dissipative pathways, is critical in the physiology of the plant cell. The ubiquitous small electron transfer protein delivers electrons to key enzymes including FNR, Fd-nitrite and sulphite reductases, gluoxoglutarate aminotransferase, fatty acid desaturase and FTR (Balmer *et al.*, 2003; Hanke *et al.*, 2004; Knaff, 1996; Sétif, 2001; Zurbriggen *et al.*, 2008). As such, the steady supply of reduced Fd in plant cell is essential for the function and regulation of many important

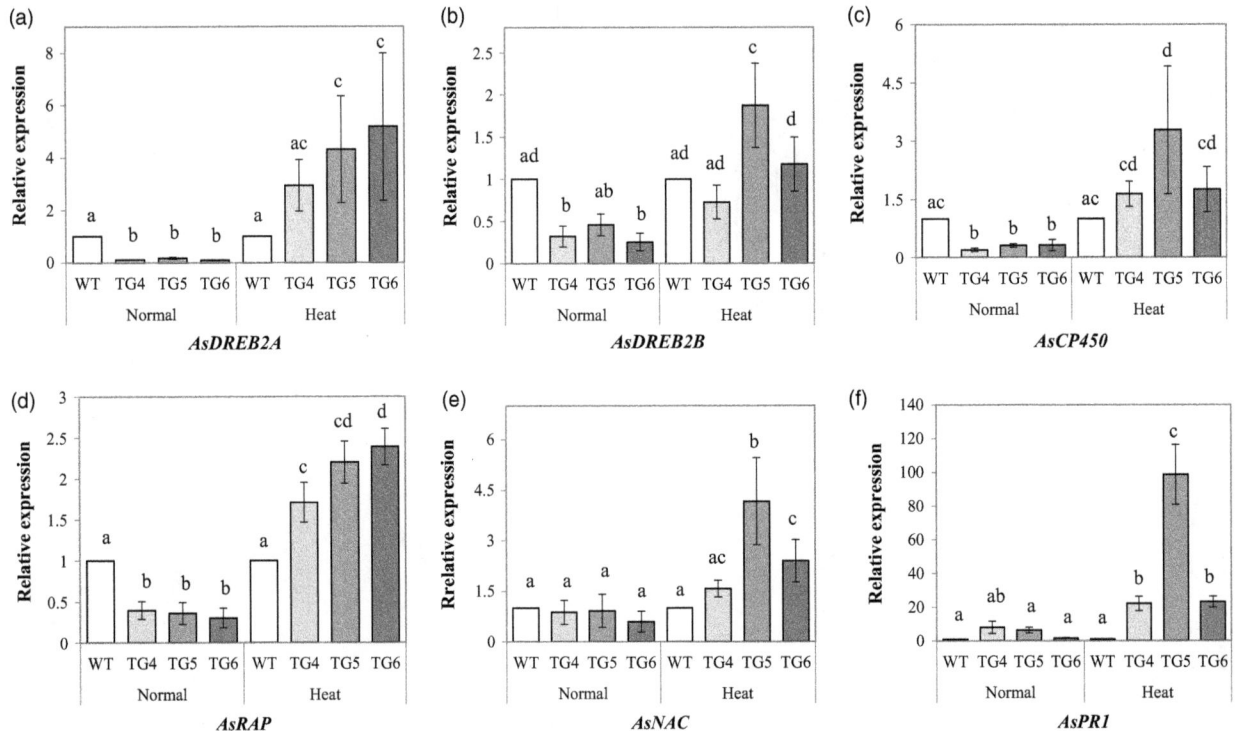

Figure 9 Expression profiles of six stress-related genes, *AsDREB2A* (a), *AsDREB2B* (b), *AsCP450* (c), *AsRAP* (d), *AsNAC* (e) and *AsPR1* (f) in transgenic (TG) and wild-type (WT) control plants under normal growth and heat stress conditions. ΔΔCt method was used for real-time RT-PCR analysis. Two reference genes, As*ACT1* and As*UBQ*, were used as the internal controls (Zhou *et al.*, 2013). Three biological replicates and three technical replicates were used for statistic analysis. Error bars indicate SE (n = 9). The statistically significant difference between groups was determined by one-way ANOVA. Means not sharing the same letter are statistically significantly different (*P* < 0.05).

cellular pathways and metabolisms including carbon fixation and allocation, N and sulphur assimilation, amino acid synthesis and fatty acid desaturation. This multiplicity of functions of Fd in plants suggests its potential role in various biological processes critical for plant development and plant response to environmental cues. Indeed, diminished leaf Fd content by antisense suppression of the *Fd* gene expression in TG potato plants altered electron distribution and photosynthesis, resulting in a pleiotropic phenotype in TG plants such as lower CO_2 assimilation rates, progressive loss of chlorophyll in a dynamic process, and therefore pale green or yellowish leaves over time, perturbed distribution of electrons, lower Trx reduction and photoinhibition and decreased growth rates (Holtgrefe *et al.*, 2003).

Fld plays a vital role in photosynthetic microorganisms as an alternative electron carrier flavoprotein under adverse environmental conditions. Cyanobacterial Fld can efficiently substitute Fd of higher plants in most electron transfer processes under stressful conditions and has been demonstrated in TG tobacco plants conferring increased tolerance to different abiotic stresses (Tognetti *et al.*, 2006, 2007a,b; Zurbriggen *et al.*, 2008). However, Fld-mediated improvement in plant stress response was not observed in *M. truncatula* although compared to WT controls, Fld-expressing TG *M. truncatula* plants exhibited less-affected N fixation in nodules by salt stress (Coba de la Peña *et al.*, 2010). In this study, we investigated the impact of Fld on creeping bentgrass, an important perennial crop species. Our data revealed that overexpression of Fld in creeping bentgrass resulted in altered plant growth and development, and most significantly, Fld-expressing transgenics exhibited enhanced tolerance to multiple sources of adverse environmental conditions, including

oxidative stress, water deficit, heat stress and N starvation. Our results suggest that like in model species, tobacco, Fld is also functional in perennial crop plants to compensate for the decline and impaired operation of Fd in the chloroplast of plants subjected to stressful conditions. It should be noted that FNR transit peptide has previously been demonstrated to efficiently target the FNR-Fld fusion protein into the tobacco chloroplast. In TG plants harbouring *FNR-Fld* fusion gene, Fld was successfully targeted to chloroplast, and the TG tobacco exhibited enhanced abiotic stress tolerance compared with WT controls and the TG plants harbouring *Fld* gene alone, in which, Fld was only cytoplasm-localized (Tognetti *et al.*, 2006). The same strategy used in our study proves to be successful. TG creeping bentgrass produced expresses *FNR-Fld* fusion gene and exhibits altered plant growth, development and response to abiotic stress in comparison with non-TG WT control plants.

In this study, Fld transgenics exhibited significantly altered plant development under normal growth conditions such as reduced biomass, delayed plant growth and altered inflorescence (Figure 1). Sakakibara *et al.* (2012) studied the protein–protein interaction of Fd and NiR by NMR spectroscopy and found that although Fds from higher plant (maize) and cyanobacterium (*Leptolyngbya boryana*) share high structural similarities, they differ significantly in the interaction with cyanobacterial NiR, highlighting the different molecular interaction between Fd and partner enzyme. Similarly, we speculate that in Fld TG creeping bentgrass, both Fld and Fd bind to partner enzymes under normal growth conditions, but the different interactions of these two flavoproteins with the same partners may alter the way the reducing equivalents generated in the PETC are distributed to various essential metabolic, regulatory and dissipative pathways therefore impacting plant growth and development.

In photosynthesizing chloroplast, rapid transients of photon capture, electron fluxes and redox potentials cause ROS to be released. As has previously been demonstrated that in Fld-expressing TG tobacco plants, the introduced flavoprotein exhibited antioxidant activity under stressed conditions with lower ROS accumulation (Tognetti et al., 2006). The Fld-mediated ROS dissipation and scavenging resulted in reduced oxidative damage to sensitive enzymes, membranes, pigments and photosynthesis (Tognetti et al., 2006). As the toxic by-products of aerobic metabolism, ROS also functions as one of a network of diverse signals. ROS signalling is a central component of the retrograde signalling network from the photosynthesizing chloroplast to the cytosol, mitochondrion and nucleus (Chan et al., 2016; Dietz, 2015). As key members in the complex signalling network of cells, ROS play a multitude of signalling roles in different organisms (D'Autréaux and Toledano, 2007; Dietz et al., 2016; Mittler et al., 2011). Hence, it is conceivable that overexpression of Fld in TG creeping bentgrass may have changed cell ROS homeostasis, which probably impacted related ROS-mediated signalling pathways, resulting in altered plant growth and development compared to WT controls (Figure 1).

It is noteworthy that Fld-mediated significant change in plant development in TG creeping bentgrass (Figure 1) was not observed in Fld-expressing tobacco plants (Tognetti et al., 2006, 2007a,b; Zurbriggen et al., 2008). These contrasting Fld effects on plant growth may imply a potentially differential regulation of the ROS signalling between dicot and monocot plant species. The regulation of plant growth and development in evolutionarily advanced monocot species may be more fine-tuned than that in dicot species therefore prone to being impacted by any altered regulation machinery. It would be interesting to find out whether the similar phenomenon could also be observed in other monocot crops, and we are currently conducting research in rice to test this hypothesis.

Consistent with the observations in Fld-expressing TG tobacco plants (Tognetti et al., 2006, 2007a,b; Zurbriggen et al., 2008), the Fld-expressing TG creeping bentgrass also exhibited enhanced tolerance to multiple adverse environmental conditions including oxidative, heat and drought stress and N starvation. The enhanced stress tolerance mediated by Fld was associated with changes in various factors known to be involved in plant stress response. For instance, sHSPs present in virtually all organisms are stress-induced molecular chaperones. They bind and stabilize their client proteins that have become denatured under stress conditions. Most of them are highly up-regulated in response to heat and have a clear role in thermotolerance (Atkinson and Urwin, 2012; Merino et al., 2014; Sun et al., 2016; Waters, 2012). Three AsHSP genes examined in this study, AsHSP17 (Sun et al., 2016), AsHSP26.7 and AsHSP26.8 (Wang and Luthe, 2003), were all significantly induced, but differentially regulated in WT and the TG plants (Figure 5), suggesting Fld implication in triggering one of the important stress response mechanisms, contributing to an enhanced plant tolerance to heat.

Trx, a class of small, ubiquitous redox proteins, plays a central role in various important biological processes, including distribution of reducing equivalents generated during photosynthesis. In algae, Fd has been proven to be a more efficient electron carrier than Fld in most reactions assayed in vitro, including Trx reduction by FTR in reconstituted systems, and a strongly preferred electron donor over Fld for nitrate reduction via nitrite reductase (NiR) and

glutamine synthetase (Meimberg and Mühlenhoff, 1999; Vigara et al., 1998). Our study indicates that overexpression of Fld in the TG creeping bentgrass led to an enhanced electron delivery efficiency in the PETC, producing more reduced Trx beneficial to plant stress response (Figure 6b).

Fld-mediated resistance to multiple stresses in a perennial grass species further confirmed the versatility of Fld in being fully operational in plant cell as an alternative intermediate for the PETC, strongly suggesting that under harsh environmental conditions, Fld could replace Fd in many, if not all, chloroplast-based pathways to maintain important metabolism within the cell and protect plants from damages elicited by the adverse environmental conditions. It should be noted that as observed in Fld-expressing M. truncatula (Coba de la Peña et al., 2010), Fld-expressing creeping bentgrass did not show significant difference from WT controls in plant response to salinity stress (Fig. S7), indicating that Fld is ineffective in prevention of all the alterations produced by oxidative damage in plants. The salt-triggered ROS might induce both oxidative and toxic damages that Fld may not be able to counteract.

Taking together, the broad-range stress tolerance exhibited in TG creeping bentgrass overexpressing Fld highlights the important role of this flavoprotein in protecting plants from multiple sources of adverse environmental conditions. The results obtained strongly suggest that using Fld, similar strategies could also be developed in important food crops to improve plant performance under adverse environmental conditions for enhanced agricultural production.

Experimental procedures

Plasmid construction

A 669-bp DNA fragment containing the coding sequence of a pea FNR chloroplast-targeting transit signal peptide (Newman and Gray, 1988; Serra et al., 1995; X12446) translationally fused to the cyanobacterial Fld gene (S68006) was chemically synthesized by Integrated DNA Technology (Coraville, IA) and used to produce a chimeric gene construct, pUbi:FNR-Fld/p35S:bar. The vector contains the corn ubiquitin (Ubi) promoter driving FNR-Fld and the cauliflower mosaic virus 35S (CaMV35S) promoter driving the bar gene for glufosinate (phosphinothricin) resistance as a selectable marker. The vector was delivered into the A. tumefaciens strain, LBA4404 for plant transformation.

Plant transformation, propagation, maintenance and stress treatments

A commercial creeping bentgrass cultivar Penn A-4 (supplied by HybriGene, Hubbard, OR) was employed for genetic transformation. The generation and nursing of TG plants were carried out as previously described (Luo et al., 2004a,b). The FNR-Fld TG and WT control plants initially maintained in glasshouse were moved to a growth room with a 14-h photoperiod for propagation. Both TG and WT control plants were clonally propagated from a single tiller and grown in cone-tainers (4.0 × 20.3 cm; Dillen Products, Middlefield, OH), 4-inch or 6-inch pots (Dillen Products) using pure silica sand or commercial soil (Fafard 3-B Mix, Fafard Inc., Anderson, SC). Shoots were trimmed weekly to maintain uniform plant growth. Growth room conditions were set up as previously described (Li et al., 2010).

To study plant growth and development, individual plants of both TG and WT controls were developed from a single tiller or 15 tillers for 12–22 weeks in cone-tainers, 6-inch or Elite 1200 pots (27.9 cm × 24.6 cm, Middlefield, OH) without/with clipping. To characterize vegetative-to-reproductive transition, 10-week-old unclipped plants were moved into a cold room for vernalization (8-h photoperiod with a light supply of 120–170 µmol m^{-2} s^{-1} at 5 °C). After 30 weeks of continuous cold treatment, the grasses were transferred into a growth room with long-day (LD) light regime (at 17–25 °C with a 16-h photoperiod of 350–450 µmol m^{-2} s^{-1} light supply) for flowering.

For stress treatments, TG and WT control plants were propagated from stolons as previously described (Li et al., 2010). For water stress, three to eight replicates of both TG and WT control plants grown in trays (57 × 48 × 11 cm^3) and in cone-tainers with pure sand were maintained in growth room for 6–14 weeks. The plants were then subjected to drought stress by water withholding after a saturated watering.

For heat and oxidative stress treatments, four replicates of both TG and WT control plants grown in cone-tainers with pure sand were maintained in growth room for 8 weeks. The plants were then subjected to heat stress as previously described (Li et al., 2013). The plant response to oxidative stress was assessed by daily spray of 30 µM of methyl viologen (MV, Sigma-Aldrich Co. LLC) with 0.02% Triton X-100 for 3 days.

To test the performance of WT and TG plants under different concentrations of N, five replicates of both TG and WT control plants were grown in cone-tainers with pure sand and developed in growth room for 9 weeks. The plants and the sand were flushed using sufficient water to remove residual nutrients and then natured using modified 1× Murashige and Skoog (MS) solution supplemented with N at different concentrations (0, 0.4, 2, 10 or 40 mM). The preparation of MS solution and N addition was as previously described (Yuan et al., 2015). Three and five weeks after N starvation treatment, the shoots were harvested for further analysis to examine expression of the genes of interest and measure various physiological parameters. Plants were recovered from N stress by nurturing with 200-ppm fertilizer and photographed for documentation.

RNA isolation, cDNA synthesis, qPCR, RT-PCR and northern blot

One hundred milligrams of young leaf tissues was employed for total RNA isolation using Trizol reagent (Invitrogen, Carlsbad, CA), and 2 µg of total RNAs was reverse-transcribed using ProtoScript® II Reverse Transcriptase (New England Biolabs, GA) following manufacturer's instructions. Fourfold diluted first-strand cDNAs were stored at −20 °C for future use.

RT-PCR for FRN-Fld expression determination was conducted using the first-strand cDNAs and gene-specific primers (Table S1). qPCR was carried out on an iCycler iQ system (Bio-Rad, Hercules, CA) in 25 µL of PCR solution containing 10 ppm SYBR Green I, 200 µM dNTPs, 1× PCR buffer, 1.5 mM MgCl$_2$, Taq DNA polymerase and 40 nM each primer. There were three technical replicates for each of the three biological replicates. PCR was conducted with the following program: an initial DNA polymerase activation at 95 °C for 180 s followed by 40 cycles of 95 °C for 30 s, 60 °C for 20 s and 72 °C for 20 s. Finally, a melting curve was performed, and the PCR products were checked with 2% agarose gel in 0.5× TBE buffer with ethidium bromide. The ΔΔCt method was used for real-time PCR analysis. Two reference genes, AsACT1 and AsUBQ (Zhou et al., 2013), were used as endogenous controls. Relative expression level was calculated using the 2$^{-\Delta\Delta Ct}$ formula. All primer pairs used for examining the expression levels of grass endogenous genes were designed

based on the cloned creeping bentgrass cDNA sequences and listed in Table S1. For northern blot, a 513-bp Fld gene fragment amplified from plasmid was used as probe. Probe was labelled with the [α-^{32}P] dCTP using the Prime-It II Random Primer Labeling Kit (Stratagene, La Jolla, CA). RNA blot (10 µg of total RNAs) hybridizations were performed in Church buffer at 68°C. Hybridization signals were detected by exposure on a phosphor screen at room temperature overnight and scanning on a Typhoon 9400 phosphorimager (GE Healthcare Bio-Sciences Corp., Piscataway, NJ).

Measurement of leaf relative water content (RWC), electrolyte leakage (EL) and chlorophyll content

Plant leaf RWC, EL and total chlorophyll content were measured as previously described (Li et al., 2010).

Measurement of reduced Trx content and NiR activities

Protein was extracted using Tris buffer of pH7.5 containing 100 mM Tris, 1 mM EDTA, 2 mM MgCl$_2$, 20 mM DTT, 0.1 mM phenylmethanesulfonyl fluoride (PMSF). DTT and PMSF were added before use. Briefly, leaf tissues (0.1 g of fresh weight) from untreated and treated WT and TG plants were ground in ten volumes of ice-cold extraction buffer and then centrifuged at 4 °C for 20 min (16 000g). The supernatants were transferred into new tubes and centrifuged again under the same conditions for another 20 min and then removed and kept on ice. Protein content was determined by a commercial Bradford assay (Bio-Rad) using BSA as a standard following manufacturer's instruction.

The Trx assay was carried out following a previous protocol (Holmgren, 1979), the ratio of free Trx content under heat stress versus that under normal conditions was used to determine the electron transfer ability of TG and WT controls.

A spectrophotometric assay (Hagenman and Hucklesby, 1971; Losada and Paneque, 1971) was used to measure NiR activity. The reaction of NiR activity assay was conducted at 30 °C for 30 min followed by vigorously vortexing to stop the reaction. Each reaction mixture contained 0.3 mL of 0.5 M Tris buffer at pH 8.0, 0.2 mL of 20 mM potassium nitrite, 0.3 mL of 5 mM methyl viologen, 0.1 mL of crude enzyme, 0.3 mL of sodium dithionite solution (25 mg of sodium dithionite in 1 mL of 0.29 M NaHCO$_3$) and 0.8 mL of H$_2$O. After stopping the reaction, 1 mL each of Diazo-coupling reagents was added to 2 mL of 100-fold dilution of the reaction mixture, and the volume was made up to 5 mL with 1 mL of H$_2$O. After 10 min, the optical density of the solution was determined at 540 nm. The nitrite content was calculated from a KNO$_2$ standard curve. NiR activity was determined following Yuan et al. (2015).

Statistical analysis

Summarized data (the counts, means and standard errors for each group) from three or more groups were subjected to a one-way ANOVA and the Tukey's honestly significant difference post hoc tests. Means not sharing the same letter are statistically significantly different ($P < 0.05$).

Acknowledgements

We thank Andrew Morris and Stephen Bolus for help on nitrate starvation treatments; Dr. Haibo Liu and Nick Menchyk for help on materials propagation. This project was supported by Biotechnology Risk Assessment Grant Program competitive grant no.

2010-33522-21656 from the USDA National Institute of Food and Agriculture as well as the USDA grant CSREES SC-1700450. Technical Contribution No. 6448 of the Clemson University Experiment Station.

References

Apse, M.P. and Blumwald, E. (2002) Engineering salt tolerance in plants. *Curr. Opin. Biotechnol.* **13**, 146–150.

Arizmendi, J.M. and Serra, J.L. (1990) Purification and some properties of the nitrite reductase from the cyanobacterium *Phormidium laminosum*. *Biochim. Biophys. Acta*, **1040**, 237–244.

Atkinson, N.J. and Urwin, P.E. (2012) The interaction of plant biotic and abiotic stresses: from genes to the field. *J. Exp. Bot.* **63**, 3523–3543.

Aubert, Y., Widemann, E., Miesch, L., Pinot, F. and Heitz, T. (2015) CYP94-mediated jasmonoyl-isoleucine hormone oxidation shapes jasmonate profiles and attenuates defence responses to *Botrytis cinerea* infection. *J. Exp. Bot.* **66**, 3879–3892.

Balmer, Y., Koller, A., del Val, G., Manieri, W., Schürmann, P. and Buchanan, B.B. (2003) Proteomics gives insight into the regulatory function of chloroplast thioredoxins. *Proc. Natl Acad. Sci. USA*, **100**, 370–375.

Blanchard, J.L., Wholey, W.-Y., Conlon, E.M. and Pomposiello, P.J. (2007) Rapid changes in gene expression dynamics in response to superoxide reveal SoxRS-dependent and independent transcriptional networks. *PLoS ONE*, **11**, e1186. doi:10.1371/journal.pone.0001186.

Blanco, N.E., Ceccoli, R.D., Segretin, M.E., Poli, H.O., Voss, I., Melzer, M., Bravo-Almonacid, F.F. *et al.* (2011) Cyanobacterial flavodoxin complements ferredoxin deficiency in knocked-down transgenic tobacco plants. *Plant J.* **65**, 922–935.

Buchanan, B.B. (2016) The path to thioredoxin and redox regulation in chloroplasts. *Annu. Rev. Plant Biol.* **67**, 1–24.

Ceccoli, R.D., Blanco, N.E., Medina, M. and Carrillo, N. (2011) Stress response of transgenic tobacco plants expressing a cyanobacterial ferredoxin in chloroplasts. *Plant Mol. Biol.* **76**, 535–544.

Ceccoli, R.D., Blanco, N.E., Segretin, M.E., Melzer, M., Hanke, G.T., Scheibe, R., Hajirezaei, M.R. *et al.* (2012) Flavodoxin displays dose-dependent effects on photosynthesis and stress tolerance when expressed in transgenic tobacco plants. *Planta*, **236**, 1447–1458.

Chan, K.X., Phua, S.Y., Crisp, P., McQuinn, R. and Pogson, B.J. (2016) Learning the languages of the chloroplast: retrograde signalling and beyond. *Annu. Rev. Plant Biol.* **47**, 25–53.

Coba de la Peña, T., Redondo, F.J., Manrique, E., Lucas, M.M. and Pueyo, J.J. (2010) Nitrogen fixation persists under conditions of salt stress in transgenic *Medicago truncatula* plants expressing a cyanobacterial flavodoxin. *Plant Biotechnol. J.* **8**, 954–965.

Coba de la Peña, T., Redondo, F.J., Fillat, M.F., Lucas, M.M. and Pueyo, J.J. (2013) Flavodoxin overexpression confers tolerance to oxidative stress in beneficial soil bacteria and improves survival in the presence of the herbicides paraquat and atrazine. *J. Appl. Microbiol.* **115**, 236–246.

D'Autréaux, B. and Toledano, M.B. (2007) ROS as signalling molecules: mechanisms that generate specificity in ROS homeostasis. *Nat. Rev. Mol. Cell Biol.* **8**, 813–824.

Dietz, K.J. (2015) Efficient high light acclimation involves rapid processes at multiple mechanistic levels. *J. Exp. Bot.* **66**, 2401–2414.

Dietz, K.J., Turkan, I. and Krieger-Liszkay, A. (2016) Redox- and reactive oxygen species-dependent signaling into and out of the photosynthesizing chloroplast. *Plant Physiol.* **171**, 1541–1550.

Flowers, T.J. (2004) Improving crop salt tolerance. *J. Exp. Bot.* **55**, 307–319.

Hagenman, R.H. and Hucklesby, D.P. (1971) Nitrate reductase from higher plants. *Methods Enzymol.* **23**, 491–503.

Hanke, G.T., Kimata-Ariga, Y., Taniguchi, I. and Hase, T. (2004) A post genomic characterization of *Arabidopsis* Ferredoxins. *Plant Physiol.* **134**, 255–264.

Hinz, M., Wilson, I.W., Yang, J., Buerstenbinder, K., Llewellyn, D., Dennis, E.S., Sauter, M. *et al.* (2010) *Arabidopsis* RAP2.2: an ethylene response transcription factor that is important for hypoxia survival. *Plant Physiol.* **153**, 757–772.

Holmgren, A. (1979) Thioredoxin catalyzes the reduction of insulin disulfides by dithiothreitol and dihydrolipoamide. *J. Biol. Chem.* **254**, 9627–9632.

Holtgrefe, S., Bader, K.P., Horton, P., Scheibe, R., von Schaewen, A. and Backhausen, J.E. (2003) Decreased content of leaf ferredoxin changes electron distribution and limits photosynthesis in transgenic potato plants. *Plant Physiol.* **133**, 1768–1778.

Karlusich, P., Lodeyro, A.F. and Carrillo, N. (2014) The long goodbye: the rise and fall of flavodoxin during plant evolution. *J. Exp. Bot.* **65**, 5161–5178.

Knaff, D.B. (1996) Ferredoxin and ferredoxin-dependent enzymes. In *Oxygenic Photosynthesis: The Light Reactions* (Ort, D.R. and Yocum, C.F., eds), pp. 333–361. Dordrecht, The Netherlands: Kluwer Academic Publishers.

Kotur, Z., Siddiqi, Y.M. and Glass, A.D. (2013) Characterization of nitrite uptake in *Arabidopsis thaliana*: evidence for a nitrite-specific transporter. *New Phytol.* **200**, 201–210.

Kramer, D.M., Avenson, T.J. and Edwards, G.E. (2004) Dynamic flexibility in the light reactions of photosynthesis governed by both electron and proton transfer reactions. *Trends Plant Sci.* **9**, 349–357.

Li, Z., Baldwin, C.M., Hu, Q., Liu, H. and Luo, H. (2010) Heterologous expression of Arabidopsis H$^+$-PPase enhances salt tolerance in transgenic creeping bentgrass (*Agrostis stolonifera* L.). *Plant, Cell Environ.* **33**, 272–289.

Li, Z., Hu, Q., Zhou, M., Vandenbrink, J., Li, D., Menchyk, N., Reighard, S. *et al.* (2013) Heterologous expression of OsSIZ1, a rice SUMO E3 ligase, enhances broad abiotic stress tolerance in transgenic creeping bentgrass. *Plant Biotechnol. J.* **11**, 432–445.

Lodeyro, A.F., Ceccoli, R.D., Pierella, K.J.J. and Carrillo, N. (2012) The importance of flavodoxin for environmental stress tolerance in photosynthetic microorganisms and transgenic plants. Mechanism, evolution and biotechnological potential. *FEBS Lett.* **586**, 2917–2924.

Losada, M. and Paneque, A. (1971) Nitrite reductase. *Methods Enzymol.* **23**, 487–491.

Luo, H., Hu, Q., Nelson, K., Longo, C., Kausch, A.P., Chandlee, J.M., Wipff, J.K. *et al.* (2004a) *Agrobacterium tumefaciens*-mediated creeping bentgrass (*Agrostis stolonifera* L.) transformation using phosphinothricin selection results in a high frequency of single-copy transgene integration. *Plant Cell Rep.* **22**, 645–652.

Luo, H., Hu, Q., Nelson, K., Longo, C. and Kausch, A. (2004b) Controlling transgene escape in genetically modified grasses. In *Molecular Breeding of Forage and Turf*, vol. **11** (Hopkins, A., Wang, Z.-Y., Mian, R., Sledge, M. and Barker, R., eds), pp. 245–254. Dordrecht, The Netherlands: Springer.

Matsukura, S., Mizoi, J., Yoshida, T., Todaka, D., Ito, Y., Maruyama, K., Shinozaki, K. *et al.* (2010) Comprehensive analysis of rice DREB2-type genes that encode transcription factors involved in the expression of abiotic stress-responsive genes. *Mol. Genet. Genomics*, **283**, 185–196.

Meimberg, K. and Mühlenhoff, U. (1999) Laser-flash absorption spectroscopy study of the competition between ferredoxin and flavodoxin photoreduction by Photosystem I in Synechococcus sp. PCC 7002: evidence for a strong preference for ferredoxin. *Photosynth. Res.* **61**, 253–267.

Merino, I., Contreras, A., Jing, Z.P., Gallardo, F., Cánovas, F.M. and Gómez, L. (2014) Plantation forestry under global warming: hybrid poplars with improved thermotolerance provide new insights on the in vivo function of small heat shock protein chaperones. *Plant Physiol.* **164**, 978–991.

Mittler, R. (2006) Abiotic stress, the field environment and stress combination. *Trends Plant Sci.* **11**, 15–19.

Mittler, R., Vanderauwera, S., Suzuki, N., Miller, G., Tognetti, V.B., Vandepoele, K., Gollery, M. *et al.* (2011) ROS signaling: the new wave? *Trends Plant Sci.* **16**, 300–309.

Mizoi, J., Ohori, T., Moriwaki, T., Kidokoro, S., Todaka, D., Maruyama, K., Kusakabe, K. *et al.* (2013) GmDREB2A;2, a canonical DEHYDRATION-RESPONSIVE ELEMENT-BINDING PROTEIN2-type transcription factor in soybean, is posttranslationally regulated and mediates dehydration-responsive element-dependent gene expression. *Plant Physiol.* **161**, 346–361.

Munekage, Y., Hashimoto, M., Miyake, C., Tomizawa, K., Endo, T., Tasaka, M. and Shikanai, T. (2004) Cyclic electron flow around PSI is essential for photosynthesis. *Nature*, **429**, 579–582.

Newman, B.J. and Gray, J.C. (1988) Characterization of a full-length cDNA clone for pea ferredoxin-NADP⁺ reductase. *Plant Mol. Biol.* **10**, 511–520.

Nogués, I., Tejero, J., Hurley, J.K., Paladini, D., Frago, S., Tollin, G., Mayhew, S.G. *et al.* (2004) Role of the C-terminal tyrosine of ferredoxin-nicotinamide adenine dinucleotide phosphate reductase in the electron transfer processes with its protein partners ferredoxin and flavodoxin. *Biochemistry,* **43**, 6127–6137.

Olsen, A.N., Ernst, H.A., Leggio, L.L. and Skriver, K. (2005) NAC transcription factors: structurally distinct, functionally diverse. *Trends Plant Sci.* **10**, 79–87.

Petrack, M.E., Dickey, L.F., Nguyen, T.T., Gatz, C., Sowinski, D.A., Allen, G.C. and Thompson, W.F. (1998) Ferredoxin-1 mRNA is destabilized by changes in photosynthetic electron transport. *Proc. Natl Acad. Sci. USA*, **95**, 9009–9013.

Sakakibara, Y., Kimura, H., Iwamura, A., Saitoh, T., Ikegami, T., Kurisu, G. and Hase, T. (2012) A new structural insight into differential interaction of cyanobacterial and plant ferredoxins with nitrite reductase as revealed by NMR and X-ray crystallographic studies. *J. Biochem.* **151**, 483–492.

Seki, M., Kamei, A., Yamaguchi-Shinozakim, K. and Shinozakim, K. (2003) Molecular responses to drought, salinity and frost: common and different paths for plant protection. *Curr. Opin. Biotechnol.* **14**, 194–199.

Sels, J., Mathys, J., De Coninck, B.M., Cammue, B.P. and De Bolle, M.F. (2008) Plant pathogenesis-related (PR) proteins: a focus on PR peptides. *Plant Physiol. Biochem.* **46**, 941–950.

Semenov, A.Y., Mamedov, M.D. and Chamorovsky, S.K. (2003) Photoelectric studies of the transmembrane charge transfer reactions in photosystem I pigment–protein complexes. *FEBS Lett.* **553**, 223–228.

Serra, E.C., Krapp, A.R., Ottado, J., Feldman, M.F., Ceccarelli, E.A. and Carrillo, N. (1995) The precursor of pea ferredoxin-NADP⁺ reductase synthesized in *Escherichia coli* contains bound FAD and is transported into chloroplasts. *J. Biol. Chem.* **270**, 19930–19935.

Sétif, P. (2001) Ferredoxin and flavodoxin reduction by photosystem I. *Biochim. Biophys. Acta*, **1507**, 161–179.

Singh, A.K., Li, H. and Sherman, L.A. (2004) Microarray analysis and redox control of gene expression in the cyanobacterium *Synechocystis* sp. PCC 6803. *Physiol. Plant.* **120**, 27–35.

Sun, X., Sun, C., Li, Z., Hu, Q., Han, L. and Luo, H. (2016) AsHSP17, a creeping bentgrass small heat shock protein modulates plant photosynthesis and ABA-dependent and independent signaling to attenuate plant response to abiotic stress. *Plant, Cell & Environ Plant Cell Environ.* **39**, 1320–1337.

Tognetti, V.B., Palatnik, J.F., Fillat, M.F., Melzer, M., Hajirezaei, M.R., Valle, E.M. and Carrillo, N. (2006) Functional replacement of ferredoxin by a cyanobacterial flavodoxin in tobacco confers broad-range stress tolerance. *Plant Cell*, **18**, 2035–2050.

Tognetti, V.B., Monti, M.R., Valle, E.M. and Carrillon, N.S.A. (2007a) Detoxification of 2,4-dinitrotoluene by transgenic tobacco plants expressing a bacterial flavodoxin. *Environ. Sci. Technol.* **41**, 4071–4076.

Tognetti, V.B., Zurbriggen, M.D., Morandi, E.N., Fillat, M.F., Valle, E.M., Hajirezaei, M.-R. and Carrillo, N. (2007b) Enhanced plant tolerance to iron starvation by functional substitution of chloroplast ferredoxin with a bacterial flavodoxin. *Proc. Natl Acad. Sci. USA*, **104**, 11495–11500.

Vigara, A.J., Inda, L.A., Vega, J.M., Gómez-Moreno, C. and Peleato, M.L. (1998) Flavodoxin as an electronic donor in photosynthetic inorganic nitrogen assimilation by iron-deficient Chlorella fusca cells. *Photochem. Photobiol.* **67**, 446–449.

Vinocur, B. and Altman, A. (2005) Recent advances in engineering plant tolerance to abiotic stress: achievements and limitations. *Curr. Opin. Biotechnol.* **16**, 123–132.

Wang, D. and Luthe, D.S. (2003) Heat sensitivity in a bentgrass variant. Failure to accumulate a chloroplast heat shock protein isoform implicated in heat tolerance. *Plant Physiol.* **133**, 319–327.

Wang, W., Vinocur, B. and Altman, A. (2003) Plant responses to drought, salinity and extreme temperatures: towards genetic engineering for stress tolerance. *Planta*, **218**, 1–14.

Waters, E.R. (2012) The evolution, function, structure, and expression of the plant sHSPs. *J. Exp. Bot.* **64**, 391–403.

Yao, D., Wei, Q., Xu, W., Syrenne, R.D., Yuan, J.S. and Su, Z. (2012) Comparative genomic analysis of NAC transcriptional factors to dissect the regulatory mechanisms for cell wall biosynthesis. *BMC Bioinform.* **13**(Suppl. 15), S10. doi: 10.1186/1471-2105-13-S15-S10.

Yousef, N., Pistorius, E.K. and Michel, K.P. (2003) Comparative analysis of *idiA* and *isiA* transcription under iron starvation and oxidative stress in *Synechococcus elongates* PCC 7942 wild-type and selected mutants. *Arch. Microbiol.* **180**, 471–483.

Yuan, S., Li, Z., Li, D., Yuan, N., Hu, Q. and Luo, H. (2015) Constitutive expression of rice MicroRNA528 alters plant development and enhances tolerance to salinity stress and nitrogen starvation in creeping bentgrass. *Plant Physiol.* **69**, 576–593.

Zhang, J.Z., Creelman, R.A. and Zhu, J.K. (2004) From laboratory to field. Using information from Arabidopsis to engineer salt, cold and drought tolerance in crops. *Plant Physiol.* **135**, 615–621.

Zheng, M., Doan, B., Schnelder, T.D. and Storz, G. (1999) *OxyR* and *SoxRS* regulation of fur. *J. Bacteriol.* **180**, 471–483.

Zhou, M. and Luo, H. (2013) MicroRNA-mediated gene regulation: potential applications for plant genetic engineering. *Plant Mol. Biol.* **83**, 59–75.

Zhou, M., Li, D., Li, Z., Hu, Q., Yang, C., Zhu, L. and Luo, H. (2013) Constitutive expression of a miR319 gene alters plant development and enhances salt and drought tolerance in transgenic creeping bentgrass. *Plant Physiol.* **161**, 1375–1391.

Zurbriggen, M.D., Tognetti, V.B. and Carrillo, N. (2007) Stress-inducible flavodoxin from photosynthetic microorganisms. The mystery of flavodoxin loss from the plant genome. *IUBMB Life*, **58**, 355–360.

Zurbriggen, M.D., Tognetti, V.B., Fillat, M.F., Hajirezaei, M.R., Valle, E.M. and Carrillo, N. (2008) Combating stress with flavodoxin: a promising route for crop improvement. *Trends Biotechnol.* **26**, 531–537.

4

Golden bananas in the field: elevated fruit pro-vitamin A from the expression of a single banana transgene

Jean-Yves Paul[1], Harjeet Khanna[1,†], Jennifer Kleidon[1], Phuong Hoang[1], Jason Geijskes[1,‡], Jeff Daniells[2], Ella Zaplin[1,§], Yvonne Rosenberg[3], Anthony James[1], Bulukani Mlalazi[1], Pradeep Deo[1], Geofrey Arinaitwe[4], Priver Namanya[1,4], Douglas Becker[1], James Tindamanyire[1], Wilberforce Tushemereirwe[4], Robert Harding[1] and James Dale[1,*]

[1]Centre for Tropical Crops and Biocommodities, Queensland University of Technology, Brisbane, Qld, Australia

[2]Agri-Science Queensland, Department of Agriculture and Fisheries, South Johnstone, Qld, Australia

[3]PlantVax Inc, Rockville, MD, USA

[4]National Agricultural Research Laboratories, National Agricultural Research Organization, Kampala, Uganda

*Correspondence
email j.dale@qut.edu.au
Present addresses: †Sugar Research Australia, Brisbane, Qld, Australia.
‡Syngenta Asia Pacific, Singapore, Singapore.
§Charles Sturt University, Wagga Wagga, NSW, Australia.

Keywords: Vitamin A deficiency, Uganda, pro-vitamin A, staple food crop, banana, biofortification, genetic modification.

Summary

Vitamin A deficiency remains one of the world's major public health problems despite food fortification and supplements strategies. Biofortification of staple crops with enhanced levels of pro-vitamin A (PVA) offers a sustainable alternative strategy to both food fortification and supplementation. As a proof of concept, PVA-biofortified transgenic Cavendish bananas were generated and field trialed in Australia with the aim of achieving a target level of 20 µg/g of dry weight (dw) β-carotene equivalent (β-CE) in the fruit. Expression of a Fe'i banana-derived phytoene synthase 2a (MtPsy2a) gene resulted in the generation of lines with PVA levels exceeding the target level with one line reaching 55 µg/g dw β-CE. Expression of the maize phytoene synthase 1 (ZmPsy1) gene, used to develop 'Golden Rice 2', also resulted in increased fruit PVA levels although many lines displayed undesirable phenotypes. Constitutive expression of either transgene with the maize polyubiquitin promoter increased PVA accumulation from the earliest stage of fruit development. In contrast, PVA accumulation was restricted to the late stages of fruit development when either the banana 1-aminocyclopropane-1-carboxylate oxidase or the expansin 1 promoters were used to drive the same transgenes. Wild-type plants with the longest fruit development time had also the highest fruit PVA concentrations. The results from this study suggest that early activation of the rate-limiting enzyme in the carotenoid biosynthetic pathway and extended fruit maturation time are essential factors to achieve optimal PVA concentrations in banana fruit.

Introduction

Micronutrient deficiency, often referred to as hidden hunger, occurs when intake and absorption of vitamins and minerals are too low to sustain good health and development. The World Health Organization (WHO) estimates that 190 million pre-school children are deficient in one of the major micronutrients, vitamin A. Vitamin A deficiency (VAD) alone is responsible for almost 6% of child deaths under the age of 60 months in Africa and 8% in South-East Asia (WHO, 2011). For the vast majority of these children, VAD is almost exclusively the result of inadequate intake of dietary vitamin A or pro-vitamin A (PVA) although exacerbated by other health conditions. Similar levels of VAD are also evident in women of childbearing age in these same regions (WHO, 2011). These levels of VAD continue despite the implementation over many years of extensive alleviating strategies such as supplements and food fortification. These strategies have been demonstrably successful, but there remain persistently high and unacceptable levels of VAD particularly in sub-Saharan Africa and south Asia (Stevens et al., 2015).

In an effort to significantly reduce VAD in these regions, strategies aimed at increasing the dietary intake of particularly α- and β-carotene together as PVA are being developed or implemented. These include programmes to encourage growing and consuming staple foods with high levels of PVA. In some instances, such foods or crops with the desired agronomic and consumer traits are already available and can therefore be easily deployed (HarvestPlus, 2012). However, the majority of accepted cultivars and landraces of staple crops are low in micronutrients such as PVA and iron, and therefore, it is necessary to develop new varieties with enhanced levels of these micronutrients. This can be achieved either through conventional breeding or by genetic modification where the traits are not available within the accessible germplasm or cannot be easily introgressed into acceptable cultivars. These two approaches are known as biofortification.

The best-known example of biofortification by genetic modification, and the most advanced in terms of development, is 'Golden Rice'. In Ye et al., 2000 and colleagues reported the generation of transgenic rice expressing the daffodil phytoene synthase (Psy) gene under the control of an endosperm-specific rice glutelin promoter together with the bacterial (Pantoea ananatis formerly known as Erwinia uredovora) phytoene desaturase (CrtI) gene under the control of the constitutive CaMV 35S promoter. The endosperm of selected lines was yellow, and one heterozygous line contained 1.6 µg/g dry weight (dw) total carotenoids. Paine et al. (2005) subsequently reported the development of the second generation of 'Golden Rice', which was engineered with a maize (*Zea mays*) *Psy* gene and the *CrtI* gene under the control of the glutelin promoter. One Golden Rice 2 elite event had a 23-fold increase in total carotenoids over the original Golden Rice with a total carotenoid level of up to 37 μg/g dw in the endosperm of which 31 μg/g was β-carotene. A number of other important food crops have been or are being developed to enhance the level of PVA through genetic modification.

Bananas are the world's most important fruit crop and one of the top 10 crops by production. They are widely grown in the wet tropics and subtropics forming an important dietary component both raw as a dessert fruit and cooked often as the major source of carbohydrate. In a number of countries, bananas are the principal staple food including Uganda where consumption levels average 0.5 kg per person per day rising to around 1 kg per person per day in some regions (Komarek, 2010; Smale and Tushemereirwe, 2007). In East Africa, the staple cultivar is East African highland banana (EAHB) (*M. acuminata* AAA-EA) prepared primarily by steaming or boiling whereas, in West Africa, plantains are dominant and are usually fried or roasted (Fungo and Pillay, 2011). In both regions, the level of VAD is high. In Uganda, it varies from 15% to 33% in children under 60 months with similar levels in women of childbearing age (UDHS - Uganda Demographic and Health Survey, 2006). Unlike rice endosperm, banana fruit contains PVA and, in some instances, very high levels particularly Fe'i bananas of Micronesia and Papua New Guinea (Englberger *et al.*, 2003). Bananas with β-carotene equivalent (β-CE) levels of 340 μg/g dw have been reported whereas the dominant dessert banana cv 'Cavendish' has between 1 and 4 μg/g dw β-CE and the EAHB clone, 'Nakitembe', has approximately 10 μg/g β-CE dw (Englberger *et al.*, 2006a; Mbabazi, 2015). Unfortunately, domesticated bananas have very low male and female fertility rendering conventional breeding extremely difficult. Thus, the introgression of the high PVA traits of Fe'i bananas for instance into farmer preferred EAHB selections would be practically impossible. However, genetic modification of bananas is well established.

Here, we report the 'proof-of-concept' technology required towards the generation of PVA-biofortified EAHB varieties in Uganda. The 'Cavendish' dessert banana was genetically modified, and greatly enhanced PVA levels were demonstrated in the fruit of plants grown in the field in Australia.

Results

The target

At the outset, it was important to identify a target fruit level of β-carotene equivalents necessary to help alleviate VAD in Uganda. The target was set at delivering 50% of the estimated average requirement (EAR) of vitamin A in vulnerable populations which for children under 60 months is 120 μg/day and for females ranging from 235 μg/day up to 445 μg/day for lactating mothers. An estimated bioconversion 6:1 ratio of β-carotene equivalents (β-CE) to vitamin A from cooked banana pulp was used (Bresnahan *et al.*, 2012) with an estimated consumption of cooked bananas of 300 g/day for children and 500 g/day for women. The α- and β-carotene retention after steaming or boiling was also estimated at 70% (Mbabazi, 2015). Using these parameters, banana fruit needed to contain β-CE levels of at least 20 μg/g dw to achieve 50% of the EAR.

There were three major technical constraints at the commencement of this project that influenced the research strategy:

(i) very little information was available regarding the expression of transgenes in bananas generally and more specifically in banana fruit, (ii) the time from transformation to harvestable fruit ranges from 2 to 2½ years, and (iii) it was clearly impractical to take large numbers of transgenic bananas through to fruit in the greenhouse. Therefore, a large number of independent transgenic lines were generated to enable the testing of a wide range of promoter and transgene combinations. Initially, a single plant per transgenic line was planted in the field with, in most instances, between 10 and 30 transgenic lines per construct. For more information, refer to the 'History of the project' section of the Supplementary information document.

Promoter characterization

Three promoters were selected as possible candidates for expressing PVA-related transgenes in banana fruit, and these were characterized for levels and patterns of expression in transgenic bananas. The promoters included the constitutive maize polyubiquitin promoter (Ubi) and two promoters isolated from banana, the expansin 1 promoter (Exp1) and the ACC oxidase promoter (ACO) which were predicted to be fruit specific. These promoters were fused to the β-glucuronidase reporter gene (*uidA*), the cassettes transformed into bananas and the transgenic plants established in the field. β-glucuronidase (GUS) protein levels were measured in the fruit pulp using ELISA; however, this approach could not be used for leaf or peel material because of very high background levels. Therefore, for leaf and peel samples, MUG fluorometric assays were used to estimate enzyme activity rather than protein levels.

GUS activity was measured in the leaves of six independent transgenic lines for each promoter. As expected, there were high but variable levels of GUS activity in the leaves of all six plants where *uidA* was under the control of the Ubi promoter (Figure 1a). In the leaves of the wild-type control plant and plants where *uidA* was under the control of either the Exp1 or ACO promoter, there was undetectable to negligible GUS activity (Figure 1b and c).

Pulp samples from the fruit of the same lines described above were collected at 3, 6, 9 and 12 weeks post-bunch emergence (S3, S6, S9 and S12) and also at 'full green' (FG), when the bunches were harvested, and 'full ripe' (FR). The FG stage is equivalent to the stage when cooking bananas are harvested in Uganda. GUS protein was not detected in the wild-type at any fruit development stage. In contrast, appreciable but varying levels of GUS protein were detected in the pulp of banana fruit from S3 through to FR in the six Ubi-*uidA* lines (Figure 1d) confirming the constitutive nature of the Ubi promoter. No reproducible trend in GUS protein levels across the six lines was observed as fruit matured from S3 to FR except that average protein accumulation was lowest at the earliest stage, S3. In the six Exp1-*uidA* lines examined, no appreciable GUS expression was detected in fruit pulp from S3 through to FG with the exception of FR fruit from lines FT258 and FT263 (Figure 1e). This indicated that the Exp1 promoter is activated very late during fruit development. Very low levels of GUS expression were detected in the fruit pulp of the ACO-*uidA* lines from S3 to S9 (Figure 1f). With the exception of line FT736 which peaked fourfold higher than any other line, the overall trend was that GUS expression slowly increased from S9 through to FG and plateaued at FR. These results indicated that the ACO promoter was activated earlier than Exp1 during fruit

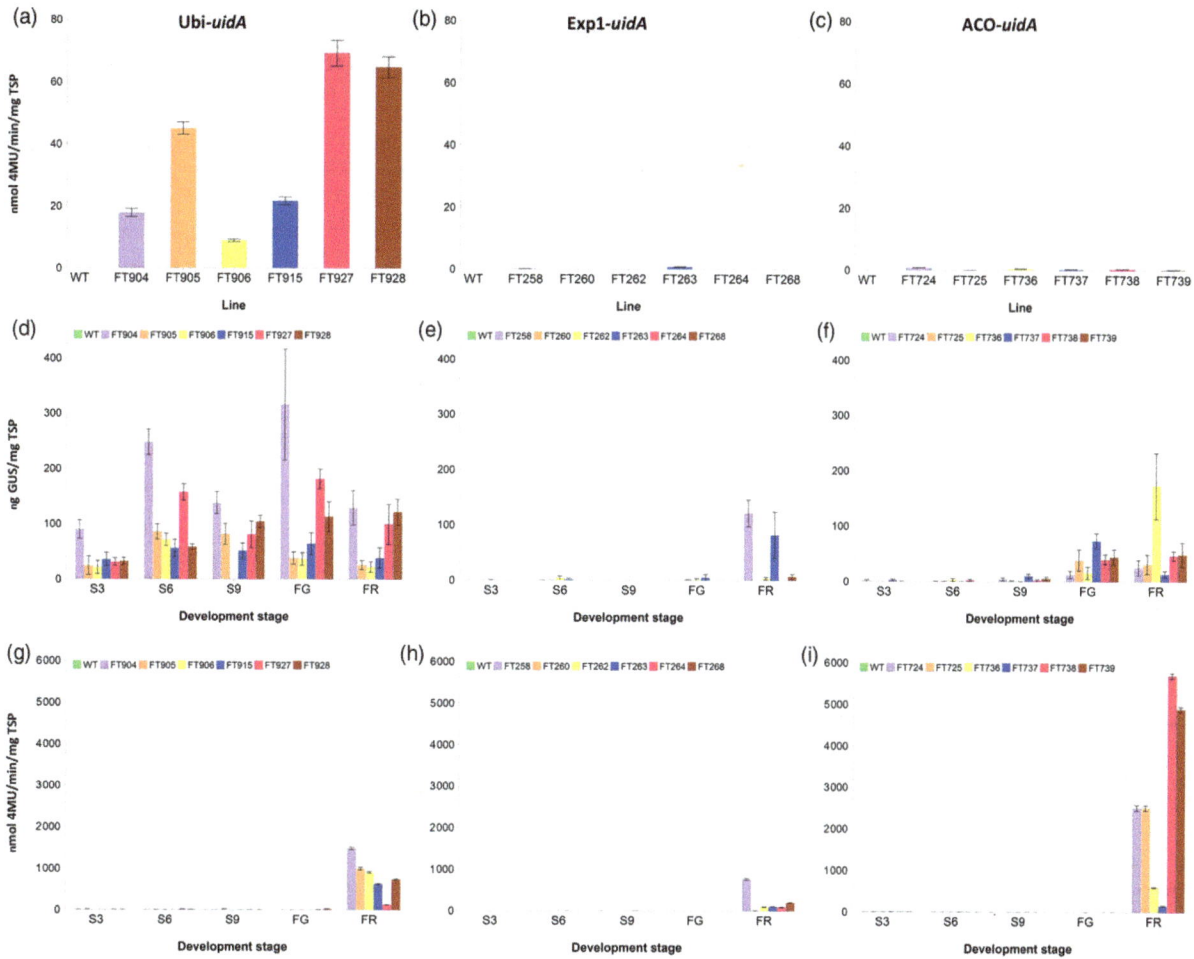

Figure 1 Analysis of promoter activity in wild-type and transgenic Cavendish banana lines. GUS activity was measured in leaf (a, b, c), and peel (g, h, i), while GUS protein concentration was measured in pulp tissue (d, e, f). Ubi promoter (a, d, g); Exp1 promoter (b, e, h); and ACO promoter (c, f, i). WT, wild-type FT432. S3, S6 and S9 represent 3, 6 and 9 weeks post-bunch emergence, respectively. FG, full green and FR, full ripe. Error bars: ±SD.

development and more consistently at the FG and FR stages (Figure 1e and f).

The pattern of Ubi driven GUS expression in fruit peel was unexpectedly different from fruit pulp with low expression from S3 to FG followed by a substantial increase at FR (Figure 1g). In the peel of Exp1-uidA lines, the pattern of GUS expression was similar to that observed in the fruit pulp (Figure 1h). ACO-uidA lines had very little GUS expression in the peel up to FG; however, there was a dramatic increase to levels higher than either the Ubi or Exp1 lines at FR (Figure 1i).

PVA analysis: plant and ratoon crops

Domesticated bananas grow as a perennial crop. The initial plant, the plant crop, develops a corm and a pseudostem from which the original bunch is produced. After the bunch is harvested, the pseudostem dies and is replaced by a second pseudostem producing the first ratoon crop which develops from a sucker originating from the corm. Similarly, a second ratoon crop is produced and so on.

The same three promoters used to assess GUS expression were used to drive the expression of three carotenoid biosynthesis transgenes. These transgenes were the phytoene synthase 1 gene from maize (ZmPsy1) used in Golden Rice 2, a phytoene synthase

2a gene isolated from the Fe'i banana cultivar Asupina (MtPsy2a) (Mlalazi et al., 2012) and the bacterial phytoene desaturase gene PaCrtI also used in Golden Rice 2. ZmPsy1 and MtPsy2a were transformed into Cavendish banana singly and in combination with PaCrtI.

A total of 244 transgenic lines were confirmed to contain the respective PVA transgene(s) by PCR. Southern blot analyses showed that the transgene copy number varied from one to more than 10 copies (Figure S1 and S2). For each line, a single plant was established in the field together with 50 non-transgenic control plants. This initial randomized trial was designated Field Trial 1 (FT-1), and the plant crop was assessed over a period of 16 months. Forty-eight transgenic lines either died or were stunted and did not produce fruit. Therefore, fruit was harvested from a total of 196 transgenic lines of which 153 samples were selected for the initial plant crop fruit analysis (Table 1). Fruit was harvested at FG, ripened to FR and sampled at both stages. The sample used for the initial PVA level screen consisted of a single fruit taken from the middle of the bunch from each of 153 transgenic lines together with the equivalent sample from the fruit of 15 non-transgenic control plants. Following lyophilization and total carotenoids extraction, each sample was analyzed by HPLC and β-CE levels calculated. The most important outcomes

Table 1 Data summary from the transgenic banana field trial

Promoter-transgene	Number of plants in the field	Number of plants harvested	Number of plants analyzed (first cut)*	β-CE in FG fruit (µg/g dw)		β-CE in FR fruit (µg/g dw)	
				Range	Average	Range	Average
Wild-type	50	50	15	0.6–3.8	1.5	0.8–5.8	3.1
Exp1-*MtPsy2a*	33	28	28	1.2–8.6	3.3	0.8–10.0	2.8
Exp1-*MtPsy2a* + Ubi-*PaCrtl*	13	8	5	0.2–2.3	1.0	1.3–3.4	2.1
Exp1-*ZmPsy1*	32	26	26	1.2–4.6	2.6	1.3–15.2	4.5
Exp1-*ZmPsy1*+ Ubi-*PaCrtl*	18	11	8	0.2–1.8	1.0	1.5–3.6	2.4
Exp1-*ZmPsy1*+ Exp1-*PaCrtl*	7	7	2	–	–	2.1–11.0	6.5
Exp1-*PaCrtl*	29	26	3	–	–	1.7–2.6	2
ACO-*MtPsy2a*	30	29	28	11.2–15.3	13.3	3.4–13.4	8.3
ACO-*ZmPsy1*	30	18	18	5.4–10.9	7.6	2.5–24.6	9.0
ACO-*ZmPsy1*+ Exp1-*PaCrtl*	6	4	3	10.6–25.7	17.1	7.8–16.7	13.2
Ubi-*MtPsy2a*	9	7	7	1.4–19.1	6.8	3.5–16.1	7.4
Ubi-*ZmPsy1*	10	5	5	0.3–13.6	6.1	1.0–16.1	7.9
Ubi-*PaCrtl*	27	27	20	–	–	1.0–3.7	1.8
Total	294	246	168	–	–	–	–

*First cut relates to an initial screening of the fruit of transgenic banana lines done by HPLC and using only a single fruit collected from the middle position of the bunch. FG, full green and FR, full ripe.

of this initial screening included: (i) there was obvious variation between individual control plants and also variation within lines with the same promoter-transgene combination, (ii) the highest expressing transgenic line for each of the different transgenes or combinations of transgenes contained higher levels of β-CE than the highest control plants except those lines containing Exp1-*PaCrtl* or Ubi-*PaCrtl* alone or where Ubi-*PaCrtl* was combined with Exp1-*MtPsy2a* or Exp1-*ZmPsy1*, (iii) in all transgenic lines and controls, the fruit contained higher levels of α-carotene than β-carotene (Figure S3), and (iv) *PaCrtl* alone had little effect on fruit PVA levels when driven by either the Exp1 promoter or the Ubi promoter; as such, none of the single transgene *PaCrtl* lines were progressed through for further analysis.

From this initial plant crop screen, 63 transgenic lines were selected for more comprehensive analysis. The selection included lines representing high, average and low PVA accumulation. As preliminary data revealed considerable variation in PVA accumulation across the bunch (data not shown), carotenoids were extracted from a composite sample including equal amounts of fruit taken from the top, middle and bottom of the bunch. Further, although fruit samples from the selected lines were to be analyzed across three generations (plant crop followed by first and second ratoon crops), the trial was hit by a severe cyclone in February 2011. As a consequence, all lines were blown over resulting in fruit from some lines in either the first or the second ratoon crops not being available for analysis.

For this project, the PVA levels at FG were considered more important as cooking bananas are harvested at this stage in Uganda. However, FR data were also collected as cooking bananas are usually consumed over a number of days post-harvest. The β-CE levels in FG and FR fruit from each of the selected 63 transgenic lines are presented in Table 2. Where expression of *MtPsy2a* or *ZmPsy1* was controlled by the Exp1

promoter, the level of β-CE in the plant crop and first ratoon increased from FG to FR in 43 of 53 samples analyzed (81%) (Table 2). A similar increase was seen in 82% (14 of 17 samples) of the analyzed samples where the Ubi promoter was used to drive the expression of the same transgenes. However, when ACO was used as a promoter, this number reduced to 50% (13 of 26 samples).

In the plant crop, there were no lines that met the target level of 20 µg/g dw β-CE where the composite sample of three fruit per bunch was analyzed. The highest β-CE level at FG was 18.7 µg/g dw in Ubi-*MtPsy2a* line FT328 that subsequently died. In addition, only eight lines had PVA levels greater than 10.0 µg/g dw β-CE at FG. These were two Ubi-*ZmPsy1* lines (FT287 and FT309 with 13.4 and 11.9 µg/g dw β-CE, respectively), two Ubi-*MtPsy2a* lines (FT328 and FT324 with 18.7 and 11.7 µg/g dw β-CE, respectively), two ACO-*MtPsy2a* lines (FT504 and FT518 with 16.6 and 15.9 µg/g dw β-CE, respectively) and two ACO-*ZmPsy1* + Exp1-*PaCrtl* lines (FT584 and FT587 with 17.1 and 11.5 µg/g dw β-CE, respectively). To investigate whether a correlation existed between transgene expression levels and accumulation of carotenoids, the expression of the *ZmPsy1* and *MtPsy2a* transgenes was determined in the FG fruit of a selection of lines using RT-PCR (Figure S4 and S5) and qRT-PCR (Figure 2). *MtPsy2a* line FT246 had the highest relative expression of the transgene followed by line FT324, FT518 and FT295 (Figure 2a). Expression was considerably lower in the other three lines tested. Expression of *ZmPsy1* was highest in line FT584 followed by FT309 while similar, but lower expression was seen in lines FT287, FT467, FT475, FT479 and FT585 (Figure 2b). Lines FT187 and FT192 had low expression.

When FG fruit from the next generation (first ratoon crop) were analyzed, 68% (23 of 34) of samples showed an increase in β-CE from the plant crop (Table 2). Interestingly, every line

Table 2 PVA carotenoid concentration in the fruit pulp of selected wild-type and transgenic Cavendish banana lines across four generations

Promoter-transgene	Line #	Plant crop		1st ratoon crop		2nd ratoon crop		Sucker crop	
		FG	FR	FG	FR	FG	FR	FG	FR
		β-carotene equivalents (µg/g dw)							
Wild-type	Average (n≥ = 6)	2.6	3.1	2.2	2.2	1.7	2.3	6.0	7.4
Exp1-*MtPsy2a*	FT246	7.3	9.3	8.3	8.5	NA	NA	18.2	19.6
	FT544	3.4	4.6	NA	NA	NA	NA	NA	NA
	FT545	3.4	4.7	NA	NA	NA	NA	NA	NA
	FT342	2.8	4.2	7.4	4.6	NA	NA	10.6	9.0
	FT335	2.5	4.0	2.4	3.3	NA	NA	NA	NA
	FT343	2.3	3.8	2.9	3.4	NA	NA	NA	NA
	FT242	2.2	3.7	2.0	2.0	NA	NA	8.6	8.0
	FT233	2.0	3.3	1.9	2.6	NA	NA	NA	NA
	FT341	1.4	3.1	1.8	2.0	NA	NA	9.3	8.1
Exp1-*MtPsy2a* + Ubi-*PaCrtI*	FT244	1.5	2.8	2.6	1.9	NA	NA	NA	NA
	FT245	0.9	1.8	1.8	4.6	NA	NA	NA	NA
	FT220	0.7	1.4	1.7	1.9	NA	NA	NA	NA
	FT232	0.2	1.9	1.8	1.5	NA	NA	NA	NA
Exp1-*ZmPsy1*	FT534	9.3	11.9	NA	NA	NA	NA	NA	NA
	FT536	9.1	8.9	NA	NA	NA	NA	NA	NA
	FT317	6.6	7.9	NA	NA	NA	NA	11.4	15.6
	FT192	4.3	6.4	NA	NA	NA	NA	20.3	31.2
	FT538	3.8	9.4	NA	NA	NA	NA	10.9	14.8
	FT187	3.5	5.3	1.8	1.8	9.5	9.9	7.0	11.0
	FT311	2.9	6.1	3.1	2.5	NA	NA	NA	NA
	FT201	2.3	3.2	1.9	3.2	NA	NA	9.1	15.0
	FT319	1.9	3.2	3.5	2.6	NA	NA	NA	NA
	FT318	1.5	2.7	1.9	2.4	NA	NA	NA	NA
	FT210	1.3	2.8	1.2	2.5	NA	NA	NA	NA
Exp1-*ZmPsy1* + Ubi-*PaCrtI*	FT217	1.8	2.4	1.5	2.4	NA	NA	NA	NA
	FT196	1.6	4.2	2.9	2.5	NA	NA	NA	NA
	FT207	1.5	3.5	2.4	2.7	NA	NA	NA	NA
	FT195	1.0	2.6	NA	NA	NA	NA	NA	NA
	FT208	0.8	2.7	2.1	2.1	NA	NA	NA	NA
	FT205	0.7	2.6	1.0	7.4	NA	NA	NA	NA
	FT213	0.4	1.8	NA	NA	NA	NA	NA	NA
ACO-*MtPsy2a*	FT504	16.6	12.0	12.1	11.7	NA	NA	20.0	24.7
	FT518	15.9	10.7	NA	NA	NA	NA	23.1	35.9
	FT511	9.4	7.3	5.3	6.0	NA	NA	14.4	13.6
	FT508	9.2	7.1	4.2	3.8	NA	NA	15.6	19.0
	FT498	8.9	9.9	8.6	8.4	NA	NA	NA	NA
	FT506	6.0	5.9	NA	NA	NA	NA	NA	NA
	FT516	4.2	7.4	NA	NA	NA	NA	NA	NA
	FT497	4.1	10.5	NA	NA	NA	NA	13.4	12.9
ACO-*ZmPsy1*	FT492	9.4	8.5	NA	NA	NA	NA	NA	NA
	FT467	7.3	10.4	NA	NA	NA	NA	10.5	13.4
	FT468	NA	NA	7.3	6.6	NA	NA	NA	NA
	FT475	4.3	10.5	NA	NA	NA	NA	20.7	22.0
	FT493	4.3	18.7	7.0	6.0	NA	NA	NA	NA
	FT476	2.9	5.4	4.5	4.7	NA	NA	NA	NA
	FT487	2.7	15.1	NA	NA	NA	NA	NA	NA
	FT479	2.7	13.6	NA	NA	NA	NA	20.5	16.2
	FT483	1.7	9.1	NA	NA	NA	NA	18.8	16.9
ACO-*ZmPsy1* + Exp1-*PaCrtI*	FT584	17.1	11.2	NA	NA	NA	NA	27.0	32.8
	FT587	11.5	NA	NA	NA	NA	NA	21.5	22.2
	FT588	7.7	10.1	NA	NA	NA	NA	NA	NA
	FT585	7.3	5.6	NA	NA	NA	NA	12.2	11.5
Ubi-*MtPsy2a*	FT328	18.7	18.9	NA	NA	NA	NA	NA	NA
	FT324	11.7	16.1	NA	NA	26.6	33.4	55.0	50.1
	FT294	6.6	9.7	13.5	9.4	13.5	12.0	29.0	25.2
	FT295	5.4	6.1	6.6	6.6	4.8	4.5	10.5	12.0
	FT296	4.2	6.0	5.6	4.1	12.5	13.1	NA	NA
	FT327	3.0	4.9	NA	NA	NA	NA	NA	NA
	FT330	2.5	4.9	3.6	4.0	NA	NA	11.9	15.1
Ubi-*ZmPsy1*	FT287	13.4	15.8	NA	NA	24.3	21.8	39.7	60.9
	FT309	11.9	14.7	NA	NA	40.4	39.3	46.9	18.5
	FT298	0.7	2.5	1.1	1.3	NA	NA	NA	NA
	FT302	0.6	1.5	1.4	2.0	NA	NA	NA	NA

FG, full green and FR, full ripe.

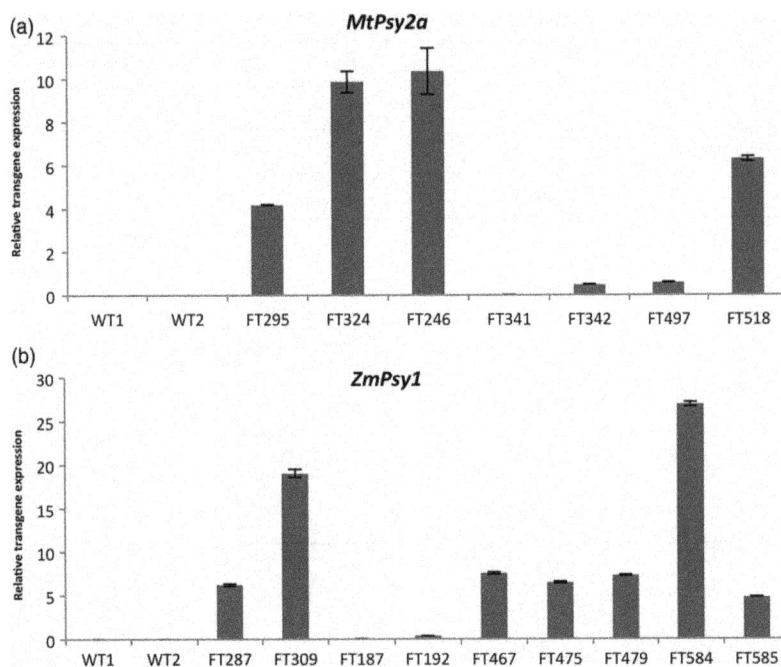

Figure 2 Analysis of mRNA expression levels in the FG fruit pulp of selected transgenic Cavendish banana lines by qRT-PCR. (a) *MtPsy2a* lines and (b) *ZmPsy1* lines. FG, full green. WT, wild-type with WT1 = FT167 and WT2 = FT430. Values are normalized expression levels ± SEM.

analyzed that contained either Ubi-*ZmPsy1* or Ubi-*MtPsy2a* showed an increased accumulation in β-CE from the plant crop. However, again, no line accumulated over the 20 µg/g dw β-CE target level. FG fruit from two lines had PVA levels above 10 µg/g dw β-CE in the ratoon crop: ACO-*MtPsy2a* line FT504 with 12.1 µg/g dw β-CE down from 16.6 µg/g dw β-CE in the plant crop and Ubi-*MtPsy2a* line FT294 with 13.5 µg/g dw β-CE up from 6.6 µg/g dw β-CE in the plant crop. Due to the impact of the 2011 cyclone, only seven lines could be assessed in the second ratoon. FG fruit from three of these lines accumulated above target levels of PVA. Ubi-*ZmPsy1* line FT309 had FG fruit with 40.4 µg/g dw β-CE, more than double the target, while Ubi-*MtPsy2a* line FT324 and Ubi-*ZmPsy1* FT287 accumulated 26.6 and 24.3 µg/g dw β-CE, respectively (Table 2).

Carotenoid accumulation throughout fruit development was also monitored in selected lines from each of the single promoter-*Psy* combinations in the second ratoon crop at 3, 6 and 9 weeks post-bunch emergence as well as FG and FR (Figure 3). For the four lines where the transgene was under the control of the Ubi promoter, PVA levels were elevated (above 15 µg/g dw β-CE) from the earliest fruit collection time point (S3) irrespective of whether the transgene was *ZmPsy1* or *MtPsy2a* (Figure 3a and d). For the two lines with the highest PVA levels, the general trend was increasing PVA during fruit development to a maximum at FG for Ubi-*ZmPsy1* line FT309 or at FR for Ubi-*MtPsy2a* line FT324. In contrast, accumulated PVA levels in the fruit of Exp1-*ZmPsy1* and Exp1-*MtPsy2a* lines remained below 5 µg/g dw β-CE from the emergence of the bunch all the way through to S9 (with the exception of line FT342 which accumulated 6.6 µg/g dw β-CE at S9) followed by an increase towards maturity with a maximum of 9.9 µg/g dw β-CE at FR in Exp1-*ZmPsy1* line FT187 (Figure 3b and e). During fruit development of lines containing ACO-*ZmPsy1*, PVA levels were lowest at S3 and S6 for two of the lines but moderately higher than the Exp1-*ZmPsy1* lines at those stages (Figure 3c). PVA levels peaked for those same two lines at either S9 or FG. In contrast, the highest PVA level in line FT476 was at

S3. For the ACO-*MtPsy2a* lines, again one line, FT511, had maximum PVA accumulation at S3, while the other two lines had maximums at S9 or FG (Figure 3f). Overall, the PVA accumulation pattern during fruit development reflected the expression profiles previously observed in transgenic lines where the same three promoters were used to drive the expression of *uidA* (Figure 1). The constitutive Ubi promoter provided consistent stronger expression throughout fruit development followed by the ACO promoter and finally Exp1.

During the plant and ratoon crops, the phenotype of each plant was recorded at regular intervals from planting to bunch harvest. None of the 50 wild-type control plants showed altered phenotypes and fruit developed normally (Figure 4a–c). The presence of the *PaCrtI* transgene did not appear to affect phenotype. However, three categories of altered phenotypes were observed in the transgenic lines: stunting, 'golden leaf' and 'golden bunch'. For the 'golden leaf' phenotype, the youngest leaf would consistently unfurl with a bright yellow colour ('golden') and progressively turn to green as it matured (Figure 4g). Fruit on the 'golden bunch' emerged bright orange instead of green (Figure 4d). As the fruit matured and filled, it progressively turned greener to a mixture of green and orange at harvest (Figure 4e and f). Fruits with increase PVA levels displayed a pulp ranging from deep yellow to bright orange (Figure 4h and i). Of the original 244 transgenic lines planted in the trial, 65 had the 'golden leaf' phenotype of which 29 were also stunted; 29 had the 'golden bunch' phenotype. The 'golden leaf' and 'golden bunch' phenotypes were highly transgene dependent where lines with those phenotypes invariably contained the *ZmPsy1* transgene in contrast to lines containing the *MtPsy2a* transgene. The 'golden leaf' phenotype was observed in 27 of 57 (50%) lines observed carrying Exp1-*ZmPsy1* alone or together with either Exp1-*PaCrtI* or Ubi-*PaCrtI*. More importantly, the 'golden bunch' phenotype which only occurred in ACO-*ZmPsy1* or ACO-*ZmPsy1* + Exp1-*PaCrtI* lines was recorded in 94% (29 of the 31) of the lines assessed. In contrast, across all 85 lines containing

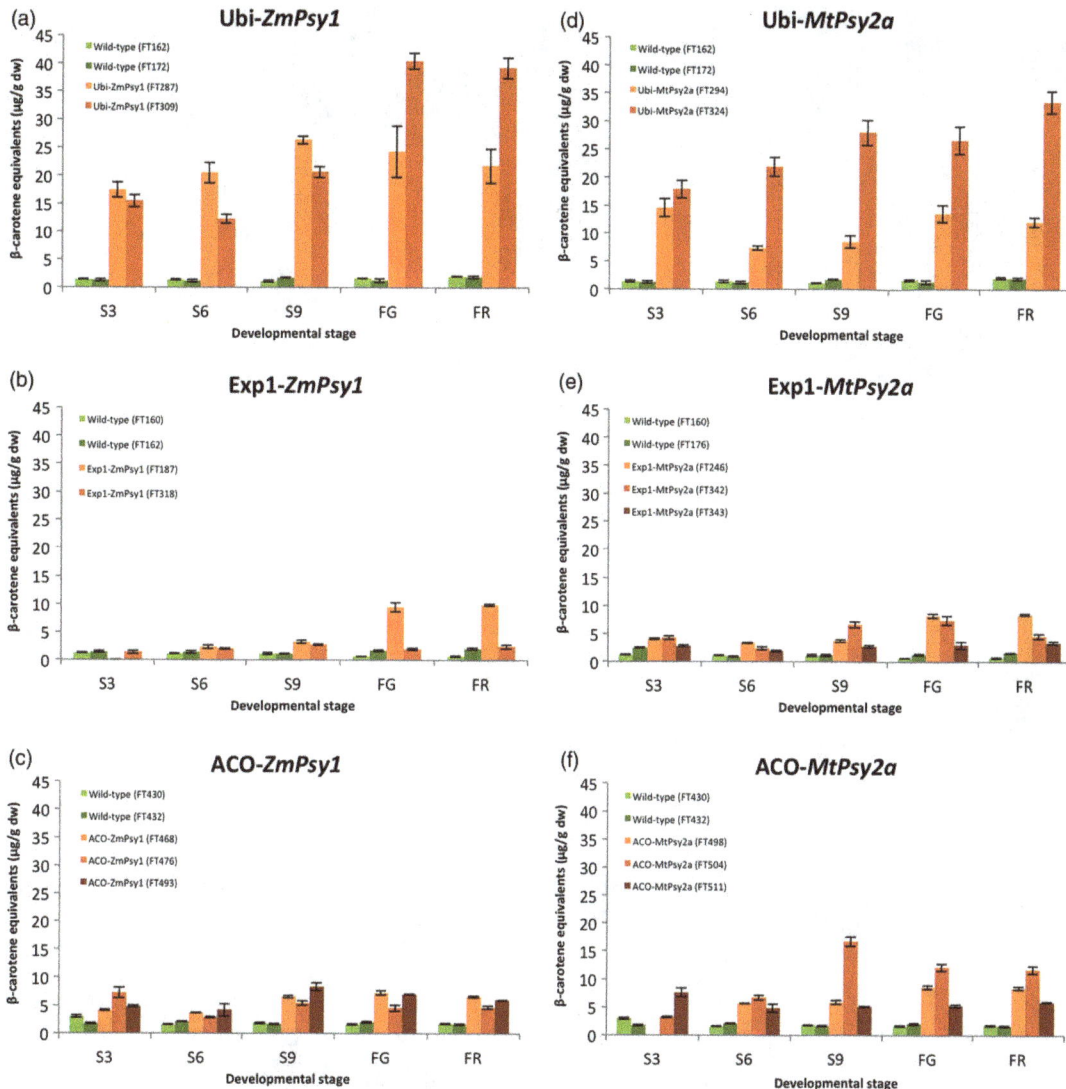

Figure 3 PVA carotenoid accumulation in the pulp of wild-type and selected transgenic Cavendish banana lines during fruit development. S3, S6 and S9 represent 3, 6 and 9 weeks post-bunch emergence, respectively. FG, full green and FR, full ripe. Error bars = ±SD.

MtPsy2a, only 7 (8%) had the 'golden leaf' phenotype and none displayed a 'golden bunch'.

The β-CE levels in the pulp and peel of FG and FR fruit from lines with the 'golden bunch' phenotype were analyzed and compared with the levels in fruit from phenotypically normal lines. In the FG fruit from four lines of ACO-*MtPsy2a*, the levels of β-CE in the peel were similar to those in the three non-transgenic controls (Table 3). However, the peel β-CE levels in four ACO-*ZmPsy1* containing lines were more than ninefold higher than in fruit peel from the control lines (Table 3). Importantly, the peel β-CE levels in the ACO-*ZmPsy1* lines did not influence fruit pulp β-CE levels as the four ACO-*ZmPsy1* lines had a similar range of β-CE levels in their fruit pulp at FG to fruit from the four ACO-*MtPsy2a* lines (Table 3).

PVA analysis: sucker crop

Following the initial plant and ratoon crop assessment with single plants per line, 30 lines selected from seven promoter/transgene combinations and five wild-type lines were multiplied through suckering to a maximum of 10 replicates per line. A total of 239

transgenic and 48 wild-type plants derived from suckers were planted in a second field trial (FT-2). Each plant was harvested, and PVA levels were measured in the fruit at FG and FR by HPLC.

In all transgenic and wild-type lines, the averaged fruit PVA level was higher in the sucker crop than in either the plant or ratoon crops. The highest average PVA level was 55.0 µg/g dw β-CE found in the FG fruit of Ubi-*MtPsy2a* line FT324 (Table 2) with one individual plant of this line reaching 73.8 µg/g dw β-CE (data not shown). Fruit from this plant had bright orange pulp compared with the fruit of non-transgenic control plants (Figure 4h). Of the 27 transgenic lines shown in Table 2, 11 had fruit PVA levels equal to or greater than the 20 µg/g dw β-CE target. However, the fruit PVA levels in the wild-type controls were also higher than in the plant and ratoon crops with an average of 6.0 µg/g dw β-CE. Interestingly, the top four sucker crop lines all contained an Ubi-*Psy* promoter/transgene combination. Furthermore, six of 12 ACO promoter lines had PVA levels equal to or greater than target level, compared to only one Exp1 promoter line of 9 (Table 2). Analysis of the carotenoid composition of fruit pulp from wild-type as well as transgenic fruit in the

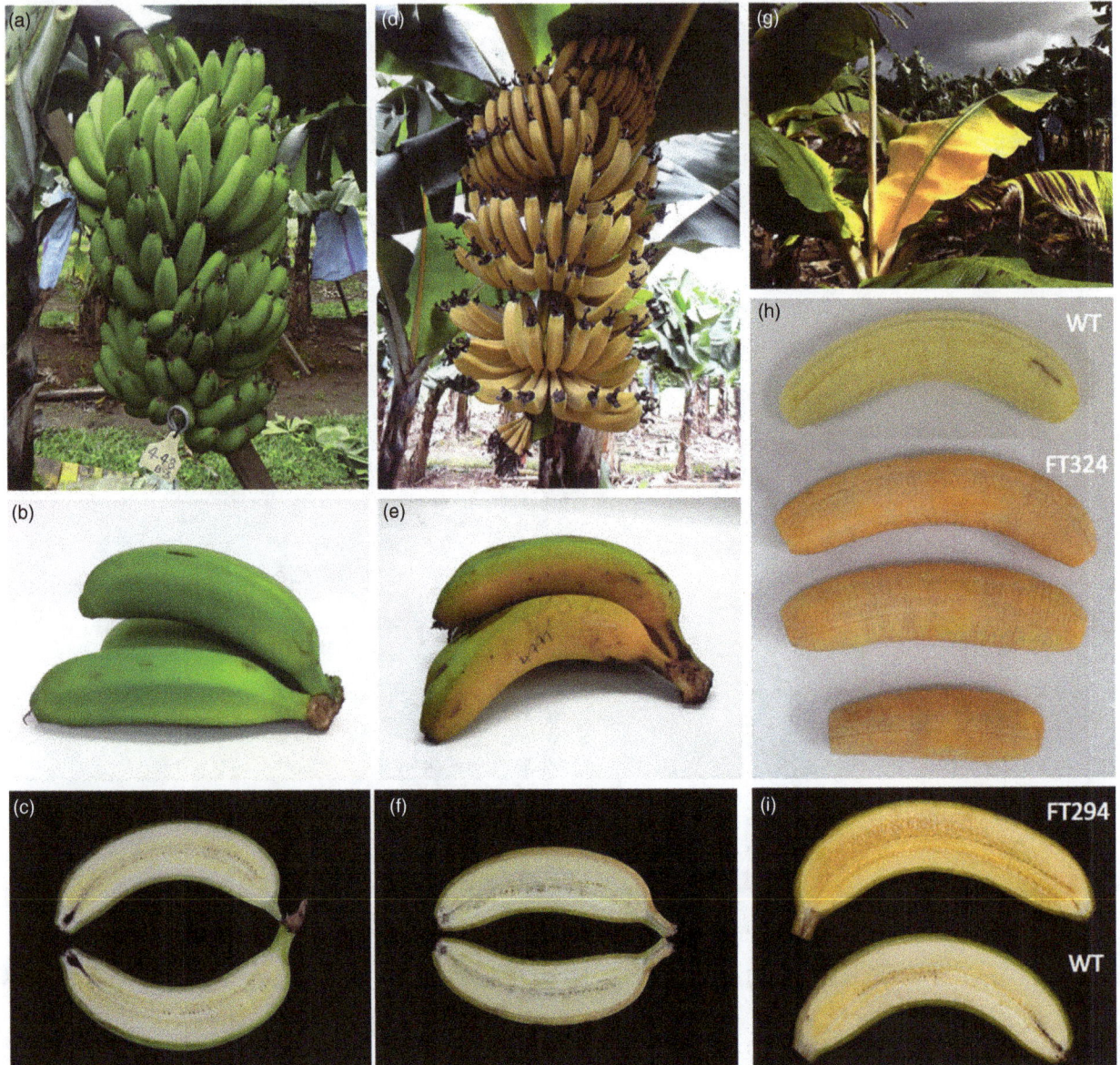

Figure 4 Characteristic phenotypes observed in wild-type and transgenic Cavendish banana lines. (a) Bunch from wild-type line FT448; (b) fruit from wild-type line FT448; (c) longitudinal section of fruit from wild-type line FT448; (d) immature bunch from ACO-*ZmPsy1* line FT477; (e) fruit from ACO-*ZmPsy1* line FT477; (f) longitudinal section of fruit from ACO-*ZmPsy1* line FT477; (g) Exp1-*ZmPsy1* line FT192; (h) Ubi-*MtApsy2a* line FT324; and (i) Ubi-*MtApsy2a* line FT294. WT, wild-type.

sucker crop revealed that, like the plant crop fruit, samples contained higher levels of α-carotene than β-carotene and in similar proportions (Figure 5).

Variation in fruit PVA levels in wild-type banana

The average PVA levels in the FG fruit of non-transgenic control plants varied considerably throughout the two field trials from a low of 1.0 µg/g dw β-CE in March-harvested fruit in the plant crop of FT-1 to 8.1 µg/g dw β-CE in September-harvested fruit of FT-2 (Figure 6). When analyzed together, a strong correlation was observed between the level of accumulated PVA in the fruit and the number of days from bunch emergence to harvest (bunch filling time). Indeed, fruit harvested in March had the shortest bunch filling time (94–97 days) and the lowest accumulated PVA level in the fruit compared with 142 days for September-

harvested fruit which had the highest fruit PVA levels (Figure 6). Longer bunch filling time and associated increased levels of PVA also appeared temperature dependent where fruit maturing during the cooler months had higher levels of accumulated PVA.

Discussion

Micronutrient deficiencies remain a substantial burden on the public health of populations particularly in developing countries. The 'poorest of the poor' are particularly impacted by micronutrient deficiencies because of their increasing dependence on nutrient poor single staple crops for the majority of their calorific intake (Muthayya *et al.*, 2013). There have been significant inroads into reducing VAD in children aged 6 to 59 months where worldwide prevalence has fallen from 39% in 1991 to

Table 3 PVA carotenoid concentration in the fruit pulp and peel of wild-type and selected transgenic Cavendish banana lines

Promoter-transgene	Line	β-CE in pulp (µg/g dw)		β-CE in peel (µg/g dw)	
		FG	FR	FG	FR
Wild-type	FT166	6.6	7.3	74.2	32.1
Wild-type	FT430	6.5	8.3	126.5	45.4
Wild-type	FT448	8.0	9.4	99.5	38.8
ACO-*MtPsy2a*	FT504	20.0	24.7	108.3	72.7
ACO-*MtPsy2a*	FT508	15.6	19.0	68.3	72.6
ACO-*MtPsy2a*	FT511	14.4	13.6	105.7	75.0
ACO-*MtPsy2a*	FT518	23.1	35.9	129.2	131.2
ACO-*ZmPsy1*	FT467	10.5	13.4	742.1	703.2
ACO-*ZmPsy1*	FT475	20.7	22.0	1112.2	1132.6
ACO-*ZmPsy1*	FT479	20.5	16.2	740.9	833.0
ACO-*ZmPsy1*	FT483	18.8	16.9	1155.1	837.3

FG, full green and FR, full ripe. All samples were collected from the sucker crop.

29% in 2013 in low- and middle-income countries (Stevens *et al.*, 2015). However, VAD prevalence in children age 6–59 months in Uganda has increased from 20% in 2006 to 38% in 2011 and a similar increase was seen from 19% to 36% in women age 15–49 years (UDHS - Uganda Demographic and Health Survey, 2006, 2011). In Uganda, the staple crop is cooking bananas, more specifically East African highland bananas (EAHB). These bananas are a group of very similar triploid clones of *Musa acuminata* with low male and female fertility levels.

Therefore, a metabolic engineering strategy that could provide the basis for elevating fruit PVA levels in EAHB was developed. Previously, the most common strategy used to achieve transgenic

elevated PVA has been the Golden Rice 2 strategy where seed expression of both *ZmPsy1* and *PaCrtI* has been reported to significantly increase seed PVA in rice (Paine *et al.*, 2005), maize (Naqvi *et al.*, 2009; Zhu *et al.*, 2008), wheat (Cong *et al.*, 2009) and sorghum (Lipkie *et al.*, 2013) with levels up to 31 µg/g dw in rice and 59.3 µg/g dw in maize. A range of other transgenes have also been tested either alone or in combinations including *PaCrtB* in potato (Ducreux *et al.*, 2005), tomato (Fraser *et al.*, 2002), canola (Shewmaker *et al.*, 1999), maize (Aluru *et al.*, 2008), soya bean (Schmidt *et al.*, 2015) and cassava (Sayre *et al.*, 2011), *AtDXS* in cassava and sorghum (Lipkie *et al.*, 2013; Sayre *et al.*, 2011) and the brassica Or gene, *BaOr,* in potato (Li *et al.*, 2012).

In the present study, the dominant dessert banana, Cavendish, was used as a model for EAHB as it is also a triploid *M. acuminata*. Two phytoene synthase transgenes were tested with and without co-expression of *PaCrtI*. Interestingly, over-expression of phytoene synthase as a strategy to increase PVA had previously only been reported in cereal crops for which the transgene used was always *ZmPsy1* except in Golden Rice 1. In this work, *ZmPsy1* was tested as well as *MtPsy2a*, a banana phytoene synthase gene previously cloned from the fruit of a naturally high PVA Fe'i banana called Asupina (Mlalazi *et al.*, 2012). This approach allowed the generation of eleven transgenic lines which produced fruit that contained greater than the target level of 20 µg/g dw β-CE. The highest level of PVA in the fruit of a single banana plant was from line FT324 (73.8 µg/g dw β-CE) which averaged at 55.0 µg/g dw β-CE. This line contained the banana phytoene synthase gene (*MtPsy2a*) under the control of the constitutive ubiquitin promoter. The highest PVA levels with the phytoene synthase transgene under the control of an apparently fruit-specific promoter with or without co-expression of *PaCrtI* were also above target and included one ACO-*ZmPsy1*+ Exp1-*PaCrtI* line with 27.0 µg/g dw β-CE, one ACO-*ZmPsy1* line with 20.7 µg/g dw β-CE and one ACO-*MtPsy2a* line with 23.1 µg/g dw β-CE. The levels of PVA in Line 324 are equivalent to the highest levels obtained in any other crop where a plant phytoene synthase gene has been over-expressed. In maize

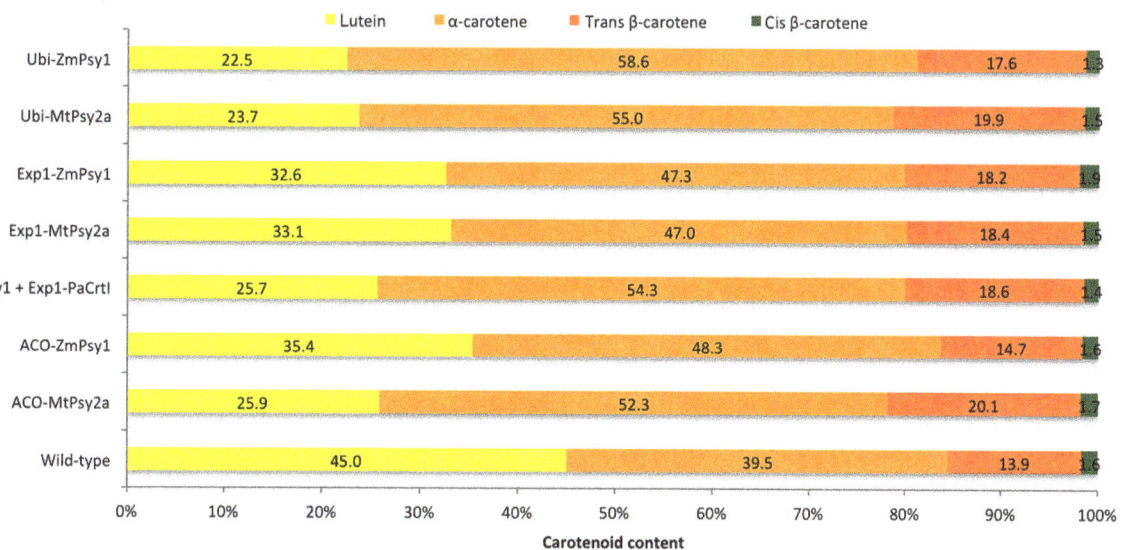

Figure 5 Percentage accumulation of individual carotenoids in the fruit pulp of wild-type and transgenic bananas. Percentage (%) carotenoid content calculated based on total carotenoid content measured in the pulp of full green fruit collected from the sucker crop. Biological replicates: wild-type (n = 5), ACO-*MtPsy2a* (n = 5), ACO-*ZmPsy1* (n = 4), ACO-*ZmPsy1* + Exp1-*PaCrtI* (n = 3), Exp1-*MtPsy2a* (n = 4), Exp1-*ZmPsy1* (n = 5), Ubi-*MtPsy2a* (n = 4) and Ubi-*ZmPsy1* (n = 2). All samples were analyzed in three technical replicates.

FT-1	Jan	Feb	Mar	Apr	May	Jun	Jul	Aug	Sep	Oct	Nov	Dec
β-CE (μg/g dw)	1.2	1.2	1.0	2.4	-	-	2.0	3.5	-	1.4	2.2	1.7
Bunch filling time (days)	108	100	97	130	-	-	117	126	-	124	118	110
FT-2	Jan	Feb	Mar	Apr	May	Jun	Jul	Aug	Sep	Oct	Nov	Dec
β-CE (μg/g dw)	1.6	2.2	1.7	1.7	-	5.7	6.8	7.8	8.1	7.7	4.4	1.7
Bunch filling time (days)	105	98	94	98	-	111	123	131	142	132	118	112
Temperature (°C)	31.2	30.7	29.8	28.2	26.2	24.4	23.9	25.2	27.0	29.0	30.5	31.2

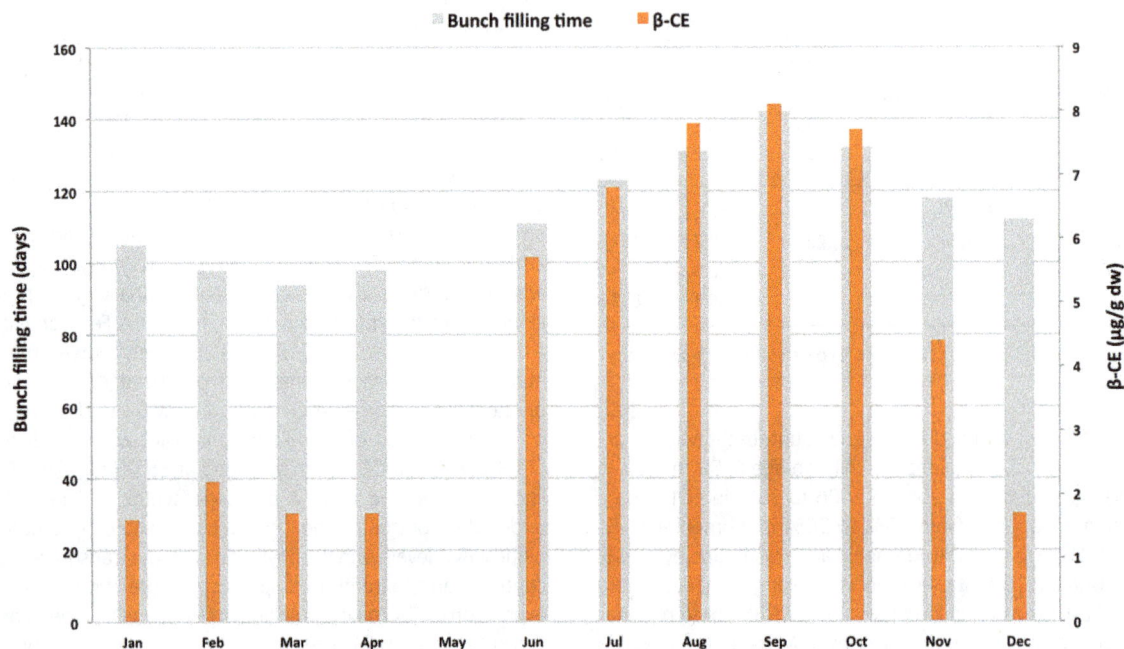

Figure 6 Influence of temperature and time to fruit maturity on the concentration of PVA carotenoids in the FG pulp of wild-type Cavendish banana. For each month, β-CE levels and time from bunch emergence to harvest were averaged from all samples collected. FG, full green.

transformed with the seed-expressed *ZmPsy1* and *PaCrtI* transgenes, Zhu *et al.* (2008) and Naqvi *et al.* (2009) both reported levels of 57.4 and 59.3 μg/g dw β-CE, respectively. However, Schmidt *et al.* (2015) recently reported that soya beans transformed with *PaCrtB* alone under the control of the seed-specific Le1 promoter accumulated up to 845 μg/g dw β-carotene in the seed.

Although two banana lines containing ACO-*ZmPsy1* had above target levels of PVA, nearly all lines transformed with this construct had a 'golden bunch' phenotype which was never observed in ACO-*MtPsy2a* transformed plants. That ACO-*ZmPsy1* expression should result in a 'golden bunch' phenotype while Exp1-*ZmPsy1* expression resulted invariably in a 'golden leaf' phenotype suggests that these two promoters express in tissue other than fruit pulp and most likely in either leaf primordia for Exp1 or 'bunch' primordia for ACO. The molecular basis for both the phenotypes has not yet been determined and is under investigation.

An important outcome from this study was that the levels of PVA in most banana lines increased through the 'generations' from the plant crop through the first ratoon, second ratoon to finally the sucker crop. Line FT324 had only 11.7 μg/g dw β-CE in the FG plant crop to 55.0 μg/g dw β-CE in the sucker crop. Bananas are vegetatively propagated and do not go through a seed phase, and thus, all generations are T_0. The explanation for the phenomenon of increasing PVA levels with successive vegetative generations is not entirely understood but it did demonstrate that, rather than there being a reduction of expression with successive vegetative generations as a result of transgene silencing, the trait was stable. Further, it is possible that this phenomenon might also occur in other vegetatively propagated crops demonstrating the importance of monitoring transgenic traits in vegetative propagated crops in the field through multiple 'generations'.

Two outcomes from this study indicated that the final level of PVA was a result of accumulation through the development of the banana fruit. Firstly, the levels of PVA in non-transgenic Cavendish fruit were found to vary considerably. Our observations indicated that these variations correlate with time to fruit maturity. Indeed, the highest PVA levels were obtained in fruit with the longest maturity time. This probably accounts for the varying reported levels of fruit PVA levels in cultivars such as Cavendish (Davey *et al.*, 2007; Englberger *et al.*, 2006b; Fungo and Pillay, 2011) as well as much of the variation observed within transgenic lines in different seasons. This variable was controlled for in the 'sucker generation' where all suckers were planted on the same day at the same location. Secondly, the two lines that accumulated the highest PVA concentration contained the phytoene synthase genes under the control of the Ubi promoter. This promoter was previously demonstrated to be active from the earliest stages of banana fruit development in contrast with the late fruit expression from the ACO and Exp1 promoters.

Two constructs containing Ubi-*MtPsy2a* and ACO-*MtPsy2a* have been transferred to the National Agricultural Research Organization (NARO) in Uganda and have been transformed into two Ugandan cultivars including EAHB.

Experimental procedures

Vector construction

The coding regions of *Musa troglodytarum* x *acuminata* cultivar Asupina phytoene synthase 2a (*MtPsy2a*; GenBank #: JX195659), *Zea mays* cultivar B73 phytoene synthase 1 (*ZmPsy1*; GenBank #: U32636) and *Pantoea ananatis* phytoene desaturase (*PaCrtI*; Genbank #: D90087) genes were used to facilitate PVA enhancement in banana. The *uidA* gene from *Escherichia coli*, encoding the enzyme β-glucuronidase (GUS), containing a catalase intron (Khanna *et al.*, 2004), was used to assess promoter activity. Each of these genes was fused to the *Agrobacterium tumefaciens* nopaline synthase (Nos) 3′ transcription termination regulatory sequence (Depicker *et al.*, 1982). The banana expansin (Exp1; GenBank #: JN172931) and 1-aminocyclopropane-1-carboxylate oxidase (ACO; GenBank #: AF221107) promoters, and the maize polyubiquitin1 promoter (Ubi, Dugdale *et al.*, 2000) where characterized in banana using each of them to drive *uidA* expression, and the resulting expression cassettes: Exp1-*uidA*-nos, ACO-*uidA*-nos and Ubi-*uidA*-nos were assembled in the pBIN-19 (GenBank #: U09365) binary vector backbone. These three promoters were also used to regulate the expression of the *MtPsy2a* and *ZmPsy1* genes, resulting in the production of six expression cassettes in the pCAMBIA-2300 (GenBank #: AF234315) binary vector backbone. Expression of the *PaCrtI* gene was driven by the Ubi or Exp1 promoter, and the resulting Ubi-*PaCrtI*-nos and Exp1-*PaCrtI*-nos expression cassettes were assembled in pBIN-19. Selection of transgenic plants was mediated by the neomycin phosphotransferase II gene (*nptII*; Beck *et al.*, 1982) in both pCAMBIA-2300 and pBIN-19.

Plant transformation and regeneration

Transgenic *Musa acuminata* (AAA Group) 'Dwarf Cavendish' lines were generated via *Agrobacterium*-mediated transformation of embryogenic cell suspensions (ECS) using *A. tumefaciens* strain AGL1 (Khanna *et al.*, 2004). Binary vectors containing *MtPsy2a* or *ZmPsy1* were used for transformation of banana ECS either alone or in combination with one of the vectors containing *PaCrtI*, while vectors containing *uidA* were all used individually for transformation.

Plant material and field trials

Transgenic and wild-type banana lines established in tissue culture were transported to the field trial site at the Department of Agriculture and Fisheries (DAF) South Johnstone Research Facility (Queensland, Australia) according to conditions on the Office of the Gene Technology Regulator (OGTR) licence number DIR109. Plants were acclimatized in soil and grown in a glasshouse for 12 weeks before planting and maintenance in the field according to standard agronomic procedures. The initial field trial (FT-1) was conducted between 2009 and 2012. Promising transgenic lines and selected wild-type control lines (up to 10 sucker plants per line) were established for 12 weeks in the glasshouse before being transferred to a second field trial (FT-2) at the same facility, which commenced in September 2012.

Field sample collection and processing

Mature green (full green—FG) fruit was harvested from each plant and sent to the Centre for Tropical Crops and Biocommodities (CTCB) laboratory in Brisbane, Australia, within 48 h of harvest. When analysis of developing fruit was required, fruit was taken at 3, 6 and 9 weeks post-bunch emergence (S3, S6 and S9) prior to harvesting the entire bunch at FG. Fruit was handled under low-light conditions and was processed either immediately (for FG analysis) or at 7 days post-exposure (24 h) to ethylene for full ripe (FR) analysis. Representative fruit from the top, middle and bottom of the bunch was received at all stages of fruit development from plants transformed with *Psy* or *Crtl* genes, while fruit from the top of the bunch only was obtained from lines transformed with the *uidA* gene. Leaf samples from the field were collected from the first fully expanded leaf prior to bunch emergence. Prior to further analysis, all samples were freeze-dried in a Benchtop 4K Freeze Dryer (VirTis®) and homogenized in a Mini-Beadbeater-8™ (Biospec Products) tissue disruptor.

Nucleic acid isolation

Genomic DNA (gDNA) for PCR and Southern blot analysis was isolated from 50 mg of homogenized freeze-dried leaf tissue using a modified CTAB method (Stewart and Via, 1993). Isolation of plasmid DNA (pDNA) was done using the Wizard® Plus SV Minipreps DNA Purification System (Promega) according to the manufacturer's instructions. Total RNA was extracted from 50 mg of homogenized freeze-dried banana fruit tissue essentially as described by Valderrama-Cháirez *et al.* (2002).

DNase treatment and complementary DNA (cDNA) synthesis

For cDNA synthesis, 3 µg of total RNA was DNase treated using an RQ1 RNase-free DNase Kit (Promega). DNA-free RNA samples (1.8 µg) were reverse-transcribed to cDNA using an oligo(dT20) primer and the GoScript™ Reverse Transcription System (Promega) in 25 µL reactions according to the manufacturer's instructions.

Polymerase chain reaction (PCR) and reverse-transcription PCR (RT-PCR)

Putatively transgenic tissue culture plants were tested under standard PCR conditions using oligonucleotide primers designed to amplify gene fragments spanning the *MtPsy2a*, *ZmPsy1* and *PaCrtl* genes and their respective promoter region and using GoTaq® Green master mix (Promega). Transgene-positive plants were further PCR tested for the presence of contaminating *Agrobacterium* using *VirC* gene primers (Haas *et al.*, 1995). Complementary DNA (cDNA) from PCR-positive plants was then used in RT-PCR to verify gene expression.

Quantitative real-time PCR (qRT-PCR)

Quantitative RT-PCRs were done in a CFX384 Touch™ Real-Time PCR Detection System (Bio-Rad) using GoTaq qPCR Master Mix (Promega). Each sample was analyzed in three technical replicates in addition to the inclusion of 'no template' and 'RT-negative' controls. The following amplification parameters were used: Hot-Start polymerase activation at 95 °C for 2 min, followed by 45 cycles of 10 s denaturation at 95 °C and 30 s annealing/extension at 60 °C. At the end of the reaction, a dissociation curve was produced from 65 to 95 °C to confirm the specificity of the amplicon from each primer set. Fluorescence was recorded in real time and detected at 470 nm. Relative expression levels were calculated using the CFX Manager 3.1 (Bio-Rad) software and the ΔCT method (Schmittgen and Livak, 2008). Ct data obtained from target gene of interest (GOI) were normalized using Ct values from the two stable reference genes cyclophilin (*CYP*) and ribosomal protein S2 (*RPS2*). All primers were designed using the Primer3Plus freeware (http://www.bioinformatics.nl/cgi-bin/primer3plus/primer3plus.cgi) (Table S1).

Southern hybridization

For determination of transgene copy number integration by Southern analysis (Southern, 1975), genomic DNA (10 μg) as well as positive control plasmid DNA (20 ng) were digested for 16 h and 1 h, respectively, by incubation at 37 °C with 20 U of restriction enzyme. Restriction enzymes were selected to only cut once within the binary vector T-DNA region without cutting the region to which the probe would hybridize. PCR based DIG-labelled probes (Roche) were designed targeting the coding regions of *MtPsy2a* and *ZmPsy1*. Digested DNA was electrophoresed, blotted and detected under standard Southern blotting conditions (Sambrook and Russell, 2001).

Carotenoid content quantification

Carotenoids were extracted from banana pulp (200 mg) or peel (25 mg) tissue and analyzed by HPLC as previously described (Buah *et al.*, 2016). Total carotenoids and β-carotene equivalents (β-CE) were expressed in μg/g dry weight (dw).

Assessment of promoter activity

β-glucuronidase (GUS) content was measured in banana pulp tissue (40 mg) by ELISA as per Dugdale *et al.* (2013), while fluorometric quantification of GUS activity was determined in leaf and peel tissue as described by Jefferson *et al.* (1987).

Acknowledgements

The Banana21 project is supported by the Bill & Melinda Gates Foundation and the Department for International Development (United Kingdom).

References

Aluru, M., Xu, Y., Guo, R., Wang, Z., Li, S., White, W., Wang, K. *et al.* (2008) Generation of transgenic maize with enhanced provitamin A content. *J. Exp. Bot.* **59**, 3551–3562.

Beck, E., Ludwig, G., Auerswald, E.A., Reiss, B. and Schaller, H. (1982) Nucleotide sequence and exact localization of the neomycin phosphotransferase gene from transposon Tn5. *Gene*, **19**, 327–336.

Bresnahan, K.A., Arscott, S.A., Khanna, H., Arinaitwe, G., Dale, J., Tushemereirwe, W., Mondloch, S. *et al.* (2012) Cooking enhances but the degree of ripeness does not affect provitamin A carotenoid bioavailability from bananas in Mongolian gerbils. *J. Nutr.* **142**, 2097–2104.

Buah, S., Mlalazi, B., Khanna, H., Dale, J.L. and Mortimer, C.L. (2016) The quest for golden bananas: investigating carotenoid regulation in a Fe'i group *Musa* cultivar. *J. Agric. Food Chem.* **64**, 3176–3185.

Cong, L., Wang, C., Chen, L., Liu, H., Yang, G. and He, G. (2009) Expression of phytoene synthase 1 and carotene desaturase *crtI* genes result in an increase

in the total carotenoids content in transgenic elite wheat (*Triticum aestivum* L.). *J. Agric. Food Chem.* **57**, 8652–8660.

Davey, M., Stals, E., Ngoh-Newilah, G., Tomekpe, K., Lusty, C., Markham, R., Swennen, R. *et al.* (2007) Sampling strategies and variability in fruit pulp micronutrient contents of West and Central African bananas and plantains (*Musa* species). *J. Agric. Food Chem.* **55**, 2633–2644.

Depicker, A., Stachel, S., Dhaese, P., Zambryski, P. and Goodman, H.M. (1982) Nopaline synthase: transcript mapping and DNA sequence. *J Mol Appl Genet.* **1**(6), 561–573.

Ducreux, L.J.M., Morris, W.L., Hedley, P.E., Shepherd, T., Davies, H.V., Millam, S. and Taylor, M.A. (2005) Metabolic engineering of high carotenoid potato tubers containing enhanced levels of β-carotene and lutein. *J. Exp. Bot.* **56**, 81–89.

Dugdale, B., Becker, D.K., Beetham, P.R., Harding, R.M. and Dale, J.L. (2000) Promoters derived from banana bunchy top virus DNA-1 to -5 direct vascular-associated expression in transgenic banana (*Musa* spp.). *Plant Cell Rep.* **19**, 810–814.

Dugdale, B., Mortimer, C.L., Kato, M., James, T.A., Harding, R.M. and Dale, J.L. (2013) In Plant Activation: an inducible, hyperexpression platform for recombinant protein production in plants. *Plant Cell*, **25**, 2429–2443.

Englberger, L., Darnton-Hill, I., Coyne, T., Fitzgerald, M.H. and Marks, G.C. (2003) Carotenoid-rich bananas: a potential food source for alleviating vitamin A deficiency. *Food Nutr. Bull.* **24**, 303–318.

Englberger, L., Schierle, J., Aalbersberg, W., Hofmann, P., Humphries, J., Huang, A., Lorens, A. *et al.* (2006a) Carotenoid and vitamin content of Karat and other Micronesian banana cultivars. *Int. J. Food Sci. Nutr.* **27**, 399–418.

Englberger, L., Wills, R.B.H., Blades, B., Dufficy, L., Daniells, J.W. and Coyne, T. (2006b) Carotenoid content and flesh color of selected banana cultivars growing in Australia. *Food Nutr. Bull.* **27**, 281–291.

Fraser, P.D., Romer, S., Shipton, C.A., Mills, P.B., Kiano, J.W., Misawa, N., Drake, R.G. *et al.* (2002) Evaluation of transgenic tomato plants expressing an additional phytoene synthase in a fruit-specific manner. *Proc. Natl Acad. Sci. USA*, **99**, 1092–1097.

Fungo, R. and Pillay, M. (2011) β-Carotene content of selected banana genotypes from Uganda. *Afr. J. Biotechnol.* **10**, 5423–5430.

Haas, J.H., Moore, L.W., Ream, W. and Manulis, S. (1995) Universal PCR primers for detection of phytopathogenic *Agrobacterium* strains. *Appl. Environ. Microbiol.* **61**, 2879–2884.

HarvestPlus (2012) *Disseminating Orange-Fleshed Sweet Potato: Uganda Country Report.* pp. 16. Washington DC, USA: HarvestPlus.

Jefferson, R.A., Kavanagh, T.A. and Bevan, M.W. (1987) GUS fusions: beta-glucuronidase as a sensitive and versatile gene fusion marker in higher plants. *Embo j* **6**(13), 3901–3907.

Khanna, H., Becker, D., Kleidon, J. and Dale, J. (2004) Centrifugation assisted *Agrobacterium tumefaciens*-mediated transformation (CAAT) of embryogenic cell suspensions of banana (*Musa* spp. Cavendish AAA and Lady finger AAB). *Mol Breed.* **14**, 239–252.

Komarek, A. (2010) The determinants of banana market commercialisation in Western Uganda. *Afr. J. Agric. Res.* **5**, 775–784.

Li, L., Yang, Y., Xu, Q., Owsiany, K., Welsch, R., Chitchumroonchokchai, C., Lu, S. *et al.* (2012) The *Or* gene enhances carotenoid accumulation and stability during post-harvest storage of potato tubers. *Mol. Plant*, **5**, 339–352.

Lipkie, T.E., De Moura, F.F., Zhao, Z.-Y., Albertsen, M.C., Che, P., Glassman, K. and Ferruzzi, M.G. (2013) Bioaccessibility of carotenoids from transgenic provitamin A biofortified sorghum. *J. Agric. Food Chem.* **61**, 5764–5771.

Mbabazi, R. (2015) *Molecular characterisation and carotenoid quantification of pro-vitamin A biofortified genetically modified bananas in Uganda.* PhD thesis, Queensland University of Technology.

Mlalazi, B., Welsch, R., Namanya, P., Khanna, H., Geijskes, R.J., Harrison, M.D., Harding, R. *et al.* (2012) Isolation and functional characterisation of banana phytoene synthase genes as potential cisgenes. *Planta*, **236**, 1585–1598.

Muthayya, S., Rah, J.H., Sugimoto, J.D., Roos, F.F., Kraemer, K. and Black, R.E. (2013) The global hidden hunger indices and maps: an advocacy tool for action. *PLoS ONE*, **8**, e67860.

Naqvi, S., Zhu, C., Farre, G., Ramessar, K., Bassie, L., Breitenbach, J., Perez Conesa, D. *et al.* (2009) Transgenic multivitamin corn through biofortification of endosperm with three vitamins representing three distinct metabolic pathways. *Proc. Natl Acad. Sci. USA*, **106**, 7762–7767.

Paine, J.A., Shipton, C.A., Chaggar, S., Howells, R.M., Kennedy, M.J., Vernon, G., Wright, S.Y. *et al.* (2005) Improving the nutritional value of Golden Rice through increased pro-vitamin A content. *Nat. Biotechnol.* **23**, 482–487.

Sambrook, J. and Russell, D.W. (2001) *Molecular Cloning: A Laboratory Manual.* Cold Spring Harbor, N.Y.: Cold Spring Harbor Laboratory Press.

Sayre, R., Beeching, J.R., Cahoon, E.B., Egesi, C., Fauquet, C., Fellman, J., Fregene, M. *et al.* (2011) The BioCassava Plus program: biofortification of cassava for Sub-Saharan Africa. *Annu. Rev. Plant Biol.* **62**, 251–272.

Schmidt, M.A., Parrott, W.A., Hildebrand, D.F., Berg, R.H., Cooksey, A., Pendarvis, K., He, Y. *et al.* (2015) Transgenic soya bean seeds accumulating β-carotene exhibit the collateral enhancements of oleate and protein content traits. *Plant Biotechnol. J.* **13**, 590–600.

Schmittgen, T.D. and Livak, K.J. (2008) Analyzing real-time PCR data by the comparative CT method. *Nat. Protocols* **3**(6), 1101–1108.

Shewmaker, C.K., Sheehy, J.A., Daley, M., Colburn, S. and Ke, D.Y. (1999) Seed-specific overexpression of phytoene synthase: increase in carotenoids and other metabolic effects. *Plant J.* **20**, 401–412.

Smale, M. and Tushemereirwe, W.K. (2007) *An economic assessment of banana genetic improvement and innovation in the Lake Victoria region of Uganda and Tanzania.* Research reports 155 from the International Food Policy Research Institute (IFPRI), Washington, D.C.

Southern, E.M. (1975) Detection of specific sequences among DNA fragments separated by gel electrophoresis. *J. Mol. Biol.* **98**, 503–517.

Stevens, G.A., Bennett, J.E., Hennocq, Q., Lu, Y., De-Regil, L.M., Rogers, L., Danaei, G. *et al.* (2015) Trends and mortality effects of vitamin A deficiency in children in 138 low-income and middle-income countries between 1991 and 2013: a pooled analysis of population-based surveys. *Lancet Glob. Health*, **3**, e528–e536.

Stewart, C.N. and Via, L.E. (1993) A rapid CTAB DNA isolation technique useful for RAPD fingerprinting and other PCR applications. *Biotechniques*, **14**, 748–750.

UDHS - Uganda Demographic and Health Survey (2006) *Uganda Bureau of Statistics (UBOS), Kampala.* Calverton, Maryland, USA: Uganda and Macro International Inc..

UDHS - Uganda Demographic and Health Survey (2011) *Uganda Bureau of Statistics (UBOS), Kampala, Uganda and MEASURE DHS ICF International, Calverton, Maryland, USA.*

Valderrama-Cháirez, M.L., Cruz-Hernández, A. and Paredes-López, O. (2002) Isolation of functional RNA from cactus fruit. *Plant Mol. Biol. Rep.* **20**, 279–286.

WHO (2011) *Guideline: Vitamin A Supplementation in Infants and Children 6–59 months of Age.* Geneva: WHO

Ye, X., Al-Babili, S., Klöti, A., Zhang, J., Lucca, P., Beyer, P. and Potrykus, I. (2000) Engineering the pro-vitamin-A (β-carotene) biosynthetic pathway into (carotenoid-free) rice endosperm. *Science*, **287**, 303–305.

Zhu, C., Naqvi, S., Breitenbach, J., Sandmann, G., Christou, P. and Capell, T. (2008) Combinatorial genetic transformation generates a library of metabolic phenotypes for the carotenoid pathway in maize. *Proc. Natl Acad. Sci. USA*, **105**, 18232–18237.

5

A collection of enhancer trap insertional mutants for functional genomics in tomato

Fernando Pérez-Martín[1,†], Fernando J. Yuste-Lisbona[1,†], Benito Pineda[2], María Pilar Angarita-Díaz[2], Begoña García-Sogo[2], Teresa Antón[2], Sibilla Sánchez[2], Estela Giménez[1], Alejandro Atarés[2], Antonia Fernández-Lozano[1], Ana Ortíz-Atienza[1], Manuel García-Alcázar[1], Laura Castañeda[1], Rocío Fonseca[1], Carmen Capel[1], Geraldine Goergen[2], Jorge Sánchez[2], Jorge L. Quispe[1], Juan Capel[1], Trinidad Angosto[1], Vicente Moreno[2] and Rafael Lozano[1,*]

[1]Centro de Investigación en Biotecnología Agroalimentaria (BITAL), Universidad de Almería, Almería, Spain
[2]Instituto de Biología Molecular y Celular de Plantas (UPV-CSIC), Universidad Politécnica de Valencia, Valencia, Spain

*Correspondence
email rlozano@ual.es †
These authors contributed equally to this work.

Keywords: T-DNA, enhancer trapping, *Solanum lycopersicum*, functional genomics, insertional mutagenesis, GUS expression.

Summary

With the completion of genome sequencing projects, the next challenge is to close the gap between gene annotation and gene functional assignment. Genomic tools to identify gene functions are based on the analysis of phenotypic variations between a wild type and its mutant; hence, mutant collections are a valuable resource. In this sense, T-DNA collections allow for an easy and straightforward identification of the tagged gene, serving as the basis of both forward and reverse genetic strategies. This study reports on the phenotypic and molecular characterization of an enhancer trap T-DNA collection in tomato (*Solanum lycopersicum* L.), which has been produced by *Agrobacterium*-mediated transformation using a binary vector bearing a minimal promoter fused to the *uidA* reporter gene. Two genes have been isolated from different T-DNA mutants, one of these genes codes for a UTP-glucose-1-phosphate uridylyltransferase involved in programmed cell death and leaf development, which means a novel gene function reported in tomato. Together, our results support that enhancer trapping is a powerful tool to identify novel genes and regulatory elements in tomato and that this T-DNA mutant collection represents a highly valuable resource for functional analyses in this fleshy-fruited model species.

Introduction

Tomato (*Solanum lycopersicum* L.) is not only an important commercial crop because of its high nutritive value for both fresh market and processing industries, but it is also a model system for dicots, especially for fleshy fruit biology (Lozano *et al.*, 2009; Meissner *et al.*, 1997). Due to its numerous advantages, tomato is recognized as a representative Solanaceae species for agronomical and fundamental research (Gillaspy *et al.*, 1993; Klee and Giovannoni, 2011; Ranjan *et al.*, 2012; Tanksley, 2004). These advantages include its being easy to cultivate, short life cycle, high multiplication rate, self-pollination and ease of mechanical crossing, together with a suitable transformation via *Agrobacterium tumefaciens* and the availability of its full genome sequence (The Tomato Genome Consortium, 2012).

Once the tomato genome sequence project has been completed, the challenge of the postgenome era is to determine the functions of the great number of genes annotated by the International Tomato Annotation Group (ITAG). The tomato nuclear genome has an estimated size of 950 Mb and consists of 12 chromosomes; its euchromatic portion contains ~220 Mb (Peterson *et al.*, 1996), including more than 90% of the genes (Wang *et al.*, 2006). Nevertheless, the majority of these genes have only been predicted by *in silico* analysis and their functions remain unknown or hypothetical (The Tomato Genome Consortium, 2009). Mutational analysis is one of the most efficient methods to isolate and understand gene functions. Thus, many spontaneous mutants have been preserved and characterized by the Tomato Genetic Resource Center (Chetelat, 2005). Furthermore, several chemical and physical mutagens have been used to generate new mutant populations (exhaustive data can be found on http://tgrc.ucdavis.edu/ and http://zamir.sgn.cornell.edu/mutants/). Nevertheless, the main disadvantage for both spontaneous and induced mutants is the difficulty to identify the mutated gene, which requires positional cloning and/or mapping-by-sequencing strategies (Schneeberger *et al.*, 2009). Insertional mutagenesis using a transposon or T-DNA insertion arises to solve this problem as the inserted element acts as a tag for gene identification. Although the potential of the maize *Ac/Ds* and *En/Spm* transposon systems has been demonstrated in different species such as Arabidopsis (Parinov *et al.*, 1999; Raina *et al.*, 2002; Speulman *et al.*, 1999; Tissier *et al.*, 1999), rice (Enoki *et al.*, 1999; Greco *et al.*, 2003) and tomato (Meissner *et al.*, 2000), the T-DNA insertional mutagenesis approach offers some advantages as T-DNA integration is stable through generations and appears to be completely random (Tinland, 1996; Tzfira *et al.*, 2004). In addition, the development of binary vectors has led to the generation of different T-DNA insertional mutagenesis methods such as activation tagging (Memelink, 2003) or several 'trapping' systems like gene trapping, promoter trapping and enhancer trapping (Springer, 2000; Stanford *et al.*, 2001). Thus, numerous T-DNA mutant collections have been developed in

Arabidopsis (Alonso et al., 2003; Campisi et al., 1999; Feldmann, 1991; Krysan et al., 1999; Qin et al., 2003; Sessions et al., 2002) and other crops like rice (Hsing et al., 2007; Jeon et al., 2000; Jeong et al., 2002; Wan et al., 2009; Wu et al., 2003). In tomato, two activation tagging collections have been generated in the cultivars Micro-Tom (Mathews et al., 2003), a dwarf genotype bearing several mutations affecting plant development (Carvalho et al., 2011; Martí et al., 2006), and M82 (Carter et al., 2013), a processing tomato variety with determinate growth habit.

Enhancer trap system is a valuable tool for identifying regulatory elements. In the enhancer trap vectors, the reporter gene is fused to a minimal promoter, which is unable to drive the reporter gene expression alone but can be activated by neighbouring cis-acting chromosomal enhancer elements (Springer, 2000; Stanford et al., 2001). Additionally, the enhancer trap system allows for the study of essential genes, as T-DNA acts as a dominant element, whose expression pattern can be detected in hemizygous state (Campisi et al., 1999). Thus, enhancer trap lines could be selected by expression profiling and/or mutant phenotype. The enhancer trap system was first described in Drosophila (O'Kane and Gehring, 1987), and since then, it has been successfully used in several plant species such as Arabidopsis (Geisler et al., 2002; He et al., 2001; Sundaresan et al., 1995) and rice (Johnson et al., 2005; Peng et al., 2005; Sallaud et al., 2004; Wu et al., 2003). In this work, the previously described pD991 binary vector (Campisi et al., 1999) was used to produce more than 7800 enhancer trap lines, which make up the first tomato enhancer trap mutant collection. Furthermore, phenotypic and molecular characterization of trans-formed lines, as well as histochemical localization of β-glucuronidase (GUS) activity in different plant tissues, proved the usefulness of enhancer trap mutagenesis as genomic tool for the identification of novel regulators of plant growth and reproductive development in tomato.

Results

A large number of enhancer trap lines have been produced with the aim to develop an insertion-based gene discovery system for tomato. The phenotypic characterization of these transgenic lines has made possible to identify mutants affected in plant growth and reproductive development. The main steps followed for the characterization of the enhancer trap mutant collection are described below (Figure S1) together with the genetic and molecular characterization of two T-DNA mutants.

Development of enhancer trap lines

The pD991 enhancer trap vector used in this work includes, at the 5′ end and close to the right border (RB), the uidA gene coding for GUS enzyme preceded by a minimal promoter, the latter being insufficient to drive GUS expression. In addition, the NEOMYCIN PHOSPHOTRANSFERASE II (NPTII) gene conferring kanamycin resistance is near the left border (LB) at the 3′ end of the T-DNA (Figure S1a), and it is used as selection marker gene. Enhancer trap lines were generated from cocultured young leaf explants of tomato cultivars P73 and Moneymaker with the Agrobacterium strain LBA4404 carrying the binary vector pD991. Ploidy-level analysis by flow cytometry showed that both diploid- and tetraploid-independent transformants were generated; however, the percentage of diploid transgenic plants was higher in cv. Moneymaker (75.3%) than in cv. P73 (56.2%), despite the fact that transformation frequency was 32.6% and 43.2%, respectively (Table S1). For this reason, cv. Moneymaker was used as main genotype to increase the number of T-DNA lines integrated in our functional genomic programme. Finally, a total of 7842 transgenic plants were generated, of which 5560 T0 lines were diploid, 1021 and 4539 T0 lines from P73 and Moneymaker tomato cultivars, respectively. Diploid T0 plants were then acclimated and subsequently grown under standard glasshouse conditions for further analysis so as to obtain their T1 progenies by selfing.

Phenotypic screening of enhancer trap lines

A total of 4189 T1 transgenic plants were screened under glasshouse conditions (Figure S1b) to detect T-DNA mutants affected in plant growth and reproductive development. Among them, 205 T0 lines displayed variations with respect to wild-type (WT) untransformed plants. The inheritance pattern of the mutant phenotypes was confirmed by a T1 progeny analysis, which showed that the phenotype segregation fitted the expected ratio for a dominant mutation (3:1 for mutant and WT phenotypes) in most cases. In addition, 1858 T1 families were also characterized to identify recessive mutations. For this purpose, sixteen T1 plants from each family were cultivated under glasshouse conditions and screened for developmental alterations. Three hundred and seventeen of 1858 T1 families (17.1%) were found to display a mutant phenotype, and no differences in the relative frequency of mutants were found between Moneymaker and P73 cultivars. Mutant phenotypes observed in most T1 families (274 out of 317) segregated according to a monogenic recessive inheritance (3:1 for WT: mutant phenotypes), whereas 43 mutant lines showed complex inheritance patterns. Thus, it was found that enhancer trap lines displaying an altered vegetative development were affected in seedling development, shoot apex morphogenesis, plant size, leaf colour and morphology, and trichome density (Figure 1). Likewise, enhancer trap lines affected in reproductive traits were detected, such as flowering time, inflorescence architecture, flower colour and morphology, fruit pigmentation, fruit morphology and parthenocarpy (Figure 2). Among the phenotypic classes (Table 1), a high percentage of mutant lines were grouped in 'plant size' and 'parthenocarpic fruit' categories (31.2% and 21.1%, respectively), whereas the less frequent phenotype classes corresponded to flowering time (0.4%), flower abscission zone (1.2%) and cuticle/cracked fruit (1.2%).

The GUS expression of uidA reporter gene was analysed in 836 T1 lines in order to provide a first overview about the organ and tissue expression specificity of genomic regions tagged by T-DNA insertions (Figure 3). Results showed histochemical localization of GUS activity in vegetative and reproductive structures of almost all T0 lines (97.7%). Moreover, organ-specific signals were exclusively found in vegetative organs (49 lines; Figure 3a,b), flowers (269 lines; Figure 3c-f) or fruits (189 lines; Figure 3g,h). Interestingly, a significant number of mutant lines with organ-specific GUS expression displayed a marked tissue or cell type specificity like that observed in vascular bundles of leaves (Figure 3b), and in several floral tissues (Figure 3d-f).

Characterization of T-DNA integration sites

The number of T-DNA insertions in mutant lines was analysed by Southern blot hybridization using a chimeric probe composed by the NPTII (kanamycin resistance) and the tomato FA genes (Figures 4 and S1c). The hybridization generated FA fragments representing a positive control of the hybridization, that is a 10-kb EcoRI-FA fragment and a 1.9-kb HindIII-FA fragment, which were

Figure 1 Representative phenotypes of enhancer trap lines altered in vegetative development. (a) Dwarf phenotype of the 102ET73 mutant. (b) 2372ETMM mutant showed chlorophyll deficiency in cotyledons. (c) Compared to wild type (left), T-DNA mutants displayed dark green leaves, likely due to a high amount of chlorophyll (797ET73) and higher number of leaflets (713ETMM). (d) 2297ETMM mutant was defective in shoot apex growth and morphogenesis. (e) Leaves of the 62ET73 mutant showed higher density of trichomes. Scale bar = 10 cm in (a) and (c); and 1 cm in (b), (d) and (e).

found in both transformed and WT plants (Figure 4a). In addition, a single 1-kb EcoRI-NPTII fragment was found in transformed plants, while the number of HindIII-NPTII fragments indicated the number of T-DNA insertions occurring in each line. Of 170 transgenic lines examined, 73 lines (42.9%) carried a single T-DNA copy and the remaining had two or more T-DNA copies (Figure 4b). The average number of T-DNA insertions per T-DNA line was 2.01 ± 0.9, and no significant differences were found in the average number of T-DNA insertions between P73 and Moneymaker cultivars (t-Student, $P < 0.05$; Figure 4c).

DNA genomic fragments flanking the T-DNA LB and RB sequences were identified by anchor PCR in 77 transgenic lines. After the sequencing of PCR products, sequence homology was firstly analysed using BLAST against the sequence of pD991 vector contained in the A. tumefaciens strain used for genetic transformation experiments. Results showed deletions of variable size affecting LB and RB sequences (Figure 4e). Indeed, LB was especially sensitive to T-DNA integration as none of the T-DNA lines analysed bore the complete LB sequence and almost all of them (99.8%) showed deletions larger than 40 bp. In addition, 15.6% of mutants were found to have aberrant T-DNA insertions, which were due to rearrangements either within T-DNA fragment or involving vector backbone. Secondly, once the vector sequences were removed from the amplified flanking region, the homology search of the trimmed sequences was carried out using BLAST against the tomato genome database (http://solge nomics.net). Results showed that T-DNA insertions were distributed over all chromosomes; however, a bias towards chromosomes 1 and 11 was detected despite that no correlations with the euchromatin ratio or gene content of these chromosomes were found (Figure 4d and S2).

The genomic sequences flanking the T-DNA were analysed to further characterize the chromosome regions where the T-DNA was inserted in the tomato genome. Thus, 37.7% of insertions were located in either the coding or the promoter region of annotated genes, which was arbitrarily defined up to 2 kb

upstream from the transcription start codon. Among them, 16.9% and 14.3% were positioned in exons and introns, respectively, while 6.5% were found in promoter regions. The remaining 62.3% of T-DNA insertions were placed in intergenic regions (Figure 4f). Furthermore, most of T-DNA insertions were found to be located in euchromatic DNA (75.4%, Figure S2).

The nucleotide composition of the sequences surrounding the insertion sites (SSIS) was also ascertained to determine whether there was a preference for insertions in particular regions. The analysis of 100-bp sequences, 50 bp upstream and downstream of each insertion site, displayed a GC content of 34.9% in the SSIS. An additional analysis performed with WebLogo software using 20-bp SSIS revealed a nonconsensus sequence in the T-DNA integration site, as well as a rich AT content (65.6%) in both RB and LB flanking sequences (Figure 4g). As expected, the genes tagged by T-DNA encoded a wide variety of proteins such as transcription factors, plant metabolism enzymes or membrane receptors. Examples of T-DNA locations and other relevant information about tagged flanking sequences are shown in Table 2.

Molecular isolation of two T-DNA tagged mutants: proof of concept

As proof of concept, here we describe the molecular characterization of two selected T-DNA mutant lines named 1381ETMM and 2477ETMM. The segregation ratio observed in the T1 progeny of the line 1381ETMM was consistent with a monogenic recessive inheritance for the mutant phenotype (16 WT: 8 mut; $\chi^2 = 0.89$, $P = 0.35$), which is characterized by a significant reduction in leaf size, giving rise to only one or two secondary leaflets (Figure 5a,c). In addition, flower development was severely altered as mutant plants produced flowers with reduced petals that opened prematurely (Figure 5b). These flowers rarely yielded fruits, and when they did, fruits were parthenocarpic (seedless) and smaller compared with WT fruits (Figure 5d). Southern blot analysis showed that 1381ETMM line only bore a

Figure 2 Representative phenotypes of enhancer trap lines affected in reproductive traits. (a) Inflorescences of wild-type tomato plants were normally composed by 7–10 flowers (top), while the 162ET73 mutant line developed a single flower inflorescence (bottom). (b) From left to right, flowers from wild-type and T-DNA mutant lines showing alterations in the colour of petals and stamens (651ET73), homeotic conversions of floral whorls (248ET73) and an increased number of floral organs (637ET73). (c) From left to right, wild-type fruit and T-DNA mutant lines displaying yellow fruit (478ET73), orange fruit (651ET73) and intense red fruit (745ETMM). (d) From left to right, wild-type fruit and fruits of three T-DNA mutant lines (12ET73, 989ET73 and 85ET73) developing parthenocarpic (seedless) fruits with altered size and morphology. (e) Longitudinal sections of the same fruits showed in (d). Scale bar = 3 cm in (a); and 1 cm in (b), (c), (d) and (e).

single T-DNA copy. Cloning of T-DNA flanking genomic sequences revealed that T-DNA was inserted at position 3 537 861 on chromosome 5 (ITAG2.4), in the sixth exon of the *LYRATE* gene (*Solyc05g009390*), which codes for a lipase-like protein (Figure 5e) involved in leaf development (David-Schwartz *et al.*, 2009). The effects of T-DNA integration on gene expression were determined by quantitative RT-PCR, which showed that *LYRATE* was significantly down-regulated in 1381ETMM mutant tissues compared with WT (Figure 5f).

Regarding the line 2477ETMM, a segregating population of 20 plants was evaluated, which segregated according to a monogenic recessive inheritance for the mutant phenotype (14 WT: 6 mut; $\chi^2 = 0.27$, $P = 0.61$). At early stages of development, leaves of mutant plants showed evident necrosis symptoms, which affected all leaf tissues, and led to a loss of photosynthetic tissue and a reduction in plant growth (Figure 6a,b). This mutant phenotype was observed in young plants developed under both *in vitro* and glasshouse conditions. Later in the development, necrosis increased and the affected leaves became curled and senescent (Figure 6b). Southern blot experiments displayed a single T-DNA insertion in the mutant plants of 2477ETMM line, and the analysis of T-DNA flanking sequences revealed that T-DNA was integrated at position 4 916 541 on chromosome 11 (ITAG2.4), in the fifth exon of the *Solyc11g011960*, a gene coding for a UTP-glucose-1-phosphate uridylyltransferase (Figure 6c). Expression analysis showed that T-DNA integration led to a decreased level of transcripts of the tagged gene in 2477ETMM mutant tissues (Figure 6d).

Table 1 Catalogue of tomato mutant phenotypes

	Category	Dominants	Recessives	Complex inheritance*	Total	Frequency (%)
i.	Seedling lethality/albinism	1	16	0	17	3.3
ii.	Root development	2	19	1	22	4.2
iii.	Plant size	39	112	12	163	31.2
iv.	Leaf morphology and colour	27	25	1	53	10.2
v.	Senescence	5	19	3	27	5.2
vi.	Flowering time	1	1	0	2	0.4
vii.	Inflorescence architecture	9	13	0	22	4.2
viii.	Flower morphology and colour	6	10	2	18	3.5
ix.	Flower abscission zone	6	0	0	6	1.2
x.	Fruit set rate	26	8	0	34	6.5
xi.	Fruit morphology and colour	12	13	3	28	5.4
xii.	Seedless (parthenocarpic) fruit	61	31	18	110	21.1
xiii.	Ripening	5	6	3	14	2.7
xiv.	Cuticle/cracked fruit	5	1	0	6	1.2
	TOTAL	205	274	43	522	

*Complex inheritance: traits that do not follow strict Mendelian inheritance.

With the aim to support the insertional nature of the mutant phenotypes above described, a cosegregation analysis of the T-DNA insertion with the mutant phenotype was assessed in T1 segregating populations (Figures 5g and 6e). In both cases, all mutant plants bore T-DNA insertion in the homozygous state, whereas WT plants were azygous or heterozygous for T-DNA, which suggested that mutant phenotypes were caused by the T-DNA insertion occurring in each line. Nevertheless, the evaluation of 77 selected T-DNA lines showed no cosegregation between the mutant phenotype and the T-DNA insertion in 26 of these lines (33.8%). These results suggested that somaclonal variation events, partial T-DNA or vector backbone fragment insertions and chromosomal rearrangements may have occurred during the *in vitro* genetic transformation, similarly to that reported in other model plant species (Feldmann, 1991; Miyao *et al.*, 2007).

In addition, different strategies were developed to further confirm that the tagged genes in the 1381ETMM and 2477ETMM lines were responsible for the mutant phenotypes observed. For the 1381ETMM line, a complementation test was carried out by crossing wild-type heterozygous plants, one bearing the 1381ETMM mutation (female parent) and the other carrying the *lyrate* mutation (*lyr2*, accession number LA2923, male parent), as homozygous mutant plants for each mutant allele rarely developed fertile flowers. The evaluation of the F1 offspring (Figure S3) showed the expected 3:1 segregation of wild-type and mutant phenotypes (18 WT: 8 mut; $\chi^2 = 0.50$, $P = 0.46$), which confirmed that the 1381ETMM mutation is a new allele of the *LYRATE* gene. Regarding the 2477ETMM line, a RNA interference (RNAi) strategy was carried out to silence the expression of the *Solyc11g011960* tagged gene (Figure S4). Thus, 10 independent transformants were obtained and used for phenotypic characterization, three of which were selected by their diploid nature and their reduced expression levels (less than 0.1-fold change relative to WT plants). These T0 RNAi lines displayed a similar mutant phenotype of that reported for the 2477ETMM line, particularly leaves with evident necrosis symptoms and a reduction of plant growth (Figure S4). These results supported that the T-DNA insertion located at the *Solyc11g011960* gene is responsible for the mutant phenotype observed in the 2477ETMM line.

Discussion

Considerable progress has been made in developing genomic resources for tomato, including the release of the complete genome sequence (The Tomato Genome Consortium, 2012). As a result, 34 727 protein-coding genes were annotated by the ITAG consortium, most of them with unknown functions. Therefore, a key research priority is to develop a set of tools to assign functions to the predicted gene sequences, thus facilitating that this genomic information can be applied in tomato genomics-assisted breeding. Insertional mutagenesis is one of the most suitable and direct approaches to define gene functions. A tomato activation tagging insertional mutant collection was developed by Mathews *et al.* (2003) using Micro-Tom, a miniature variety originally bred for ornamental purposes (Scott and Harbaugh, 1989). The dwarf phenotype of Micro-Tom plants is determined by a combination of hormonal and photomorphogenetic mutations (altered sensitivity or endogenous levels of auxin, ethylene, abscisic acid, gibberellin, brassinosteroid and light response) into its genetic background (Carvalho *et al.*, 2011; Martí *et al.*, 2006), which may make this genotype unsuitable for the identification of genetic factors controlling important developmental traits like those related to plant vigour and fruit size. In this study, two commercial tomato cultivars, that is Moneymaker and P73, with adequate agronomic performance have been used to develop a large-scale insertional mutagenesis approach. Thus, more than 5500 diploid T0 lines have been generated using the *Agrobacterium*-mediated transformation protocol with the pD991 enhancer trap vector. The average transformation frequency was 32.4% in cv. Moneymaker and 42.9% in cv. P73; in both cases, it was higher than previously described by Hu and Phillips (2001) for the industrial processing cultivar UC82 (25%), although lower than previously reported by Dan *et al.* (2006) for the Micro-Tom variety (57%). Both this work and the two previously mentioned reports used an *Agrobacterium tumefaciens*-mediated transformation procedure, which corroborates that T-DNA integration into tomato genomes highly depends on the genotype (Ellul *et al.*, 2003).

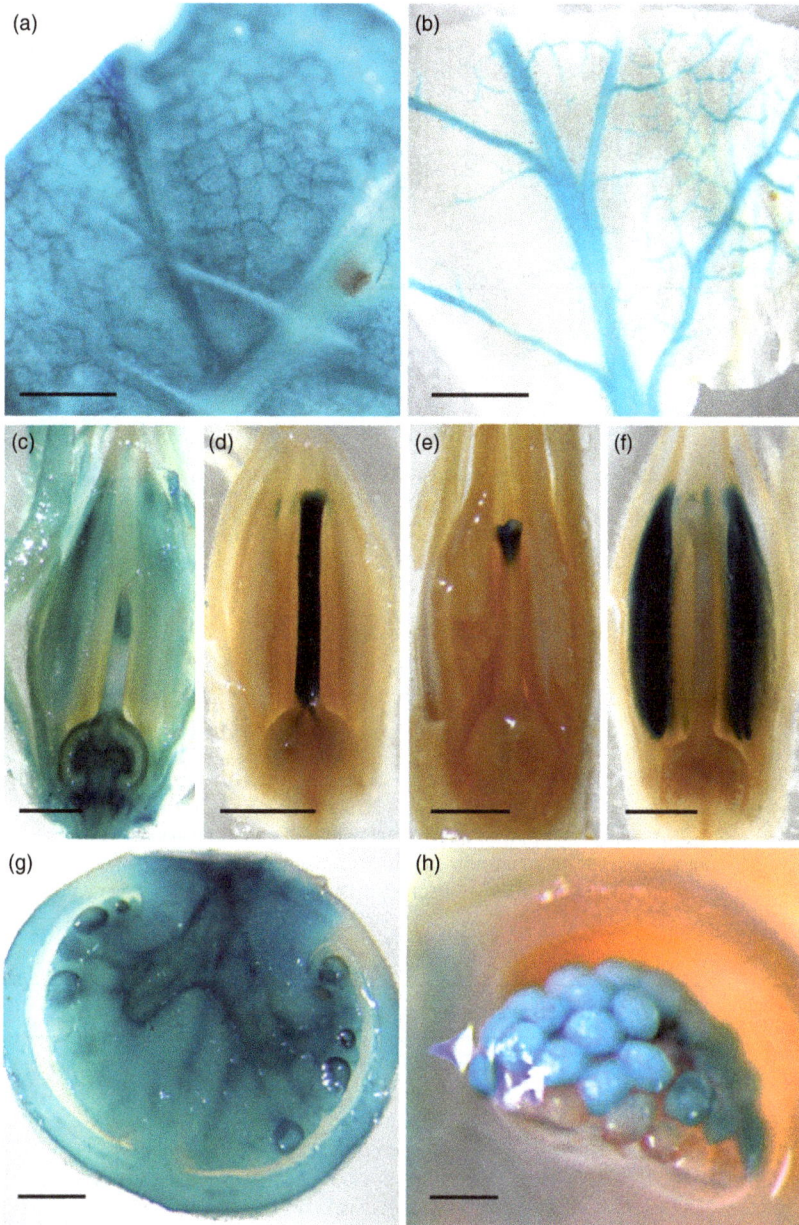

Figure 3 GUS expression patterns in enhancer trap lines. (a), (c) and (g) show organ-specific GUS staining in leaf, flower and fruit, respectively. Tissue-specific GUS expression was detected in vascular bundles of leaves (b), and in style (d), stigma (e), stamens (f) and ovules (h) of flowers. Scale bar = 1 cm in (a), (b) and (g); 0.25 cm from (c) to (f); and 50 μm in (h).

The *in vivo* screening of an insertional mutant collection is a space- and time-consuming process, particularly in tomato cultivars like Moneymaker and P73, which show an indeterminate growth habit and have 4- to 6-month-long life cycles, as it is characteristic of tomato varieties for fresh consumption market. Thus, for more than 6 years (two seasons per year, i.e. autumn–winter and spring–summer), a total of 4189 T0 plants and 1858 T1 progenies (16 plants of each T1 line) were evaluated under glasshouse conditions. Based on this evaluation, mutant lines with defective vegetative (Figure 1) and reproductive (Figure 2) development were found; most of them belonged to 'plant size' (31.2%) and 'parthenocarpic fruit' (21.1%) categories (Table 1). Two hundred and five of the 522 mutant lines identified had an autosomal dominant mode of inheritance, which means that a novel dominant mutated allele was generated in 4.9% of the evaluated T0 lines. This percentage was smaller than that previously reported by Mathews *et al.* (2003) in the Micro-Tom activation tagging collection, where 12.8% of T0 lines evaluated showed a dominant mutant phenotype. Such differences might be due to the transformation method or the tomato genetic background used in each study. However, it should be taken into account that Mathews *et al.* (2003) characterized a small number of T1 progenies to confirm the mutant phenotypes observed in T0 selected plants; hence, it is not possible to compare the percentage of recessive mutant lines detected in both T-DNA collections. On the other hand, the present study evaluated a limited number of discrete traits; hence, if enhancer trap lines were screened under other conditions such as drought or temperature stress conditions or under pathogen pressure, phenotypic description of this collection would be much more enriched, which would allow for the identification of new mutant phenotypes. In fact, the same pD991 vector-based gene construct has been used to generate a T-DNA mutant collection in the wild-related species *S. pennellii* whose screening has provided useful information regarding regulatory genes involved in salt stress response (Atarés *et al.*, 2011).

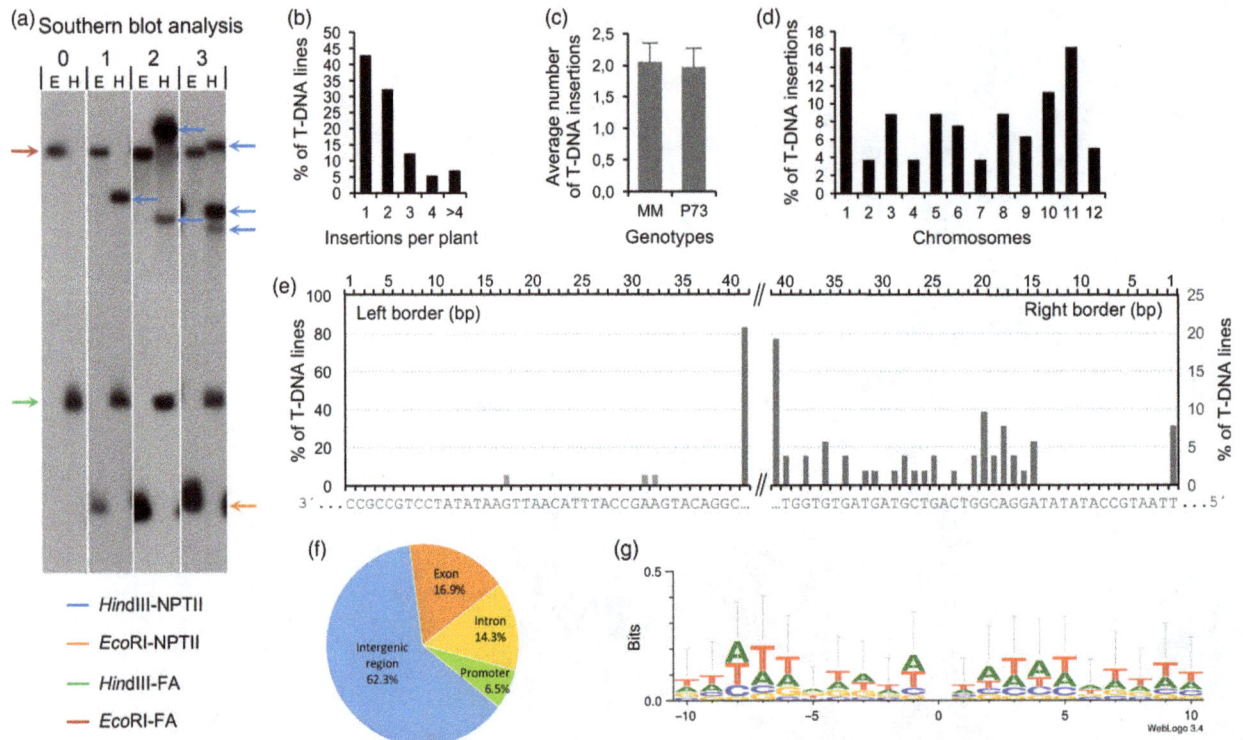

Figure 4 Molecular characterization of enhancer trap lines. (a) Southern blot analysis of genomic DNA digested by restriction enzymes EcoRI (e) and HindIII (h) and hybridized with the NPTII-FA probe (for details, see Methods). (b) Number of T-DNA insertions per T0 plant. (c) Average number of T-DNA insertions in Moneymaker (MM) and P73 cultivars. (d) Distribution of T-DNA insertions on tomato chromosomes. (e) Percentage of enhancer trap lines with deletions in the sequence of the integrated right (RB) and left (LB) borders (the last 40 bp are only shown). (f) Distribution of T-DNA insertions on intergenic and genic regions. (g) WebLogo analysis of 20-bp sequences surrounding the T-DNA insertion sites. Zero position represents the insertion site.

Table 2 Examples of insertion sites of enhancer trap T-DNAs in the tomato genome.

Line	RB/LB*	Ch.†	Region	Gene	Protein function
746ET73	RB	1	Intron	Solyc01g010500	Ein3-binding f-box protein 3
374ETMM	RB	1	Exon	Solyc01g095030	MYB Transcription factor
282ET73	LB	3	Exon	Solyc03g005580	Legumin 11S-globulin
1381ETMM	RB	5	Exon	Solyc05g009390	Lipase-like protein
515ETMM	RB	5	Promoter	Solyc05g012020	MADS-box transcription factor
386ETMM	RB	5	Exon	Solyc05g013480	ATP-dependent protease
136ETMM	RB	5	Exon	Solyc05g013530	Octicosapeptide
832ETMM	RB	6	Promoter	Solyc06g008020	Zinc Finger Transcription factor
1336ETMM	RB	6	Promoter	Solyc06g068090	Phospholipase PLDa1
390ETMM	RB	6	Exon	Solyc06g068980	Serine/threonine-protein kinase B-raf
1635ETMM	LB	8	Intron	Solyc08g007380	Histidine triad protein
365ET73	RB	8	Exon	Solyc08g061240	Catalytic/hydrolase
51ET73	RB	10	Exon	Solyc10g049460	Transposon Ty1-A Gag-Pol polyprotein
740ET73	RB	10	Intron	Solyc10g083250	RNA-binding protein
1527ETMM	RB	11	Intron	Solyc11g008620	Phosphoglycolate phosphatase
2477ETMM	RB	11	Exon	Solyc11g011960	UTP-glucose-1-phosphate uridylyltransferase
651ET73	RB	11	Exon	Solyc11g069740	Nitrate transporter

*T-DNA flanking genomic sequences were amplified from RB: right border or LB: left border.
†Ch: Chromosome.

In addition, the expression of the reporter *uidA* gene was evaluated in vegetative structures, flowers and immature fruits. While different GUS patterns were detected in the enhancer trap lines (Figure 3 and Table S2), a significant percentage of these lines (60.6%) displayed organ- or tissue-specific GUS activity. The combined use of T-DNA-based mutagenesis and GUS

Figure 5 Phenotypic and molecular characterization of the 1381ETMM line. Mutant plants of the 1381ETMM line were affected in the development of leaves (a, c), flowers (b) and fruits (d). (e) Schematic representation of T-DNA localization in the 1381ETMM line. (f) Relative expression of *LYRATE* (*Solyc05g009390*) in wild-type and 1381ETMM mutant plants. Asterisk denotes significant differences at *P* < 0.05. (g) Cosegregation analysis of T-DNA insertion and 1381ETMM mutant phenotype. Red numbers indicate plants displaying mutant phenotype. Scale bar = 1 cm in (a), (b) and (d); and 5 cm in (c).

histochemical detection has also been carried out in other Solanaceae species, such as *Nicotiana tabacum* and *S. tuberosum* (Goldsbrough and Bevan, 1991; Lindsey *et al.*, 1993; Topping *et al.*, 1991). Using the weak (−90 bp) CaMV35S promoter, Topping *et al.* (1991) analysed patterns of GUS gene expression in a collection of 184 tobacco T-DNA lines, from which 73% displayed GUS activity with different organ and tissue specificities. Comparable results were observed by Goldsbrough and Bevan (1991) in potato T-DNA lines, using a similar enhancer trap vector. Thus, different patterns of GUS expression were detected at high frequency. Likewise, similar findings were reported by Lindsey *et al.* (1993) in T-DNA lines of tobacco, tomato and Arabidopsis. The percentage of lines showing GUS activity was high for all three species; however, the frequencies of GUS activity detected

in a given organ were different among species, which ranged from 25% in stems of potato and 30% in roots of Arabidopsis, up to 92% in flowers of tobacco T-DNA lines. Therefore, the set of T-DNA lines here reported showing specific GUS expression in the flower and fruit tissues (269 and 189 lines, respectively) could be used to further studies of functional genomics in tomato. Among them, the high percentage of lines (26.2%) displaying a stamen-specific GUS staining pattern is remarkable. As defects during pollen ontogeny produce parthenocarpic (seedless) fruits, such percentage is in agreement with that of mutant lines with parthenocarpic fruit (21.1%) identified during phenotypic screening under glasshouse conditions. In fleshy fruit plants like tomato, parthenocarpy is considered to be of commercial importance as seedless fruits usually have increased fruit quality traits, and

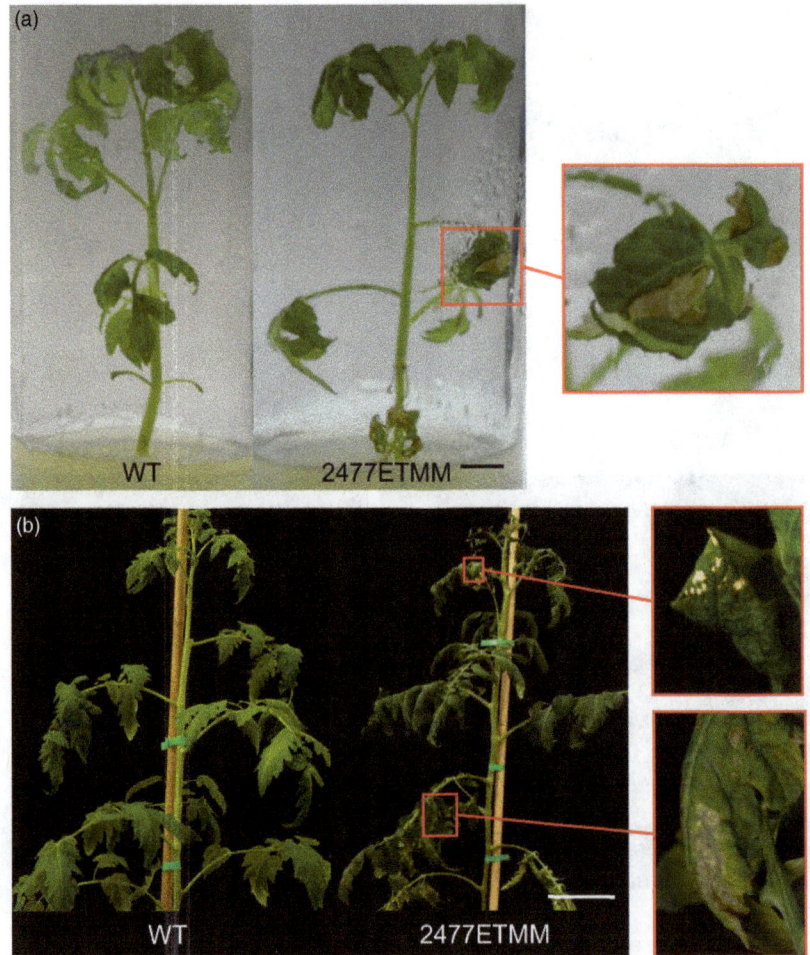

Figure 6 Phenotypic and molecular characterization of the 2477ETMM line. Necrosis of leaf tissues observed in the mutant phenotype of 2477ETMM line when plants grew either under *in vitro* (a) or glasshouse (b) conditions (magnification pictures of necrotic tissues are shown in right panels). (c) Schematic representation of T-DNA integration site in the 2477ETMM line. (d) Relative expression of the gene coding the UTP-glucose-1-phosphate uridylyltransferase (*Solyc11g011960*) in wild-type and 2477ETMM mutant genotypes. Asterisk denotes significant differences at $P < 0.05$. (e) Cosegregation analysis of T-DNA insertion and 2477ETMM mutant phenotype. Red numbers indicate plants displaying mutant phenotype. Scale bar = 1 cm in (a) and 5 cm in (b).

parthenocarpic varieties can provide tomato yield under unfavourable climatic conditions (Gorguet *et al.*, 2005; Pandolfini, 2009). Therefore, these mutant lines could help to uncover novel genes, which may exert a fundamental role during pollen and fruit developmental processes. Moreover, enhancer trapping is suitable for isolating regulatory genes involved in developmental traits which may be difficult to address from mutants showing highly pleiotropic or lethal phenotypes. In this case, gene discovery mostly depends on the reporter gene expression rather than the mutant phenotype (Groover *et al.*, 2004).

As regards molecular characterization, Southern blot analysis revealed that enhancer trap lines contain an average of 2.01 T-DNA insertions although 43% of assessed lines bear a single T-DNA copy. This result is in accordance with that reported by Wu *et al.* (2003) in a rice enhancer trap collection; however, it differed from the 1.4 T-DNA insertions found as average in Arabidopsis and rice mutant collections developed by other trapping systems (Feldmann, 1991; Jeon *et al.*, 2000). Examination of the junctions between the T-DNA borders and tomato genomic DNA revealed that right and left borders were not completely integrated (Figure 4e). The deletions in the left border junction were even more severe (deletions larger than 40 bp). Nevertheless, this phenomenon seems to be common in *Agrobacterium*-mediated T-DNA-transferring processes as it had also been previously found in Arabidopsis and rice (Hiei *et al.*, 1994; Tinland, 1996; Wu *et al.*, 2003). The rationale of this phenomenon is that T-DNA integration into plant genome is usually achieved by a form of illegitimate recombination, which is initiated by a break in the DNA involved in the mutational process (Gheysen *et al.*, 1991). Recombination of only a few identical nucleotides preferentially occurs at the base where VirD2 protein nicks the right border, as T-DNA transfer is a polar process, which is initiated at the right border and ends at the left border. However, the left junction between bacterial and plant DNA frequently does not occur within the left border sequence, which results in the commonly found deletion of left border sequences (Rossi *et al.*, 1996; Tinland *et al.*, 1994, 1995).

In the present study, 75.4% of T-DNA insertion occurred in these large gene-rich euchromatic regions, where 62.3% were located in intergenic regions (Figure 4f). This result further supports the significant percentage of enhancer trap lines showing GUS activity and agrees with previous studies which have reported that T-DNA integration favours intergenic regions over genic regions (Alonso *et al.*, 2003; Krysan *et al.*, 1999; Pan *et al.*, 2005; Rosso *et al.*, 2003). In the Solanaceae family, fluorescence *in situ* hybridization (FISH) experiments carried out in Petunia indicated that T-DNA insertions occur preferentially in distal chromosome regions, where gene density is higher and chromatin is loosely packed and transcriptionally active (Ten Hoopen *et al.*, 1996; Wang *et al.*, 1995). Likewise, comparative analysis in Arabidopsis and rice revealed that T-DNA insertions were randomly found in the Arabidopsis genome, which contains little repetitive DNA and is globally rich in gene concentration whereas in the rice genome, T-DNA fragments were inserted in gene-dense euchromatic regions (Barakat *et al.*, 2000; Zhang *et al.*, 2007). Furthermore, GC content (34.9%) in sequences surrounding the T-DNA insertion sites was similar to that previously reported by Barakat *et al.* (2000) and Qin *et al.* (2003) in Arabidopsis and rice insertional collections, suggesting that T-DNA integration events most likely occur in genome sequences having a moderate GC content.

As proof of concept, we have reported the isolation of the genes tagged in two T-DNA lines. Firstly, a new T-DNA allele of the *LYRATE* gene has been identified from the 1381ETMM line. *LYRATE* was found to be the tomato homologue of the *Arabidopsis JAGGED* gene, and the functional analysis proved that it functions as crucial regulator of leaf development and patterning by interacting with other transcription factors (David-Schwartz *et al.*, 2009). Mutations at the *LYRATE* locus also affected the proper development of floral organs, mainly stamens and carpels, as well as fruit formation, which were likely due to pleiotropic effects. Our results corroborated the functional analysis of *LYRATE* and provided a new allele for further insight into the molecular and physiological mechanisms underlying complex biological processes such as vegetative and reproductive development. In addition, the gene coding for a UTP-glucose-1-phosphate uridylyltransferase, an enzyme involved in the biosynthesis of carbohydrate cell components, such as cellulose and callose, was isolated from the 2477ETMM line. Phenotypic characterization of the T-DNA mutant suggests that this gene should play an important role in regulating cell death during leaf development of tomato. In *A. thaliana*, an *UTP-GLUCOSE-1-PHOSPHATE*

URIDYLYLTRANSFERASE homologue (*UGP1*) gene has been reported as a crucial regulator of programmed cell death (Chivasa *et al.*, 2013), which supports our hypothesis on the role of the tagged gene. Furthermore, *UGP1* and *UGP2* seem to act redundantly in plant growth and reproduction in *A. thaliana* (Park *et al.*, 2010), suggesting that UTP-glucose-1-phosphate uridylyltransferase may have overall housekeeping functions during plant development. Noteworthy, our results in tomato also provide a suitable scenario for further functional and evolutionary studies on the *UTP-GLUCOSE-1-PHOSPHATE URIDYLYLTRANSFERASE* genes. Likewise, the screening of this T-DNA mutant collection has allowed us to identify other tagged mutants. Among them are *vegetative inflorescence* (*mc-vin*) and *altered response to salt stress 1* (*ars1*) mutants (Campos *et al.*, 2016; Yuste-Lisbona *et al.*, 2016). All together, these results strongly support the usefulness of enhancer trapping as an efficient strategy for functional genomics, allowing for the discovery of novel genes and regulatory elements.

Experimental procedures

Generation of enhancer trap lines

The enhancer trap vector used for transformation was pD991 (kindly supplied by Dr. Thomas Jack; Department of Biological Sciences, Dartmouth College, USA), which was described by Campisi *et al.* (1999). Young leaf explants were transformed with *A. tumefaciens* strain LBA4404 following the protocol described by Ellul *et al.* (2003). The transformed plants (T0) were selected by growing the explants in the salt medium reported by Murashige and Skoog (1962), sucrose (10 g/L) and kanamycin (100 mg/L). To ensure that each regenerated plant represented an independent transgenic event, only one regenerated plant from a single poked area of an inoculated leaf explant was selected. Transformation frequency was estimated as the number of independent transgenic events divided by the total number of inoculated leaf explants and then multiplied by 100. Furthermore, the ploidy level in transgenic plants was evaluated according to the protocol described by Atarés *et al.* (2011). Thus, the diploid plants from the T-DNA insertion lines were selected and labelled with a consecutive number and the tag 'ET73' or 'ETMM', depending on whether the callus was originated from P73 or Moneymaker cultivars, respectively. Seeds of Moneymaker (accession LA2706) were obtained from Tomato Genetics Resource Center (TGRC, http://tgrc.ucdavis.edu/), whereas P73 seeds were kindly provided by Dr. M.J. Díez (COMAV-UPV, Valencia, Spain). Several clonal replicates for each T0 line were obtained by culturing axillary buds in rooting medium. These replicates were used to maintain the T-DNA collection under *in vitro* growth conditions as well as to acclimatize a sufficient number of replicates under glasshouse conditions to identify dominant insertion mutants and obtain T1 seeds by selfing. The collected T1 seeds were dried and catalogued in a temperature- and humidity-controlled chamber. Furthermore, in order to detect recessive insertion mutants, sixteen T1 plants from each progeny were cultivated under glasshouse conditions for two seasons each year (autumn–winter and spring–summer) from 2009 to 2015.

Phenotypic characterization

Vegetative and reproductive relevant traits were considered for phenotypic characterization (depicted in Table 1). Consequently, the mutant lines were classified into 14 phenotypic categories according to criteria described by Menda *et al.* (2004) with several modifications: (i) seedling lethality/albinism, mutations affecting embryo survival and absence or deficiency of chlorophyll during seedling; (ii) root development, that is altered root morphogenesis; (iii) plant size, from the soil surface to the apex at the fifteenth leaf stage; (iv) leaf morphology and colour, reflected by alterations in size, colour and complexity of leaf and leaflet (margin, venation, shape), as well as an increase or decrease in the number of trichomes; (v) senescence, that is premature death of the plant; (vi) flowering time, measured as the number of leaves before flowering; (vii) inflorescence architecture, comprised by variations in the number of inflorescences and the number of flowers per inflorescence; (viii) flower morphology and colour, including any mutants with homeotic changes, as well as alterations in size and colour; (ix) flower abscission zone, mutations affecting abscission layer development that cause alteration in flower dropping; (x) fruit set rate, measured as the proportion of flowers that yielded fruits compared with the wild type; (xi) fruit morphology and colour, reflected by variations in size and shape (rounded, elliptic, heart shape, among others) and colour (variation not due to late ripening, e.g. orange, yellow, green); (xii) seedless (parthenocarpic) fruit, comprising those mutants with any case of partial or full sterility which gave rise to parthenocarpy (absence of seeds) or stenospermocarpy (contain only rudiments of seeds) fruits; (xiii) fruit ripening, measured as fruit firmness compared with the wild type; and (xiv) cuticle/cracked fruit, mutations affecting fruit cuticle, epidermis and pericarp properties. Measurements were taken in centimetres and weight in grams.

GUS assay

A histochemical GUS assay was conducted as described by Atarés et al. (2011). Different tissues of T0 transformed plants were placed in GUS staining solution [100 mM sodium phosphate at pH 7.0, 10 mM ethylenediaminetetraacetic acid (EDTA), 0.1% Triton X-100, 0.5 mg/mL X-Gluc, 0.5 mM potassium ferricyanide, 0.5 mM potassium ferrocyanide and 20% methanol] and incubated at 37 °C for 20–24 h. Subsequently, the GUS-stained tissues were washed with 70% ethanol and examined under a zoom stereomicroscope (MZFLIII, Leica). Three replicates of each sample were analysed.

DNA isolation

Tomato genomic DNA was isolated according to Dellaporta et al. (1983). Genomic DNA was quantified by fluorometry using SYBR Green I (Sigma-Aldrich) as fluorophore. Fluorescence measurements were made at room temperature using Synergy MX (Biotek) fluorometer.

Southern blot analysis

The number of T-DNA insertions existing in selected mutants was determined by Southern blot. DNA blot hybridization was performed as described by Ausubel et al. (1995) using 10 μg of genomic DNA digested by restriction enzymes EcoRI and HindIII, electrophoresed throughout 0.8% agarose gel in 1X TBE buffer (100 mM Tris-borate, 1 mM EDTA, pH 8.3), and blotted onto Hybond N+ membranes (GE Healthcare). Hybridization was carried out with a chimeric probe, fusing the complete coding sequence of the NPTII gene to 811 bp of coding sequence of endogenous tomato FALSIFLORA (FA) gene, which was employed as hybridization positive control. Finally, the chimeric FA-NPTII probe (1635 bp) was labelled with [α-32P]dCTP using High Prime random priming kit (Roche Applied Science) following the

manufacturer's instructions. Nylon membranes were exposed to Hyperfilms (GE Healthcare).

Identification of T-DNA flanking sequences

The T-DNA flanking sequences were isolated by anchor PCR according to the protocol previously established by Schupp et al. (1999) and Spertini et al. (1999) with some modifications: (i) genomic DNA (500 ng) was digested with blunt-end restriction enzymes EcoRV, DraI, ScaI, StuI, AluI, HincII, PvuII or SmaI; (ii) additional third nested PCR was employed to avoid nonspecific amplification products; and (iii) new specific primers were designed for the RB, LB and Adapter. The sequence of primers used is listed in Table S3. PCR products were sequenced using the BigDye Terminator Cycle Sequencing Ready Reaction Kit (Applied Biosystems) following the manufacturer's instructions. The cloned sequences were compared with SGN Database (http://solgenomics.net/tools/blast/) to assign the T-DNA insertion site on tomato genome. Furthermore, flanking sequences tags were examined to search for an integration pattern sequence using the WebLogo v3.4 software (http://weblogo.threeplusone.com/) described in Crooks et al. (2004).

PCR genotyping

Cosegregation of the T-DNA insertion site with the mutant phenotype in the T1 progeny for selected mutants was checked by PCR using (i) the specific genomic forward and reverse primers to amplify the wild-type allele (without T-DNA insertion) and (ii) one specific genomic primer and the specific T-DNA border primer (from RB or LB) to amplify the mutant allele (carrying the T-DNA insertion). The primers located upstream and downstream of the T-DNA insertional sites in each line were designed based on sequence information available from SGN Database (http://solgenomics.net/). The sequence of genotyping primers used is listed in Table S3.

Tomato RNA isolation and qRT-PCR analysis

Total RNA was isolated using TRIzol (Invitrogen) following the manufacturer's instructions from young leaves. The DNA-freeTM kit (Ambion) was used to remove contaminating DNA from each sample. The cDNA was synthetized by M-MuLV reverse transcriptase (Fermentas Life Sciences) with a mixture of random hexamer and oligo(dT)$_{18}$ primers. Specific primer pairs for each evaluated gene were described in Supplementary Table S3. Gene expression analysis was performed with three biological and two technical replicates using SYBR Green PCR Master Mix (Applied Biosystems) kit and the 7300 Real-Time PCR System (Applied Biosystems). The ΔΔCt calculation method (Winer et al., 1999) was used to express the results in arbitrary units by comparison with a data point from the wild-type samples. The housekeeping gene Ubiquitine3 (Solyc01g056940) was used as a control. Means of WT and mutant samples were compared using a least significant difference (LSD) test ($P < 0.05$).

Generation of silencing lines

A RNA interference (RNAi) approach was followed to generate Solyc11g011960 silencing lines. A 117-bp fragment of the Solyc11g011960 cDNA was amplified using the primers RNAi-F (5′-TCTAGACTCGAGGGTTTGGATCTAGCGTTACCC-3′) and RNAi-R (5′-ATCGATGGTACCCCTCAGGTCCATTGATGTCC-3′), and the PCR product was cloned in sense and antisense orientation separated by intronic sequences into the pKannibal vector, which was then digested with NotI and cloned into the binary vector pART27, according to Campos *et al.* (2016). The binary plasmids generated were transformed with *A. tumefaciens* strain LBA4404 following the protocol described by Ellul *et al.* (2003).

Acknowledgements

This work was supported by research grants from the Spanish Ministerio de Economía y Competitividad (AGL2012-40150-C02-01, AGL2012-40150-C02-02, AGL2015-64991-C3-1-R and AGL2015-64991-C3-3-R), Junta de Andalucia (P10-AGR-6931) and UE-FEDER. B.P. received a JAE-Doc research contract from the CSIC (Spain). PhD fellowships were funded by the FPU (M.G-A. and R.F.) and the FPI (M.P.A., S.S. A.F-L., A.O-A and L.C.) Programmes of the Ministerio de Ciencia e Innovación, the JAE predoc Programme of the Spanish CSIC (G.G.), the CONACYT and Universidad de Sinaloa of México (J.S.) and the LASPAU (J.L.Q.). The authors thank research facilities provided by the Campus de Excelencia Internacional Agroalimentario (CeiA3).

References

Alonso, J.M., Stepanova, A.N., Leisse, T.J., Kim, C.J., Chen, H., Shinn, P., Stevenson, D.K. *et al.* (2003) Genome-wide insertional mutagenesis of *Arabidopsis thaliana*. *Science*, **301**, 653–657.

Atarés, A., Moyano, E., Morales, B., Schleicher, P., García-Abellan, J.O., Antón, T., García-Sogo, B. *et al.* (2011) An insertional mutagenesis programme with an enhancer trap for the identification and tagging of genes involved in abiotic stress tolerance in the tomato wild-related species *Solanum pennellii*. *Plant Cell Rep.* **30**, 1865–1879.

Ausubel, F.M., Brent, R., Kingston, R.E., Moore, D.D., Seidman, J.G., Smith, J.A. and Struhl, K. (1995) *Current Protocols in Molecular Biology*. New York: John Wiley and Sons Inc.

Barakat, A., Gallois, P., Raynal, M., Mestre-Ortega, D., Sallaud, C., Guiderdoni, E., Delseny, M. *et al.* (2000) The distribution of T-DNA in the genomes of transgenic Arabidopsis and rice. *FEBS Lett.* **471**, 161–164.

Campisi, L., Yang, Y., Yi, Y., Heiling, E., Herman, B., Cassista, A.J., Allen, D.W. *et al.* (1999) Generation of enhancer trap lines in Arabidopsis and characterization of expression patterns in the inflorescence. *Plant J.* **17**, 699–707.

Campos, J.F., Cara, B., Pérez-Martín, F., Pineda, B., Egea, I., Flores, F.B., Fernández-García, N. *et al.* (2016) The tomato mutant *ars1* (*altered response to salt stress 1*) identifies an R1-type MYB transcription factor involved in stomatal closure under salt acclimation. *Plant Biotechnol. J.* **14**, 1345–1356.

Carter, J.D., Pereira, A., Dickerman, A.W. and Veileux, R.E. (2013) An active Ac/Ds transposon system for activation tagging in tomato cultivar M82 using clonal propagation. *Plant Physiol.* **162**, 145–156.

Carvalho, R.F., Campos, M.L., Pino, L.E., Crestana, S.L., Zsögön, A., Lima, J.E., Benedito, V.A. *et al.* (2011) Convergence of developmental mutants into a single tomato model system: 'Micro-Tom' as an effective toolkit for plant development research. *Plant Methods*, **7**, 18.

Chetelat, R.T. (2005) Revised list of miscellaneous stocks. *TGC Rep.* **55**, 48–69.

Chivasa, S., Tomé, D.F. and Slabas, A.R. (2013) UDP-glucose pyrophosphorylase is a novel plant cell death regulator. *J. Proteome Res.* **5**, 1743–1753.

Crooks, G.E., Hon, G., Chandonia, J.M. and Brenner, S.E. (2004) WebLogo: A sequence logo generator. *Genome Res.* **14**, 1188–1190.

Dan, Y., Yan, H., Munyikwa, T., Dong, J., Zhang, Y. and Armstrong, C.L. (2006) Micro-Tom–a high-throughput model transformation system for functional genomics. *Plant Cell Rep.* **25**, 432–441.

David-Schwartz, R., Koenig, D. and Sinha, N.R. (2009) *LYRATE* is a key regulator of leaflet initiation and lamina outgrowth in tomato. *Plant Cell*, **21**, 3093–4104.

Dellaporta, S.L., Wood, J. and Hicks, J.B. (1983) A plant DNA minipreparation: Version II. *Plant Mol. Biol. Rep.* **1**, 19–21.

Ellul, P., García-Sogo, B., Pineda, B., Ríos, G., Roig, L.A. and Moreno, V. (2003) The ploidy level of transgenic plants in Agrobacterium-mediated transformation of tomato cotyledons (*Lycopersicon esculentum* L. Mill.) is genotype and procedure dependent. *Theor. Appl. Genet.* **106**, 231–238.

Enoki, H., Izawa, T., Kawahara, M., Komatsu, M., Koh, S., Kyozuka, J. and Shimamoto, K. (1999) *Ac* as a tool for the functional genomics of rice. *Plant J.* **19**, 605–613.

Feldmann, K.A. (1991) T-DNA insertion mutagenesis in Arabidopsis: mutational spectrum. *Plant J.* **1**, 71–82.

Geisler, M., Jablonska, B. and Springer, P.S. (2002) Enhancer trap expression patterns provide a novel teaching resource. *Plant Physiol.* **130**, 1747–1753.

Gheysen, G., Villaroel, R. and Van Montagu, M. (1991) Illegitimate recombination in plants: a role for T-DNA integration. *Genes Dev.* **5**, 287–297.

Gillaspy, G., Ben-David, H. and Gruissem, W. (1993) Fruits: A developmental perspective. *Plant Cell*, **5**, 1439–1451.

Goldsbrough, A. and Bevan, M. (1991) New patterns of gene activity in plants detected using an *Agrobacterium* vector. *Plant Mol. Biol.* **16**, 263–269.

Gorguet, B., van Heusden, A.W. and Lindhout, P. (2005) Parthenocarpic fruit development in tomato. *Plant Biol.* **7**, 131–139.

Greco, R., Ouwerkerk, P.B.F., de Kam, R.J., Sallaud, C., Favalli, C., Colombo, L., Guiderdoni, E. *et al.* (2003) Transpositional behaviour of an *Ac/Ds* system for reverse genetics in rice. *Theor. Appl. Genetics*, **108**, 10–24.

Groover, A., Fontana, J.R., Dupper, G., Ma, C., Martienssen, R., Strauss, S. and Meilan, R. (2004) Gene and enhancer trap tagging of vascular-expressed genes in poplar trees. *Plant Physiol.* **134**, 1742–1751.

He, Y., Tang, W., Swain, J.D., Green, A.L., Jack, T.P. and Gan, S. (2001) Networking senescence-regulating pathways by using Arabidopsis enhancer trap lines. *Plant Physiol.* **126**, 707–716.

Hiei, Y., Ohta, S., Komari, T. and Kumashiro, T. (1994) Efficient transformation of rice (*Oryza sativa* L.) mediated by Agrobacterium and sequence analysis of the boundaries of the T-DNA. *Plant J.* **6**, 271–282.

Hsing, Y.I., Chem, C.G., Fan, M.J., Lu, P.C., Chen, K.T., Lo, S.F., Sun, P.K. *et al.* (2007) A rice gene activation/knockout mutant resource for high throughput functional genomics. *Plant Mol. Biol.* **63**, 351–364.

Hu, W. and Phillips, G.C. (2001) A combination of overgrowth-control antibiotics improves *Agrobacterium tumefaciens*-mediated transformation efficiency for cultivated tomato (*L. esculentum*). *In Vitro Cell. Dev. Biol.* **37**, 12–18.

Jeon, J.S., Lee, S., Jung, K.H., Jun, S.H., Jeong, D.H., Lee, J., Kim, C. *et al.* (2000) T-DNA insertional mutagenesis for functional genomics in rice. *Plant J.* **22**, 561–570.

Jeong, D.H., An, S., Kang, H.G., Moon, S., Han, J.J., Park, S., Lee, H.S. *et al.* (2002) T-DNA insertional mutagenesis for activation tagging in rice. *Plant Physiol.* **130**, 1636–1644.

Johnson, A.A.T., Hibberd, J.M., Gay, C., Essah, P.A., Haseloff, J., Tester, M. and Guiderdoni, E. (2005) Spatial control of transgene expression in rice (*Oryza sativa* L.) using the GAL4 enhancer trapping system. *Plant J.* **41**, 779–789.

Klee, H.J. and Giovannoni, J.J. (2011) Genetics and control of tomato fruit ripening and quality attributes. *Annu. Rev. Genet.* **45**, 41–59.

Krysan, P.J., Young, J.C. and Sussman, M.R. (1999) T-DNA as an insertional mutagen in Arabidopsis. *Plant Cell*, **11**, 2283–2290.

Lindsey, K., Wei, W., Clarke, M.C., McArdle, H.F., Rooke, L.M. and Topping, J.F. (1993) Tagging genomic sequences that direct transgene expression by activation of a promoter trap in plants. *Transgenic Res.* **2**, 33–47.

Lozano, R., Giménez, E., Cara, B., Capel, J. and Angosto, T. (2009) Genetic analysis of reproductive development in tomato. *Int. J. Dev. Biol.* **53**, 1635–1648.

Martí, E., Gisbert, C., Bishop, G.J., Dixon, M.S. and García-Martínez, J.L. (2006) Genetic and physiological characterization of tomato cv. *Micro-Tom*. *J. Exp. Bot.* **57**, 2037–2047.

Mathews, H., Clendennen, S.K., Caldwell, C.G., Liu, X.L., Connors, K., Matheis, N., Schuster, D.K. *et al.* (2003) Activation tagging in tomato identifies a transcriptional regulator of anthocyanin biosynthesis, modification, and transport. *Plant Cell*, **15**, 1689–1703.

Meissner, R., Jacobson, Y., Melamed, S., Levyatuv, S., Shalev, G., Ashri, A., Elkind, Y. *et al.* (1997) A new model system for tomato genetics. *Plant J.* **12**, 1465–1472.

Meissner, R., Chague, V., Zhu, Q., Emmanuel, E., Elkind, Y. and Levy, A.A. (2000) A high throughput system for transposon tagging and promoter trapping in tomato. *Plant J.* **22**, 265–274.

Memelink, J. (2003) T-DNA activation tagging. *Methods Mol. Biol.* **236**, 345–362.

Menda, N., Semel, Y., Peled, D., Eshed, Y. and Zamir, D. (2004) *In silico* screening of a saturated mutation library of tomato. *Plant J.* **38**, 861–872.

Miyao, A., Iwasaki, Y., Kitano, H., Itoh, J., Maekawa, M., Murata, K., Yatou, O. *et al.* (2007) A large-scale collection of phenotypic data describing an insertional mutant population to facilitate functional analysis of rice genes. *Plant Mol. Biol.* **63**, 625–635.

Murashige, T. and Skoog, F. (1962) A revised medium for rapid growth and bioassays with tobacco tissue cultures. *Physiol. Plant.* **15**, 473–497.

O'Kane, C.J. and Gehring, W.J. (1987) Detection in situ of genomic regulatory elements in Drosophila. *Proc. Natl Acad. Sci. USA*, **84**, 9123–9127.

Pan, X., Li, Y. and Stein, L. (2005) Site preferences of insertional mutagenesis agents in Arabidopsis. *Plant Physiol.* **137**, 168–175.

Pandolfini, T. (2009) Seedless fruit production by hormonal regulation of fruit set. *Nutrients*, **1**, 168–177.

Parinov, S., Sevugan, M., Ye, D., Yang, W.C., Kumaran, M. and Sundaresan, V. (1999) Analysis of flanking sequences from *Dissociation* insertion lines: a database for reverse genetics in Arabidopsis. *Plant Cell*, **11**, 2263–2270.

Park, J.I., Ishimizu, T., Suwabe, K., Sudo, K., Masuko, H., Hakozaki, H., Nou, I.S. *et al.* (2010) UDP-glucose pyrophosphorylase is rate limiting in vegetative and reproductive phases in *Arabidopsis thaliana*. *Plant Cell Physiol.* **51**, 981–996.

Peng, H., Huang, H.M., Yang, Y.Z., Zhai, Y., Wu, J.X., Huang, D.F. and Lu, T.G. (2005) Functional analysis of GUS expression patterns and T-DNA integration characteristics in rice enhancer trap lines. *Plant Sci.* **168**, 1571–1579.

Peterson, D.G., Stack, S.M., Price, H.J. and Johnston, J.S. (1996) DNA content of heterochromatin and euchromatin in tomato (*Lycopersicon esculentum*) pachytene chromosomes. *Genome*, **39**, 77–82.

Qin, G., Kang, D., Dong, Y., Shen, Y., Zhang, L., Deng, X., Zhang, Y. *et al.* (2003) Obtaining and analysis of flanking sequences from T-DNA transformants of Arabidopsis. *Plant Sci.* **165**, 941–949.

Raina, S., Mahalingam, R., Chen, F. and Fedoroff, N. (2002) A collection of sequenced and mapped *Ds* transposon insertion sites in *Arabidopsis thaliana*. *Plant Mol. Biol.* **50**, 91–108.

Ranjan, A., Ichihashi, Y. and Sinha, N.R. (2012) The tomato genome: implications for plant breeding, genomics and evolution. *Genome Biol.* **13**, 167.

Rossi, L., Hohn, B. and Tinland, B. (1996) Integration of complete transferred DNA units is dependent on the activity of virulence E2 protein of *Agrobacterium tumefaciens*. *Proc. Natl Acad. Sci. USA*, **93**, 126–130.

Rosso, M.G., Li, Y., Strizhov, N., Reiss, B., Dekker, K. and Weisshaar, B. (2003) An *Arabidopsis thaliana* T-DNA mutagenized population (GABI-Kat) for flanking sequence tag-based reverse genetics. *Plant Mol. Biol.* **53**, 247–259.

Sallaud, C., Gay, C., Lamande, P., Bès, M., Piffanelli, P., Piégu, B., Droc, G. *et al.* (2004) High throughput T-DNA insertion mutagenesis in rice: a first step towards in silico reverse genetics. *Plant J.* **39**, 450–464.

Schneeberger, K., Ossowski, S., Lanz, C., Juul, T., Petersen, A.H., Nielsen, K.L., Jorgensen, J.E. *et al.* (2009) SHOREmap: simultaneous mapping and mutation identification by deep sequencing. *Nat. Methods*, **6**, 550–551.

Schupp, J.M., Price, L.B., Klevytska, A. and Keim, P. (1999) Internal and flanking sequence from AFLP fragments using ligation-mediated suppression PCR. *Biotechniques*, **26**, 905–912.

Scott, J.W. and Harbaugh, B.K. (1989) Micro-Tom. A miniature dwarf tomato. *Fl. Agric. Exp. Stn. Circular*, **S-370**, 1–6.

Sessions, A., Burke, E., Presting, G., Aux, G., McElver, J., Patton, D., Dietrich, B. *et al.* (2002) A high-throughput Arabidopsis reverse genetics system. *Plant Cell*, **14**, 2985–2994.

Spertini, D., Béliveau, C. and Bellemare, G. (1999) Screening of transgenic plants by amplification of unknown genomic DNA flanking T-DNA. *Biotechniques*, **27**, 308–314.

Speulman, E., Metz, P.L.J., van Arkel, G., te Lintel, H.B., Stiekema, W.J. and Pereira, A. (1999) A two-component enhancer-inhibitor transposon mutagenesis system for functional analysis of the Arabidopsis genome. *Plant Cell*, **11**, 1853–1866.

Springer, P.S. (2000) Gene traps: tools for plant development and genomics. *Plant Cell*, **12**, 1007–1020.

Stanford, W.L., Cohn, J.B. and Cordes, S.P. (2001) Gene-trap mutagenesis: past, present and beyond. *Nat. Rev. Genet.* **2**, 756–768.

Sundaresan, V., Springer, P., Volpe, T., Haward, S., Jones, J.D., Dean, C., Ma, H. *et al.* (1995) Patterns of gene action in plant development revealed by enhancer trap and gene trap transposable elements. *Genes Dev.* **9**, 1797–1810.

Tanksley, S.D. (2004) The genetic, developmental, and molecular bases of fruit size and shape variation in tomato. *Plant Cell*, **16**(Suppl), S181–S189.

Ten Hoopen, R., Robbins, T.P., Fransz, P.F., Montijn, B.M., Oud, O., Gerats, A.G.M. and Nanninga, N. (1996) Localization of T-DNA insertions in Petunia by Fluorescence in Situ Hybridization: physical evidence for suppression of recombination. *Plant Cell*, **8**, 823–830.

The Tomato Genome Consortium. (2009) A snapshot of the emerging tomato genome sequence. *Plant Genome*, **2**, 78–92.

The Tomato Genome Consortium. (2012) The tomato genome sequence provides insights into fleshy fruit evolution. *Nature*, **485**, 635–641.

Tinland, B. (1996) The integration of T-DNA into plant genomes. *Trends Plant Sci.* **1**, 178–184.

Tinland, B., Hohn, B. and Puchta, H. (1994) *Agrobacterium tumefaciens* transfers single-stranded transferred DNA (T-DNA) into the plant cell nucleus. *Proc. Natl Acad. Sci. USA*, **91**, 8000–8004.

Tinland, B., Schoumacker, F., Gloedkler, V., Bravo-Angel, A.M. and Hohn, B. (1995) The *Agrobacterium tumefaciens* virulence D2 protein is responsible for precise integration of T-DNA into the plant genome. *EMBO J.* **14**, 3585–3595.

Tissier, A.F., Marillonnet, S., Klimyuk, V., Patel, K., Torres, M.A., Murphy, G. and Jones, J.D. (1999) Multiple independent defective suppressor-mutator transposon insertions in Arabidopsis: a tool for functional genomics. *Plant Cell*, **11**, 1841–1852.

Topping, J.F., Wei, W. and Lindsey, K. (1991) Functional tagging of regulatory elements in the plant genome. *Development*, **112**, 1009–1019.

Tzfira, T., Vaidya, M. and Citovsky, V. (2004) Involvement of targeted proteolysis in plant genetic transformation by *Agrobacterium*. *Nature*, **431**, 87–92.

Wan, S., Wu, J., Zhang, Z., Sun, X., Lv, Y., Gao, C., Ning, Y. *et al.* (2009) Activation tagging, an efficient tool for functional analysis of the rice genome. *Plant Mol. Biol.* **69**, 69.

Wang, J., Lewis, M.E., Whallon, J.H. and Sink, K.C. (1995) Chromosomal mapping of T-DNA inserts in transgenic Petunia by *in situ* hybridization. *Transgenic Res.* **4**, 241–246.

Wang, Y., Tang, X., Cheng, Z., Mueller, L., Giovannoni, J.J. and Tanksley, S.D. (2006) Euchromatin and pericentromeric heterochromatin: comparative composition in the tomato genome. *Genetics*, **172**, 2529–2540.

Winer, J., Jung, C.K., Shackel, I. and Williams, P.M. (1999) Development and validation of real-time quantitative reverse transcriptase-polymerase chain reaction for monitoring gene expression in cardiac myocytes in vitro. *Anal. Biochem.* **15**, 41–49.

Wu, C., Li, X., Yuan, W., Chen, G., Kilian, A., Li, J., Xu, C. *et al.* (2003) Development of enhancer trap lines for functional analysis of the rice genome. *Plant J.* **35**, 418–427.

Yuste-Lisbona, F.J., Quinet, M., Fernández-Lozano, A., Pineda, B., Moreno, V., Angosto, T. and Lozano, R. (2016) Characterization of *vegetative inflorescence* (*mc-vin*) mutant provides new insight into the role of *MACROCALYX* in regulating inflorescence development of tomato. *Sci. Rep.* **6**, 18796.

Zhang, J., Guo, D., Chang, Y.X., You, C.J., Li, X.W., Dai, X.X., Weng, Q.J. *et al.* (2007) Non-random distribution of T-DNA insertions at various levels of the genome hierarchy as revealed by analyzing 13 804 T-DNA flanking sequences from an enhancer-trap mutant library. *Plant J.* **49**, 947–959.

6

Overexpression of an *Arabidopsis thaliana* galactinol synthase gene improves drought tolerance in transgenic rice and increased grain yield in the field

Michael Gomez Selvaraj[1], Takuma Ishizaki[2], Milton Valencia[1], Satoshi Ogawa[1,3], Beata Dedicova[1], Takuya Ogata[4], Kyouko Yoshiwara[4], Kyonoshin Maruyama[4], Miyako Kusano[5,6,7], Kazuki Saito[5,6,8], Fuminori Takahashi[5,6], Kazuo Shinozaki[5,6], Kazuo Nakashima[4] and Manabu Ishitani[1,*]

[1]*International Center for Tropical Agriculture (CIAT), Cali, Colombia*

[2]*Tropical Agriculture Research Front (TARF), Japan International Research Center for Agricultural Sciences (JIRCAS), Ishigaki, Okinawa, Japan*

[3]*Japan Society for the Promotion of Science, The University of Tokyo, Bunkyo-ku, Tokyo, Japan*

[4]*Biological Resources and Post-harvest Division, Japan International Research Center for Agricultural Sciences (JIRCAS), Tsukuba, Ibaraki, Japan*

[5]*RIKEN Center for Sustainable Resource Science, Yokohama, Kanagawa, Japan*

[6]*RIKEN Center for Sustainable Resource Science, Tsukuba, Ibaraki, Japan*

[7]*Graduate School of Life and Environmental Sciences, University of Tsukuba, Tsukuba, Ibaraki, Japan*

[8]*Department of Molecular Biology and Biotechnology, Graduate School of Pharmaceutical Sciences, Chiba University, Chiba, Japan*

*Correspondence
email m.ishitani@CGIAR.ORG

Keywords: galactinol synthase, drought, transgenic rice, grain yield, confined field trial.

Summary

Drought stress has often caused significant decreases in crop production which could be associated with global warming. Enhancing drought tolerance without a grain yield penalty has been a great challenge in crop improvement. Here, we report the Arabidopsis thaliana galactinol synthase 2 gene (*AtGolS2*) was able to confer drought tolerance and increase grain yield in two different rice (*Oryza sativa*) genotypes under dry field conditions. The developed transgenic lines expressing *AtGolS2* under the control of the constitutive maize ubiquitin promoter (*Ubi:AtGolS2*) also had higher levels of galactinol than the non-transgenic control. The increased grain yield of the transgenic rice under drought conditions was related to a higher number of panicles, grain fertility and biomass. Extensive confined field trials using *Ubi:AtGolS2* transgenic lines in Curinga, tropical japonica and NERICA4, interspecific hybrid across two different seasons and environments revealed the verified lines have the proven field drought tolerance of the Ubi:*AtGolS2* transgenic rice. The amended drought tolerance was associated with higher relative water content of leaves, higher photosynthesis activity, lesser reduction in plant growth and faster recovering ability. Collectively, our results provide strong evidence that *AtGolS2* is a useful biotechnological tool to reduce grain yield losses in rice beyond genetic differences under field drought stress.

Introduction

Drought is a major abiotic stress condition critically limiting crop production and yield (Edmeades, 2008). Climate prediction models suggest that abiotic stresses will increase in the near future because of global climate change (Ahuja *et al.*, 2010). The ever-rising world population and recurrent global climate change challenge the agricultural system to produce sufficient food to feed the world (Godfray *et al.*, 2010). As the world's second-largest crop, rice plays a critical role in food security for more than half of the world's population (FAO, 2016: http://faostat3.fao.org/browse/Q/QC/E).

Rice accounts for about 27% of total cereal production, with a worldwide production of roughly 738.2 million tons (FAO, 2016). By 2035, a 26% increase in rice production will be required to feed the growing population (Cassman *et al.*, 2003; Seck *et al.*, 2012). Global water shortage is a major issue for cultivated rice, which needs large quantities of water (Manavalan *et al.*, 2012). It was reported that the global reduction in rice production due to drought averages 18 million tons annually (O'Toole, 2004). Worldwide, drought affects approximately 23 million ha of rice production under rainfed conditions. Drought is particularly frequent in unbunded uplands, bunded uplands and shallow rainfed lowland fields in many parts of South and South-East Asia, sub-Saharan Africa and Latin America (Serraj *et al.*, 2011). To resolve those global problems, it is important to improve crop yields especially within staple food crops like rice (*Oryza sativa* L.) through breeding-improved stress tolerance.

Transgenic technologies are one of the numerous tools available to plant breeding programmes, which help to open new avenues for crop improvement by developing crop cultivars resistant to various biotic and abiotic stresses (Younis *et al.*, 2014). Around 175.2 million hectares of biotech crops were grown globally and transgenic acreage grew 3% in 2013, representing 35% of the global seed market (Marshall, 2014). In rice, progress has been made in the generation and evaluation of transgenic rice events against drought tolerance (Todaka *et al.*, 2015).

Plants have evolved several mechanisms to accustom to abiotic stresses through changes at the physiological levels and molecular levels (Todaka *et al.*, 2012; Yamaguchi-Shinozaki and Shinozaki, 2006). It is suggested that overexpression of stress-related genes could improve drought tolerance in rice (reviewed by Nakashima *et al.*, 2014 and Todaka *et al.*, 2015). Despite such efforts to develop drought-tolerant rice plants, very few have been shown to improve grain yields under the field environments (Gaudin *et al.*, 2013). Encouraging results include transgenic rice plants expressing *OsNAC5* (Jeong *et al.*, 2013), *OsNAC9/SNAC1* (Redillas *et al.*, 2012) or *OsNAC10* (Jeong *et al.*, 2010), which was shown to improve grain yield under field drought conditions. Many genes that may play an important role under drought have been mostly tested on a single model rice genetic background (Nipponbare) under laboratory conditions, but very few have been tested vigorously in a natural target environment using different commercial rice genetic backgrounds. For improved rice to be accepted by consumers, it is necessary to consider both adaptation to the target environments and fulfilment of local grain quality and taste preferences. This is predominantly important in transgenic studies in which the recipient genetic background is often chosen according to its ability to be transformed rather than agronomic or cultural considerations (Gaudin *et al.*, 2013).

The accumulations of metabolite or osmoprotectants are one of the key adaptive mechanisms for plants to handle with dehydration stress and cellular injury (Hare *et al.*, 1998). Soluble sugars, including those in the sucrose, trehalose and raffinose families also known as oligosaccharides (RFOs), have been found to accumulate during drought stress in many plants (Collett *et al.*, 2004; Farrant, 2007; Peters *et al.*, 2007; Taji *et al.*, 2002). Galactinol synthase (GolS), a key enzyme in the metabolic pathway leading to RFOs, synthesizes galactinol (from UDP-Gal and myoinositol), which serves as a galactosyl donor to form raffinose, stachyose and verbascose (Panikulangara *et al.*, 2004). It has been reported that the production of enzymes involving the biosynthesis of RFOs, and the resulting accumulation of RFOs, plays critical roles in acquired tolerance of *Arabidopsis thaliana* to drought and heat stresses (Taji *et al.*, 2002; Nishizawa *et al.*, 2008; reviewed by Sengupta *et al.*, 2015). In *Arabidopsis*, seven *GolS* genes and three putative *GolS* genes have been identified, and intricate induction patterns were reported (Nishizawa *et al.*, 2008). *AtGolS1* was inducible by drought, salinity (Taji *et al.*, 2002) and temperature stresses (Panikulangara *et al.*, 2004); *AtGolS2* was induced only by drought and salinity stresses; and *AtGolS3* induction was detected solely after cold stress (Taji *et al.*, 2002). Overexpression of *AtGolS2* caused the increase in galactinol and raffinose contents in leaves and exhibited enhanced drought tolerance of transgenic *Arabidopsis* (Taji *et al.*, 2002).

Here, we describe the production of transgenic rice events that overexpress the *AtGolS2* cDNA driven by the maize ubiquitin promoter (*Ubi:AtGolS2*) in the background of Curinga (a Brazilian local upland rice variety) and NERICA4 (a popular upland rice variety in African countries) and present the results of multiple confined field trials over two different environmental conditions. Our extensive field test over different seasons and environmental conditions using multiple rice genetic backgrounds clearly demonstrated that *AtGolS2* overexpression consistently increased biomass and grain yield under drought stress conditions. These findings implied that the *Ubi:AtGolS2* transgene played an important role in improving agronomic traits and yield characteristics of rice and that overexpression would be an efficient way to accelerate the rice breeding programme for drought tolerance.

Results

Generation and molecular analyses of rice events expressing *Ubi:AtGolS2*

Our goal was to produce and select best-performing Curinga and NERICA4 transgenic rice lines for drought tolerance by expressing *AtGolS2* from the constitutive maize ubiquitin (*Ubi*) promoter. We accomplished this using *Agrobacterium*-mediated transformation (Ishizaki and Kumashiro, 2008; Zuniga-Soto *et al.*, 2015). At least 20 independent transgenic events were produced from each variety. T_3 or T_4 seeds that possessed genetically fixed single copy of transgene were used for further analyses.

To link field performance of the transgenic lines to the transgene expression levels and metabolite accumulation levels. Therefore, quantitative PCR expression of *Ubi:AtGolS2* transgenic Curinga and *Ubi:AtGolS2* transgenic NERICA4 events was analysed by quantitative real-time PCR (RT-PCR). All of the transgenic Curinga lines expressed the transgene (Figure 1a). Galactinol synthase (GolS) catalyses the first committed step in the biosynthesis of raffinose family oligosaccharides (RFOs) including galactinol and raffinose and plays a key regulatory role in carbon partitioning between sucrose and RFOs (Saravitz *et al.*, 1987; Taji *et al.*, 2002). We measured galactinol content in the promising transgenic plants grown in glasshouse under unstressed conditions. The NT Curinga and NERICA4 were used as the control. Under normal growth conditions, each transgenic plant showed significantly higher accumulation of galactinol as compared with NT rice plants (Figure 1b). All of the transgenic Curinga lines expressed the transgene and the expression levels of the transgene in lines #3025 and #3214 were higher than in the other lines (#2580, #2590, #2783 and #3020). In case of NERICA4, the *AtGolS2* gene was overexpressed and expression level of the gene and accumulation level of galactinol in line #1577 (NERICA4) were higher than in other lines tested (Figure 1a, b). We also analysed the expression of representative drought marker genes in the *Ubi:AtGolS2* transgenic rice. Expression of genes for a transcription factor OsNAC6, isocitrate lyase (ICL) and late embryogenesis-abundant protein LEA3 (Maruyama *et al.*, 2014; Nakashima *et al.*, 2014) was not induced in the events for *Ubi:AtGolS2* without drought stress (Figure S1). These results indicate that the accumulation of galactinol is not related to the expression of these drought-inducible genes.

Drought tolerance of *Ubi:AtGolS2* Curinga lines at vegetative-stage stress

Drought tolerance during the seedling growth period was important for rice plant establishment in areas where early-season drought overlapped with the vegetative stage. We conducted vegetative-stage drought experiments using homozygous transgenic rice lines overexpressing *Ubi:AtGolS2* in Curinga. The experiment was conducted in rainout shelter facility at CIAT, Colombia, and biomass was used as a main criterion to select promising lines.

To evaluate the growth performance of the *Ubi:AtGolS2*-overexpressing Curinga under drought conditions at the vegetative stage, three-week-old transgenic and nontransgenic (NT) Curinga control plants were subjected to drought stress for up to 3 weeks in November-December 2011 (Figure 2a). Agronomic

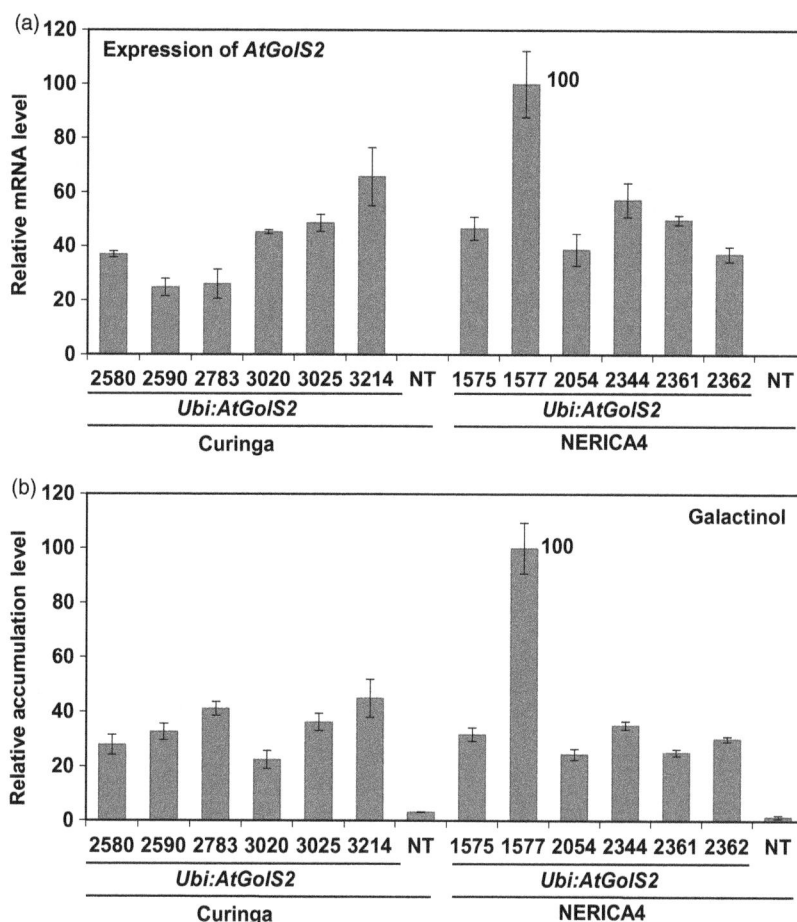

Figure 1 Ectopic overexpression of *AtGolS2* gene confers higher galactinol accumulation on transgenic Curinga and NERICA4 rice. (a) Expression of *AtGolS2* in *Ubi:AtGolS2* transgenic rice was analysed by quantitative real-time PCR. Expression of *OsUbi1* was analysed as an internal control to normalize the expression of *AtGolS2*. (b) Accumulation of galactinol in *Ubi:AtGolS2* transgenic rice was analysed using GC-TOF-MS. The highest average value of the samples was set as 100, and the relative values were shown in (a) and (b). Five plants were combined into one sample for each line, and the relative values are mean ± SD of three technical replicates. Nontransgenic (NT) Curinga and NERICA4 were used as the control.

data collected before stress treatment showed that there was no significant variation among the lines, which helps to explain the uniformity in the experiment.

During peak stress, the transgenic lines significantly maintained more plant height than NT Curinga (Figure 2e). The NT Curinga started to show visual symptoms of drought-induced leaf rolling at an earlier stage than the transgenic plants (Figure 2b). The transgenic lines showed low leaf rolling score, 2-3 compared with the NT Curinga plants with (6.5) during peak drought stress (Figure 2b). Only the event #2590 accumulated significantly higher dry biomass than NT Curinga at the end of drought stress (Figure 2c). However, dry biomasses of the transgenic plants measured after rewatering were significantly higher than those of NT Curinga in all but one instance (Figure 2d).

Drought tolerance of the *Ubi:AtGolS2* Curinga lines under Managed Drought Stress Environment (MDSE)

In order to confirm the drought tolerance of Curinga transgenic lines at the reproductive stage, three consecutive confined field trials were conducted under the removable rainout shelter facility at CIAT, Palmira. As mentioned above, we had conducted two drought stress trials in the year 2012 over two cultivating seasons (2012-rainy season-MDSE-Trial-1 and 2012-dry season-MDSE-Trial-2) and one in 2014 (2014-rainy season-MDSE-Trial) (Figure S2). Up to eight independent T_4 single-copy homozygous transgenic and NT Curinga lines were used to conduct the dry-down experiments. In addition to the field drought experiments, one normal well-watered paddy field trial (WW-field trial) was also

conducted during the dry season of 2012 (Table S1). In the WW-field trial, single plant grain yield of all the tested transgenic lines and NT Curinga was not significantly different (Table S1), which suggests no yield penalty in these transgenic lines. Based on these results, grain yield was also used as main parameter to compare yield in dry fields between these transgenic lines and NT Curinga.

Drought intensity varied among drought trials from mild to severe (Figure S2). Although similar levels of stress duration and environmental conditions occurred in the rainy season drought experiments, it was observed that average single plant yield of NT Curinga was sharply reduced compared to the other two rainy season experiments (Figure 3a, b and c). This suggests a dry season effect. Additionally, we found that soil moisture rapidly decreased within the 0- to 40-cm soil layer compared to other two rainy season experiments (Figure S2b). The level of drought stress imposed under upland rainout shelter conditions was equivalent to that which caused an average reduction of around 60%–70% in the single plant grain yield obtained in the NT Curinga under well-watered paddy field conditions (Table S1 and Figure 3a, b and c).

Statistical analysis of yield and yield-related parameters scored for three rainout shelter drought experiments revealed that the *Ubi:AtGolS2* overexpression consistently produced grain yield compared to NT across the season. Interestingly, in the *Ubi:AtGolS2* Curinga lines, the morphophysiological trait performance was significantly better than NT Curinga (Table S2 and Figure 3). In each trial, we found some transgenic lines that had significantly higher yield: three of eight transgenic

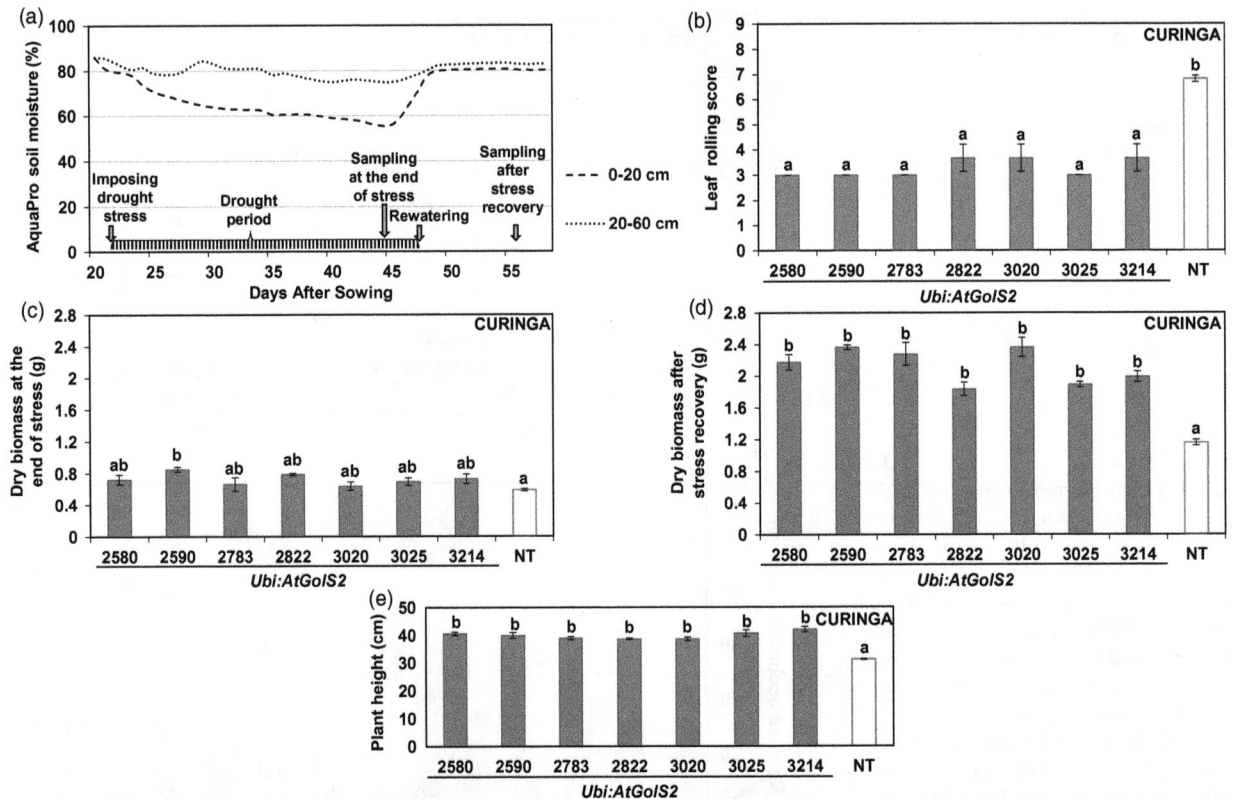

Figure 2 *Ubi:AtGolS2* improves vegetative drought tolerance in transgenic Curinga. (a) Soil moisture profile during the vegetative drought stress rainout shelter experiment. Fragmented line and dotted line indicate upper (0–20 cm) and average lower (20–60 cm) soil moisture, respectively. Arrowheads indicate drought stress scheduling and sampling time. (b) Variation in leaf rolling score among the transgenic lines during peak stress (three weeks after stress). (c) Variation in plant dry biomass among the transgenic lines at the end of the stress. (d) Variation in plant dry biomass among the transgenic lines at the end of the harvest after stress recovery. (e) Variation plant height among the transgenic lines at the end of the stress. Each dry biomass and plant height value represents the mean ± SE (n = 3), in each replication data point derived from three individual uniform plants; leaf rolling score based on whole plot performance and represents the mean ± SE (n = 3) from three replications. Different letters in each column denote significant differences at $P < 0.05$ by Tukey–Kramer method.

lines in 2012-rainy-MDSE-Trial-1, four of six transgenic lines in 2012-dry-MDSE-Trial-2 and two of five transgenic lines in 2014-rainy-MDSE-Trial, respectively (Figure 3a, b and c). We also found two promising transgenic lines (#2580 and #2590), which consistently outperformed NT Curinga in terms of grain yield over the drought experiments (Figure 3). In these lines, the higher grain yield under severe drought stress (2012-dry-MDSE-Trial-2) was associated with a significantly higher number of panicles, higher accumulation of biomass, low leaf rolling and leaf drying score, faster recovering ability, early flowering (Table S2), panicle length and grain fertility. Furthermore, under moderate stress (drought experiments in rainy season), no significant difference in panicle number and biomass between transgenic and nontransgenic lines was observed. Altogether, these results demonstrate that the *Ubi: AtGolS2* expression increased grain yield under drought stress conditions imposed at the reproductive stage through a mechanism that involves the maintenance of early flowering, increased vegetative biomass, higher numbers of panicles and enhanced grain fertility (Table S2).

Analysis of physiological parameters in *Ubi:AtGolS2* Curinga lines

Based on the results of the rainout shelter experiments in 2012 (the 2012-MDSE-Trial-1 and 2), five promising T4 *Ubi:AtGolS2* Curinga lines were selected for further analysis of physiological

parameters and field gene expression analysis in the 2014-rainy-MDSE-Trial. During the dry-down experiment, the most uniform transgenic lines for physiological analysis were tagged and repeated measurements were taken during drought stress.

Relative water content (RWC) was measured before and after subjecting the transgenic lines and NT Curinga to the drought stress treatment. Before stress, there were no obvious differences in the leaf RWC between NT Curinga and transgenic lines, and the RWC was within the range of 92%–95% (Figure 4a). After the lines were subjected to water stress for one week, the RWC of the NT Curinga leaves reduced sharply with respect to their first reading (before stress) from 95% to 89 %, whereas the RWC of most of the transgenic lines declined very slowly (#2580, 94%; #2590, 94%; #2783, 94%; #3020, 94%; and #3214, 93%). After three weeks of drought stress, the RWC of the transgenic lines had declined by just 18%–22% as compared to 30% in the NT Curinga. Lines #2580, #2590, #2783, #3020 and #3214 maintained RWC very well even three weeks after drought stress, with RWC percentages of 73, 77, 77, 72 and 76, respectively. The rapid decline of the RWC (average of 66%) was observed in the NT Curinga after three weeks of drought stress.

To further verify the mechanism of drought tolerance, we measured F_v/F_m values of the transgenic and NT Curinga during stress period using FluorPen-FP100, (Photon Systems Instruments, spol. s r.o., Czech Republic) (Figure 4b). The F_v/F_m values represent the maximum photochemical efficiency of photo

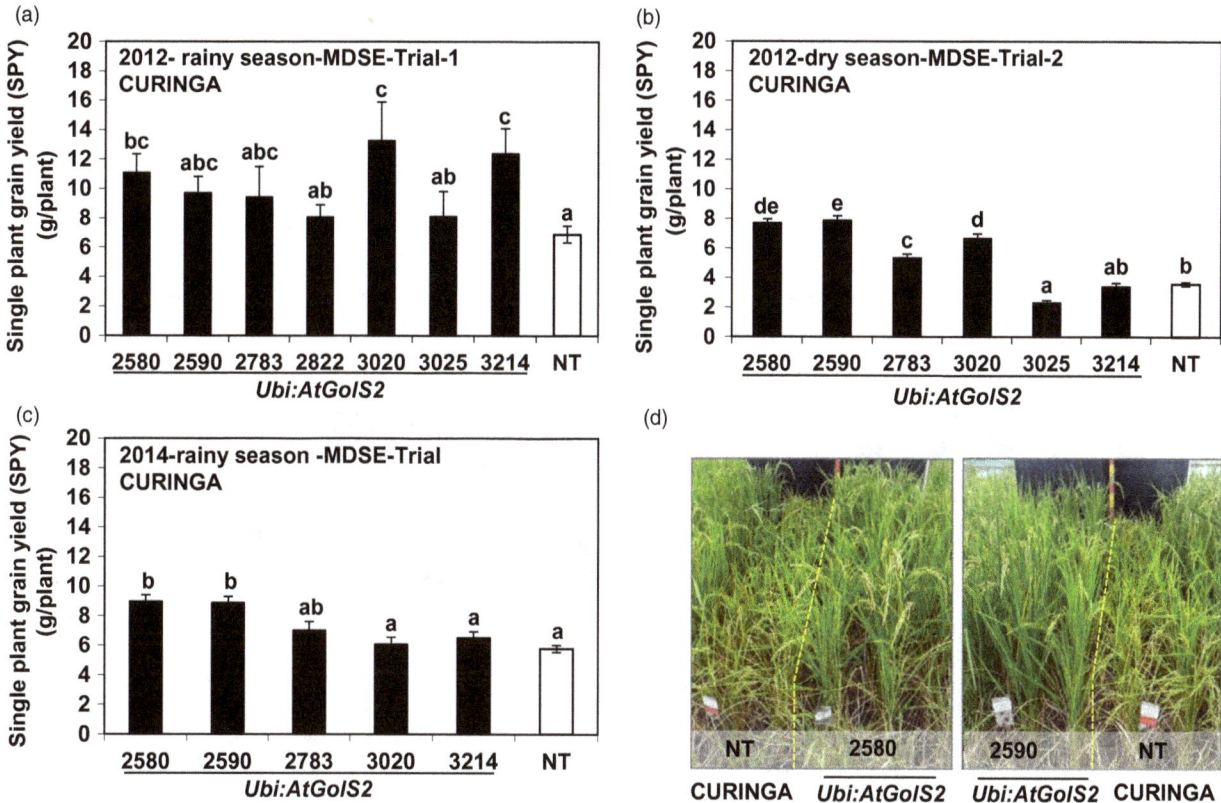

Figure 3 *Ubi:AtGolS2* improves rice grain yield in Managed Drought Stress Environment (MDSE) over the growing seasons—rainout shelter experiments, CIAT, Palmira. Single plant grain yield performance of Curinga transgenic lines in 2012-rainy season-MDSE-Trial-1 (a), 2012-dry season-MDSE-Trial-2 (b) and 2014-rainy season-MDSE-Trial (c). (d) Field performance of NT Curinga and promising transgenic lines in 2014-rainy-MDSE-Trial. Photographs were taken next day after rewatering for recovery after drought stress at the flowering stage. Each single plant yield value represents the mean ± SE (*n* = 9–24). Different letters in each column denote significant differences at *P* < 0.05 by Tukey–Kramer method.

system (PS) II in a dark-adapted state, where F_v stands for variable fluorescence and F_m stands for maximum fluorescence. Initially under unstressed conditions, the F_v/F_m values of both NT and transgenic plants were similar, ranging from 0.70 to 0.74. After one week of drought stress, the F_v/F_m value of the NT Curinga slightly decreased (0.68), but we did not find any significant differences between NT Curinga and the *Ubi:AtGolS2* transgenic lines. However, after three weeks of drought stress, the F_v/F_m value for the transgenic lines #3214, #2580, #2590, #2783 and #3020 was maintained at 0.59, 0.67, 0.65, 0.65 and 0.61, respectively, while the NT Curinga value significantly decreased to 0.56 (Figure 4b). The promising transgenic lines #2580 and #2590 showed significantly higher F_v/F_m values compared to NT Curinga even after three weeks of stress.

Chlorophyll content was measured using a SPAD-502 Chlorophyll Meter (Konica Minolta Inc., Tokyo, Japan). Transgenic and NT Curinga plants were measured for their chlorophyll content before and during peak stress (Figure 4c). Before stress, the SPAD values of the transgenic and NT Curinga plants were not significantly different ranging from 37 to 40. At peak stress (three weeks after stress), the chlorophyll values of the NT Curinga plants were reduced (average of 34) compared to the initial reading; in contrast, promising lines like #2580 and #2590 maintained a similar chlorophyll content after three weeks of the stress (Figure 4c). A RT-PCR analysis was also performed on the tested promising lines in the 2014-rainy-MDSE-Trial during different stages of the stress development to confirm the

expression of the *Ubi:AtGolS2* transgene under field conditions (Figure 4d).

Drought tolerance of *Ubi:AtGolS2* Curinga in the different environments—Target Environment (TE) Trial

Based on the initial vegetative and reproductive drought stress experiments, up to six potential transgenic Curinga lines along with the NT Curinga were chosen for upland rainfed field trials at CIAT Santa Rosa station, Villavicencio, Colombia. To test the hypothesis of gene × environment interactions of *Ubi:AtGolS2*, the three consecutive TE field trials were carried out from 2012 to 2015. The field trial conditions, design and the plot size were well described in the experimental procedure section. The rainfall and temperature pattern of this site during trial period is shown in Figure 5. Ten years of rainfall data from this site revealed that natural rainfall failure events usually occur in the months of January–February, which coincides with the reproductive stage of the crop. For instance, trial years TE-2012-13 and TE-2013-14 were very dry with continuous rain-free days of 31 and 39, respectively (Figure 5a and b). However, trial year TE-2014-15 had rainfall on and off during the reproductive stage (Figure 5c).

In the first two rainfed trials (TE-2012-13 and TE-2013-14), Curinga transgenic lines reached 50% flowering significantly earlier than NT Curinga (4 and 5 days earlier), indicating that *Ubi:AtGolS2* overexpression induced earliness in Curinga lines (Table S3). The continuous rain-free period during the reproductive stage caused a marked reduction in soil moisture in the 0- to

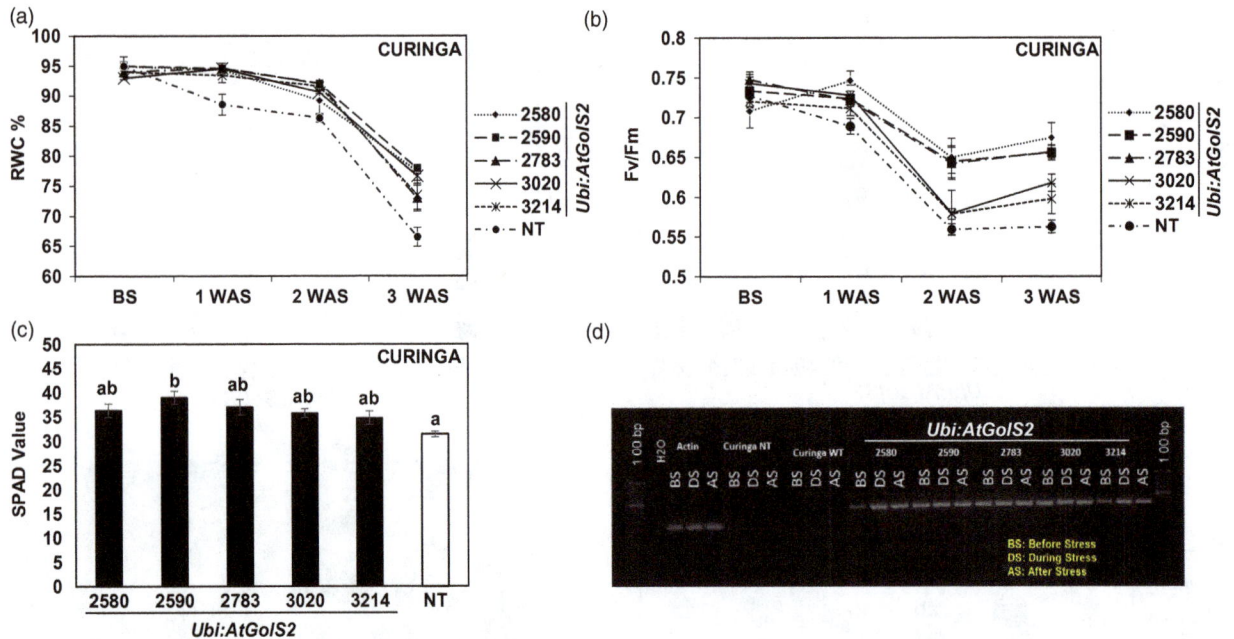

Figure 4 Variation in physiological parameters among the Curinga *Ubi:AtGolS2* transgenic lines in 2014-rainy-MDSE-Trial. (a) Percentage of relative water content (RWC) values of transgenic and NT lines at before, first, second and third weeks after stress (WAS). (b) Changes in chlorophyll fluorescence (F_V/F_M) of transgenic and NT lines at before (BS), first, second and third weeks after stress (WAS). (c) SPAD chlorophyll values of transgenic and NT lines at peak drought stress. (d) RT-PCR analysis of *Ubi:AtGolS2* transgenic and NT lines evaluated at different timing points: before stress (BS), during peak stress (DS) and after stress (AS). RT-PCR analyses were performed using RNAs from leaf tissue at different points of drought development using *Ubi:AtGolS2* gene-specific primers. Each physiological parameter value represents the mean ± SE ($n = 9$), three individual plants from three replications. Different letters in each figure denote significant differences at $P < 0.05$ by Tukey–Kramer method.

40-cm soil layer (Figure S3). Under these severe stress conditions, lines #2580 and #2590 maintained higher panicle number and showed significantly higher grain yield (GY) with relative gains of 49%, 18%, 34% and 17%, respectively, compared to NT Curinga (Figure 6a and Table S3) in the first trial (TE-2012-13). We also observed yield gains continued in promising transgenic lines in the second (TE-2013-14) and third (TE-2014-15) field trials (Figure 6b and c). In the third rainfed trial (TE-2014-15), drought stress was mild; there was about 19 days rain-free period that coincided the grain-filling stage (Figure 5c). Field vegetative performance of *Ubi:AtGolS2* transgenic Curinga (#2580) was much better than NT Curinga as shown in Figure 6d. Thus, the average grain yields of promising *Ubi:AtGolS2* transgenic lines were significantly increased compared to NT Curinga in three field trials (Figure 6).

Drought tolerance in the different genetic background expressing *Ubi:AtGolS2*

To understand how the *Ubi:AtGolS2* ubiquitously works in a rice genotype, we conducted drought tolerance experiments in homozygous transgenic lines overexpressing *Ubi:AtGolS2* in the interspecific hybrid, NERICA4. We first tested seedling survival rate in a glasshouse pot experiment at Japan International Research Center for Agriculture Sciences (JIRCAS) in Japan. Seedling survival of NERICA4 transgenic lines was evaluated through a previously reported method (Ishizaki *et al.*, 2013). Before drought stress treatment, no obvious phenotypic differences were observed between the NT NERICA4 plants and the *Ubi:AtGolS2* transgenic NERICA4 lines. After nine days of drought treatment and subsequent recovery for seven days, the majority of NT NERICA4 never recovered and only 11.5% survived. By

contrast, four of seven *Ubi:AtGolS2* transgenic NERICA4 lines exhibited a significantly higher survival ratio, ranging from 26.2% to 34.5% (Table S4). These results demonstrate that *Ubi:AtGolS2* can significantly improve seedling survival under drought in NERICA4.

The *Ubi:AtGolS2*-NERICA4 lines were also evaluated during the second (TE-2013-14) and third (TE-2014-15) rainfed reproductive trials in Colombia along with Curinga lines; NERICA4 lines were not included in the trial 1 (TE-2012-13) (Figure 7 and Table S5). AquaPro soil moisture profiles indicated that the conditions of NERICA4 plots were similar to those of Curinga plots in TE-2013-14-Trial and TE-2014-15-Trial (Figure S4a and b). Under severe drought stress conditions in the second trial (TE-2013-14), the *Ubi:AtGolS2*-NERICA4 lines #1577 and #2344 showed significantly higher GY with relative gains of 34% and 49%, respectively, compared to NT NERICA4 (Figure 7a). In the third trial (TE-2014-15), #1577, #2361 and #2362 showed significantly higher grain yield than NT NERICA4 (Figure 7b). Interestingly, line 1577 consistently performed well in both vegetative and reproductive experiments (Table S4 and Figure 7).

Correlation between accumulation level of galactinol and grain yield

To link field performance of the transgenic lines to the expression level of transgene *AtGolS2* and the galactinol accumulation levels, we calculated Pearson's coefficient of correlation between accumulation level of galactinol and mRNA level of *AtGolS2*, SPY and GY in Curinga and NERICA4 evaluated under field (Figure S5). The expression level of *AtGolS2* correlated with the accumulation level of galactinol: Pearson's coefficient of correlation between those factors was 0.72 in Curinga and 0.94 in NERICA4, revealing

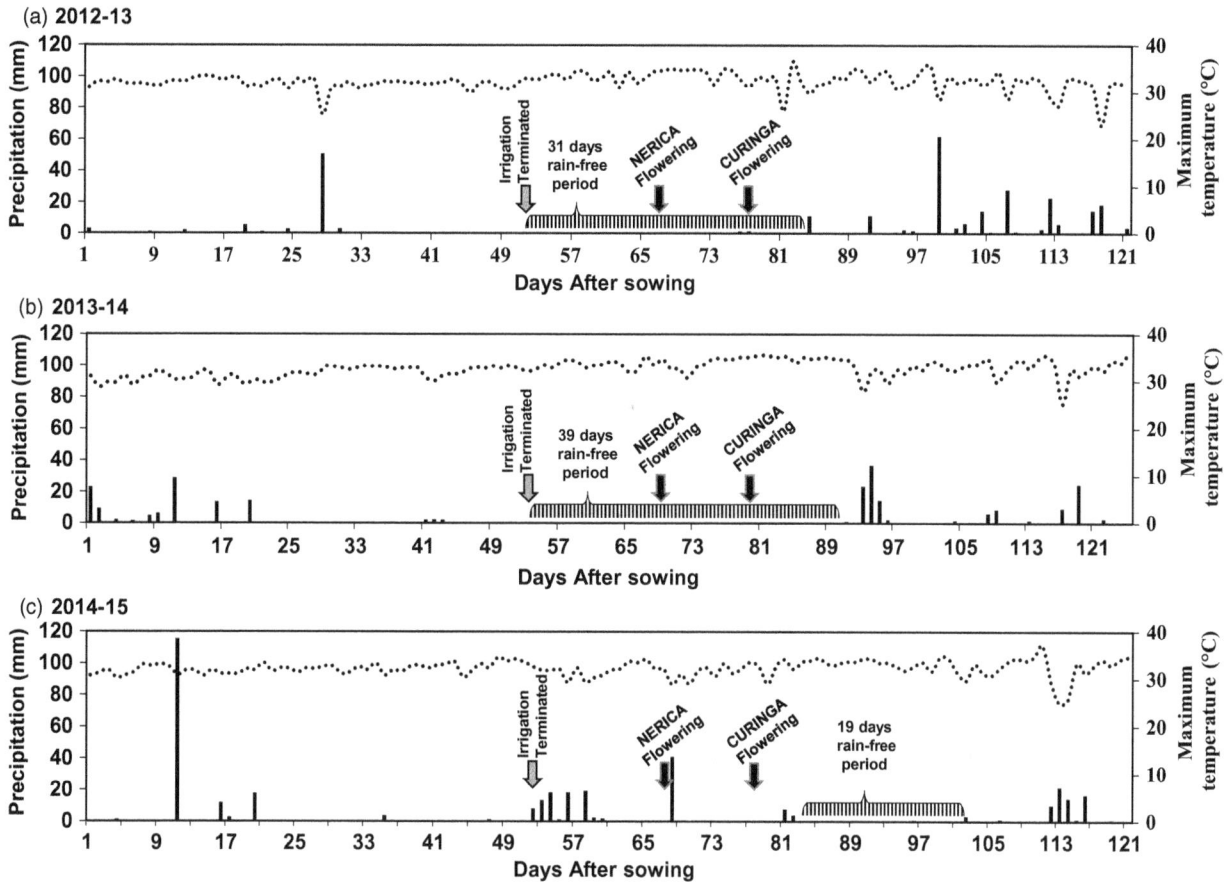

Figure 5 Rainfall and temperature pattern during crop period in upland confined rainfed field trial, CIAT, Santa Rosa upland station. (a) Climatic profile of TE-2012-13-Trial. (b) Climatic profile of TE-2013-14-Trial. (c) Climatic profile of TE-2014-15-Trial. Black bar shows amount of rainfall received during crop period, and dotted line graph shows the maximum daily temperature during trial period. Temperature data are daily averages and rainfall is daily total.

that the expression of *AtGolS2* certainly conferred the accumulation of galactinol in rice plant. However, SPY and GY did not always correlate with the accumulation level of galactinol: Pearson's coefficient of correlation between those factors ranged from -0.05 to 0.65, suggesting the galactinol did not have dose effects on yield under field.

Discussion

Ubi:AtGolS2 is versatile: improving drought tolerance across stages of rice, genetic background, drought intensity and environments

While developing drought-tolerant crops, plant productivity should be taken into consideration. Plant productivity is widely affected by natural drought incidences under field conditions (Todaka *et al.*, 2015). Droughts are random events and dry spells can occur at virtually any time during the rice growing period in drought-prone areas, leading to drought stress of varying intensity. Although rice is highly sensitive to drought stress during the reproductive stage (Venuprasad *et al.*, 2007), drought at early vegetative stage of rice growth can considerably affect plant performance. Commonly, drought survival test during the vegetative stage is obtained under laboratory or glasshouse conditions and is therefore not perfectly comparable to

interpretations made under real-field conditions. Extensive field trials are thus critical for the appropriate evaluation of stress-tolerant transgenic crops (Todaka *et al.*, 2015). In this paper, we carried out field trials in CIAT, Colombia, and demonstrated that *Ubi:AtGolS2* overexpression in rice was effective at conferring drought tolerance during both the vegetative and the reproductive stages (Figures 2 and 3). It is a very rare phenomenon when the results obtained from vegetative screening experiments concur with reproductive stage, indicating that the *Ubi:AtGolS2* overexpression can be exploited for both early- and mid-season drought. This is very important in the perspective of targeting rice varieties to rainfed environments where rainfall uncertainty is expected.

In rice, several reports are available that examine field drought tolerance caused by overexpression of transgenes (Xiao *et al.*, 2009; You *et al.*, 2012; Yu *et al.*, 2013). However, most of these studies were conducted on plants that were grown under glasshouse conditions. There have been instances where a transgene-mediated trait expressed in the glasshouse was unstable under field conditions (Brandle *et al.*, 1995). As it was reported that the effect of transgene expression in wheat varied from year to year based on the climatic conditions of a particular growing season (Bahieldin *et al.*, 2005), it was considered essential to explore the yield stability of the *Ubi:AtGolS2*

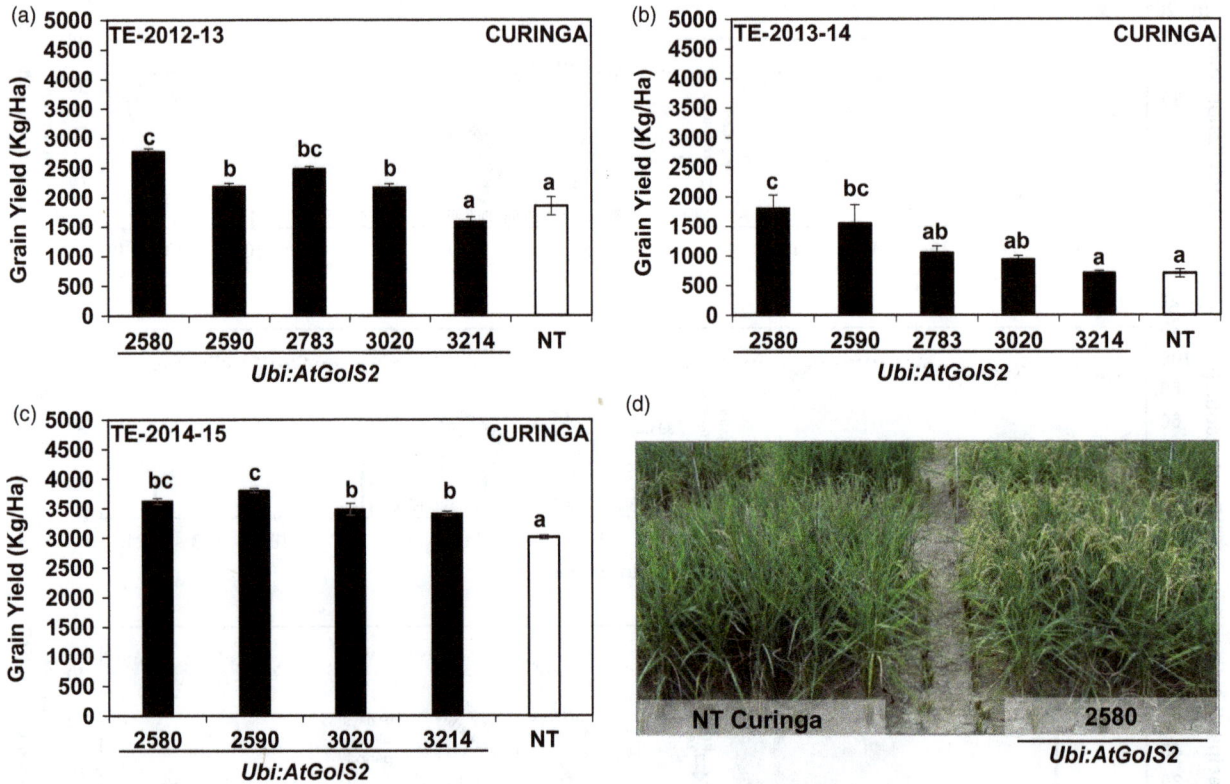

Figure 6 *Ubi:AtGolS2* improves Curinga grain yield in Target Environment (TE)—Santa Rosa rainfed Trial, CIAT upland rainfed station, Villavicencio. (a) Grain yield performance of Curinga transgenic lines in TE-2012-13-Trial. (b) Grain yield performance of Curinga transgenic lines in TE-2013-14-Trial. (c) Grain yield performance of Curinga transgenic lines in TE-2014-15-Trial. (d) Field performance of NT Curinga and promising transgenic event 2580 at TE-2013-14-Trial. Photographs were taken during stress at the grain-filling stage. Estimated grain yield (kg/ha) was derived from plot yield from three replications, and value represents the mean ± SE (*n* = 3). Different letters in each column denote significant differences at *P* < 0.05 by Tukey–Kramer method.

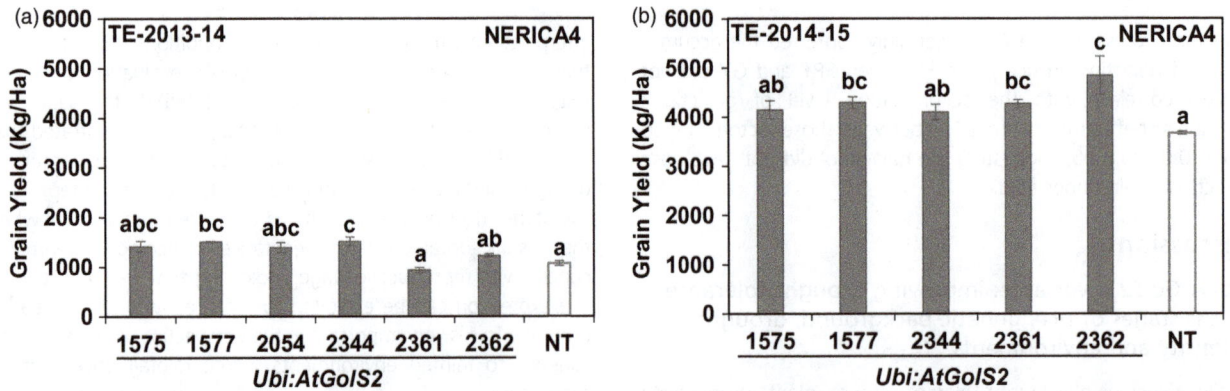

Figure 7 *Ubi:AtGolS2* improves NERICA4 grain yield in rainfed upland trials. (a) Grain yield performance of NERICA4 lines in Target Environment (TE)-2013-14-Trial at Santa Rosa rainfed trial. (b) Grain yield performance of NERICA4 lines in TE-2014-15-Trial at Santa Rosa rainfed trial. Estimated grain yield (kg/ha) derived from plot yield. Estimated grain yield (kg/ha) was derived from plot yield from three replications, and value represents the mean ± SE (*n* = 3). Different letters in each column denote significant differences at *P* < 0.05 by Tukey–Kramer method.

transgenic rice in different seasons, for example rainy and dry and under several environmental conditions such as rainfed and well-watered conditions.

In this study, higher GY in the *Ubi:AtGolS2* transgenic lines was consistently observed over the season and environments (Figures 3, 6 and 7). The promising *Ubi:AtGolS2* transgenic lines

showed significantly enhanced drought tolerance in the field across different genetic backgrounds, Curinga and NERICA4, with a grain yield of 17%–100% higher than NT Curinga under mild to severe drought stress, whereas the transgenic lines displayed no significant differences under normal growth conditions (Figures 3, 6, 7 and Table S1). These improvements of grain yields under drought can be considered greater than what has been reported for other transgenic rice lines expressing genes conferring field drought tolerance, which have often been challenged with milder stresses as demonstrated by the grain yield reduction under drought of the control checks (Oh et al., 2009). However, in this study, the drought intensity was mild because plants were irrigated to evade leaf rolling, which resulted in a yield loss of around 32% in the WT. In another study, rice plants overexpressing *OsNAC10* showed enhanced drought tolerance during the flowering stage and increased grain yield by 25%–42% compared to WT, but again milder drought stress conditions were applied (Jeong et al., 2010).

A limited number of studies applied field drought conditions similar to our work (Todaka et al., 2015). Under severe stress, rice transgenic lines expressing *OsCPI1* showed 2.5- to 3-fold greater GY over the control, for which yield dropped 90% (Huang et al., 2007). Likewise, plants overexpressing *LOS5* and *ZAT10* exhibited gains between 11% and 36% compared to their controls which suffered 82% yield reduction (Xiao et al., 2009).

Through this international collaborative project, we realized the importance of conducting the initial screening efforts in a farmer-adapted variety, because these are popular over large growing areas, locally adapted and because relatively quick introgression of the transgene into other megavarieties is possible (Gaudin et al., 2013). In our study, we attempted to improve two farmer-adapted varieties, one from Latin America (Curinga) and another one from Africa (NERICA4). The phenotype might be controlled by genes (**G**) including transgenes and genetic background (genotypes) in transgenic plants, and plant responses to drought are also influenced by environment (**E**) including intensity, duration and frequency of the stress as well as by diverse plant–soil–atmosphere interactions (Saint Pierre et al., 2012). It is always suggested to test **G** × **G** and **G** × **E** interactions/stability before recommending a potential transgene into the breeding pipeline. In this study, we have conducted drought experiments in two contrasting field seasons (rainy and dry). We found the response of the *Ubi:AtGolS2* lines to be different than the NT control based on the season. However, regardless of the season, the promising lines #2580 and #2590 had significantly greater plant biomass and panicle numbers. A similar result was found within the target environment.

Our results provide strong evidence that overexpression of *AtGolS2* is a useful biotechnological tool to reduce yield losses under field drought conditions under different environmental conditions (**E**) and in different rice genetic backgrounds (**G**), which suggests that *AtGolS2* is an essential gene to improve drought tolerance in rice regardless of **G** × **G** and **G** × **E** interactions.

Mechanism of drought tolerance offered by the *Ubi: AtGolS2* transgene

Even though many stress resistance genes have been identified in noncrop species such as *Arabidopsis*, evaluation of the effect of these genes on improving field drought tolerance in a given crop has seldom been reported (Xiao et al., 2009). GolS plays a key role in the accumulation of galactinol under abiotic stress conditions, conferring drought stress tolerance to plants, because galactinol may function as osmoprotectants and scavenger of hydroxyl radicals (Nishizawa et al., 2008; Taji et al., 2002). Based on our extensive evaluation of transgenic lines, the *Ubi:AtGolS2* transgene improved grain yield of rice under drought conditions. Although high expression of *AtGolS2* and accumulation of

galactinol were confirmed in *Ubi:AtGolS2* events, no significant change in the expression of drought marker genes was induced (Figure 1 and Figure S1). These results suggest that drought tolerance of *Ubi:AtGolS2* transgenic lines with the accumulation of galactinol was not correlated with the expression of drought-responsive genes. This could be contributed through the following mechanisms. First, the transgenic lines are more tolerant to drought and gain yield over the NT because they are protected by elevated galactinol (RFOs) that can act as osmoprotectants and scavenger of hydroxyl radicals (Figure 1b). The increased transcription of *GolS* genes during drought has been reported in many plants and crops. In *Cucumis melo*, it was observed that GolS activates accumulation of RFO in plants submitted to drought stresses (Volk et al., 2003).

As second physiological perspective, the promising Curinga transgenic lines have a better maximum photochemical efficiency (F_v/F_m) and leaf chlorophyll content than NT under drought stress (Figure 4b). The decrease in F_v/F_m and SPAD chlorophyll values under drought stress could be an indicator of oxidative stress and damage in PSII (Farooq et al., 2009). Under severe drought stress, we observed high SPAD chlorophyll and F_v/F_m values in the *Ubi: AtGolS2* Curinga than in NT Curinga, and the *Ubi:AtGolS2* leaves were greener than those of the NT, which confirmed normal photosynthesis in transgenic *Ubi:AtGolS2* rice. In addition to photosynthetic-related traits, stress-related traits such as RWC, leaf rolling and drying of transgenic lines were significantly better than NT Curinga (Tables S2, S3 and S5). Maintenance of high plant water status, as expressed in high RWC of the *Ubi:AtGolS2* rice, was an good indicator of drought tolerance (as shown by Babu et al., 2004), and capacity of transgenic lines maintained higher leaf RWC compared with NT Curinga under drought stress, which was consistent with their ability to postpone dehydration (as indicated by Castonguay and Markhart, 1992). In this study, we observed that the *Ubi:AtGolS2* transgenic lines had higher RWC values than NT rice during peak drought stress and had a lower leaf rolling and leaf drying score (Figure 4a; Tables S2, S3 and S5). By contrast, we observed that NT rice quickly wilted and dried as compared to those transgenic lines with higher RWC values under drought stress.

Third, the *AtGolS2* Curinga transgenic lines showed earlier flowering than NT Curinga (Table S3) under drought stress, but displayed no difference in growth under normal growth conditions (Table S1). In rice, early maturation is an escape mechanism to ensure production under conditions of stress (Gur et al., 2010). In this study, the *Ubi:AtGolS2* Curinga lines showed the earliest flowering and exhibited higher grain fertility (82%) than NT (70%), which may be due to the drought escape mechanism of transgenic lines (data not shown). However, early flowering was not observed in the *Ubi:AtGolS2* NERCA4 background (Table S5). As NERICA4 is a short-duration variety, early flowering of the *Ubi: AtGolS2* lines may be profound in Curinga due to the long duration nature.

Transcription levels of the *AtGolS2* gene of the transgenic lines correlated with the accumulation level of galactinol; however, the accumulation level of galactinol did not correspond with their field performance (Figures 1 and S5). These results suggest that good field performance might not always be associated with levels of gene expression and of accumulation of galactinol. The complexities of environments and other factors influencing performance of rice plants under drought may be reasons for no dosage effects of galactinol on grain yield under field.

Conclusions and prospects

Our study reported extensive field evaluation of transgenic rice plants expressing the *Ubi:AtGolS2* transgene under drought stress environments under field conditions in Colombia. We clearly observed that the *Ubi:AtGolS2* expression and the accumulation of galactinol significantly enhanced grain yield under drought field conditions, but did not affect either grain yield or plant growth under well-watered paddy field conditions. Improved grain yield under stress was associated with early flowering, higher biomass accumulation, higher number of panicles and lower panicle sterility.

In this study, the same gene construct, the *Ubi:AtGolS2,* transgene was tested on two different commercial genetic backgrounds, Curinga and NERICA4, and contrasting different seasons and different environments. We presented the results of extensive confined field testing of transgenic rice overexpressing *AtGolS2* and the responses of these transgenic rice plants to contrasting environments. Notably, we evaluated the agronomic traits of these transgenic lines at all stages of plant growth in the field as a function of the environment and genetic background. As the *Ubi:AtGolS2* transgene was tested in the commercial rice genetic backgrounds of Latin America (Curinga) and Africa (NERICA4), we think it is easy to pyramid the *Ubi:AtGolS2* transgene into ongoing transgenic rice breeding programmes in Latin America and Africa. The promising NERICA4 transgenic lines selected from this study can be integrated into ongoing NEWEST —the NERICA4 (Nitrogen-use Efficient, Water-use Efficient and Salt Tolerant) rice project where extensive transgenic field trials are currently being implemented in Ghana, Uganda and Nigeria through USAID feed the future programme. Development of this drought-tolerant rice through the *Ubi:AtGolS2* transgene should have significant economic and environmental benefits in low-input agricultural systems like Latin America and Africa.

Experimental procedures

Generation of *Ubi:AtGolS2* Plants

To generate transgenic rice plants overexpressing *AtGolS2* encoding galactinol synthase 2 of *Arabidopsis thaliana* (Taji *et al.*, 2002), the pBIG-ubi vector was used (Becker, 1990; Ito *et al.*, 2006). *AtGolS2* cDNA was amplified using *Bam*HI linker primers. The resulting DNA fragment carrying *Bam*HI sites at the 5′ and 3′ termini was inserted into pBIG-ubi at the *Bam*HI site. The construct was introduced into rice cv. Curinga and NERICA4 by *Agrobacterium*-mediated transformation as described previously (Ishizaki and Kumashiro, 2008; Zuniga-Soto *et al.*, 2015). The molecular characterization of putative transgenic events involved PCR and Southern blot analysis. The primers used for this study are reported in Table S6.

Expression analysis of rice plants expressing *Ubi:AtGolS2*

The transgenic and nontransgenic (NT) rice plants (Curinga and NERICA4) were grown in soil-filled, open-bottomed 50-mL plastic tubes in the glasshouse. After the drought treatment, the leaves from five plants were collected, frozen in liquid nitrogen and stored at −80 °C. Total RNA was isolated from the leaf samples using RNAiso Plus reagent (Takara Bio, Shiga, Japan). Extracted RNA was subjected to a DNase treatment using a RQ1 DNase (Promega, WI), and complementary DNA was synthesized using a PrimeScript RT Master Mix (Takara Bio). Real-time quantitative RT-PCR was performed with the QuantStudio 7 Flex real-time PCR

system (Thermo Fisher Scientific, MA) using SYBR Premix Ex Taq (Takara Bio). Primers used for qRT-PCR are listed in Table S6.

Sugar metabolite analysis of rice plants expressing *Ubi: AtGolS2*

The transgenic and control rice lines were grown in soil-filled, open-bottomed 50-ml plastic tubes in the glasshouse. After the drought treatment, the leaves from five plants were collected, frozen in liquid nitrogen and stored at −80 °C. Sugar metabolites were analysed using GC-TOF-MS as described previously by Maruyama *et al.* (2014).

Seedling survival test of NERICA4 transgenic events

The ability of NERICA4 transgenic events to survive under rapid drying was evaluated by the reported method (Ishizaki *et al.*, 2013).

Vegetative drought stress experiment in confined field

To evaluate the drought tolerance of transgenic rice plants at vegetative stage, single-copy independent homozygous lines of *Ubi:AtGolS2* Curinga transgenic lines, together with nontransgenic (NT) Curinga controls, were direct-seeded in confined field conditions under a rainout shelter at the International Center for Tropical Agriculture (CIAT), Palmira, Colombia, in the dry season, November-December 2011. A randomized block design was employed with three replicates with each event sown in two rows placed 16 cm apart in a rainout shelter where the depth of restructured soil was 85 cm. Each row was 1 m long and 40 plants were accommodated in each row with equal spacing (5 cm) between plants. Drought stress was imposed by withholding irrigation at initial tillering stage (21 days after direct sowing) and rewatered after 21 days (3 weeks) until severe wilting symptoms appeared in NT Curinga (Figure 2a). The intensity of drought was monitored through AquaPro soil moisture probes (AquaPro sensors Inc, California, USA). Plant height, the number of tillers and destructive plant dry biomass data were measured from uniform tagged plants at the before, during peak stress and at the end of the harvest after rewatering. Leaf rolling was determined at the time of peak drought stress.

Managed Drought Stress Environment (MDSE) Trial— Rainout Shelter (RS) reproductive Stage trial at CIAT, Palmira, Colombia

All rainout shelter reproductive stress experiments were carried out at our confined field facility at CIAT, Palmira, Colombia. For Curinga, three confined field drought trials (from 2012 to 2014) over two contrasting seasons were conducted under the movable semi-automatic rainout shelter facility. All three experiments were conducted using same protocol with respect to designs and field drought characterization. A randomized block design with three replications was followed to layout the experiment under the rainout shelter facility at CIAT. The seeds of T_4 homozygous lines were sown in the dry soil of the experimental plots in rows (20 cm spacing between rows). Each event was sown in two rows placed 20 cm apart where the depth of restructured soil was 85 cm. Each row was 2 m long and had 20 plants with equal spacing. The recommended fertilizer application for upland rice was used. Drought was imposed by withholding irrigation when panicle initiation was around 10 mm long (60–66 days after sowing in the case of Curinga) for 3–4 weeks (or) until severe leaf rolling and drying appeared in the NT control. Then, the plants were rewatered to 90% field capacity until physiological maturity. The intensity of drought was monitored through AquaPro soil moisture probes that were installed to measure moisture in the soil profile to a depth of 0.85 m.

Leaf rolling (LR), leaf drying (LD) and drought recovery scores were recorded on a 1-9 IRRI scale standardized for rice (IRRI, 2002). The following agronomic traits were measured based on the criteria established in the Standard Evaluation System for Rice (SES) (IRRI, 2002): flowering date, plant height (cm), single plant dry biomass (g), panicle length (cm), the number of tillers, the number of fully emerged panicles and grain fertility (%). Single plant yield (from five more uniform tagged plants from each block with three replications) and plot yield were also recorded.

The degree of relative chlorophyll content in the fully expanded flag leaf was determined while the plant was under stress, using a SPAD-502 Chlorophyll Meter (Konica Minolta Co., Tokyo, Japan). Chlorophyll a fluorescence parameters were also measured using a FluorPen FP100 chlorophyll (Photon Systems Instruments, spol. s r.o., Czech Republic). Relative water content was calculated using the protocol based on Schonfeld et al. (1988).

Well-watered experiment in confined field conditions

To evaluate the yield components of the transgenic Curinga lines under normal well-watered field conditions, selected independent T_4 homozygous lines of *Ubi:AtGolS2* transgenic plants together with NT controls were transplanted to a rice paddy confined field at CIAT, Palmira (dry season, September–January 2013). A randomized design was employed with three replicates of two 2-m-long rows per plot. For each plot, 20 seedlings per line were randomly transplanted with a 20×10 cm spacing 25 days after sowing. The recommended fertilizer application for lowland rice was used. Yield parameters were scored for five tagged uniform plants per plot.

RT-PCR analysis of field samples—Rainout shelter Drought trials

Total RNA was extracted from the flag leaves of tested promising transgenic lines at before stress, peak stress and after rewatering from the 2014-rainy-MDSE-rainout shelter trial using the Trizol reagent (Invitrogen). Reverse transcription was carried out using DNase (Promega) and SuperScript III (Invitrogen). Endpoint PCR was conducted with primers listed in Table S6, using standard protocols and an annealing temperature of 55 °C. PCR products were checked on 1% agarose gel with SYBR-safe stain.

Target Environment (TE) Trial—Confined Field Evaluation of Transgenic lines at Santa Rosa, Colombia

To evaluate yield components of transgenic plants under rainfed upland conditions with natural drought condition, the most promising independent T_4 homozygous lines of Curinga and NERICA4 from the previous drought experiments along with their NT controls were evaluated in a replicated plot trial with randomized block design from 2013 to 2015 (three consecutive field trials) at CIAT Santa Rosa rainfed upland station, Colombia. Promising NERICA4 lines were selected based on the survival test results from Japan and rainout shelter trials. An upland field trial was laid out in a random complete block design with three replicates. The transgenic events along with NT rice were sown in 2×2 m plots with 25×10 cm spacing. Seeds were sown directly by hand at the rate of 120 kg/ha when soil moisture was about 80% of field capacity. The recommended fertilizer application for upland rice was used. The plants were established in dry soil and irrigation was provided until 50 days after sowing (DAS) via sprinklers to establish the crop. After plant establishment, irrigation was stopped and plants were totally dependent on rainfall. The soil moisture was monitored throughout the cropping period by Aquapro soil moisture device. Plant growth and development of each of the transgenic events relative to NT rice was monitored regularly and plot yield (g) was recorded. Grain yield (kg/ha) was estimated from plot yield based on plant density.

Data analysis

Data were analysed by one-factor ANOVA at $P < 0.05$. When significant differences were found, multiple comparisons by the Tukey–Kramer method ($P < 0.05$) were made.

Acknowledgements

We are grateful for technical support provided by Emiko Kishi and Miho Kishimoto of JIRCAS. We also thank Dr. Jagadish Ranne, AL.Chavez, Cesar Zuluaga, Santiago Jaramillo and Maria Recio of CIAT for the technical assistance and Dr. Joe Tohme, Director of Agrobiodiversity Research Area of CIAT, for critical discussion on field phenotyping. We also thank Richard Bruton, Texas A&M, and Angela Fernando, CIAT, for their critical English edition and suggestions of the manuscript. This work was supported by the grants from the Ministry of Agriculture, Forestry and Fisheries (MAFF) of Japan for a project: Development of Drought-Tolerant Crops for Developing Countries. The authors declare no conflict of interest.

Author contributions

K.S., K.N. and M.I designed the total experiments. M. G. S., S.O. and M.O.V. planned, conducted and analysed drought field experiments. B.D. and M.I. conducted transformation experiments at CIAT. T.I. conducted transformation experiments and glasshouse experiments at JIRCAS. T.O. and K.N. conducted gene expression experiments. K.Y., K. M., M. K. and K. S. conducted sugar analysis. F.T. made the construct for transformation. M.G.S. wrote the manuscript, and all the authors checked it.

Ethical standards

The authors declare that the transgenic experiments comply with the current biosafety laws of the country in which they were performed.

References

Ahuja, I., de Vos, R.C., Bones, A.M. and Hall, R.D. (2010) Plant molecular stress responses face climate change. *Trends Plant Sci.* **15**, 664–674.

Babu, R.C., Zhang, J., Blum, A., Ho, T.-H.D., Wu, R. and Nguyen, H. (2004) HVA1, a LEA gene from barley confers dehydration tolerance in transgenic rice (*Oryza sativa* L.) via cell membrane protection. *Plant Sci.* **166**, 855–862.

Bahieldin, A., Mahfouz, H.T., Eissa, H.F., Saleh, O.M., Ramadan, A.M., Ahmed, I.A., Dyer, W.E. et al. (2005) Field evaluation of transgenic wheat plants stably expressing the HVA1 gene for drought tolerance. *Physiol. Plantarum*, **123**, 421–427.

Becker, D. (1990) Binary vectors which allow the exchange of plant selectable markers and reporter genes. *Nucleic Acids Res.* **18**, 203.

Brandle, J., McHugh, S., James, L., Labbe, H. and Miki, B. (1995) Instability of transgene expression in field grown tobacco carrying the csr1-1 gene for sulfonylurea herbicide resistance. *Nat. Biotechnol.* **13**, 994–998.

Cassman, K.G., Dobermann, A., Walters, D.T. and Yang, H. (2003) Meeting cereal demand while protecting natural resources and improving environmental quality. *Annu. Rev. Environ. Resour.* **28**, 315–358.

Castonguay, Y. and Markhart, A.H. (1992) Leaf gas exchange in water-stressed common bean and tepary bean. *Crop Sci.* **32**, 980–986.

Collett, H., Shen, A., Gardner, M., Farrant, J.M., Denby, K.J. and Illing, N. (2004) Towards transcript profiling of desiccation tolerance in *Xerophyta humilis*: construction of a normalized 11 k *X. humilis* cDNA set and microarray expression analysis of 424 cDNAs in response to dehydration. *Physiol. Plantarum*, **122**, 39–53.

Edmeades, G.O. (2008) Drought tolerance in maize: an emerging reality: a feature in James, Clive. 2008. Global Status of Commercialized Biotech/GM Crops: 2008. In: *ISAAA Brief No. 39*. (Greg O., Edmeades, ed), pp. 1–11. Ithaca, NY: ISAAA.

FAO (2016) *Rice Market Monitor*, April 2016, Volume XIX Issue No. 1 available at http://www.fao.org/fileadmin/templates/est/COMM_MARKETS_MONITORING/Rice/Images/RMM/RMM_APR16.pdf.

Farooq, M., Wahid, A., Kobayashi, N., Fujita, D. and Basra, S.M.A. (2009) Plant drought stress: effects, mechanisms and management. *Agron. Sustain. Dev.* **29**, 185–212.

Farrant, J.M. (2007) Mechanisms of desiccation tolerance in angiosperm resurrection plants. In *Plant Desiccation Tolerance* (Jenks, M.A. and W.A.J., eds), pp. 51–90. Oxford, UK: Blackwell.

Gaudin, A.C., Henry, A., Sparks, A.H. and Slamet-Loedin, I.H. (2013) Taking transgenic rice drought screening to the field. *J. Exp. Bot.* **64**, 109–117.

Godfray, H.C.J., Beddington, J.R., Crute, I.R., Haddad, L., Lawrence, D., Muir, J.F., Pretty, J. *et al.* (2010) Food security: the challenge of feeding 9 billion people. *Science*, **327**, 812–818.

Gur, A., Osorio, S., Fridman, E., Fernie, A.R. and Zamir, D. (2010) *hi2-1*, A QTL which improves harvest index, earliness and alters metabolite accumulation of processing tomatoes. *Theor. Appl. Genet.* **121**, 1587–1599.

Hare, P., Cress, W. and Van Staden, J. (1998) Dissecting the roles of osmolyte accumulation during stress. *Plant, Cell Environ.* **21**, 535–553.

Huang, Y., Xiao, B. and Xiong, L. (2007) Characterization of a stress responsive proteinase inhibitor gene with positive effect in improving drought resistance in rice. *Planta*, **226**, 73–85.

IRRI. (2002) *Standard Evaluation System (SES)*, pp. 11–30. Manila, Philippines: International Rice Research Institute.

Ishizaki, T. and Kumashiro, T. (2008) Genetic transformation of NERICA, interspecific hybrid rice between *Oryza glaberrima* and *O. sativa*, mediated by *Agrobacterium tumefaciens*. *Plant Cell Rep.* **27**, 319–327.

Ishizaki, T., Maruyama, K., Obara, M., Fukutani, A., Yamaguchi-Shinozaki, K., Ito, Y. and Kumashiro, T. (2013) Expression of *Arabidopsis DREB1C* improves survival, growth, and yield of upland New Rice for Africa (NERICA) under drought. *Mol Breed*, **31**, 255–264.

Ito, H., Iwamoto, I., Morishita, R., Nozawa, Y., Asano, T. and Nagata, K.-I. (2006) Identification of a PDZ protein, PIST, as a binding partner for Rho effector Rhotekin: biochemical and cell-biological characterization of Rhotekin–PIST interaction. *Biochem J.* **397**, 389–398.

Jeong, J.S., Kim, Y.S., Baek, K.H., Jung, H., Ha, S.-H., Do Choi, Y., Kim, M. *et al.* (2010) Root-specific expression of *OsNAC10* improves drought tolerance and grain yield in rice under field drought conditions. *Plant Physiol.* **153**, 185–197.

Jeong, J.S., Kim, Y.S., Redillas, M.C., Jang, G., Jung, H., Bang, S.W., Choi, Y.D. *et al.* (2013) *OsNAC5* overexpression enlarges root diameter in rice plants leading to enhanced drought tolerance and increased grain yield in the field. *Plant Biotech. J.* **11**, 101–114.

Manavalan, L.P., Chen, X., Clarke, J., Salmeron, J. and Nguyen, H.T. (2012) RNAi-mediated disruption of squalene synthase improves drought tolerance and yield in rice. *J. Exp. Bot.* **63**, 163–175.

Marshall, A. (2014) Drought-tolerant varieties begin global march. *Nat. Biotechnol.* **32**, 308.

Maruyama, K., Urano, K., Yoshiwara, K., Morishita, Y., Sakurai, N., Suzuki, H., Kojima, M. *et al.* (2014) Integrated analysis of the effects of cold and dehydration on rice metabolites, phytohormones, and gene transcripts. *Plant Physiol.* **164**, 1759–1771.

Nakashima, K., Yamaguchi-Shinozaki, K. and Shinozaki, K. (2014) The transcriptional regulatory network in the drought response and its crosstalk in abiotic stress responses including drought, cold, and heat. *Front. Plant Sci.* **5**, 170.

Nishizawa, A., Yabuta, Y. and Shigeoka, S. (2008) Galactinol and raffinose constitute a novel function to protect plants from oxidative damage. *Plant Physiol.* **147**, 1251–1263.

Oh, S.J., Kim, Y.S., Kwon, C.W., Park, H.K., Jeong, J.S. and Kim, J.K. (2009) Overexpression of the Transcription Factor AP37 in Rice Improves Grain Yield

under Drought Conditions. *Plant Physiol.* **150**, 1368–1379.

O'Toole, J.C. (2004) Rice and water: the final frontier. In *First International Conference on Rice for the Future*, (The University, 2004, ed), pp. 1–26. Bangkok, Thailand: Rockefeller Foundation.

Panikulangara, T.J., Eggers-Schumacher, G., Wunderlich, M., Stransky, H. and Schöffl, F. (2004) *Galactinol synthase1*. A novel heat shock factor target gene responsible for heat-induced synthesis of raffinose family oligosaccharides in Arabidopsis. *Plant Physiol.* **136**, 3148–3158.

Peters, S., Mundree, S.G., Thomson, J.A., Farrant, J.M. and Keller, F. (2007) Protection mechanisms in the resurrection plant *Xerophyta viscosa* (Baker): both sucrose and raffinose family oligosaccharides (RFOs) accumulate in leaves in response to water deficit. *J. Exp. Bot.* **58**, 1947–1956.

Redillas, M.C., Jeong, J.S., Kim, Y.S., Jung, H., Bang, S.W., Choi, Y.D., Ha, S.H. *et al.* (2012) The overexpression of *OsNAC9* alters the root architecture of rice plants enhancing drought resistance and grain yield under field conditions. *Plant Biotech. J.* **10**, 792–805.

Saint Pierre, C., Crossa, J.L., Bonnett, D., Yamaguchi-Shinozaki, K. and Reynolds, M.P. (2012) Phenotyping transgenic wheat for drought resistance. *J. Exp. Bot.* **63**, 1799–1808.

Saravitz, D.M., Pharr, D.M. and Carter, T.E. (1987) Galactinol synthase activity and soluble sugars in developing seeds of four soybean genotypes. *Plant Physiol.* **83**, 185–189.

Schonfeld, M.A., Johnson, R.C., Carver, B.F. and Mornhinweg, D.W. (1988) Water relations in winter wheat as drought resistance indicators. *Crop Sci.* **28**, 526–531.

Seck, P.A., Diagne, A., Mohanty, S. and Wopereis, M.C. (2012) Crops that feed the world 7: rice. *Food Secur.* **4**, 7–24.

Sengupta, S., Mukherjee, S., Basak, P. and Majumder, A.L. (2015) Significance of galactinol and raffinose family oligosaccharide synthesis in plants. *Front. Plant Sci.* **6**, 656.

Serraj, R., McNally, K.L., Slamet-Loedin, I., Kohli, A., Haefele, S.M., Atlin, G. and Kumar, A. (2011) Drought resistance improvement in rice: an integrated genetic and resource management strategy. *Plant Prod. Sci.* **14**, 1–14.

Taji, T., Ohsumi, C., Iuchi, S., Seki, M., Kasuga, M., Kobayashi, M., Yamaguchi Shinozaki, K. *et al.* (2002) Important roles of drought-and cold-inducible genes for galactinol synthase in stress tolerance in Arabidopsis thaliana. *Plant J.* **29**, 417–426.

Todaka, D., Nakashima, K., Maruyama, K., Kidokoro, S., Osakabe, Y., Ito, Y., Matsukura, S. *et al.* (2012) Rice phytochrome-interacting factor-like protein OsPIL1 functions as a key regulator of internode elongation and induces a morphological response to drought stress. *Proc. Natl Acad. Sci. USA*, **109**, 15947–15952.

Todaka, D., Shinozaki, K. and Yamaguchi-Shinozaki, K. (2015) Recent advances in the dissection of drought-stress regulatory networks and strategies for development of drought-tolerant transgenic rice plants. *Front. Plant Sci.* **6**, 84.

Venuprasad, R., Lafitte, H. and Atlin, G. (2007) Response to direct selection for grain yield under drought stress in rice. *Crop Sci.* **47**, 285–293.

Volk, G.M., Haritatos, E.E. and Turgeon, R. (2003) Galactinol synthase gene expression in melon. *J. Am. Soc. Hortic. Sci.* **128**, 8–15.

Xiao, B.-Z., Chen, X., Xiang, C.-B., Tang, N., Zhang, Q.-F. and Xiong, L.-Z. (2009) Evaluation of seven function-known candidate genes for their effects on improving drought resistance of transgenic rice under field conditions. *Mol. Plant*, **2**, 73–83.

Yamaguchi-Shinozaki, K. and Shinozaki, K. (2006) Transcriptional regulatory networks in cellular responses and tolerance to dehydration and cold stresses. *Annu. Rev. Plant Biol.* **57**, 781–803.

You, J., Hu, H. and Xiong, L. (2012) An ornithine δ-aminotransferase gene *OsOAT* confers drought and oxidative stress tolerance in rice. *Plant Sci.* **197**, 59–69.

Younis, A., Siddique, M.I., Kim, C.-K. and Lim, K.-B. (2014) RNA interference (RNAi) induced gene silencing: a promising approach of hi-tech plant breeding. *Int. J. Biol. Sci.* **10**, 1150.

Yu, L., Chen, X., Wang, Z., Wang, S., Wang, Y., Zhu, Q., Li, S. *et al.* (2013) Arabidopsis *enhanced drought tolerance1/HOMEODOMAIN GLABROUS11* confers drought tolerance in transgenic rice without yield penalty. *Plant Physiol.* **162**, 1378–1391.

Zuniga-Soto, E., Mullins, E. and Dedicova, B. (2015) Ensifer-mediated transformation: an efficient non-Agrobacterium protocol for the genetic modification of rice. *SpringerPlus*, **4**, 600.

7

Indel-seq: a fast-forward genetics approach for identification of trait-associated putative candidate genomic regions and its application in pigeonpea (*Cajanus cajan*)

Vikas K. Singh[1,†], Aamir W. Khan[1,†], Rachit K. Saxena[1,†], Pallavi Sinha[1,†], Sandip M. Kale[1], Swathi Parupalli[1], Vinay Kumar[1], Annapurna Chitikineni[1], Suryanarayana Vechalapu[1], Chanda Venkata Sameer Kumar[1], Mamta Sharma[1], Anuradha Ghanta[2], Kalinati Narasimhan Yamini[2], Sonnappa Muniswamy[3] and Rajeev K. Varshney[1,4,*]

[1]*International Crops Research Institute for the Semi-Arid Tropics, Patancheru, Telangana State, India*

[2]*Agricultural Research Station (ARS)-Tandur, Professor Jayashankar Telangana State Agricultural University (PJTSAU), Hyderabad, Telangana State, India*

[3]*Agricultural Research Station (ARS)-Gulbarga, University of Agricultural Sciences (UAS), Raichur, Karnataka, India*

[4]*School of Plant Biology and Institute of Agriculture, The University of Western Australia, Crawley, WA, Australia*

*Correspondence

email r.k.varshney@cgiar.org
[†]Authors contributed equally to this work.

Summary

Identification of candidate genomic regions associated with target traits using conventional mapping methods is challenging and time-consuming. In recent years, a number of single nucleotide polymorphism (SNP)-based mapping approaches have been developed and used for identification of candidate/putative genomic regions. However, in the majority of these studies, insertion–deletion (Indel) were largely ignored. For efficient use of Indels in mapping target traits, we propose Indel-seq approach, which is a combination of whole-genome resequencing (WGRS) and bulked segregant analysis (BSA) and relies on the Indel frequencies in extreme bulks. Deployment of Indel-seq approach for identification of candidate genomic regions associated with fusarium wilt (FW) and sterility mosaic disease (SMD) resistance in pigeonpea has identified 16 Indels affecting 26 putative candidate genes. Of these 26 affected putative candidate genes, 24 genes showed effect in the upstream/downstream of the genic region and two genes showed effect in the genes. Validation of these 16 candidate Indels in other FW- and SMD-resistant and FW- and SMD-susceptible genotypes revealed a significant association of five Indels (three for FW and two for SMD resistance). Comparative analysis of Indel-seq with other genetic mapping approaches highlighted the importance of the approach in identification of significant genomic regions associated with target traits. Therefore, the Indel-seq approach can be used for quick and precise identification of candidate genomic regions for any target traits in any crop species.

Keywords: bulked segregant analysis, fusarium wilt, Indels, sterility mosaic disease, whole-genome resequencing.

Introduction

Conventional trait mapping methods are generally expensive and take much time in generating and analysing genotyping data on segregating populations. Trait mapping becomes more time-consuming if genotyping is performed using low-throughput marker systems such as simple sequence repeat (SSR) markers. Visual scoring in such marker systems also adds to the possibility of discovering spurious marker–trait associations (MTAs). High-throughput marker systems such as single nucleotide polymorphism (SNP) in combination with automated genotyping platforms (SNP arrays, KASpar assays, GoldenGate assays, etc.) have provided better options in generation of genotyping data. However, downstream analysis of such large volume data (quality assessment, identification of parental polymorphism and subsequently assessment of informative SNPs in population) takes time to provide meaningful information, which can be used for MTAs. This limits the rapid deployment of high probability MTAs in genomics-assisted breeding (GAB) and, subsequently, delays development of new

breeding lines (Varshney *et al.*, 2007). Additionally, meeting the increasing demand of nutritious food under anticipated climate change scenario along with ever-decreasing agricultural lands and limited water resources is a challenging task (Khoury *et al.*, 2014). It requires sophisticated rapid genome mapping and targeted GAB approaches to produce better and high-yielding crop varieties in faster manner (Godfray, 2010; Varshney *et al.*, 2005).

The rapid development of next-generation sequencing (NGS) technologies has enabled generation of genomic resources at large scale with faster pace during the last decade (Pazhamala *et al.*, 2015; Pandey *et al.*, 2016). NGS-based approaches have also provided rapid ways to establish relationship between genotype and phenotype at higher resolution (Varshney *et al.*, 2014). Nevertheless, despite the decreasing sequencing cost, development of individual reference-based assembly for each accession in a given species or progeny of mapping populations is still a challenging task. To overcome this bottleneck and to identify genomic segments responsible for phenotypic traits using NGS, many alternative approaches such as SHOREmap

(Schneeberger *et al.*, 2009), Next-generation mapping (NGM) (Austin *et al.*, 2011), MutMap (Abe *et al.*, 2012), Isogenic mapping-by-sequencing (Hartwig *et al.*, 2012), SNP-ratio mapping (SRM) (Lindner *et al.*, 2012), MutMap+ (Fekih *et al.*, 2013), MutMap-Gap (Takagi *et al.*, 2013a) have been used. Abovementioned studies rely on a number of different principles, which can handle mainly qualitative traits (traits governed by 1-2 genes). In contrast, QTL-seq approach was proposed primarily to deal with quantitative traits, based on Δ SNP index to map the target genomic region(s) for blast resistance and seedling vigour in rice (Takagi *et al.*, 2013b). Similarly, whole-genome resequencing (WGRS)-based BSA was applied to calculate G' statistics to identify the QTLs for cold tolerance in rice seedling (Yang *et al.*, 2013). Recently, genome resequencing of contrasting parents together with identification of nonsynonymous SNP (nsSNP) substitution was utilized for identification of candidate genes in defined QTL regions or new genic regions in many crops (Silva *et al.*, 2012; Singh *et al.*, 2016a; Xu *et al.*, 2014). To list a few, nsSNP substitution approach has been successfully utilized in mapping the candidate genes for sheath blight resistance in rice (Silva *et al.*, 2012), drought tolerance in maize (Xu *et al.*, 2014) fusarium wilt (FW) and sterility mosaic disease (SMD) resistance in pigeonpea (Singh *et al.*, 2016a). In all these studies, SNP genotyping data were used for establishing MTAs. However, Indels in the genomic regions based on bulked segregant sequencing have not yet been targeted for trait mapping. Evidence of involvement of Indels in altering the gene functions has been reported in different crops (see Kage et al., 2015). Further, in comparison with other markers, Indels have a number of inherent advantages such as abundance in the genome, multiallelic and codominant, ease in marker conversion and amenable to low-cost genotyping.

In view of above, this study reports a novel approach called 'Indel-seq', which is a combination of WGRS and BSA, for the identification of Indels associated with target traits. An example of application of Indel-seq has been provided in pigeonpea with FW and SMD resistance as target traits. In this context, the extreme bulks (resistant and susceptible) along with the resistant parents of recombinant inbred lines (RILs) segregating for FW and SMD resistance were sequenced. Candidate genomic regions/genes were identified for FW and SMD resistance in pigeonpea using Indel-seq approach. Further, the identified Indels were validated on a set of FW- and SMD-resistant and FW- and SMD-susceptible genotypes. In summary, Indel-seq seems to be a suitable approach for coarse as well as fine mapping of quantitative traits in a rapid and precise manner.

Results

Principle of Indel-seq

Indel-seq combines WGRS and BSA to identify the genomic regions associated with the target traits. To initiate Indel-seq approach, any segregating population (F$_2$/RILs/back-cross) for the target traits could be utilized. Based on the phenotypic data of segregating population, 15–20 lines of extreme classes can be selected to constitute DNA pools in high trait bulk (HTB) and low trait bulk (LTB). Subsequently, two bulks (HTB and LTB) along with the high trait parent (HTP) are subjected to WGRS with high genome coverage (~10×) (Figure 1). WGRS data, subsequently, can be analysed in a proposed manner to detect trait(s)-associated Indels. As the first step in analysis,

high-quality WGRS data from HTP, HTB and LTB are mapped to the reference genome (RG). Mapped/aligned data are used for the identification of genomewide Indels. Identified Indels are then subjected to high-quality filtering parameters such as Q value >30, homozygous and no 'N' (missing call) in any tested sample. Further homozygous Indels supported by a minimum of seven sequencing reads in both the bulks (HTB and LTB) can be selected for establishing MTAs. In this direction, each Indel could be passed through the either (i) or (ii) of following criteria:

1. RG = HTP = HTB ≠ LTB; here RG is similar to HTP. Indel should be selected if call is similar in RG, HTP and HTB and alternative call in LTB.
2. HTP = HTB ≠ LTB = RG; here RG is similar to LTP. Indel should be selected if similar call is present in HTP and HTB and contrasting call in LTB and RG.

Further selected Indels based on above principles are subjected to chi-square (χ^2) analysis to check their goodness of fit ratio, that is 1:1 in HTB and LTB. A significant deviation from the normally expected ratio of any Indel would indicate the possible association with the target trait. Effect of significantly associated Indels on genes and genomic regions can be predicted through SnpEff (http://snpeff.sourceforge.net/) software.

Application of Indel-seq approach in pigeonpea

Extreme pools for Indel-seq

To deploy Indel-seq in pigeonpea for detecting the candidate genomic regions/genes for FW and SMD resistance, available sequencing and phenotypic data were utilized in this study (Singh *et al.*, 2016a). In brief, phenotyping data generated for resistance to FW and SMD on the RIL population, that is ICPL 20096 (resistant to FW and SMD, HTP) × ICPL 332 (susceptible to FW and SMD, LTP), were used for defining resistant bulk (HTB) and susceptible bulk (LTB) of 16 individual RILs in each group (Figures 2 and S1–S2). Using WGRS, a total of 9.27, 8.99 and 8.43 Gb data were generated for the resistant parent or HTP and HTB and LTB, respectively (Table S1). Cleaned data were aligned to the pigeonpea reference genome resulting in mapping of total 90.6% (HTP), 81.8% (HTB) and 82.5% (LTB) of the total high-quality reads. Genome coverage was found to be 89.21% in HTP, 87.72% in HTB and 87.37% in LTB with an average depth of 13.4 X in HTP, 11.4 X in HTB and 10.8 X in LTB (Table S1).

Candidate Indels

Resequencing data sets for HTP, HTB and LTB were aligned with reference genome (RG) for identification of Indels (Varshney *et al.*, 2012a). As a result, 211 603 genomewide Indels were identified. Of 211 603 Indels, 89 261 were identified on the pseudomolecules and the remaining were present on CcLG0 or floating scaffolds.

A total of 88,867 Indels with Q value >30 were selected for downstream analysis (Table S2 and Figure S3-S5). The lengths of these Indels were ranged from 1 bp to 99 bp (Figure S6). Indels with heterozygous and 'N' (missing) calls in the HTP, HTB and LTB were also discarded, and a set of 33 577 Indels was subjected to further filtration. After applying final filtering criteria, that is Indels with read depth ≥7 were selected, the number of Indels reduced to 14 408 across HTP, HTB and LTB. On pairwise analysis, a total of 1290 Indels were identified between HTB and LTB. These Indels were checked for the concept, that is RG = HTP = HTB ≠ LTB. As

(a) Mapping population development

Low trait parent (LTP)

High trait parent (HTP)

Single seed descent

F_2 ← • • → F_1 ● ● ● ● ● → RILs

(b) Phenotyping of mapping population

No of individuals

Percent disease incidence

(c) HTB   LTB

(d) Sequencing of both bulks and HTP

(e) Alignment of HTP, HTB and LTB reads on reference genome for identification of Indels

(f) BSA analysis using genome wide Indels

Indels filtering criteria

(i) Total Indels identified among HTP, HTB & LTB
(ii) Total Indels identified among HTP, HTB & LTB with Q value ≥ 30
(iii) Total homozygous Indels among HTP, HTB & LTB
(iv) Indels supported by a minimum read depth ≥7 (HTB & LTB)
(v) Presence of homozygous polymorphic Indels in both bulks (HTB & LTB)
(vi) Calls following RG=HTP=HTB≠LTB
(vii) Select position ≥ 6.63 Chi-square value in both bulks (HTB & LTB)

(g) List of putatively linked Indels

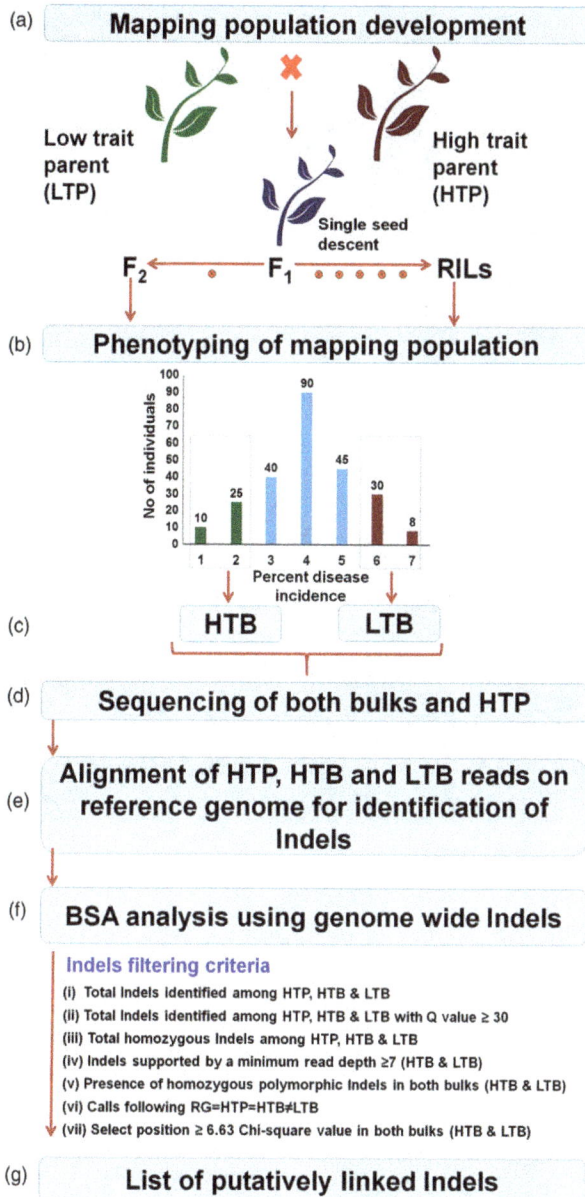

Figure 1 Pipeline of Indel-seq approach. (a) Two contrasting parents (high trait parent (HTP) and low trait parent (LTP) are crossed to develop segregating population (F_2/RILs) for target traits). (b) Based on the phenotyping of mapping population for the target traits, ~20 plants with extreme phenotype are selected for the constitution of extreme pools. (c) Low trait bulk (LTB) and high trait bulk (HTB) are constructed based on equimolar bulking of ~20 individuals of DNA for each bulk. (d) These two DNA bulks along with high trait parent (HTP) are used to whole-genome resequencing. (e) Raw reads of HTP, HTB and LTB are aligned to the reference genome (RG, which is similar to high trait parent in term of target phenotype) for the identification of Indels. (f) Bulked segregant analysis (BSA) approach is applied to identify the associated Indels with the target traits using several Indel filtering criteria to identify putatively associated Indels between resistance and susceptibility. (g) List of putatively linked Indels identified.

a result, 464 putative Indels were identified. Based on chi-square test of the 464 Indels, only 16 Indels showed chi-square values ≥6.63 depicted to have an association with traits of interest

(Figure 3). The chi-square values in HTB ranged from 7 (P-value: 0.008151) to 12 (P-value: 0.000532) and in LTB ranged from 7 (P value: 0.008151) to 14.22 (P-value: 0.000163) (Table 1). These 16 Indels were found affecting 26 genes (Table 1). Of 26 affected genes, 24 genes showed the effect in the upstream/downstream of the genic region and two genes have effect at genic level (Table S3). Few of these candidate genes have been reported to play significant role in the defence mechanisms in other plant species (Table S3).

Validation of candidate Indels

To validate and classify the identified 16 candidate Indels associated with the target genes for FW and SMD resistance, a comparative analysis based on allele frequencies in available sequence data was performed among four additional FW/SMD-resistant and FW/SMD-susceptible genotypes along with HTP, RG, HTB and LTB (Table 2). As a result, of 16 candidate Indels, five with an effect on eight candidate genes were validated (Table 2).

Indels for FW resistance

Three Indels, one each on CcLG02, CcLG07 and CcLG08, were found to be associated with FW resistance. For instance, one-bp deletion identified on CcLG02 (at position 1 253 647 bp) showed 'C' allele in FW-resistant genotypes and HTB, whereas 'CA' allele was present in LTB and FW-susceptible genotype (ICPB 2049) with a P-value <0.001. The identified one-bp insertion in susceptible genotypes ('C' to 'CA') was found to be affecting AP-1 complex subunit sigma-2 (C.cajan_05665) and L-ascorbate oxidase (C.cajan_05665) at upstream and downstream regions, respectively. At 405 527bp position on CcLG07, 'AT' allele was identified in HTB- and FW-resistant genotypes and 'A' allele identified in LTB and susceptible genotype (ICPB 2049) with a P-value of <0.001. This single-bp deletion ('AT' to 'A') in susceptible genotypes showed an effect at intronic region and targeting receptor-like protein kinase (C.cajan_17341). On CcLG08 (at position 7 106 619 bp), one-bp deletion was observed in HTB- and FW-resistant genotypes ('T' allele) in comparison with LTB and FW-susceptible genotypes ('TG' allele) (with P-value <0.001). The insertion of one bp ('T' to 'TG') in susceptible genotypes has shown the effect on two genes (C.cajan_16014; Transcriptional corepressor SEUSS and C.cajan_16015; Uncharacterized protein).

Indels for SMD resistance

For SMD resistance, Indel-seq analysis has provided two associated Indels, one each on CcLG02 and CcLG10. On CcLG02 at 14 020 849 bp position, one-bp insertion in HTB and SMD resistance genotypes ('CA' allele) was detected. In the case of LTB and susceptible genotype (ICP 8863), 'C' allele was present (with P-value <0.001). The identified one-bp deletion ('CA' to 'C') in susceptible genotypes targeting two genes (C.cajan_05815 at upstream and C.cajan_05816 at downstream region), and both the genes were annotated as conserved oligomeric Golgi complex subunit 5. Similarly, On CcLG10 (at position 18 889 276 bp) one-bp insertion was observed in HTB- and SMD-resistant genotypes ('AT' allele) in comparison with LTB and SMD-susceptible genotype ('A' allele) (with P-value <0.001). This single-bp deletion in susceptible genotypes ('AT' to 'A') showed frame-shift effect in an uncharacterized protein (C.cajan_15032).

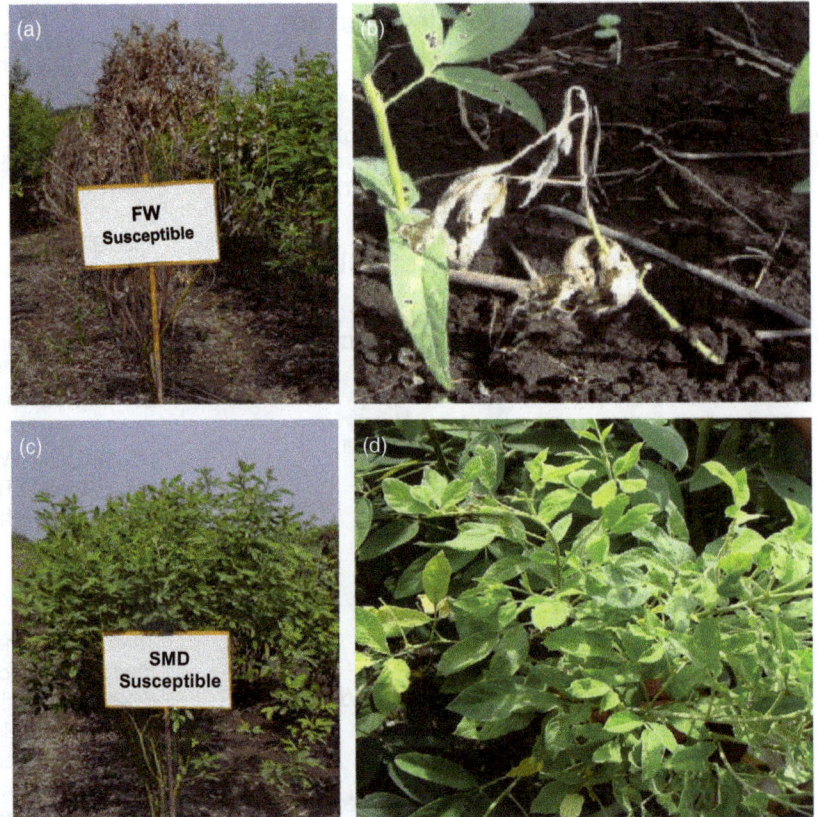

Figure 2 Phenotypic reaction of resistant and susceptible *Fusarium* wilt (FW) and sterility mosaic disease (SMD) plants. FW is a seed and soil borne fungal disease caused by *Fusairum udum.* Wilt symptoms usually appear when plants are in flowering and podding stage (a), but sometimes occur earlier when plants are 1-2-month-old (b). SMD is a viral disease caused by *Pigeonpea sterility mosaic virus (PSMV)*. This disease can be easily identified from a distance as patches of bushy, pale green plants (c) without flower or pods (d). Due to excess vegetative growth, without growing into reproductive phase, this disease is known as the *green plague* of pigeonpea.

Figure 3 Flow diagram of Indel-seq analysis for identification of candidate genes for FW and SMD resistance in pigeonpea. (a) Whole-genome resequencing of the resistant parent (HTP), resistant bulk (HTB) and susceptible bulk (LTB) was performed with more than ≥10× genome coverage. (b) The generated raw reads of HTP, HTB and LTB were aligned with the reference genome (RG) for identification of genomewide Indels. The value presented in the funnel is the number of Indels identified/selected in each step, which is further classified as insertion (I) and deletion (D) (c) Total number of Indels identified after mapping of HTP, HTB and LTB on RG. (d) Further, only those Indels were selected, which possess ≥30 quality score. (e) Only homozygous Indels among HTP, HTB and LTB bulks were selected for further analysis (f) To remove false-positive associations, only those Indels were selected which possesses reads ≥7 at both the bulk positions. (g) Homozygous polymorphic Indels were identified between both the bulks. (h) The classical concept of bulked segregant analysis (BSA) approach was implemented (RG = HTP = HTB ≠ LTB) for identification of putatively associated Indels (see text for the explanation). (i) Chi-square test at 99% probability level was performed at each selected positions based on the presence of reads at selected Indel positions to select trait-associated Indels.

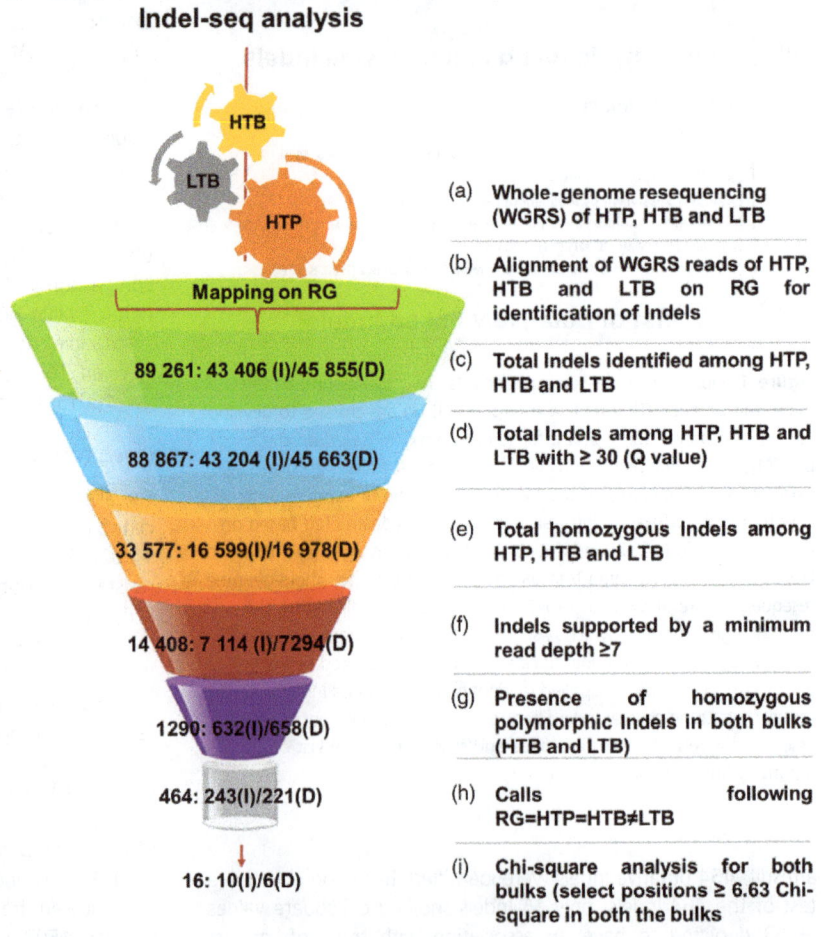

Indel-seq analysis

Mapping on RG

89 261: 43 406 (I)/45 855(D)

88 867: 43 204 (I)/45 663(D)

33 577: 16 599(I)/16 978(D)

14 408: 7 114 (I)/7294(D)

1290: 632(I)/658(D)

464: 243(I)/221(D)

16: 10(I)/6(D)

(a) **Whole-genome resequencing (WGRS) of HTP, HTB and LTB**

(b) **Alignment of WGRS reads of HTP, HTB and LTB on RG for identification of Indels**

(c) **Total Indels identified among HTP, HTB and LTB**

(d) **Total Indels among HTP, HTB and LTB with ≥ 30 (Q value)**

(e) **Total homozygous Indels among HTP, HTB and LTB**

(f) **Indels supported by a minimum read depth ≥7**

(g) **Presence of homozygous polymorphic Indels in both bulks (HTB and LTB)**

(h) **Calls following RG=HTP=HTB≠LTB**

(i) **Chi-square analysis for both bulks (select positions ≥ 6.63 Chi-square in both the bulks**

Table 1 Identification of Indels between resistant and susceptible bulks using Indel-seq approach

Gene*	Type[†]	Linkage group	Position (bp)	RG[‡] base	HTP[§] base	Resistant bulk					Susceptible bulk			
						HTB[¶] base	Read depth	χ^2 value	P value		LTB[‖] base	Read depth	χ^2 value	P value
C.cajan_05665 (d)	I	CcLG02	12 535 647	C	C	C	15	8.07	<0.001		CA	14	9.94	<0.001
C.cajan_05666 (u)														
C.cajan_05815 (d)	D	CcLG02	14 020 849	CA	CA	CA	7	7.00	<0.001		C	15	8.00	<0.001
C.cajan_05816 (u)														
C.cajan_05857 (u)	I	CcLG02	14 397 213	A	A	A	19	8.89	<0.001		AT	18	7.12	<0.001
C.cajan_05858 (d)														
C.cajan_06311 (d)	I	CcLG02	19 386 341	T	T	T	14	7.14	<0.001		TC	11	8.00	<0.001
C.cajan_09080 (u)	D	CcLG03	10 887 279	GTA	GTA	GTA	13	9.31	<0.001		G	16	9.80	<0.001
C.cajan_11099 (d)	I	CcLG06	890 690	A	A	A	12	12.00	<0.001		AT	12	9.00	<0.001
C.cajan_11101 (u)														
C.cajan_11323 (u)	I	CcLG06	3 364 388	C	C	C	8	8.00	<0.001		CT	7	8.33	<0.001
C.cajan_11324 (d)														
C.cajan_17341 (i)	D	CcLG07	405 527	AT	AT	AT	8	8.00	<0.001		A	14	7.14	<0.001
C.cajan_16014 (u)	I	CcLG08	7 106 619	T	T	T	8	8.00	<0.001		TG	20	7.00	<0.001
C.cajan_16015 (d)														
C.cajan_16060 (d)	I	CcLG08	7 820 397	C	C	C	19	11.84	<0.001		CCAACAA	11	10.29	<0.001
C.cajan_22308 (u)	I	CcLG09	2 209 342	A	A	A	14	7.14	<0.001		AT	11	11.00	<0.001
C.cajan_22309 (d)														
C.cajan_14502 (u)	I	CcLG10	13 435 965	C	C	C	13	9.31	<0.001		CA	17	8.07	<0.001
C.cajan_14503 (d)														
C.cajan_14515 (u)	D	CcLG10	13 516 086	TTA	TTA	TTA	15	8.07	<0.001		T	8	14.22	<0.001
C.cajan_14516 (d)														
C.cajan_15032 (f)	D	CcLG10	18 889 276	AT	AT	AT	18	8.00	<0.001		A	17	7.36	<0.001
C.cajan_01566 (u)	D	CcLG11	17 030 340	CA	CA	CA	19	8.89	<0.001		C	8	7.36	<0.001
C.cajan_01567 (d)														
C.cajan_02069 (u)	I	CcLG11	22 814 098	G	G	G	15	8.07	<0.001		GT	11	7.36	<0.001

*Gene: u: upstream region; d: downstream region: i, intron; f, frame shift.

[†]Type of Indels: 'I' stand for insertion and 'D' stand for deletion.

[‡]RG: Reference genome (Asha; ICPL 87119) (http://www.icrisat.org/gt-bt/iipg/genomedata.zip).

[§]HTP: Resistant parent (ICPL 20096).

[¶]HTB: Resistant bulk.

[‖]LTB: Susceptible bulk.

Discussion

NGS-based genome mapping enables identification of candidate genomic regions/genes in a rapid way, which is often difficult using traditional methods in terms of time and resources required (Varshney et al., 2012b). Recently, a number of SNP-based approaches combining BSA and WGRS have been successfully developed and implemented to identify the target candidate genes (see Zou et al. 2016). In the present study, an Indel-seq approach has been proposed for the identification of candidate genes/Indels associated with target traits. This approach has been tested in pigeonpea for rapid identification of candidate genes associated with the FW and SMD resistance.

To enable WGRS-based identification of candidate genes using mapping-by-sequencing approach, several methods have been developed and discussed in different crops (Abe et al., 2012; Austin et al., 2011; Hartwig et al., 2012; Nordström et al., 2013; Schneeberger et al., 2009; Takagi et al., 2013a; Trick et al., 2012). Based on the published literature and through large-scale simulation studies, James et al. (2013) developed user guide for mapping-by-sequencing. Among different NGS-based

approaches, QTL-seq approach provided the first successful example of mapping candidate genomic regions through NGS-based approach in crop species like rice (Takagi et al., 2013b). QTL-seq approach was found successful for identification of candidate genomic regions (SNPs) for FW and SMD resistance in pigeonpea (Singh et al., 2016a) and 100-seed weight and root trait ratio (RTR %) in chickpea (Singh et al., 2016b). However, in the majority of above-mentioned studies, Indels have been ignored. For effective applications of Indels in trait mapping, we propose here Indel-seq approach that is a combination of WGRS and BSA. Deployment of Indel-seq approach has been used for identification of candidate genomic regions associated with FW and SMD resistance in the present study.

Application of Indel-seq approach for identification of trait-associated Indels

Two types of genetic variations, namely SNPs and Indels, are the most promising variations and used in the trait mapping studies in a number of crops (Huang et al., 2012; Li et al., 2013; Thudi et al., 2014). In the recent past, NGS-based trait mapping approaches utilizing a large number of SNPs generated through

Table 2 Validation of candidate Indels in four known (resistant and susceptible) genotypes for FW and SMD resistance

Linkage group	Indel positions (bp)	RG*		HTP†		HTB‡		LTB§		ICPB 2049		ICPL 99050		ICPL 20097		ICP 8863		P-value for FW resistance	P-value for SMD resistance
		FW-R¶	SMD-R¶	FW-R¶	SMD-R¶	FW-R¶	SMD-R¶	FW-S‖	SMD-S‖	FW-S‖	SMD-R¶	FW-R¶	SMD-R¶	FW-R¶	SMD-R¶	FW-R¶	SMD-S‖		
CcLG02	12 535 647	C		C		C		CA		CA		C**		C		C		**<0.00**	0.42
CcLG02	14 020 849	CA		CA		CA		C		CA		CA		CA		C		0.42	**<0.00**
CcLG02	14 397 213	A		A		A		AT		A		A		A		A		0.08	0.08
CcLG02	19 386 341	T		T		T		TC		T		T		T		T		0.08	0.08
CcLG03	10 887 279	GTA		GTA		GTA		G		G		G		G		G		0.27	0.27
CcLG06	890 690	A		A		A		AT		A		A		A		A		0.08	0.08
CcLG06	3 364 388	C		C		C		CT		C		C		C		C		0.08	0.08
CcLG07	405 527	AT		AT		AT		A		A		AT		AT		AT		**<0.00**	0.42
CcLG08	7 106 619	T		T		T		TG		TG		T		T		T		**<0.00**	0.42
CcLG08	7 820 397	C		C		C		CCAACAA		C		C		C		C		0.08	0.08
CcLG09	2 209 342	A		A		A		AT		AT		AT		AT		AT		0.27	0.27
CcLG10	13 435 965	C		C		C		CA		C		C		C		C		0.08	0.08
CcLG10	13 516 086	TTA		TTA		TTA		T		TTA		TTA		TTA		TTA		0.08	0.08
CcLG10	18 889 276	AT		AT		AT		A		AT		AT**		AT		A**		0.04	**<0.00**
CcLG11	17 030 340	CA		CA		CA		C		C		C		C		C		0.27	0.27
CcLG11	22 814 098	G		G		G		GT		G		G		G		G		0.08	0.08

*RG: Reference genome (Asha; ICPL 87119) (http://www.icrisat.org/gt-bt/iipg/genomedata.zip).

†HTP: Resistant parent (ICPL 20096).

‡HTB: Resistant bulk.

§LTB: Susceptible bulk.

¶R: resistant reaction.

‖S: susceptible reaction.

**Heterozygous calls.

P-value <0.00 (boldface) found significant for specific disease resistance.

resequencing/genotyping have been used for trait mapping (Varshney et al., 2014). SNP-based mapping approaches identified candidate genes for the target traits in many reports but identification of a large number of cloned genes with the presence of functional Indels through map-based cloning experiments for different traits in different crops revealed the importance of Indels for trait mapping and development of functional markers (Kage et al., 2015).

Comparative analysis of Indel-seq approach with other NGS-based QTL mapping approaches combining WGRS and BSA revealed some pros and cons over other methods of trait mapping (Table S4). The additional advantage of Indel-seq mapping approach is to map the candidate genes in the population developed by crossing gamma-induced mutants with the wild types due to the presence of genomewide Indels in the genome. Another important feature of Indel-seq is the high probability of development of PCR-based markers for trait mapping. The rapid fall in the cost of sequencing will facilitate application of Indel-seq for trait mapping in diploid crops with relatively smaller genomes such as rice (389 Mb), chickpea (738 Mb), sorghum (818 Mb), pigeonpea (833 Mb). However, analysis of data sets for complex and large genome species requires some additional modification in the selection criteria of Indel for marker–trait association analysis.

Indel-seq analysis in pigeonpea for mapping FW and SMD resistance has been very effective as it overcomes many constraints like identification of polymorphic markers between parents, the time required for genotyping of the mapping population, preparation of the (low density) genetic maps, and identification of QTLs (with large intervals). WGRS data of parental line and bulks revealed a higher number of genomewide Indels; however, comparatively low mapping percentage and genome coverage was obtained after aligning the raw sequences to the reference genome. This lower mapping and coverage percentage might be due to sequencing library used, sequencing errors, structural rearrangements or insertions in the query genome or deletions in the reference, a high percentage of repetitive elements (Sims et al., 2014) and quality of the reference genome. WGRS analysis of resistant parent and both the bulks revealed 89 261 genomewide Indels and 33 577 Indels between the bulks (HTB vs LTB), which further narrowed down to 1290 Indels, based on stringent selection criteria (read depth and homozygosity of calls in the bulks). The number of Indels was further reduced to 464 based on Indel-seq principle. However, this number is comparatively higher than the previous SMD resistance mapping experiments in which only 120 and 78 SSRs were found polymorphic in two mapping populations after screening of 3000 SSR markers (Gnanesh et al., 2011). Finally, based on chi-square analysis 16 candidate Indels were identified targeting 26 different candidate genes. The Indel-seq pipeline discussed in this report is very simple and after mapping raw reads to the reference genome, analysis can be done using simple Perl Script or in Microsoft Excel program (2010 and above).

Identification of significant genomic regions for FW- and SMD-resistant breeding

To check the efficiency of Indel-seq in identifying possible candidates (markers/genes) for the target traits, we have also used identified Indels (Table S5) in a recently proposed method known as EXPLoRA-web BSA (Duitama et al., 2014). We have significant results from each of the models proposed in EXPLoRA-web BSA (Tables S6). As expected, the lowest number of QTLs

was reported in the high sensitive model ($\alpha = 5$, $\beta = 1$) and highest number of QTLs in the high specific model ($\alpha = 30$, $\beta = 1$). Interestingly, 12 of 16 candidate Indels identified through Indel-seq approach were found common in EXPLoRA-web BSA analysis (Table S7). Moreover, from the five validated Indels in the present study, four were also found in EXPLoRA-web BSA analysis. This has enhanced our confidence in proposing Indel-seq as a possible approach for fast trait mapping experiments. However, it is important to mention that EXPLoRA-web BSA has provided a large number of possible Indels' associations, which directly cannot be applied for genomics-assisted breeding (GAB) programmes, whereas Indel-seq has provided reasonable numbers of high confidence MTAs (three for FW and two for SMD) which can be converted into KASP markers. After validation of KASP markers, it can be utilized in GAB for development of FW- and SMD-resistant pigeonpea genotypes.

Conclusions

It is evident from the present study that identification of candidate genes for targeted traits based on NGS will not only increase the precision and power but also generate results in less time than the conventional methods of genome mapping. In near future due to rapid declining of sequencing cost and availability of high-quality draft genome sequences in several crops, we envisage application of Indel-seq for trait mapping and GAB for crop improvement. Identified target genes and associated Indels in the present study were validated on defined sets of genotypes for which sequence data were available. These results after validation on larger sets of genotypes will be useful in guiding diseases resistance breeding efforts in pigeonpea.

Materials and methods

Plant materials and construction of pools

Six pigeonpea genotypes were selected based on their FW and SMD responses identified from our previous experiments (Saxena et al., 2010a; Singh et al., 2016a; Varshney et al., 2012a). Among the selected genotypes, ICPL 20096, ICPL 20097, ICPL 8863, ICPL 99050 and ICPL 87119 were FW resistant and ICPL 20096, ICPL 20097, ICPL 99050, ICPB 2049 and ICPL 87119 were SMD resistant. Similarly, among the six genotypes, ICPB 2049 was FW susceptible, and ICP 8863 was SMD susceptible. Two genotypes ICPL 20096 (FW and SMD resistant) and ICP 332 (FW and SMD susceptible) with contrasting phenotypes were crossed and selfed through single seed descent method to develop 188 F_7 recombinant inbred lines (RILs).

These RILs were phenotyped for FW and SMD resistance using standard procedures as mentioned in Nene and Reddy (1976) and Singh et al. (2003). The detailed descriptions on sick plot nursery, filed design and construction of bulks have been presented in Singh et al. (2016a).

Sequencing libraries and alignment of short reads of bulks

Raw sequencing data of ICPL 20096 (resistant parent or HTP) and resistant bulk or HTB and susceptible bulk or LTB were used for Indel-seq analysis (Table S2). The generated paired end reads of 251 bp lengths were cleaned using the tool Sickle (Joshi and Fass, 2012) with minimum phred quality score of 30 and minimum read length of 70 bp. The reads containing 'Ns' were also removed. The clean data of samples were used to align to the pigeonpea reference genome (Varshney *et al.*, 2012a) using BWA: Burrows–Wheeler Aligner (Li and Durbin, 2009) to get the Sequence Alignment/Map (SAM)/BAM (Binary Alignment/Map) alignment files, which results in alignment files in BAM format. The BAM files were further processed for Indel realignment using IndelRealigner component of Genome Analysis Toolkit (GATK; McKenna *et al.*, 2010), and Picard utility was used for adding read group information. These processed BAM files were then subjected for the variants calling through GATK (DePristo *et al.*, 2011) using standard parameters for the parent and both the bulks. The identified genomewide variants were further used for Indel-seq analysis for the identification of MTAs.

Mining of resequencing data sets for validation

To validate the candidate SNPs, resequencing data sets of four genotypes, namely ICPL 20097 (R-FW and R-SMD) and ICP 8863 (R-FW and S-SMD), ICPB 2049 (S-FW and R-SMD) and ICPL 99050 (R-FW and R-SMD), were used to find out the genes/markers unique to FW and SMD (Kumar *et al.*, 2016). To test the association, *p*-value was calculated between identified Indels with the target traits using single factor ANOVA in Microsoft Excel 2013.

EXPLoRA-web BSA

EXPLoRA-web BSA works upon the principle of LD to detect QTLs using Hidden Markov Model (HMM) (Duitama *et al.*, 2014). Genomewide mapping reads of susceptible bulk (LTB) onto the reference genome (RG) was utilized to develop input files for EXPLoRA-web BSA analysis. Only those positions were selected for analyses, which were supported by a minimum of 10 reads. LTB was chosen for BSA in the present analysis because RG was resistant to both the diseases (FW and SMD). To control the EXPLoRA-web models, three different parameters were utilized for identification of QTLs (i) $\alpha = 5$; $\beta = 1$ (high sensitivity) (ii) $\alpha = 10$; $\beta = 1$ (the middle ground between sensitivity and specificity) and (iii) $\alpha = 30$; $\beta = 1$ (high specificity). The α/β ratio determines the shape of the β distribution in the models, which reflects the probability for the phenotype-linked states (Pulido-Tamayo *et al.*, 2016).

Acknowledgements

Authors are thankful to the United States Agency for International Development (USAID); Biotechnology Industry Partnership Programme (BIPP) and Department of Biotechnology of Government of India; Ministry of Agriculture, Government of Karnataka state of India for funding various projects related to pigeonpea genomics. This work has been undertaken as part of the CGIAR Research Program on Grain Legumes. ICRISAT is a member of CGIAR Consortium.

References

Abe, A., Kosugi, S., Yoshida, K., Natsume, S., Takagi, H., Kanzaki, H., Matsumura, H. *et al.* (2012) Genome sequencing reveals agronomically important loci in rice using MutMap. *Nat. Biotechnol.* **30**, 174–178.

Austin, R.S., Vidaurre, D., Stamatiou, G., Breit, R., Provart, N.J., Bonetta, D., Zhang, J. *et al.* (2011) Next-generation mapping of Arabidopsis genes. *Plant J.* **67**, 715–725.

DePristo, M.A., Banks, E., Poplin, R., Garimella, K.V., Maguire, J.R., Hartl, C., Philippakis, A.A. *et al.* (2011) A framework for variation discovery and genotyping using next-generation DNA sequencing data. *Nat. Genet.* **43**, 491–498.

Duitama, J., Sánchez-Rodríguez, A., Goovaerts, A., Pulido-Tamayo, S., Hubmann, G., Foulquié-Moreno, M.R., Thevelein, J.M. *et al.* (2014) Improved linkage analysis of Quantitative Trait Loci using bulk segregants unveils a novel determinant of high ethanol tolerance in yeast. *BMC Genom.* **15**, 1–15.

Fekih, R., Takagi, H., Tamiru, M., Abe, A., Natsume, S., Yaegashi, H., Sharma, S. *et al.* (2013) MutMap+: genetic mapping and mutant identification without crossing in rice. *PLoS ONE*, **8**, e68529.

Gnanesh, B.N., Bohra, A., Sharma, M., Byregowda, M., Pande, S., Wesley, V., Saxena, R. K., et al. (2011) Genetic mapping and quantitative trait locus analysis of resistance to sterility mosaic disease in pigeonpea [Cajanus cajan (L.) Millsp.]. *Field Crops Res.*, **123**, 56–61.

Godfray, H.C.J. (2010) Food security: the challenge of feeding 9 billion people. *Science*, **327**, 812–818.

Hartwig, B., James, G.V., Konrad, K., Schneeberger, K. and Turck, F. (2012) Fast isogenic mapping-by-sequencing of ethyl methane sulfonate-induced mutant bulks. *Plant Physiol.* **160**, 591–600.

Huang, X., Zhao, Y., Wei, X., Li, C., Wang, A., Zhao, Q., Li, W. *et al.* (2012) Genome-wide association study of flowering time and grain yield traits in a worldwide collection of rice germplasm. *Nature Genet.* **44**, 32–39.

James, G.V., Patel, V., Nordström, K.J., Klasen, J.R., Salomé, P.A., Weigel, D. and Schneeberger, K. (2013) User guide for mapping-by-sequencing in Arabidopsis. *Genome Biol.* **14**, 1.

Joshi, N. and Fass, J. N. (2011). *Sickle-A windowed adaptive trimming tool for FASTQ files using quality.* Available at: https://github.com/najoshi/sickle (accessed 5 Feb, 2016).

Kage, U., Kumar, A., Dhokane, D., Karre, S. and Kushalappa, A.C. (2015) Functional molecular markers for crop improvement. *Crit Rev Biotechnol.*, **16**, 1–14.

Khoury, C.K., Bjorkman, A.D., Dempewolf, H., Ramirez-Villegas, J., Guarino, L., Jarvis, A., Rieseberg, L.H. *et al.* (2014) Increasing homogeneity in global food supplies and the implications for food security. *Proc. Natl Acad. Sci. USA*, **111**, 4001–4006.

Kumar, V., Khan, A.W., Saxena, R.K., Garg, V. and Varshney, R.K. (2016) First-generation HapMap in Cajanus spp. reveals untapped variations in parental lines of mapping populations. *Plant Biotechnol. J.* **14**, 1673–1681.

Li, H. and Durbin, R. (2009) Fast and accurate short read alignment with Burrows-Wheeler transform. *Bioinformatics*, **25**, 1754–1760.

Li, H., Peng, Z., Yang, X., Wang, W., Fu, J., Wang, J., Han, Y. *et al.* (2013) Genome-wide association study dissects the genetic architecture of oil biosynthesis in maize kernels. *Nat. Genet.* **45**, 43–50.

Lindner, H., Raissig, M.T., Sailer, C., Shimosato-Asano, H., Bruggmann, R. and Grossniklaus, U. (2012) SNP-ratio mapping (SRM): identifying lethal alleles and mutations in complex genetic backgrounds by next-generation sequencing. *Genetics*, **191**, 1381–1386.

McKenna, A., Hanna, M., Banks, E., Sivachenko, A., Cibulskis, K., Kernytsky, A., Garimella, K. *et al.* (2010) The genome analysis toolkit: a MapReduce framework for analyzing next-generation DNA sequencing data. *Genome Res.* **20**, 1297–1303.

Nene, Y.L. and Reddy, M.V. (1976) A new technique to screen pigeonpea for resistance to sterility mosaic. *Trop. Grain Legume Bull.* **5**, 23.

Nordström, K.J., Albani, M.C., James, G.V., Gutjahr, C., Hartwig, B., Turck, F., Paszkowski, U. *et al.* (2013) Mutation identification by direct comparison of whole-genome sequencing data from mutant and wild-type individuals using k-mers. *Nature Biotechnol.* **31**, 325–330.

Pandey, M.K., Roorkiwal, M., Singh, V.K., Ramalingam, A., Kudapa, H., Thudi, M., Chitikineni, A. *et al.* (2016) Emerging genomic tools for legume breeding: current status and future prospects. *Front. Plant Sci.* **7**, 455.

Pazhamala, L., Saxena, R.K., Singh, V.K., Sameerkumar, C.V., Kumar, V., Sinha, P., Patel, K. *et al.* (2015) Genomics-assisted breeding for boosting crop improvement in pigeonpea (*Cajanus cajan*). *Front. Plant Sci.* **6**, 50.

Pulido-Tamayo, S., Duitama, J. and Marchal, K. (2016) EXPLoRA-web: linkage analysis of quantitative trait loci using bulk segregant analysis. *Nucleic Acids Res.* **44**, W142–W146.

Saxena, R.K., Saxena, K.B., Kumar, R.V., Hoisington, D.A. and Varshney, R.K. (2010a) Simple sequence repeat-based diversity in elite pigeonpea genotypes for developing mapping populations to map resistance to Fusarium wilt and sterility mosaic disease. *Plant Breed.* **129**, 135–241.

Schneeberger, K., Ossowski, S., Lanz, C., Juul, T., Petersen, A.H., Nielsen, K.L., Jørgensen, J.E. *et al.* (2009) SHOREmap: simultaneous mapping and mutation identification by deep sequencing. *Nat. Methods*, **6**, 550–551.

Silva, J., Scheffler, B., Sanabria, Y., De Guzman, C., Galam, D., Farmer, A., Woodward, J. *et al.* (2012) Identification of candidate genes in rice for resistance to sheath blight disease by whole genome sequencing. *Theor. Appl. Genet.* **124**, 63–74.

Sims, D., Sudbery, I., Ilott, N.E., Heger, A. and Ponting, C.P. (2014) Sequencing depth and coverage: key considerations in genomic analyses. *Nat. Rev. Genet.* **15**, 121–132.

Singh, I.P., Vishwadhar, and Dua, R.P. (2003) Inheritance of resistance to sterility mosaic in pigeonpea (*Cajanus cajan*). *Indian J. Agric. Sci.* **73**, 414–417.

Singh, V.K., Khan, A.W., Saxena, R.K., Kumar, V., Kale, S.M., Sinha, P., Chitikineni, A. *et al.* (2016a) Next-generation sequencing for identification of candidate genes for Fusarium wilt and sterility mosaic disease in pigeonpea (*Cajanus cajan*). *Plant Biotechnol. J.* **14**, 1183–1194.

Singh, V.K., Khan, A.W., Jaganathan, D., Thudi, M., Roorkiwal, M., Takagi, H., Garg, V. *et al.* (2016b) QTL-seq for rapid identification of candidate genes for 100-seed weight and root/total plant dry weight ratio under rainfed conditions in chickpea. *Plant Biotechnol. J.* **14**, 2110–2119.

Takagi, H., Uemura, A., Yaegashi, H., Tamiru, M., Abe, A., Mitsuoka, C., Utsushi, H. *et al.* (2013a) MutMap-Gap: wholegenome resequencing of mutant F$_2$ progeny bulk combined with de novo assembly of gap regions identifies the rice blast resistance gene Pii. *New Phytol.* **200**, 276–283.

Takagi, H., Abe, A., Yoshida, K., Kosugi, S., Natsume, S., Mitsuoka, C., Uemura, A. *et al.* (2013b) QTL-seq: rapid mapping of quantitative trait loci in rice by whole genome resequencing of DNA from two bulked populations. *Plant J.* **74**, 174–183.

Thudi, M., Upadhyaya, H.D., Rathore, A., Gaur, P.M., Krishnamurthy, L., Roorkiwal, M., Nayak, S.N. *et al.* (2014) Genetic dissection of drought and heat tolerance in chickpea through genome-wide and candidate gene-based association mapping approaches. *PLoS ONE*, **9**, e96758.

Trick, M., Adamski, N.M., Mugford, S.G., Jiang, C.-C., Febrer, M. and Uauy, C. (2012) Combining SNP discovery from next-generation sequencing data with bulked segregant analysis (BSA) to fine-map genes in polyploidy wheat. *BMC Plant Biol.* **12**, 14.

Varshney, R.K., Graner, A. and Sorrells, M.E. (2005) Genomics-assisted breeding for crop improvement. *Trends Plant Sci.* **10**, 621–630.

Varshney, R.K., Chabane, K., Hendre, P.S., Aggarwal, R.K. and Graner, A. (2007) Comparative assessment of EST-SSR, EST-SNP and AFLP markers for evaluation of genetic diversity and conservation of genetic resources using wild, cultivated and elite barleys. *Plant Sci.* **173**, 638–649.

Varshney, R.K., Chen, W., Li, Y., Bharti, A.K., Saxena, R.K., Schlueter, J.A., Donoghue, M.T. *et al.* (2012a) Draft genome sequence of pigeonpea (*Cajanus cajan*), an orphan legume crop of resource-poor farmers. *Nat. Biotechnol.* **30**, 83–89.

Varshney, R.K., Ribaut, J.M., Buckler, E.S., Tuberosa, R., Rafalski, J.A. and Langridge, P. (2012b) Can genomics boost productivity of orphan crops? *Nat. Biotechnol.* **30**, 1172–1176.

Varshney, R.K., Terauchi, R. and McCouch, S.R. (2014) Harvesting the promising fruits of genomics: applying genome sequencing technologies to crop breeding. *PLoS Biol.* **12**, e1001883.

Xu, J., Yuan, Y., Xu, Y., Zhang, G., Guo, X., Wu, F., Wang, Q. *et al.* (2014) Identification of candidate genes for drought tolerance by whole-genome resequencing in maize. *BMC Plant Biol.* **14**, 83.

Yang, Z., Huang, D., Tang, W., Zheng, Y., Liang, K., Cutler, A.J. and Wu, W. (2013) Mapping of quantitative trait loci underlying cold tolerance in rice seedlings via high-throughput sequencing of pooled extremes. *PLoS ONE*, **8**, e68433.

Zou, C., Wang, P. and Xu, Y. (2016) Bulked sample analysis in genetics, genomics and crop improvement. *Plant Biotechnol. J.* **14**, 1941–1955.

8

Mapping of homoeologous chromosome exchanges influencing quantitative trait variation in *Brassica napus*

Anna Stein[1,*], Olivier Coriton[2], Mathieu Rousseau-Gueutin[2], Birgit Samans[1], Sarah V. Schiessl[1], Christian Obermeier[1], Isobel A.P. Parkin[3], Anne-Marie Chèvre[2] and Rod J. Snowdon[1]

[1]*Department of Plant Breeding, IFZ Research Centre for Biosystems, Land Use and Nutrition, Justus Liebig University, Giessen, Germany*
[2]*IGEPP, INRA, Agrocampus Ouest, Université de Rennes 1, Le Rheu, France*
[3]*Agriculture and Agri-Food Canada, Saskatoon, Canada*

Correspondence
email anna.stein@agrar.uni-giessen.de

Keywords: Genome rearrangements, homoeologous exchange, genetic mapping, quantitative trait loci, single nucleotide polymorphism.

Summary

Genomic rearrangements arising during polyploidization are an important source of genetic and phenotypic variation in the recent allopolyploid crop *Brassica napus*. Exchanges among homoeologous chromosomes, due to interhomoeologue pairing, and deletions without compensating homoeologous duplications are observed in both natural *B. napus* and synthetic *B. napus*. Rearrangements of large or small chromosome segments induce gene copy number variation (CNV) and can potentially cause phenotypic changes. Unfortunately, complex genome restructuring is difficult to deal with in linkage mapping studies. Here, we demonstrate how high-density genetic mapping with codominant, physically anchored SNP markers can detect segmental homoeologous exchanges (HE) as well as deletions and accurately link these to QTL. We validated rearrangements detected in genetic mapping data by whole-genome resequencing of parental lines along with cytogenetic analysis using fluorescence *in situ* hybridization with bacterial artificial chromosome probes (BAC-FISH) coupled with PCR using primers specific to the rearranged region. Using a well-known QTL region influencing seed quality traits as an example, we confirmed that HE underlies the trait variation in a DH population involving a synthetic *B. napus* trait donor, and succeeded in narrowing the QTL to a small defined interval that enables delineation of key candidate genes.

Introduction

Brassica napus (rapeseed, oilseed rape, canola) is a very recent allopolyploid species that since its origin has become one of the world's most important crops. The species was formed by hybridization and genome doubling from the diploid donor genomes of *Brassica oleracea* and *Brassica rapa*, respectively. Because this cross can be readily reproduced with the help of embryo rescue techniques, *B. napus* has become a popular model for studying the genetic and genomic consequences of *de novo* allopolyploidization and how these have shaped natural and agricultural selection in a modern crop (Mason and Chèvre, 2016).

Segmental exchanges between homoeologous chromosomes are frequent throughout the *B. napus* genome. Numerous small-scale homoeologous exchanges (HE) are observed throughout the genomes of natural *B. napus* accessions, whereas large-scale HE are common in synthetic accessions (Chalhoub et al., 2014; Rousseau-Gueutin et al., 2017). In fact, resynthesized lines are specifically prone to homoeologous rearrangements, including deletions, duplications and translocations (Gaeta et al., 2007; Szadkowski et al., 2010; Xiong et al., 2011). Other than the assumption that resynthesized rapeseed displays some kind of accelerated oilseed rape evolution, it is unknown whether different mechanisms take place in natural and resynthesized oilseed rape.

Implementation of synthetic *B. napus* in breeding is an interesting, yet challenging strategy to overcome the extreme narrow genetic diversity in modern rapeseed breeding pools. Causes of the genetic bottlenecks are the small number of founder allopolyploidization events during the origin of the *B. napus* species (Allender and King, 2010), strong adaptive selection in strict eco-geographic gene pools and intensive agronomic selection during recent breeding for essential seed quality traits (Snowdon et al., 2015). The diploid progenitor species harbour important variation particularly for disease resistance (e.g. Mei et al., 2015; Rygulla et al., 2007a,b; Werner et al., 2008) and improvement of the heterotic potential (Snowdon et al., 2015). Unfortunately, the rich genetic diversity available through *de novo* resynthesis of *B. napus* carries the price of a heavy genetic load, with poor fertility and agronomic performance. Although marker-assisted backcrossing can accelerate incorporation of such exotic materials, the complex genome restructuring common in synthetic *B. napus* makes such germplasm particularly difficult to deal with in linkage mapping studies.

Furthermore, because most marker systems commonly used for genetic mapping do not assay presence–absence variation (PAV) or copy number variation (CNV) automatically, the extent to which this kind of variation underlies QTL for important agronomic traits is still completely unknown in *B. napus*. Most genetic linkage maps constructed using high-density SNP array data do not consider CNV or PAV (e.g. Delourme *et al.*, 2013; Fopa *et al.*, 2014; Liu *et al.*, 2013; Raman *et al.*, 2014) although both are inherent phenomena in chromosome regions shaped by HEs or deletions. The consequence may be that genome regions affected by genomic rearrangements including deletions, which sometimes span entire chromosomes in synthetic *B. napus* (Chalhoub *et al.*, 2014), may not be incorporated into genetic maps and result in large gaps in linkage groups.

Accurate mapping of genomic rearrangement events is essential for mapping of associated QTL and evaluation of their impact in allopolyploid crops. Mason *et al.* (2017) proposed guidelines for scoring of SNP calling results that suggest presence–absence variation (PAV), corresponding with the expected segregation ratio, in a segregating mapping population. Today, high-density SNP data array, like that generated using the Brassica 60K Illumina Infinium genotyping array (Clarke *et al.*, 2016), enable rapid, high-resolution mapping of large *B. napus* mapping populations at low cost. Due to the large numbers of polymorphic markers that can be assayed using this array, it is easy for users to generate highly dense genetic linkage maps, even when markers that show unexpected segregation or excessive quantities of failed SNP calls are excluded.

Similarly, hemi-SNPs (Trick *et al.*, 2009) are frequently encountered between homoeologous loci in *B. napus*. A hemi-SNP results from either (i) a simultaneous hybridization of the marker to two homoeologous loci or (ii) a simultaneous hybridization of the marker to duplicated fragments containing a SNP mutation, where only one locus in each instance is polymorphic. These variants generally exceed common threshold limits for tolerated heterozygosity in segregating mapping populations, meaning that hemi-SNP data tend to be automatically discarded from genetic mapping data sets. On the other hand, inclusion of expected segregation ratios between heterozygous and homozygous individuals enables genotypes to be resolved at a single-locus level and used for map calculation in a mapping population.

Physical anchoring of SNP loci provides positional information that can improve confidence in calling of deletion and duplication events based on missing SNP calls or hemi-SNPs, respectively. Although missing calls due to technical failures are reasonably common in array-based genotyping systems, and repeated failure of the same SNP may point to technical problems with the assay, it is unlikely that two or more physically adjacent SNPs will by chance show the same segregation patterns of technical failures vs. successful amplifications of one or the other SNP allele. Similarly, the calling of a hemi-SNP does not necessarily indicate presence of a duplication, but may also occur as the consequence of unspecific SNP probe hybridization. Nevertheless, using positionally anchored markers in genetic mapping indicates putatively rearranged genomic regions. Those regions may be validated by genomic resequencing.

Mapping of short-read genomic resequencing data to the recently published *B. napus* reference genome sequence (Chalhoub *et al.*, 2014) enables comprehensive analysis regarding nucleotide polymorphisms and gene content, but also and importantly identification of presence–absence variation (PAV) in *B. napus* samples. However, neither reciprocal nor nonreciprocal rearrangements between the homoeologous subgenomes can be detected from resequencing data *per se*. Nevertheless, segmental deletions that show corresponding duplications of the homoeologous chromosome segment imply that this incidence is a homoeologous exchange (HE).

Fluorescence *in situ* hybridization using bacterial artificial chromosome probes (BAC-FISH) in *B. napus* enables chromosome painting using BAC probes containing subgenome-specific repeat sequences (Leflon *et al.*, 2006), or even molecular karyotyping based on chromosome-specific probes (Xiong and Pires, 2011). Multicolour combination of chromosome-specific BACs and subgenome-specific BACs enables unambiguous chromosome identification *in situ*. Thus, comparing the combination of FISH signals from specific BAC clones within putative HE regions can distinguish individuals, carrying single copies of the exchanged region, from individuals with duplicated copies of one homoeologous locus and/or deleted copies of the other homoeologue. BAC-FISH therefore represents an interesting method for independent validation of HE events imputed from genomic sequencing reads or molecular marker segregation data. As homoeologous reciprocal translocations cannot be validated by genome resequencing, however in principle they can be detected by genetic mapping and BAC-FISH.

Results

Natural and synthetic *B. napus* parental genotypes exhibit widespread genomic rearrangements

The aligned resequencing data yielded a generally uniform coverage over the lengths of each chromosome in the respective mapping parents. This facilitated calling of putatively deleted and duplicated segments, which showed consistent patterns of coverage either lower or higher than the chromosome-wide average and a minimum length of 50 kb. As described by Chalhoub *et al.* (2014), we found widespread evidence for deletions, duplications and HEs among homoeologous A-subgenome and C-subgenome chromosomes. Also, as expected, the natural *B. napus* accession Express 617 showed the lowest degree of segmentation (685 deleted or duplicated segments), whereas the two synthetic *B. napus* parents exhibited a considerably higher segmentation degree (821 and 1,630 deletions or duplications, for 1012-98 and R53, respectively). The semisynthetic parent V8, derived from backcrossing of a synthetic *B. napus* to natural *B. napus*, showed the expected intermediate degree of segmentation (795 segments), between that of the synthetic and natural accessions. In summary, Express 617 showed genomic rearrangement events affecting 8.0% of the genome, 1012-98 16.2% of the genome, V8 12.3% of the genome and R53 41.5% of the genome. Although these figures may be biased by normalization based on chromosome mean coverage values and alignment to a European winter–oilseed rape reference sequence (Darmor-*bzh*), the overall scale of rearrangements is clearly greater in synthetic *B. napus*.

As an overview, Figure S1 displays locations of segmental deletions or duplications larger than 500 kb in the four mapping parents. Full details of all detected segmental deletions and duplications, also including those events between 50 kb and 500 kb in size, are provided in Table S1. Coverage plots for the 19 chromosomes of the four parental genotypes are given in Figures S2, S3, S4 and S5.

In Express 617, only five deletions larger than 500 kb were detected, three on chromosome C01, one on chromosome C02 and one on C08. In contrast, duplications and deletions larger than 500 kb were numerous and widespread across both subgenomes in the synthetic and semisynthetic accessions: 22 events in V8, 40 events in 1012-98 and 107 events in R53. Interestingly, deletions in five genomic regions on chromosomes

A03, C02, C05, C07 and C09 were consistently detected in all three parental accessions with synthetic *B. napus* background.

Chromosome C02 in the genotype R53 was not represented by sequence reads and is therefore assumed to be missing (completely or almost completely). On the other hand, chromosome number of R53 was confirmed cytogenetically to be 38 (data not shown). Due to the elevated coverage of sequence reads from chromosome A02 in R53, we assume that chromosome C02 has been replaced by a duplication of A02. Deletions and duplications affecting whole or nearly whole chromosomes have also been observed by Rousseau-Gueutin *et al.* (2017) in other synthetic *B. napus*.

Genetic mapping

The total amount of SNP markers used to calculate the genetic bin maps was reduced based on allele frequency and cosegregation from 35 170 preselected SNPs, with putative unique positions in the reference sequence, to sets of 2204, 3135 and 2029 markers for the populations ExV8-DH, ExR53-DH and Ex1012-98-DH, respectively. The resulting genetic linkage maps comprised 1733, 2186 and 1631 markers, respectively, covering 2512 cM, 3780 cM and 2358 cM over 20, 21 and 22 linkage groups. A total of 273 consensus markers were found in all three populations, with 704 consensus markers between ExV8-DH and Ex1012-98-DH, 729 between ExV8-DH and ExR53-DH and 677 between ExV8-DH and ExR53-DH, respectively.

High collinearity was achieved among the three linkage maps, whereby the linkage groups in the ExR53-DH map were generally larger than those in the other two populations, reflecting the considerably greater number of markers. Lengths of individual linkage groups vary among the three populations, indicating differential degrees of diversity and recombination across the chromosomes of the synthetic parents 1012-98, V8 and R53. All linkage maps are displayed in Figure S6; the map texts can be found in Table S2.

Notably, some linkage groups were found to be considerably longer in a single population than in the other two. Because all populations had mapping parent Express 617 in common, this could indicate a propensity for higher recombination frequency on these specific chromosomes conferred by the particular synthetic/semisynthetic mapping parent.

We also observed frequent incidents where multiple markers assigned to the same chromosome in the Darmor-*bzh* reference were genetically mapped onto two linkage groups that could not be combined into a single linkage group. The data suggest that this phenomenon, which is common in *B. napus* genetic maps, corresponds with the presence of either large or numerous genomic rearrangements in one of the synthetic or semisynthetic mapping parents and a consequent disruption of pairing and linkage disequilibrium in the DH lines. 'noncontiguous' linkage groups are distinguishable from translocated genomic segments, because the latter map to their new position. In this study, Ex1012-98DH exhibits two noncontinuous linkage groups representing chromosome A01. Additionally, a duplicated and translocated A01-fragment of 1012-98 maps to its new position in linkage group C01.

Genomic rearrangements can be localized by genetic mapping

Calling of 'het' and 'PA' marker types as outlined above proved extremely helpful in detecting genomic rearrangements. Segments of the genetic map carrying three or more adjacent 'het'

markers or 'PA' markers were validated by comparing read densities in the corresponding regions of the genomic sequence data. For single mapped 'PA' or 'het' markers, validation by resequencing data is very difficult. These markers were regarded with care, as their annotation may be false due to sequence similarity. Still, some examples in this study prove them to be useful rearrangement markers.

Two examples for independent validation of prominent HE and deletion events by plotting genomic resequencing coverage and genetic mapping in DH mapping populations are shown in Figures 1 and 2. In Figure 1, four homoeologous nonreciprocal translocations (HNRTs) were identified in linkage group C03 in population ExR53-DH using genomic resequencing data; each could be correlated to some extent with SNP mapping data. The first of them was mapped by a block of C03-PA-markers spanning the region 0–6.1 cM on top of linkage group C03. This corresponds to a deletion in R53 chromosome C03 of 1.16 Mb. A putatively translocated homoeologous A03 duplicate fragment was not mapped. The second was a A03-duplication and translocation to C03 mapped by a block of A03-markers spanning the region 53–108.1 cM. This corresponds to the positions 3.9–6.8 Mb on chromosome A03 and 5.5–9.0 Mb on chromosome C03. The third was a C03-deletion at the position 15.4–18.8 Mb, which was mapped by a single C03-PA-marker at 189.2 cM. The deletion corresponds to a duplication of an A03-fragment. Although the genomic segments here are large in size, the corresponding SNP markers did not meet the expected 1:1 segregation ratio and could not be mapped. The forth HE event, a C03-deletion at the position 47.9–48.8 Mb and a homoeologous A08-duplication of 0.8 Mb, was mapped by a single A08-marker at 268.7 cM.

In Figure 2, several smaller and larger segmental deletions detected by resequencing coverage analysis, spanning the same region of chromosome A05 in both V8 and R53, were genetically mapped by a block of PA-markers in both the ExV8-DH and ExR53-DH genetic maps. The corresponding regions span 20.5 cM (1.750 Mb) in ExV8-DH and 13.9 cM (4.651 Mb) in ExR53-DH. Some PA-markers are common between the two populations, whereas others are unique to one of the populations.

QTL mapping

In this study, 15 QTL for 15 traits were identified in the Ex1012-98-DH population, 61 QTL for 37 traits in the ExV8-DH population and 25 QTL for 14 traits in the ExR53-DH population. All QTL share LOD scores higher than 5, allowing an alpha-error not larger than 0.05. Table 1 gives a list of detected QTL in chromosome regions with putative genomic rearrangements for all measured traits in the three mapping populations. The list includes their genetic and inferred physical positions (unique BLAST hits) in the *B. napus* Darmor-*bzh* reference genome, R^2 values indicating the proportion of phenotypic variation explained by the QTL, and the LOD scores at QTL peaks. The colocalization of QTL with regions involved in genomic rearrangements is consistent with the hypothesis that genomic rearrangements generate significant phenotypic variation with considerable selective and evolutionary potential.

A homoeologous rearrangement on A09 causes variation in seed fibre content

In a previous study involving the same plant material, we concluded that a major QTL for seed colour and seed coat fibre

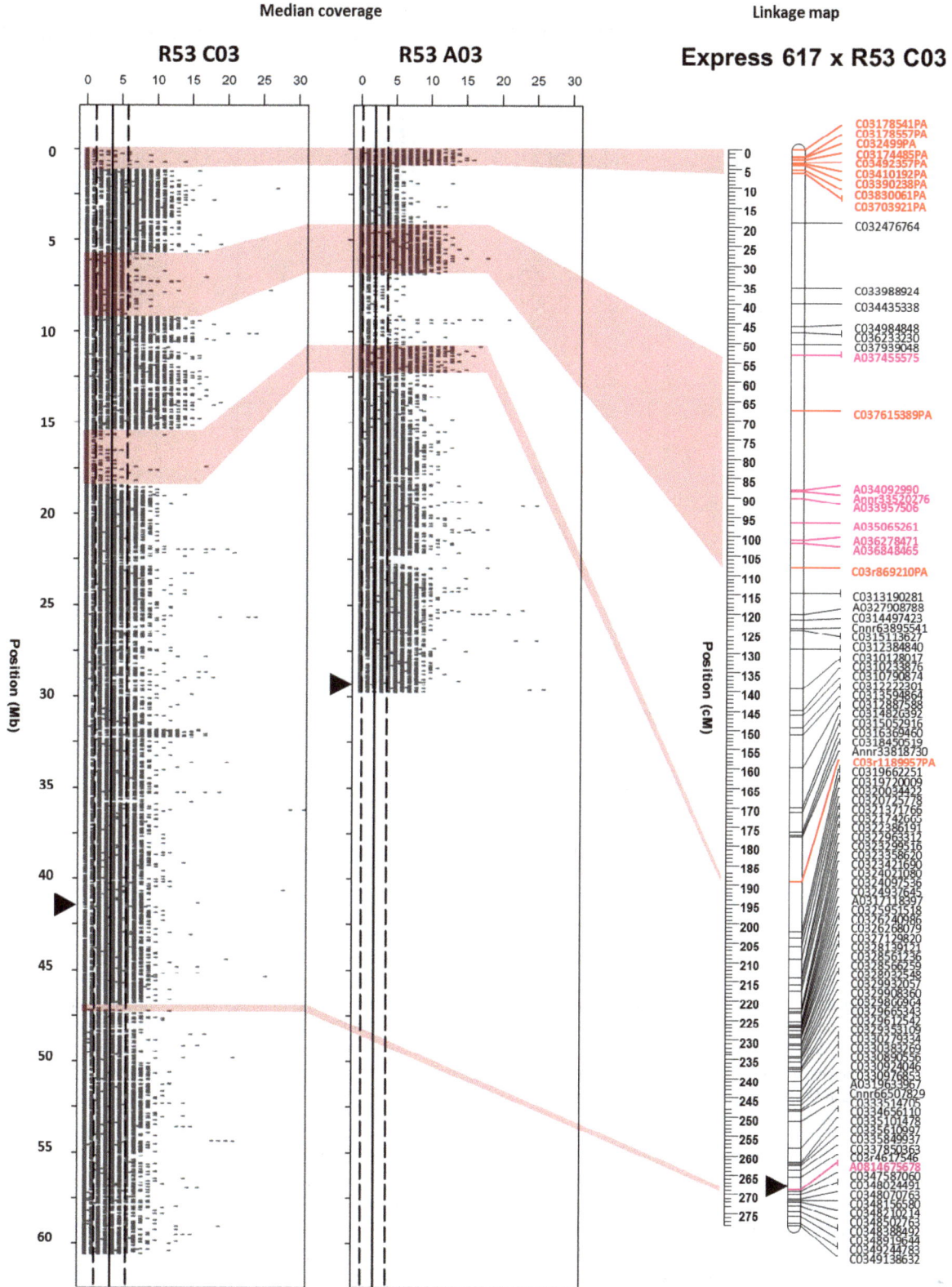

Figure 1 A03 to C03 translocation in the synthetic *B. napus* genotype R53 identified by resequencing and validated by genetic mapping. The plots show resequencing read coverage across the lengths of the respective chromosomes, calculated for segments of 1 kb. The genetic linkage maps on the right of the read maps show genetic mapping including SNPs with normally segregating, bi-allelic calls with locus names in black text. SNPs called as deletions (presence–absence markers, with suffix '–PA') are indicated by bold red marker names, whereas SNPs with heterozygous–homozygous segregation due to polymorphism in one of two duplicated copies (with suffix '–het') are indicated by bold blue marker names. Polymorphic markers in bold magenta text indicate duplicated markers mapping to their homoeologous position. Opaque red blocks link putative deletions detected in coverage blocks with the corresponding regions in the genetic maps. Centromere regions are indicated by black triangles according to (Mason *et al.*, 2013).

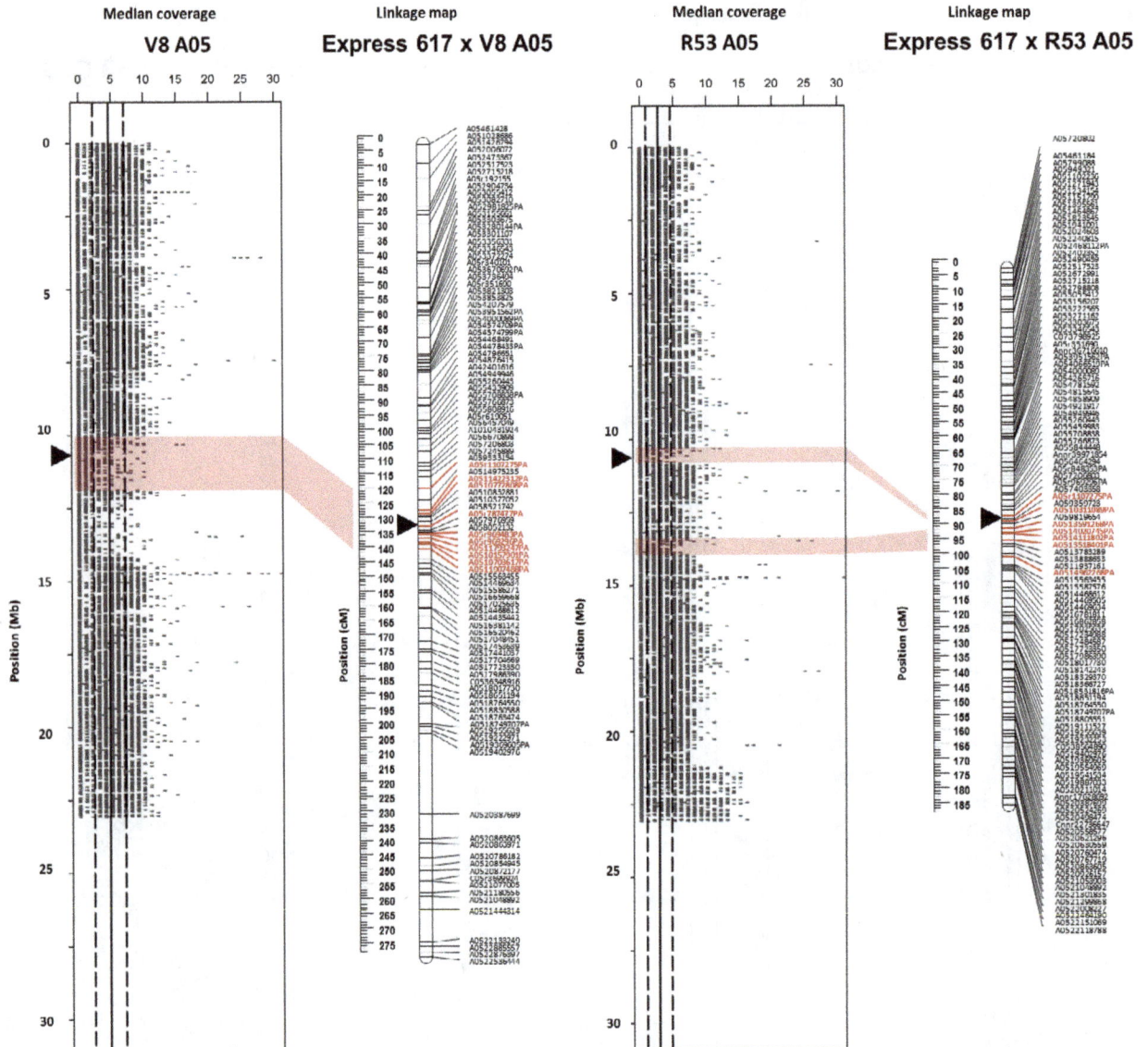

Figure 2 A05 Deletion in the synthetic *B. napus* genotypes V8 und R53 identified by resequencing and validated by genetic mapping. The plots show resequencing read coverage across the lengths of the respective chromosomes, calculated for segments of 1 kb. The genetic linkage maps on the right of the read plots show genetic mapping including SNPs with normally segregating, bi-allelic calls with locus names in black text. SNPs called as deletions (presence–absence markers, with suffix '–PA') are indicated by bold red marker names, whereas SNPs with heterozygous–homozygous segregation due to polymorphism in one of two duplicated copies (with suffix '–het') are indicated by bold blue marker names. Polymorphic markers in bold magenta text indicate duplicated markers mapping to their homoeologous position. Opaque red blocks link putative deletions detected in coverage blocks with the corresponding regions in the genetic maps. Centromere regions are indicated by black triangles according to (Mason *et al.*, 2013).

content, colocalizing on chromosome A09 in the mapping populations Ex1012-98-DH and ExV8-DH, might have derived from an HE causing deletion or conversion of important candidate genes on chromosome A09 (Stein *et al.*, 2013). Because of the comparatively low marker density in that study, it was not possible to precisely map the HE borders and correctly localize the physical position of the QTL. In this study, we confirmed the presence of this HE by sequence coverage and validated the genetic position using a considerably larger set of molecular markers than before.

The coverage plots of the homoeologous chromosomes of the synthetic parent 1012-98 (Figure 3) clearly show coincidence of a segmental deletion in A09, corresponding to a duplication of the homoeologous segment in C08. SNP markers

derived from chromosome C08 map as 'het' markers on linkage group A09, marking the position of a duplicated block. A09 markers representing a deletion ('PA') flank this block. Using the new, high-density genetic map of Ex1012-98-DH chromosome A09, including the HE-tracing markers, we narrowed the QTL to a very small interval with an extremely high LOD peak for the seed fibre QTL (Figure 3). The 173-kb interval harbours one gene from the monolignol biosynthesis pathway, *BnaCAD2/3* (*BnaA09g42930/BnaC08g35540*). The whole translocation region includes also *BnaCCR1* (*Bna A09g56490/BnaC08g38580*) and *BnaAHA10* (*BnaA09g41670/ BnaC08g34260*), confirming the previously hypothesized involvement of these genes in the seed fibre and seed colour QTL (Stein *et al.*, 2013). Cinnamoyl-CoA-reductase (CCR1)

Table 1 Major QTL for seed quality and agronomic traits located in genomic regions that are rearranged with respect to the *B. napus* Darmor-*bzh* reference sequence. Calculation was performed by composite interval mapping, considering only QTL with a LOD score of >5; QTL in bold text could also be identified in genetic mapping

Population	Trait	Linkage group	Genetic position (Marker interval) [cM]	Inferred physical position [kb]	LOD score	R^2	Structural rearrangement
ExV8-DH	Seedling volume increase	A05	161–165	A05 17 065–17 605	8.0	0.14	Deletion
	Seed sulphur	**C03**	**222–227.5**	**C03 51 659–52 628**	**5.2**	**0.09**	**Deletion**
	Seed sulphur	C09	22.5–27.5	C09 2814–3592	29.7	0.42	Deletion
	Seed glucosinolate	C09	22.5–27.5	C09 2814–3592	35.0	0.48	Deletion
Ex1012-98-DH	**Seed ADL (acid detergent lignin)**	**A09**	**152**	**C08 33 479**	**28.9**	**0.56**	**HE**
	Seed NDF (neutral detergent fibre)	**A09**	**152**	**C08 33 479**	**31.6**	**0.59**	**HE**
	Seed colour	**A09**	**152**	**C08 33 479**	**7.7**	**0.20**	**HE**
	Seed C18:3	**A09**	**176–177**	**A09 31 557**	**8.4**	**0.21**	**HE**
	Days to flowering	A09	193–194.5	A09r 3826 (~A09 30 000)	7.4	0.20	HE
	Flowering duration	**A09**	**207.5–208**	**C08 36 510**	**6.3**	**0.18**	**HE**
	Seeds per silique	**A09**	**208**	**C08 36 510**	**5.1**	**0.15**	**HE**
ExR53-DH	Seed ADL (acid detergent lignin)	A05	15–17.5	A05 1464	5.4	0.10	Deletion
	Seed ADF (acid detergent fibre)	C01	193.5–194	C01 34 145–38 105	8.6	0.16	Deletion
	Seed NDF (neutral detergent fibre)	C01	193.5–194	C01 34 145–38 105	10.4	0.19	Deletion

confers reduction in *p*-coumaroyl-CoA to *p*-coumaraldehyde, which is subsequently reduced to the monolignol *p*-coumarylalcohol by cinnamyl alcohol dehydrogenase (CAD2/3). While CCR1 acts substrate-specific, CAD2/3 has a potentially wider substrate spectrum, which allows a certain plasticity in monolignol metabolism (Bonawitz and Chapple, 2010). Impaired activity of either CCR1 or CAD2/3 does not necessarily reduce the lignin content in the plant organ, but can affect the lignin composition. Double knockout of both genes, however, has been shown to decrease lignin content (Chabannes *et al.*, 2001). Autoinhibited H+-ATPase isoform 10 (AHA10) is a seed-expressed transcription factor involved in proanthocyanidin formation in the seed coat endothelium, which has been shown to influence seed coat pigment accumulation (Baxter *et al.*, 2005). The physical annotation of these genes within a deleted segment of chromosome A09 in the genotype 1012-98 likely explains the colocalization of QTL for both seed acid detergent lignin (ADL) and seed colour (Figure 3).

A targeted sequence-capture experiment carried out for *BnaCCR1* in the mapping parents Express 617 and 1012-98 underpins the hypothesis that the large deletion described above has led to loss of an A-genome copy of *BnaCCR1* in 1012-98. Details of the experimental setup and data analysis are described by Schiessl *et al.* (2017). Figure 4 presents sequence-capture results of the two *BnaCCR1* copies *BnaA09g56490* and *BnaC08g38580* at very high average sequencing depth (approximately 1200×), allowing highly accurate estimations of copy number. Coverage for *BnaA09g56490* is considerably lower and less consistent over the length of the gene copy in 1012-98. In comparison with the normalized gene coverage across the whole experimental panel of 280 genotypes described by Schiessl *et al.* (2017), the gene coverage is reduced to 51%.

PCR and BAC-FISH validation of the QTL-associated HE between chromosomes C08 and A09

We validated the deletion of a 900-kb region (29 274–30 174 Mb) in 1012-98 using three primer pairs specific to this region. As expected, no amplification was observed in 1012-98, whereas the expected PCR product, indicating presence of the chromosome segment, was observed in Express 617, R53, V8 and Darmor-*bzh*.

The BAC KBrB043F18 and BoB014O06 were hybridized to identify the A09 and C08 chromosomes and all C-genome chromosomes, respectively, to detect the homoeologous rearrangement (green) (Figure 5b and c). To further validate the A09 chromosome fragment losses, a BAC-FISH experiment was performed using BAC clone 54 probe (isolated from Express 617) present in the rearranged region of 1012-98. Due to the high-sequence similarity between A and C genomes, two or four signals are expected in the case of a deletion or an HNRT, respectively. In 1012-98, four signals were observed with BAC clone 54 (Figure 5c), indicating the presence of an HNRT. As the limit of GISH-like resolution is 5 Mb to clearly observe rearranged chromosomes by harbouring a dual colour signal, it is not easy to observe smaller rearrangements by hybridization of the BoB014O06 BAC. However, in Figure 5b (and enlarged in Figure 5h), the C08-to-A09 translocation in 1012-98 can be observed by a faint green signal in the otherwise unstained A09 chromosome.

Discussion

Genomic rearrangements are widespread in the recent allopolyploid species *B. napus*, making it an interesting model for studying the cause and effect of *de novo* polyploidization. The presence of two additive genomes with a high degree of

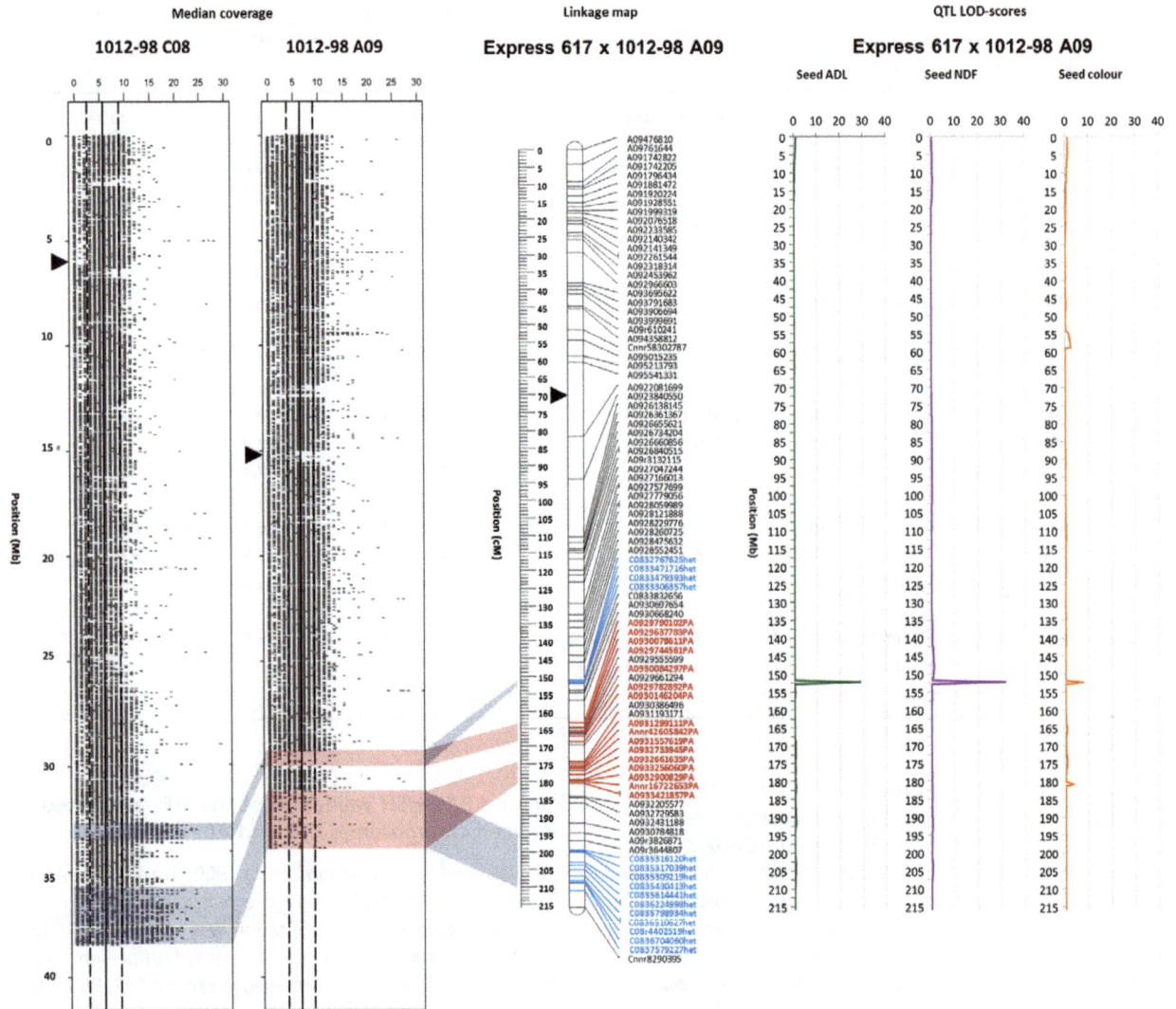

Figure 3 Localization of a major QTL on A09 influencing numerous seed quality traits in the mapping population Ex1012-98-DH, within a prominent HE between the distal ends of homoeologous chromosomes A09 and C08. The two plots on the left hand side show median read coverage across the lengths of the respective homoeologous chromosomes, calculated for segments of 1 kb. The two genetic linkage maps on the right show genetic mapping including SNPs with normally segregating, bi-allelic calls with locus names in black text. SNPs called as deletions (presence–absence markers, with suffix '–PA') are indicated by bold red marker names, whereas SNPs with heterozygous–homozygous segregation due to polymorphism in one of two duplicate copies (with suffix '–het') are indicated by bold blue marker names. Opaque red blocks link putative deletions, detected based on sequence coverage blocks, with the corresponding regions in the genetic maps, while opaque blue blocks indicate putative duplications, respectively. Centromere regions are indicated by black triangles position according to (Mason *et al.*, 2013).

homoeology in the same nucleus leads to meiotic chromosome pairing between homoeologous chromosomes during the first generations after allopolyploidization (Szadkowski *et al.*, 2010), causing considerable HE occurrence and gene conversion (Chalhoub *et al.*, 2014). Implementation of novel variation caused by genomic rearrangement events, such as HE events or pure deletions in synthetic *B. napus,* is of considerable interest for rapeseed and canola breeders. However, skewed marker segregation patterns, which occur as consequence of genomic rearrangements, prevent standard mapping procedures from accurately localizing QTL, tightly linked markers and causal genes in these regions.

Here, we present a method which enabled us to accurately map agronomic QTL to a number of genomic rearrangement events. The 60K Illumina SNP array marker data were used to generate high-density genetic maps in half-sib DH populations from three synthetic *B. napus* accessions, carrying interesting disease resistance (Obermeier *et al.*, 2013), agronomic and yield-related traits (Basunanda *et al.*, 2007, 2010; Radoev *et al.*, 2008) and seed quality characters (Badani *et al.*, 2006; Stein *et al.*, 2013). Although the putative colocalization of HE events with QTL in *B. napus* has been suggested in previous studies (Chalhoub *et al.*, 2014; Liu *et al.*, 2012; Stein *et al.*, 2013) to our knowledge, this is the first study providing independent validation of QTL-HE colocalization. Alongside genetic mapping and evidence from genomic coverage sequencing, we provide cytological indications from BAC-FISH that a chromosome fragment spanning an HE-associated QTL is indeed involved in the suspected HE. This supports our use of SNP marker data from loci spanning deletions or duplications, which normally would be

BnaA09g56490 (A09random: 3 748 661 – 3 751 552)

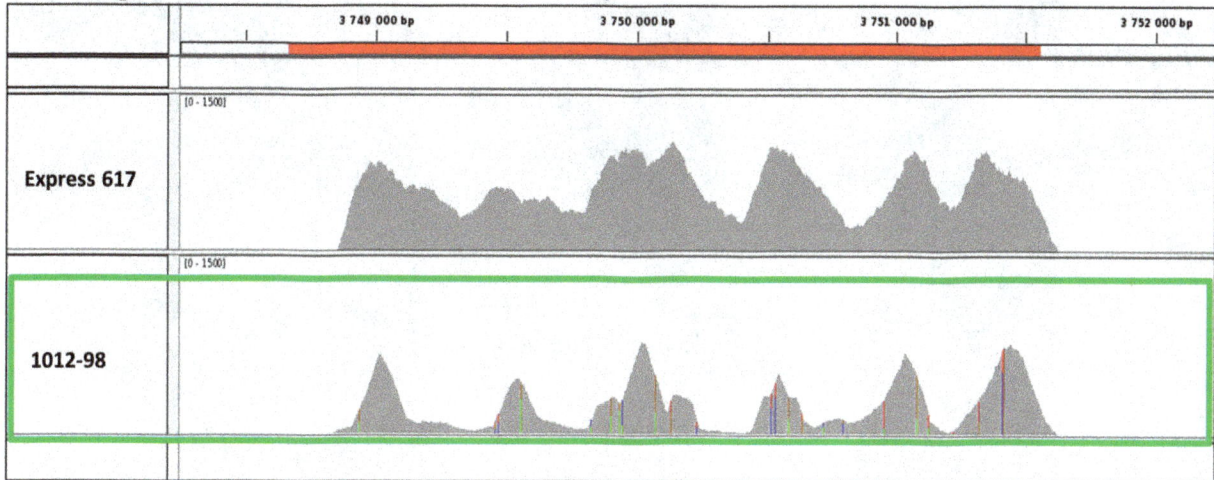

BnaC08g38580 (C08: 34 776 298 – 34 779 240)

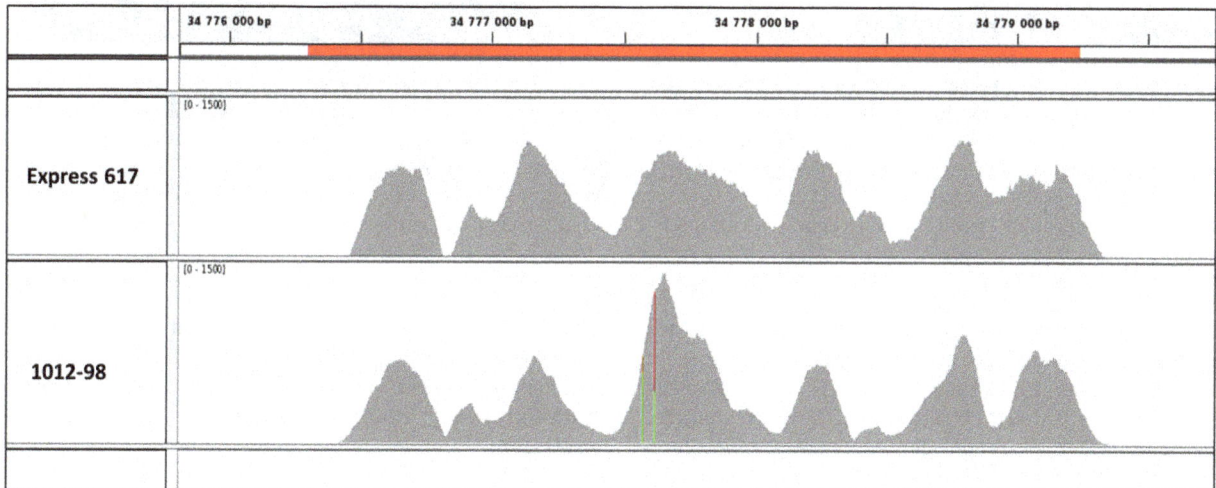

Chromosome	Start position (bp)	Stop position (bp)	Gene	Relative coverage value	Relative coverage value	Normalized mean coverage over 280 genotypes
				Express 617	1012-98	
A09r	3 748 661	3 751 552	BnaA09g56490	1.0535	0.5162	1239.8
C08	34 776 298	34 779 240	BnaC08g38580	1.0667	0.9169	1226.1

Figure 4 Mapping results from targeted sequence capture of the two *BnaCCR1* copies *BnaA09g56490* and *BnaC08g38580* in the *B. napus* genotypes Express 617 and 1012-98. Sequence read coverages were mapped against the *B. napus* reference sequence Darmor-bzh v. 4.1 and are displayed along the gene length. SNPs (indicated by coloured bars) in the homozygous doubled-haploid genotype 1012-98 suggest two variants of the *BnaC08g38580* gene copy. This may have also led to mis-alignment of some reads to the homoeologous gene *BnaA09g56490*. Both the coverage landscape for *BnaA09g56490* in 1012-98 and the according relative coverage values given in the table indicate a deletion of this gene copy (highlighted in green boxes). Relative coverage values were calculated as the ratio of the normalized mean coverage of the genotype over normalized mean coverage over 280 genotypes for each specific gene copy. Sequence-capture data were obtained from (Schiessl *et al.*, 2017).

discarded in mapping procedures. The results provide an interesting reminder about the importance of structural chromosome variation in genome mapping and QTL analysis and suggest implementation of routine screening for presence–absence variation as a resource for breeding diversity.

Using the 60K Infinium SNP array allows genetic mapping of deletions and translocations

Homoeologous translocations were first detected in mapping populations using codominant RFLP markers (Parkin *et al.*, 1995;

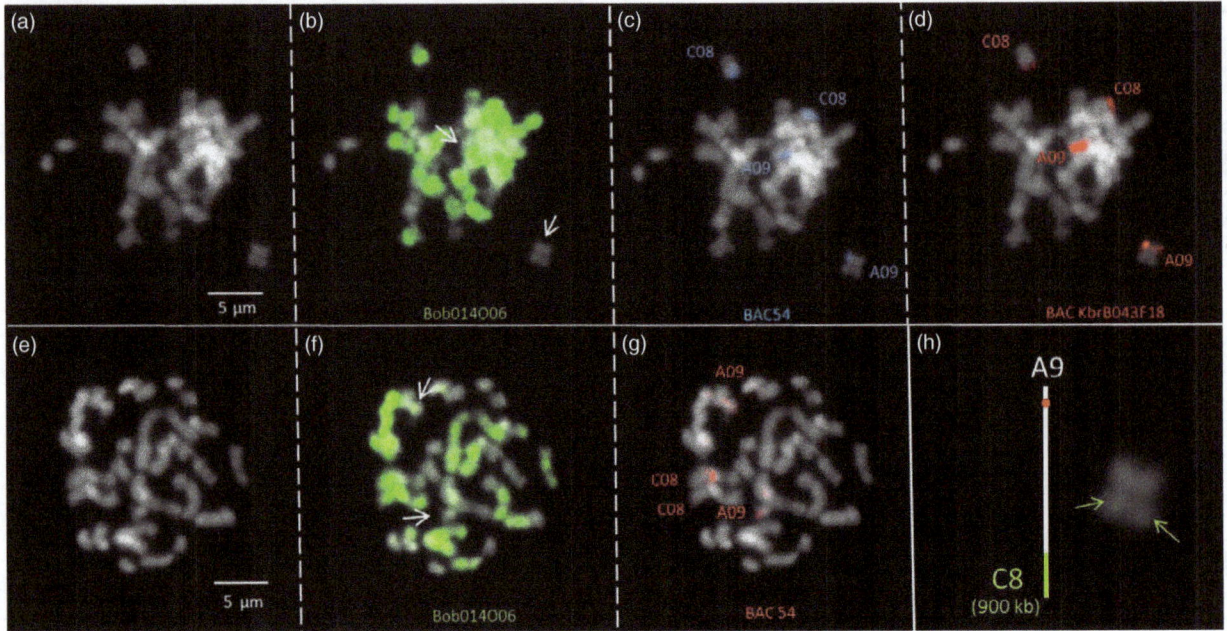

Figure 5 Fluorescence *in situ* hybridization using bacterial artificial chromosome probes to identify a putative HNRT between chromosomes A09 and C08 in the synthetic *B. napus* genotype 1012-98. (a, e) DAPI background staining of somatic metaphase chromosomes of the synthetic *B. napus* genotype 1012-98, carrying a putative HNRT between chromosomes A09 and C08. (b, f) Green FISH signals from BAC BoB014O06, which identifies all *Brassica* C-genome chromosomes. (c) Blue FISH signals and red signals (g) from BAC54, specific for chromosomes A09 and C08. (d) Red FISH signals from KBrB043F18, also specific for chromosomes A09 and C08. Arrows indicate chromosomes with putative rearrangements. (h) Schematic representation of translocation of enlarged A09 chromatids. Green arrows indicate green FISH signal, representing the presence of a C-subgenome fragment on chromosome A09.

Sharpe *et al.*, 1995), which also gave initial insights into the frequency and extent of HE events in natural and synthetic *B. napus* (Song *et al.*, 1995). In contrast to RFLP assays, rapid and cost-effective analysis with high-density SNP arrays provides considerably higher resolution of genetic recombination in mapping populations. For analyses of homoeologous rearrangements, the ability to anchor SNP markers to a reference genome sequence and to compare patterns for multiple SNPs per homoeologous chromosome segment allows a more accurate and detailed assessment of genomewide HE events. Genetic mapping of homoeologous reciprocal translocations is more challenging unless duplicated markers can be assigned to a unique locus. SNPs with multiple physical BLAST positions were discarded. This may mean that markers from highly similar homoeologous regions are neglected in the mapping process.

Applicability of HE mapping approaches

Our method to derive HE information from high-density SNP array data in segregating mapping populations provides an opportunity to include structural genome variation in genetic maps and QTL studies. Revisiting historical QTL data sets using SNP array data analysed with this technique will give a more comprehensive picture of HE associations to QTL. We are currently developing thresholds and techniques to apply HE calling from SNPs in nonrelated populations for genomewide association studies. In multiparent mapping populations, for example those described by Snowdon *et al.* (2015) for nested-association mapping in *B. napus*, we expect that capturing of HE variants will greatly increase the power to detect QTL in genome regions that were previously very difficult to accurately access by genetic mapping techniques.

Access to multiple genome assemblies per species, using new assembly methods that greatly improve genome coverage and accuracy, will further improve our ability to trace the impact of chromosome-level structural variation on quantitative trait expression.

Experimental procedures

Plant material

Three half-sib doubled-haploid (DH) winter–oilseed rape populations were used in this study, developed from crosses of the black-seeded inbred winter–oilseed rape line 'Express 617' to the synthetic *B. napus* lines 1012-98 and R53 and to the semisynthetic line V8. The origins of the parental lines, along with the DH mapping populations Express 617 × 1012-98 (Ex1012-98-DH, $n = 164$), Express 617 × R53 (ExR53-DH, $n = 248$) and Express 617 × V8 (ExV8-DH, $n = 248$), have been described previously in detail (Badani *et al.*, 2006; Basunanda *et al.*, 2007, 2010; Radoev *et al.*, 2008).

Detection of genomic rearrangements from genomic resequencing data

Genomic resequencing data were collected from the parental genotypes Express 617, 1012-98, V8 and R53 using the Illumina MiSeq and HiSeq 2000 systems. The MiSeq system delivered 250-bp paired-end reads, while 100-bp paired-end reads were collected using the HiSeq system. All reads were aligned to the *B. napus* reference sequence version 4.1 (Chalhoub *et al.*, 2014) using the short oligonucleotide alignment program (SOAP2 v2.21) (Li *et al.*, 2008), and read depths were calculated using the command *genomecov* in the bedtools package v.2.20.1 for every nucleotide position genomewide. Median coverage over 1000-bp blocks was calculated using an R script, and adjacent blocks with the same coverage value were aggregated to consecutive segments using a circular binary segmentation algorithm implemented in R package PSCBS (Bengtsson et al., 2010; Olshen et al., 2011). Adjacent segments over 50 kb in length and with the same mean coverage value were merged unless separated by gaps larger than 50 kb.

Mean read coverage and standard deviation were calculated for each individual and chromosome. Segments with a coverage value exceeding the chromosome mean by 1 standard deviation were defined as segmental duplications, while segments with a coverage value of 1 standard deviation lower than the chromosome mean were defined as deletions. Accordingly, segments not deviating by more than 1 standard deviation from the chromosome-wide average were assumed to have 'normal' coverage. The script used is provided in Data S1.

DNA extraction and high-density genetic mapping including HE markers

Total genomic DNA was extracted from 200 mg leaf material of young leaves of mapping parents and DH lines, using the Qiagen BioSprint 96 DNA Plant Kit (Qiagen GmbH, Hilden, Germany). All samples were subjected to high-density, genomewide SNP genotyping using the Brassica Illumina 60K SNP genotyping array (Clarke et al., 2016). Genotyping was outsourced to TraitGenetics GmbH (Gatersleben, Germany).

Genomic positions in the B. napus reference genome sequence were assigned to all tested SNP markers showing a BLAST hit with the SNP flanking sequences, as described previously (Qian et al., 2014; Mason and Snowdon, 2016). In brief, SNP markers were physically localized on the Brassica napus Darmor-bzh reference genome sequence assembly (version 4.1) (Chalhoub et al., 2014) using the following criteria: minimum overlap of 50-bp length, minimum identity of 95%, no sequence gaps. SNPs with only one BLAST hit were regarded as informative in terms of physical position, and only these markers were included in the genetic mapping. Linkage mapping was conducted using the software Joinmap 4 (van Ooijen, 2006), after assignment of cosegregating markers into bins with Perl to reduce the locus number. Markers were subsequently scored according to three different genotype patterns (Mason et al., 2017): (i) polymorphic simple SNP calls; (ii) polymorphic hemi-SNP calls; and (iii) polymorphic presence–absence SNP calls. The latter two allow recording of markers potentially affected by genomic rearrangement events. All of these scoring types were required to fit the expected 1:1 segregation for DH populations.

All SNP markers that were included in the mapping file were named with their predetermined physical position and a suffix indicating markers showing dominant presence–absence segregation ('PA') on the one hand, or codominantly segregating heterozygous hemi-SNPs ('het') on the other hand. The map was calculated using the maximum-likelihood algorithm and groups were formed with a cut-off at recombination frequency 0.2. Maps were subsequently joined using Mapchart 2.3 to link bins containing SNP markers shared across the three maps.

QTL mapping

Quantitative trait loci were calculated using Qgene4.3.10 (Joehanes and Nelson, 2008) based on the calculated genetic maps. Seed trait phenotyping for all populations was performed by near-infrared reflectance spectrometry (NIRS) analysis on seed

samples produced in field trials at four different locations in Germany, over multiple years from 2003 until 2015, following the seed analysis procedures (Wittkop et al., 2012). The seed colour and quality data from the first 2 years of experiments correspond with those reported by Badani et al. (2006). A fully randomized complete block design was used, with plots sizes of 10–13 m^2 depending on the standard practice at each location. A high number of locations and years were preferred rather than multiple replications of genotypes per location, as is standard practice when testing large rapeseed populations, because the large plot size reduces field homogeneity when too many test plots are included per location. Selfed seeds from 3 to 5 representative plants per line were hand-harvested at maturity and subjected to NIRS as above, with two technical replicates per sample. Germination traits in the ExV8-DH population were phenotyped at the French national seed testing laboratory (GEVES, Angers, France), with 100 seeds per genotype and repetition with seed lots from two different production environments (Hatzig et al., 2015). Morphological traits in the ExV8-DH population were collected from two locations in Germany in the years 2006 and 2007.

HE validation using specific primers and BAC-FISH

To validate the deletion of a 900-kb fragment on A09 (29 274–30 174 Mb) in 1012-98, we extracted this 900-kb region from B. napus 'Darmor-bzh' genome (Chalhoub et al., 2014) and blasted it against the whole 'Darmor-bzh' genome sequence, enabling us to identify fragments of at least 500 bp that were specific to this A09 chromosome region (i.e. absent from the homoeologous region). We then designed primer pairs from these specific A09 regions using Primer 3.0 (Rozen and Skaletsky, 2000). Only the primers (3 pairs) that gave a single band in different B. napus varieties tested and no amplification in B. oleracea were retained (primer details are given in Table S3). The PCR products obtained using B. napus 'Darmor' DNA were also directly sequenced, enabling further validation of the specificity of these primer pairs to A09. Subsequently, these primer pairs were tested using DNA from synthetic line 1012-98. Each PCR amplification was performed in a total volume of 50 μL containing 10 μL of 5× buffer (Promega), 4 μL of 25 mM MgCl2, 0.5 μL of 25 mM dNTP mix, 2.5 μL of each primer (10 mM), 0.2 μL of GoTaq® G2 Hot Start polymerase (5 U/μL) and 50 ng of DNA. For PCRs, genomic DNA was denatured at 94 °C for 2 min, followed by 30 cycles of 94 °C for 30 s, 58 °C for 30 s and 72 °C for 1 min.

The proximal 0.5–1.5 cm of young seedling roots were excised, treated in the dark with 0.04% 8-hydroxyquinoline for 2 h at 4 °C and then transferred to room temperature for 2 h to accumulate cells at metaphase. Root tips were then fixed in 3:1 ethanol–glacial acetic acid for 48 h at 4 °C and stored in 70% ethanol at −20 °C until required. After washing in 0.01 M citric acid–sodium citrate (pH 4.5) for 15 min, the tips were digested in 5% Onozuka R-10 cellulase (Sigma-Aldrich, St. Louis, MO) containing 1% Y23 pectolyase (Sigma) at 37 °C for 30 min and then washed carefully with distilled water for 30 min. Single root tips were transferred to cleaned microscope slides and macerated with a drop of 3:1 fixation solution using a preparation needle. After air-drying, slides with good metaphase chromosome spreads were stored at −20 °C until further use.

The B. napus BAC clone 54 from the parental genotype Express 617 was used to probe a chromosome spread of synthetic B. napus parental genotype 1012-98, in which a putative HE was suspected between homoeologous chromosomes C08 and A09. The BAC clone was previously identified by PCR screens with markers from a QTL region on chromosome A09 (see below). Sanger sequencing confirmed alignment of the BAC clone to *B. napus* Darmor-*bzh* chromosome A09.

BAC clone 54 was labelled by random priming with biotin-14-dUTP (Invitrogen, Life Technologies Waltham, MA). The BAC clones KBrB043F18 (from *B. rapa* chromosome A09, homoeologous to *B. oleracea* chromosome C08 (Xiong and Pires, 2011) and *B. oleracea* BoB014O06 (Howell *et al.* 2002) were labelled by random priming with Alexa 594-5-dUTP and Alexa 488-5-dUTP, respectively. BoB014O06 was used as genomic *in situ* hybridization (GISH)-like probe to specifically stain all C-genome chromosomes in *B. napus* (Suay *et al.*, 2014; Szadkowski *et al.*, 2010).

Chromosome preparations were incubated in RNAse A (100 ng/μL) and pepsin (100 μg/mL) in 0.01 M HCl and fixed with paraformaldehyde (4%). Chromosomes were denatured in a solution of 70% formamide in 2× saline-sodium citrate buffer (SSC) at 70 °C for 2 min, dehydrated in an ethanol series (70%, 90% and 100%) and air-dried. The hybridization mixture consisted of 50% deionized formamide, 10% dextran sulphate, 2xSSC and 1% SDS. Labelled probes (200 ng per slide) were denatured at 92 °C for 6 min and transferred to ice. The denatured probe was placed on the slide, and *in situ* hybridization was carried out overnight in a moist chamber at 37 °C. After hybridization, slides were washed for 5 min in 50% formamide in 2xSSC at 42 °C, followed by several washes in 4xSSC-Tween to remove nonconjugated probe. Biotinylated probe was immunodetected by Texas Red Avidin DCS (Vector Laboratories, Burlingame, CA), and the signal was amplified with biotinylated anti-avidin D (Vector Laboratories). The chromosomes were mounted and counterstained in Vectashield (Vector Laboratories) containing 2.5 μg/mL 4′,6-diamidino-2-phenylindole (DAPI). Fluorescence images were captured using a CoolSnap HQ camera (Photometrics, Tucson, AZ) on an Axioplan 2 microscope (Zeiss, Oberkochen, Germany) and analysed using MetaVueTM (Universal Imaging Corporation, Downingtown, PA).

Acknowledgements

This work was financed by grant SN 14/17-1 for the ERA-CAPS project Evo-Genapus 'Evolution of genomes: Structure-function relationships in the polyploid crop species *Brassica napus*'. The authors declare no conflict of interest.

References

Allender, C.J. and King, G.J. (2010) Origins of the amphiploid species *Brassica napus* L. investigated by chloroplast and nuclear molecular markers. *BMC Plant Biol.* **10**, 54.

Badani, A.G., Snowdon, R.J., Wittkop, B., Lipsa, F.D., Baetzel, R., Horn, R., de Haro, A. *et al.* (2006) Colocalization of a partially dominant gene for yellow seed colour with a major QTL influencing acid detergent fibre (ADF) content in different crosses of oilseed rape (*Brassica napus*). *Genome*, **49**, 1499–1509.

Basunanda, P., Spiller, T.H., Hasan, M., Gehringer, A., Schondelmaier, J., Lühs, W., Friedt, W. *et al.* (2007) Marker-assisted increase of genetic diversity in a double-low seed quality winter oilseed rape genetic background. *Plant Breed.* **126**, 581–587.

Basunanda, P., Radoev, M., Ecke, W., Friedt, W., Becker, H.C. and Snowdon, R.J. (2010) Comparative mapping of quantitative trait loci involved in heterosis for seedling and yield traits in oilseed rape (*Brassica napus* L.). *TAG. Theoret. Appl. Genet. Theoretische und angewandte Genetik*, **120**, 271–281.

Baxter, I.R., Young, J.C., Armstrong, G., Foster, N., Bogenschutz, N., Cordova, T., Peer, W.A. *et al.* (2005) A plasma membrane H+-ATPase is required for the formation of proanthocyanidins in the seed coat endothelium of Arabidopsis thaliana. *Proc. Natl Acad. Sci.* **102**, 2649–2654.

Bengtsson, H., Neuvial, P. and Speed, T.P. (2010) TumorBoost: normalization of allele-specific tumor copy numbers from a single pair of tumor-normal genotyping microarrays. *BMC Bioinform.* **11**, 245.

Bonawitz, N.D. and Chapple, C. (2010) The genetics of lignin biosynthesis: connecting genotype to phenotype. *Annu. Rev. Genet.* **44**, 337–363.

Chabannes, M., Barakate, A., Lapierre, C., Marita, J.M., Ralph, J., Pean, M., Danoun, S. *et al.* (2001) Strong decrease in lignin content without significant alteration of plant development is induced by simultaneous down-regulation of cinnamoyl CoA reductase (CCR) and cinnamyl alcohol dehydrogenase (CAD) in tobacco plants. *Plant J.* 257–270.

Chalhoub, B., Denoeud, F., Liu, S., Parkin, I.A.P., Tang, H., Wang, X., Chiquet, J. *et al.* (2014) Early allopolyploid evolution in the post-Neolithic *Brassica napus* oilseed genome. *Science*, **345**, 950–953.

Clarke, W.E., Higgins, E.E., Plieske, J., Wieseke, R., Sidebottom, C., Khedikar, Y., Batley, J. *et al.* (2016) A high-density SNP genotyping array for *Brassica napus* and its ancestral diploid species based on optimised selection of single-locus markers in the allotetraploid genome. *TAG. Theoret. Appl. Genet. Theoretische und angewandte Genetik*, **129**, 1887–1899.

Delourme, R., Falentin, C., Fomeju, B.F., Boillot, M., Lassalle, G., André, I., Duarte, J. *et al.* (2013) High-density SNP-based genetic map development and linkage disequilibrium assessment in *Brassica napus* L. *BMC Genom.* **14**, 120.

Fopa, F.B., Falentin, C., Lassalle, G., Manzanares-Dauleux, M.J. and Delourme, R. (2014) Homoeologous duplicated regions are involved in quantitative resistance of *Brassica napus* to stem canker. *BMC Genom.* **15**, 498.

Gaeta, R.T., Pires, J.C., Iniguez-Luy, F., Leon, E. and Osborn, T.C. (2007) Genomic changes in resynthesized *Brassica napus* and their effect on gene expression and phenotype. *Plant Cell*, **19**, 3403–3417.

Hatzig, S.V., Frisch, M., Breuer, F., Nesi, N., Ducournau, S., Wagner, M.-H., Leckband, G. *et al.* (2015) Genome-wide association mapping unravels the genetic control of seed germination and vigor in *Brassica napus*. *Front. Plant Sci.* **6**, 507.

Howell, E.C., Barker, G.C., Jones, G.H., Kearsey, M.J., King, G.J., *et al.* (2002) Integration of the cytogenetic and genetic linkage maps of Brassica oleracea. *Genetics* **161**, 1225–1234.

Joehanes, R. and Nelson, J.C. (2008) QGene 4.0, an extensible Java QTL-analysis platform. *Bioinformatics*, **24**, 2788–2789.

Leflon, M., Eber, F., Letanneur, J.C., Chelysheva, L., Coriton, O., Huteau, V., Ryder, C.D. *et al.* (2006) Pairing and recombination at meiosis of Brassica rapa (AA) × *Brassica napus* (AACC) hybrids. *TAG. Theoret. Appl. Genet. Theoretische und angewandte Genetik*, **113**, 1467–1480.

Li, R., Li, Y., Kristiansen, K. and Wang, J. (2008) SOAP: short oligonucleotide alignment program. *Bioinformatics*, **24**, 713–714.

Liu, L., Stein, A., Wittkop, B., Sarvari, P., Li, J., Yan, X., Dreyer, F. *et al.* (2012) A knockout mutation in the lignin biosynthesis gene CCR1 explains a major QTL for acid detergent lignin content in *Brassica napus* seeds. *Theor. Appl. Genet.* **124**, 1573–1586.

Liu, L., Qu, C., Wittkop, B., Yi, B., Xiao, Y., He, Y., Snowdon, R.J. *et al.* (2013) A high-density SNP map for accurate mapping of seed fibre QTL in *Brassica napus* L. *PLoS ONE*, **8**, e83052.

Mason, A.S. and Chèvre, A.M. (2016) Optimization of Recombination in Interspecific Hybrids to Introduce New Genetic Diversity into Oilseed Rape (Brassica napus L.). In: *Polyploidy and Hybridization for Crop Improvement* (Mason, A.S., ed), pp. 431–444. Boca Raton, FL: CRC Press.

Mason, A.S. and Snowdon, R.J. (2016) Oilseed rape: learning about ancient and recent polyploid evolution from a recent crop species. *Plant Biol (Stuttg)*, **18**, 883–892.

Mason, A.S., Rousseau-Gueutin, M., Morice, J., Bayer, P.E., Besharat, N., Cousin, A., Pradhan, A. *et al.* (2013) Centromere locations in Brassica A and C genomes revealed through half-tetrad analysis. *Genetics*, **202**, 513–523.

Mason, A.S., Higgins, E.E., Snowdon, R.J., Batley, J., Stein, A., Werner, C. and Parkin, I. (2017) A user guide to the Brassica 60K Illumina Infinium™ SNP genotyping array. *Theoret. Appl. Genet.* **130**, 621–633.

Mei, J., Liu, Y., Wei, D., Wittkop, B., Ding, Y., Li, Q., Li, J. *et al.* (2015) Transfer of sclerotinia resistance from wild relative of Brassica oleracea into Brassica napus using a hexaploidy step. *TAG. Theoret. Appl. Genet. Theoretische und angewandte Genetik*, **128**, 639–644.

Obermeier, C., Hossain, M.A., Snowdon, R.J., Knüfer, J., Tiedemann, A. and Friedt, W. (2013) Genetic analysis of phenylpropanoid metabolites associated with resistance against Verticillium longisporum in *Brassica napus*. *Mol. Breed.* **31**, 347–361.

Olshen, A.B., Bengtsson, H., Neuvial, P., Spellman, P.T., Olshen, R.A. and Seshan, V.E. (2011) Parent-specific copy number in paired tumor-normal studies using circular binary segmentation. *Bioinformatics (Oxford, England)*, **27**, 2038–2046.

van Ooijen, J.W. (2006) *JoinMap 4 Manual*.

Parkin, I., Sharpe, A.G., Keith, D.J. and Lydiate, D.J. (1995) Identification of the A and C genomes in the amphidiploid *Brassica napus* (oilseed rape). *Genome*, **38**, 1122–1131.

Qian, L., Qian, W. and Snowdon, R.J. (2014) Sub-genomic selection patterns as a signature of breeding in the allopolyploid *Brassica napus* genome. *BMC Genom.* **15**, 1.

Radoev, M., Becker, H.C. and Ecke, W. (2008) Genetic analysis of heterosis for yield and yield components in rapeseed (*Brassica napus* L.) by quantitative trait locus mapping. *Genetics*, **179**, 1547–1558.

Raman, H., Dalton-Morgan, J., Diffey, S., Raman, R., Alamery, S., Edwards, D. and Batley, J. (2014) SNP markers-based map construction and genome-wide linkage analysis in *Brassica napus*. *Plant Biotechnol. J.* **12**, 851–860.

Rousseau-Gueutin, M., Morice, J., Coriton, O., Huteau, V., Trotoux, G., Nègre, S., Falentin, C. *et al.* (2017) The impact of open pollination on the structural evolutionary dynamics Allotetraploid *Brassica napus* L. *Genes, Genomes, Genet.* **7**, 705–717.

Rozen, S. and Skaletsky, H. (2000) Primer3 on the WWW for General Users and for Biologist Programmers. In *Bioinformatics Methods and Protocols* (Misener, S. and Krawetz, S.A., eds), pp. 365–386. Totowa, N.J.: Humana Press.

Rygulla, W., Friedt, W., Seyis, F., Lühs, W., Eynck, C., von Tiedemann, A. and Snowdon, R.J. (2007a) Combination of resistance to Verticillium longisporum from zero erucic acid Brassica oleracea and oilseed Brassica rapa genotypes in resynthesized rapeseed (*Brassica napus*) lines. *Plant Breed.* **126**, 596–602.

Rygulla, W., Snowdon, R.J., Eynck, C., Koopmann, B., von Tiedemann, A., Lühs, W. and Friedt, W. (2007b) Broadening the genetic basis of Verticillium longisporum resistance in *Brassica napus* by interspecific hybridisation. *Phytopathology*, **97**, 1391–1396.

Schiessl, S., Huettel, B., Kuehn, D., Reinhard, R. and Snowdon, R.J. (2017) Post-polyploidisation morphotype diversification associates with gene copy number variation. *Sci. Rep.*, **7**, 41845.

Sharpe, A.G., Parkin, I.A., Keith, D.J. and Lydiate, D.J. (1995) Frequent nonreciprocal translocations in the amphidiploid genome of oilseed rape (*Brassica napus*). *Genome*, **28**, 1112–1121.

Snowdon, R.J., Abbadi, A., Kox, T., Schmutzer, T. and Leckband, G. (2015) Heterotic Haplotype Capture: precision breeding for hybrid performance. *Trends Plant Sci.* **20**, 410–413.

Song, K., Lu, P., Tang, K. and Osborn, T.C. (1995) Rapid genome change in synthetic polyploids of Brassica and its implications for polyploid evolution. *Proc. Natl Acad. Sci. USA*, **92**, 7719–7723.

Stein, A., Wittkop, B., Liu, L., Obermeier, C., Friedt, W. and Snowdon, R.J. (2013) Dissection of a major QTL for seed colour and fibre content in *Brassica napus* reveals colocalization with candidate genes for phenylpropanoid biosynthesis and flavonoid deposition. *Plant Breed.* **132**, 382–389.

Suay, L., Zhang, D., Eber, F., Jouy, H., Lodé, M., Huteau, V., Coriton, O. *et al.* (2014) Crossover rate between homologous chromosomes and interference are regulated by the addition of specific unpaired chromosomes in Brassica. *New Phytol.* **201**, 645–656.

Szadkowski, E., Eber, F., Huteau, V., Lodé, M., Huneau, C., Belcram, H., Coriton, O. *et al.* (2010) The first meiosis of resynthesized Brassica napus, a genome blender. *New Phytol.* **186**, 102–112.

Trick, M., Long, Y., Meng, J. and Bancroft, I. (2009) Single nucleotide polymorphism (SNP) discovery in the polyploid *Brassica napus* using Solexa transcriptome sequencing. *Plant Biotechnol. J.* **7**, 334–346.

Werner, S., Diederichsen, E., Frauen, M., Schondelmaier, J. and Jung, C. (2008) Genetic mapping of clubroot resistance genes in oilseed rape. *Theor. Appl. Genet.* **116**, 363–372.

Wittkop, B., Snowdon, R.J. and Friedt, W. (2012) New NIRS calibrations for fiber fractions reveal broad genetic variation in *Brassica napus* seed quality. *J. Agric. Food Chem.* 2248–2256.

Xiong, Z. and Pires, J.C. (2011) Karyotype and identification of all homoeologous chromosomes of allopolyploid *Brassica napus* and its diploid progenitors. *Genetics*, **187**, 37–49.

Xiong, Z., Gaeta, R.T. and Pires, C. (2011) Homoeologous shuffling and chromosome compensation maintain genome balance in resynthesized allopolyploid *Brassica napus*. *Proc. Natl Acad. Sci. USA*, **108**, 7908–7913.

9

Synthesis of bacteriophage lytic proteins against *Streptococcus pneumoniae* in the chloroplast of *Chlamydomonas reinhardtii*

Laura Stoffels[1,†], Henry N. Taunt[1,†,‡], Bambos Charalambous[2] and Saul Purton[1,*]

[1]Algal Biotechnology Group, Institute of Structural and Molecular Biology, University College London, London, UK
[2]Research Department of Infection, University College London Medical School, London, UK

*Correspondence
email: s.purton@ucl.ac.uk
†These authors contributed equally to this work.
‡Present address: Algenuity, Eden Laboratory, Broadmead Road, Stewartby, UK.

Keywords: antimicrobials, *Chlamydomonas reinhardtii*, chloroplast, Pal, Cpl-1, endolysin.

Summary

There is a pressing need to develop novel antibacterial agents given the widespread antibiotic resistance among pathogenic bacteria and the low specificity of the drugs available. Endolysins are antibacterial proteins that are produced by bacteriophage-infected cells to digest the bacterial cell wall for phage progeny release at the end of the lytic cycle. These highly efficient enzymes show a considerable degree of specificity for the target bacterium of the phage. Furthermore, the emergence of resistance against endolysins appears to be rare as the enzymes have evolved to target molecules in the cell wall that are essential for bacterial viability. Taken together, these factors make recombinant endolysins promising novel antibacterial agents. The chloroplast of the green unicellular alga *Chlamydomonas reinhardtii* represents an attractive platform for production of therapeutic proteins in general, not least due to the availability of established techniques for foreign gene expression, a lack of endotoxins or potentially infectious agents in the algal host, and low cost of cultivation. The chloroplast is particularly well suited to the production of endolysins as it mimics the native bacterial expression environment of these proteins while being devoid of their cell wall target. In this study, the endolysins Cpl-1 and Pal, specific to the major human pathogen *Streptococcus pneumoniae*, were produced in the *C. reinhardtii* chloroplast. The antibacterial activity of cell lysates and the isolated endolysins was demonstrated against different serotypes of *S. pneumoniae*, including clinical isolates and total recombinant protein yield was quantified at ~1.3 mg/g algal dry weight.

Introduction

The increase of antibiotic resistance in the last few decades has severely reduced the reliability of antibiotic treatments (Levy, 2005). In particular, the prevalence and range of multidrug resistant strains has increased alarmingly (Levy and Marshall, 2004; WHO report 2014), and there has been a simultaneous decline in the development of new antibiotics, mainly owing to the lower profitability of antibiotics compared to other drugs (Spellberg *et al.*, 2008; Wright, 2012). Health agencies, including the World Health Organization (WHO) and the Infectious Disease Society of America (IDSA), have warned that antimicrobial resistance is an 'increasingly serious threat to global public health' and that there is a real risk of falling back into a pre-antibiotic era in the 21st century (Infectious Diseases Society of America (IDSA) 2011; WHO report 2014). Furthermore, many antibiotics are broad spectrum and therefore each treatment affects the human commensal bacterial flora in addition to the target pathogen. Not only does this increase the patient's susceptibility to opportunistic bacterial and fungal pathogens, there is also increasing evidence that disruption to the natural microbiota can play a key role in conditions such as obesity, type 1 diabetes, inflammatory bowel disease, allergies and asthma (Blaser, 2011; Chen and Blaser, 2007). Repeated exposure to broad spectrum antibiotics can also result in the spread of antimicrobial resistance genes within the human microbiome, which in turn can be passed on to pathogenic bacteria (Schmelcher *et al.*, 2012). With the availability of enhanced diagnostic methods it would therefore be highly desirable to develop narrow spectrum antibiotics that specifically target the pathogen, thus reducing the collateral effects on the commensal bacterial flora and in turn the spread of antimicrobial resistance genes (Blaser, 2011).

Endolysins are bacteriophage-encoded enzymes that accumulate in the cytoplasm of infected bacterial cells throughout the lytic cycle (Loessner, 2005). Towards the end of this cycle, the next generation of phage particles is assembled, and the synthesis of a second phage protein known as a holin is initiated. The holin creates pores in the plasma membrane allowing the endolysin to exit the cytoplasm and attack the cell wall of the infected bacterium (Schmelcher *et al.*, 2012; Young *et al.*, 2000). Cleavage of specific bonds within the peptidoglycan results in the lysis of the host cell and the release of the phage progeny (Loessner, 2005). In the case of Gram-positive bacteria, where the peptidoglycan layer is exposed to the bulk solution, such enzymes have been shown to function as 'exolysins', lysing the target bacterium when applied externally (Schmelcher *et al.*, 2012). Gram-positive targeting endolysins are also typically highly specific for the cell wall of the bacterial species from which they are naturally produced (Borysowski *et al.*, 2006), which allows for their use as targeted antibacterial treatments that specifically eliminate the pathogen without harming the commensal human microbiota, in contrast to many conventional antibiotics.

Bacteriophages and their host bacterial strains have co-evolved over billions of years, with the functionality of the phage endolysin representing a key factor in the ensuing arms race: if a host strain develops effective resistance against endolytic attack, then the phage progeny will not be released and that particular phage lineage will not continue. It is thus inferred that any modern-day phage must necessarily have evolved to target components in the cell wall that are essential for bacterial viability, and as such are highly immutable. It was proposed that the development of bacterial resistance against an endolysin-based antibiotic would be less frequent than that seen for conventional antibiotics (Schmelcher *et al.*, 2012), and indeed attempts to induce resistance by treating bacterial populations to sublethal doses of endolysin over several generations (even in combination with a mutagen) have failed to generate any observable effect (Loeffler *et al.*, 2001; Schuch *et al.*, 2002). Recombinant endolysins have been successfully tested against several different pathogenic bacteria *in vitro* and in a range of animal models (Entenza *et al.*, 2005; Loeffler *et al.*, 2001; Schuch *et al.*, 2002). The high activity, specificity and low occurrence of resistance make endolysins highly promising candidates for use as novel antibacterial agents in human and veterinary medicine.

Despite the benefits associated with endolysins, they are yet to achieve clinical adoption. The reason for this is largely considered to be economic; the production costs for recombinant therapeutic proteins are significantly higher than small molecule bioactives such as traditional antibiotics (Dove, 2002). It is therefore desirable to develop an inexpensive production system for recombinant endolysins that also fulfils all safety criteria for the production of therapeutic proteins. Microalgae offer several potential advantages for the production of recombinant proteins over more established expression platforms, including a low cost of cultivation using media with simple nutrient requirements and the ability to rapidly develop and scale-up production. A number of key species have also been granted GRAS (Generally Recognised as Safe) status for human consumption – as such, the removal of any endogenous toxins and infectious agents from the product is not a concern (Dove, 2002; Rasala *et al.*, 2010; Specht *et al.*, 2010). Furthermore, the ability of photosynthetic organisms to grow using just light, inorganic substances and CO_2 has the potential to decrease the overall carbon footprint of the production process. Finally, microalgae can be grown in full containment under sterile conditions in simple photobioreactors (PBRs) (Gimpel *et al.*, 2014; Specht *et al.*, 2010), significantly reducing the potential for release of transgenes to the environment and reducing the risk of environmental contamination of the production system.

The eukaryotic microalga *Chlamydomonas reinhardtii* offers all the aforementioned advantages together with established techniques for the expression of foreign genes from the chloroplast and nuclear genomes (Almaraz-Delgado *et al.*, 2014; Fuhrmann *et al.*, 1999). Expression in the chloroplast is more attractive as the integration of transgenes into the genome occurs via homologous recombination and so the insertion site can be easily defined. In addition, the levels of recombinant proteins that can be achieved in the organelle are markedly higher compared to expression from the nuclear genome (Potvin and Zhang, 2010).

Furthermore, the chloroplast compartment offers several specific advantages for the production of endolysins given its evolution from a cyanobacterial endosymbiotic ancestor and hence its similarity to the bacterial cells in which endolysins naturally accumulate. For example, (i) recombinant proteins are retained and accumulate exclusively within the chloroplast (Tran *et al.*, 2013), but the chloroplast lacks a peptidoglycan cell wall to which the endolysin might bind or cleave during cell breakage, thereby hindering cultivation and purification; (ii) endolysins are likely to have a long-half-life in this environment as the endogenous proteases in the chloroplast are homologues of the bacterial proteases to which endolysins have evolved to be resistant (Adam *et al.*, 2006); (iii) proteins synthesized in the chloroplast are not subject to post-translational modifications such as glycosylation, unlike cytoplasmic production platforms. Indeed, the potential exploitation of chloroplasts of plants and microalgae as platforms for the production of therapeutic proteins such as antimicrobials and vaccines has been highlighted in several recent reviews (Bock, 2015; Bock and Warzecha, 2010; Daniell *et al.*, 2016; Rasala and Mayfield, 2015).

The biosynthesis of active endolysins in the chloroplast of tobacco was demonstrated by Oey *et al.* (2009a) who showed that the PlyGBS endolysin, which targets *Streptococcus* pathogens, can accumulate in the leaves to levels as high as 70% of total soluble protein. A second study by the same group similarly showed high level production in tobacco of two further *Streptococcus* endolysins, Pal and Cpl-1 (Oey *et al.*, 2009b). Additionally, this study highlighted the limitations of trying to over-express endolysin genes using an *E. coli* platform, as both of Pal and Cpl-1 had a lethal effect on the bacterium, potentially caused by unspecific peptidoglycan cleavage at high concentrations or by binding to the cell wall (Oey *et al.*, 2009b). The tobacco chloroplast has also been employed to make other types of antimicrobial proteins and peptides that are technically difficult to produce in established platforms. For example, the clinically relevant antimicrobial peptides retrocyclin and protegrin have complex secondary structures with multiple disulphide bonds, but are able to assemble into a functional form in the chloroplast (Lee *et al.*, 2011). However, there are major challenges associated with the commercial production of recombinant pharmaceuticals in plants grown in fields or greenhouses, as compared to single cells cultured in contained, sterile fermenter-based systems. These challenges include both compliance with good manufacturing practice (GMP) regulations, and the potentially high cost of purifying product from plant biomass under these GMP regulations (Fischer *et al.*, 2012).

In this study, we have investigated the production of endolysins in the chloroplast of *C. reinhardtii* to examine the suitability of this microalga as a production platform. We have focussed on the Cpl-1 and Pal endolysins which are specific to the human pathogen *Streptococcus pneumoniae*, as previous studies have demonstrated the antibacterial efficacy of these endolysins both *in vitro* and *in vivo* (Garcia *et al.*, 1983; Grandgirard *et al.*, 2008; Jado *et al.*, 2003; Loeffler *et al.*, 2001). We report the generation of transgenic *C. reinhardtii* lines expressing codon-optimized genes for Pal and Cpl-1, and the stable accumulation of the endolysins to a level of ~1% of total soluble protein (TSP). Both crude cell lysates and the isolated endolysins show antibacterial activity against *S. pneumoniae*, including clinical isolates with resistance against penicillin and co-trimoxazole.

Results

Creation of marker-less transgenic lines of *Chlamydomonas reinhardtii* expressing *pal* and *cpl-1* in the chloroplast

To investigate the synthesis of endolysins in *C. reinhardtii*, we created transgenic lines in which a synthetic gene encoding either Pal or Cpl-1 was introduced into the chloroplast genome. The coding regions of the genes were designed based on the codon preference for *C. reinhardtii* chloroplast genes and synthesized *de novo*. (Nakamura *et al.*, 2000) (Figure S1). The coding sequence for a C-terminal human influenza haemagglutinin (HA) epitope tag was also added to each gene to facilitate the detection of the proteins. These synthetic genes are referred to as *pal* and *cpl-1*. A *pal* gene without the HA-tag sequence (referred to as *pal(HA-)*) was also included to investigate whether the HA-tag has an influence on the stability or activity of the endolysin. As shown in Figure 1a, the genes were cloned into the chloroplast transformation vector pSRSapI that uses the chloroplast *psaA-1* promoter and 5′ untranslated region (UTR) to drive transgene expression (Young and Purton, 2014). To compare the efficiency of expression, *pal* and *cpl-1* were also cloned into a second vector, pASapI, that differs from pSRSapI in that it uses the promoter/5′UTR from the *atpA* chloroplast gene (Economou *et al.*, 2014). Each vector is designed to target the transgene cassette to a neutral site in the chloroplast genome downstream of the essential photosystem II gene *psbH*, with restoration of phototrophy in the Δ*psbH* recipient line TN72 used for selection of transformants (Young and Purton, 2014). Such transformants therefore lack any antibiotic-resistance marker and carry the gene of interest (GOI) as the only section of foreign DNA (Figure 1a).

We analysed colonies of putative *C. reinhardtii* transformants for the correct insertion of the endolysin genes into the chloroplast genome and homoplasmicity of the polyploid genome using a three-primer PCR assay (Economou *et al.*, 2014). As shown in Figure 1b, homoplasmic transformants were obtained for pSRSapI_cpl-1, _pal and _pal(HA-). Homoplasmic transformants were also obtained for the pASapI_pal and pASapI_cpl-1 constructs, and also for a control transformant line (TN72_control) lacking the GOI in which TN72 was transformed with the empty pSRSapI vector (data not shown). Furthermore, the correct insertion of the expression cassettes carrying *pal* and *cpl-1* into the lines TN72_pal and Tn72_cpl-1 was confirmed by Southern blot analysis using probes binding within the *pal* and *cpl-1* genes, as well as a probe that binds adjacent to the insertion site of the expression cassette (Figure S2).

Accumulation of Pal and Cpl-1 in the *C. reinhardtii* chloroplast

We analysed representative TN72_cpl-1 and TN72_pal lines generated using the pSRSapI constructs for the presence of the recombinant protein. This was carried out on crude cell extracts by Western blot analysis using antibodies against the HA-tag. The Cpl-1 protein has an expected size of 40 kDa and Pal an expected size of 36 kDa, and as seen in Figure 2a, both proteins are readily detected in the extracts, confirming the accumulation of Pal and Cpl-1 in the respective lines. For the detection of untagged Pal, antibodies to Pal were raised in rabbits using two synthetic peptides derived from the protein sequence (Figure S3) and used to compare Pal levels in the TN72_pal and TN72_pal (HA-) lines. As seen in Figure 2b, a 36-kDa band of similar

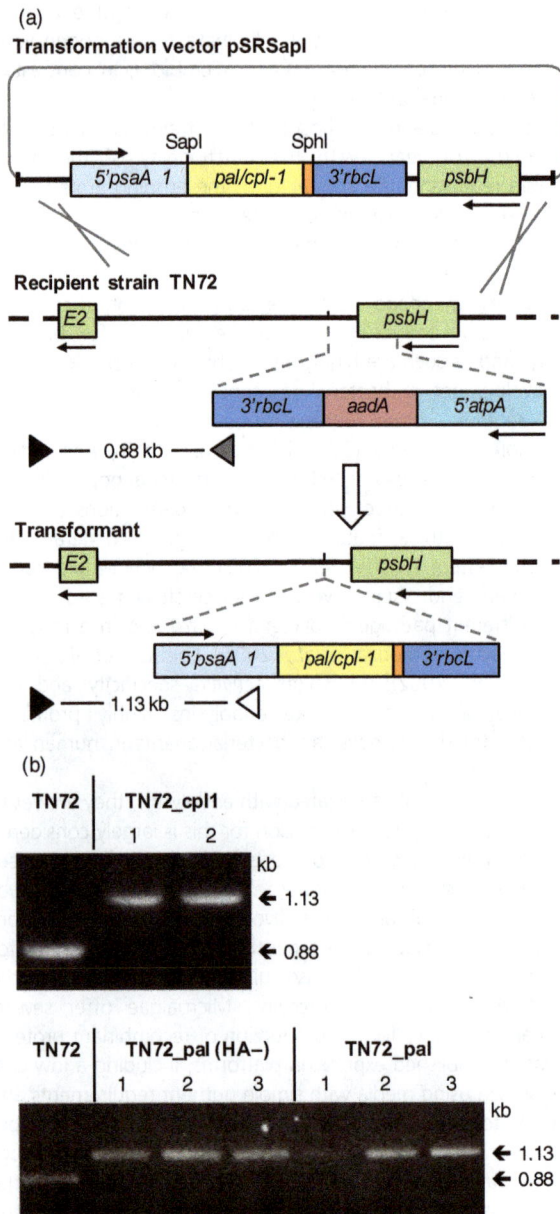

Figure 1 Schematic diagram of the transformation system used for the introduction of *cpl-1* or *pal* into the chloroplast genome. (a) Homologous recombination (grey crosses) between chloroplast sequences (thick black lines) on the transformation vector pSRSapI and the genome of the recipient strain TN72 allows the introduction of the gene (under the control of the *psaA* exon 1 promoter/5′UTR and *rbcL* 3′ UTR from *C. reinhardtii*) into a neutral locus between *psbH* and *trnE2*. Recombination also replaces the *aadA*-disrupted copy of *psbH* with a functional copy of this essential photosystem II gene, allowing transformants to be selected by restoration of phototrophic growth (Wannathong *et al.*, 2016). (b) PCR screening of putative transformants for correct insertion of *cpl-1*, *pal(HA-)* and *pal* and homoplasmicity of the polyploidy chloroplast genome using a set of three primers. Primer F1 (black triangle in (a)) binds to the genome outside of the recombination region. Primer R1 (grey triangle) binds within the *aadA* cassette of TN72 and results together with primer F1 in a product of 0.88 kb. Primer R3 (white triangle) binds within the gene cassette of the transformants and results together with primer F1 in a product of 1.13 kb. Homoplasmy is scored by a band at 1.13 kb and an absence of a 0.88 kb band arising from untransformed copies of the plastome (Wannathong *et al.*, 2016).

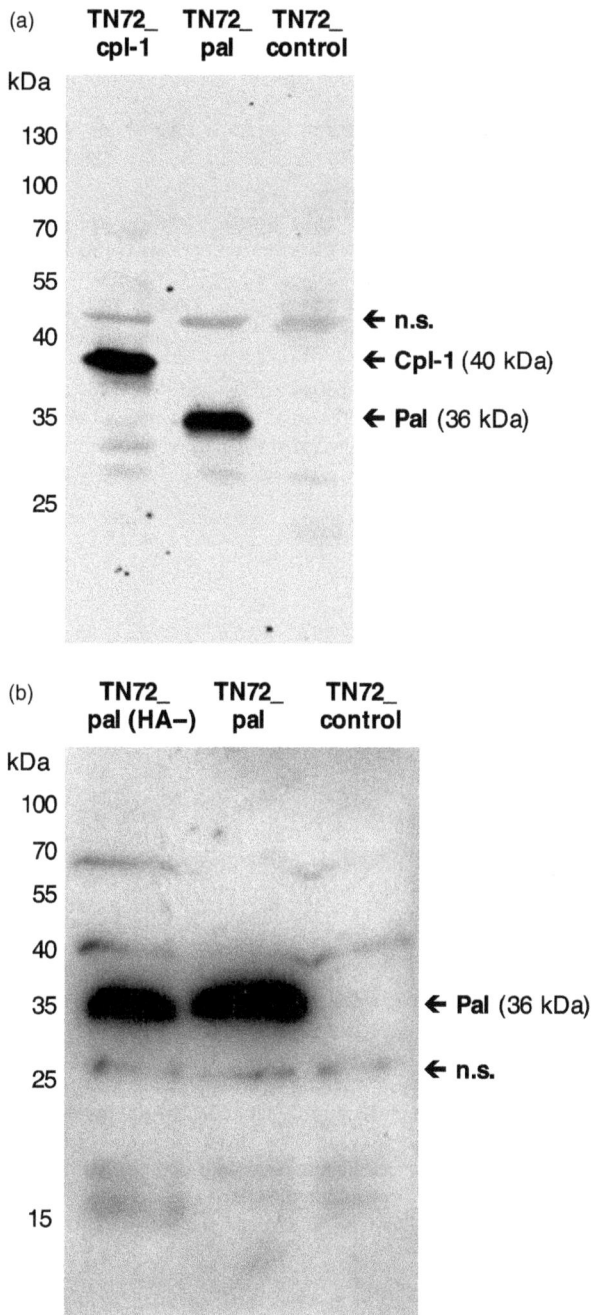

Figure 2 Western blot analysis of endolysin accumulation in transgenic lines. (a) Detection of HA-tagged Cpl-1 and HA-tagged Pal in equivalent loadings of cell extracts from two representative transgenic lines (TN72_cpl1 and TN72_pal) using anti-HA antibodies, together with the negative control transformant (TN72_control). (b) Comparison of levels of untagged Pal and HA-tagged Pal in extracts from two representative lines (TN72_pal(HA-)) using custom made anti-Pal antibodies. Binding of the different primary antibodies to nonspecific (n.s.) bands within the extracts serves as a loading control.

intensity is seen for both the HA-tagged and untagged protein indicating that the HA-tag does not interfere with the stability of this endolysin. Furthermore, we did not observe any reduction in growth rate or biomass production of the Pal and Cpl-1 producing strains in comparison with TN72_control, indicating that the production of the endolysins in the chloroplast is not

detrimental to *C. reinhardtii* growth (shown for TN72_pal in Figure S4).

When we compared the levels of Pal and Cpl-1 synthesized using the *atpA* promoter/5'UTR with those from the *psaA-1* promoter/5'UTR, a higher level was seen for both proteins when using the *psaA-1* elements (Figures S5 and S6), as has been reported previously (Michelet *et al.*, 2011). We therefore used the transgenic lines with the *psaA-1* promoter/5'UTR for all further studies.

To investigate the stability of the endolysins in the *C. reinhardtii* chloroplast, we grew cultures of TN72_pal to mid-log phase and specifically inhibited further protein synthesis in the chloroplast with the addition of chloramphenicol. As shown in Figure 3, there is very little decrease in the amount of Pal in the chloramphenicol-treated cultures compared to the untreated controls even after 71 h, and the protein was still detectable 122 h after the start of the inhibition. In contrast, the endogenous D1 protein (a component of photosystem II) decreased to undetectable levels 7 h after the start of the inhibition. In addition, we found that the stability of Pal was further increased during cultivation in the dark (Figure 3). This indicates a high stability of Pal in the *C. reinhardtii* chloroplast, as has been reported previously for both Pal and Cpl-1 when produced in the tobacco chloroplast (Oey *et al.*, 2009b).

Isolation and quantification of Pal and Cpl-1

For the development of the *C. reinhardtii* chloroplast as a production platform for therapeutic endolysins, it is important to show that the proteins can be extracted and purified in a biologically active state. Furthermore, it was necessary to produce pure standards for the quantification of Pal and Cpl-1 levels within the algal cell. We isolated the endolysins from crude cell extracts as described in the methods, using ultracentrifugation followed by chromatography with the weak anion exchanger diethylaminoethyl (DEAE) cellulose and choline as specific eluent. Pal and Cpl-1 bind to choline in the *S. pneumoniae* cell wall, and DEAE is a choline analogue which is commonly used for the affinity chromatography of choline binding proteins (Jado *et al.*, 2003; Sanz *et al.*, 1988). As the final step, we concentrated Pal and Cpl-1 by either ammonium sulphate precipitation or with centrifugal concentrators. As seen in Figure 4, the Pal and Cpl-1 isolated from the algal biomass using this procedure are in a near purified state with only trace amounts of other proteins, including a 55-kDa protein which is most likely the large subunit of the highly abundant Rubisco enzyme.

To optimize the yield of recombinant endolysin, we first analysed the level of Pal in the algal cell at different growth stages and conditions. This indicated that under mixotrophic growth conditions, the concentration of Pal on a per cell basis is highest during the logarithmic phase and then declines once the cells enter stationary phase (Figures S5, S7 and S8). However, different experiments suggested that a shift from light to dark (*i.e.* heterotrophic conditions) can limit this decline (Figures S7 and 3), possibly due to the inactivity of chloroplast proteases that are dependent on ATP generated through photo-phosphorylation as reported by Preiss *et al.* (2001). However, heterotrophic cultivation of the transformant line from the start of inoculation resulted in an overall reduction in Pal yield per culture volume because of a reduction in cell density under these conditions when compared to mixotrophic growth (Figure S8). Hence, the maximal yield per culture volume was achieved towards the end of the logarithmic phase under conditions of mixotrophic growth.

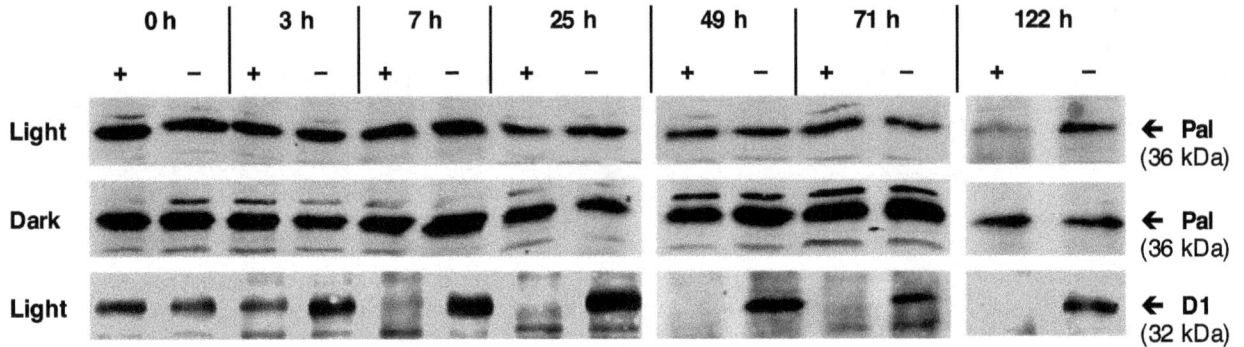

Figure 3 Stability of Pal after inhibition of chloroplast protein synthesis by chloramphenicol. Cultures of TN72_pal were grown under standard conditions to an $OD_{750\ nm}$ of 1 before treating half with 500 µg/mL chloramphenicol and incubating for a further 122 h. The experiment was performed both in the light and the dark with samples taken at different time points. Samples were normalized to the same cell density and analysed by Western blotting using anti-HA and anti-D1 antibodies. + = Chloramphenicol-treated cultures, − = untreated controls.

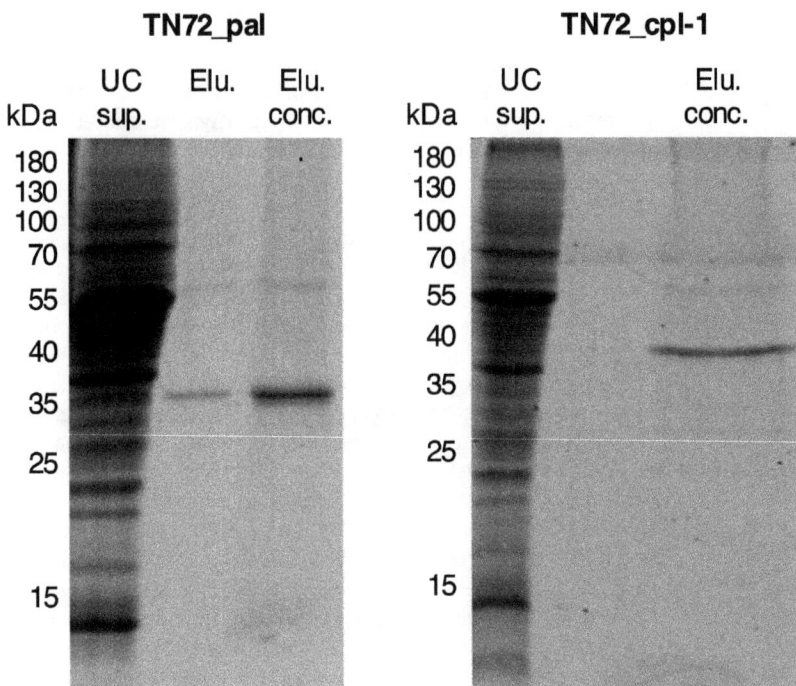

Figure 4 SDS-PAGE analysis of isolated Pal and Cpl-1. The recombinant proteins were isolated from the crude cell lysate following ultracentrifugation (=UC sup.) using diethylaminoethanol (DEAE) cellulose with choline as the specific eluent. The elution fractions with the highest amount of Pal and Cpl-1 were pooled (=Elu.) and concentrated using ammonium sulphate precipitation (=Elu. conc.). The figure shows Coomassie stained gels recorded using the Odyssey® Infrared Imaging system.

To quantitate the amount of recombinant protein, we performed Western blot analyses with a dilution series of the isolated endolysins alongside crude cell lysates of TN72_pal and TN72_cpl-1 taken from mid- and late-logarithmic phase cultures (Figure 5a). Using the Odyssey® imaging system (LI-COR) to quantify the amount of antigen in the samples (Figure 5b), we found that both Pal and Cpl-1 represent ~1% of total soluble protein (TSP) in *C. reinhardtii*. This equates to ~1 mg of recombinant protein per litre of culture volume at the end of the logarithmic phase during mixotrophic cultivation, or ~1.3 mg per gram of algal dry weight (Table 1).

Antibacterial activity of Pal and Cpl-1

To analyse whether Pal and Cpl-1 produced in the *C. reinhardtii* chloroplast are biologically active, we performed turbidity reduction assays (TRA) with the target bacterium *Streptococcus pneumoniae*. In a TRA, the lytic activity of an enzyme is measured as a decrease in optical density (OD_{595}) of a bacterial suspension.

Deoxycholate, which induces rapid autolysis of *S. pneumoniae*, was used as a positive control (Mellroth et al., 2012). As seen in Figure 6, after the addition of a crude extract from TN72_pal or TN72_cpl-1, the OD_{595} of a *S. pneumoniae* suspension decreased in comparison with assays using crude extract from TN72_control or untreated controls. In control assays, the OD_{595} showed a gradual decline, which is most likely caused by autolysis of *S. pneumoniae*. The lytic activity of Pal and Cpl-1 was confirmed in more than 10 independent TRAs. Crude extracts containing either HA-tagged or untagged Pal caused comparable rates of lysis and decreased the OD_{595} to a similar extent in five independent TRAs. This suggests that the HA-tag does not interfere with the enzymatic activity of the endolysin. Furthermore, we found that the extracts containing Pal and Cpl-1 were active against different clinical isolates of *S. pneumoniae*, with all four serotypes (6A, 6B, 6C and 19F) showing lysis including strain ST4157 that is intermediate resistant to penicillin (MIC (minimal inhibitory concentration): 0.125 µg/mL) and resistant to co-trimoxazole (MIC:

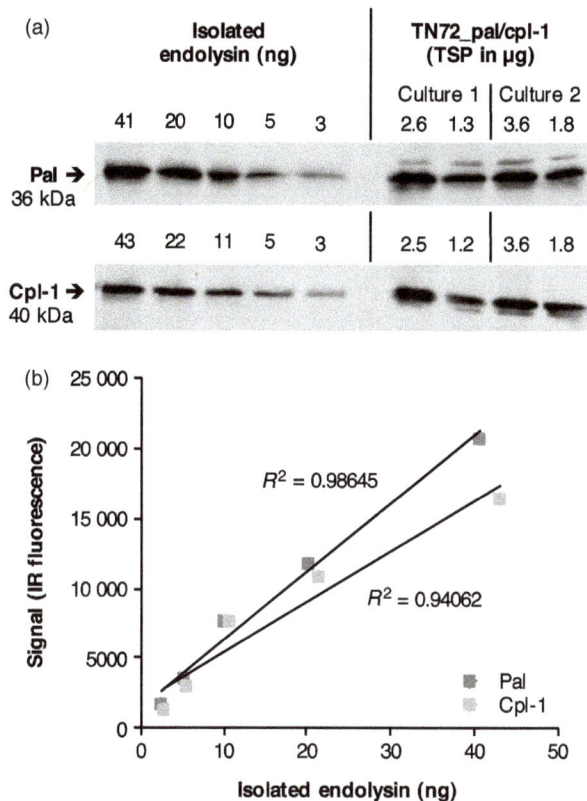

Figure 5 Quantification of Pal and Cpl-1 accumulation in *C. reinhardtii*. (a) Cell lysates of TN72_pal and TN72_cpl-1 were compared to a dilution series of known amounts of isolated Pal or Cpl-1 in Western blot analysis using anti-HA antibodies and IRDye® secondary antibodies. The amount of total soluble protein (TSP) in the cell lysates is stated above each lane. For each strain, 'Culture 1' was grown to an OD_{750} of 2.5 and 'Culture 2' to an OD_{750} of 3.0. The cultures were concentrated five times (lane 1) or 2.5 times (lane 2) in comparison with the initial culture volume before preparation of the cell lysates. (b) The IR fluorescence signals of the serial dilutions were plotted against the protein concentration of the isolated endolysin preparations (taking impurities in the isolated proteins into account) and the equations of the resulting trend lines used to calculate the concentration of Pal/Cpl-1 in the cell lysates.

Table 1 Quantification of Pal and Cpl-1 accumulation in *C. reinhardtii*

	% of total soluble protein (TSP)	mg endolysin/g of cell dry weight	mg endolysin/L culture volume
Pal	0.9–1.2%	1.3 ± 0.4	1.2 ± 0.4 (Culture $OD_{750\ nm}$ of 3.8)
Cpl-1	1.1–1.2%	1.2 ± 0.7	0.9 ± 0.3 (Culture $OD_{750\ nm}$ of 3.0)

6 µg/mL) (Figure 7a and S9). The specificity of Pal was demonstrated by performing TRAs using *Escherichia coli*, *Staphylococcus aureus* and *Streptococcus pyogenes*. As shown in Figure 7b and S10, the algal extract containing Pal, which was active against *S. pneumoniae* cells, did not have a measurable effect on the tested bacteria. This suggests that the algal-produced Pal has specificity for *S. pneumoniae* as described for bacteria-produced

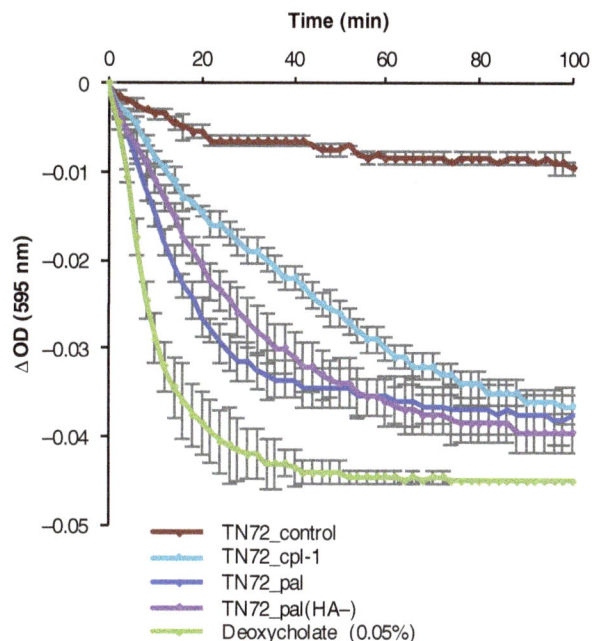

Figure 6 Turbidity reduction assay showing the lytic activity of crude cell extracts containing Pal (HA-tagged and untagged) and Cpl-1. *Streptococcus pneumoniae* (16 NP3, serotype 19F) cells were resuspended in Na-Pi buffer to an OD_{595} of 0.1. Crude extracts of equal concentration of TN72_pal, TN72_pal(HA-), TN72_cpl-1 and TN72_control were added to the *S. pneumoniae* suspension, and the cell lysis was measured as a decrease in OD_{595}. Lysis of the bacterium by 0.05% deoxycholate was used as a positive control. Error bars show ± one standard deviation ($n = 2$).

Pal and other endolysins specific to Gram-positive bacteria (Loeffler *et al.*, 2001; Schmelcher *et al.*, 2012).

To demonstrate directly that Pal and Cpl-1 kill *S. pneumoniae*, we determined the number of colony-forming units per mL (cfu/mL) remaining in a cell suspension after endolysin treatment (Figure 8). One complication we encountered in initial experiments using the crude algal extracts was the presence of an endogenous bactericidal activity in *C. reinhardtii*, as has been previously reported (Ghasemi *et al.*, 2007; Jørgensen, 1962). While this activity did not cause cell lysis (see the TN72_control in Figure 6), it did have a marked effect on the viability of *S. pneumoniae* and therefore masked the effect of the endolysins in the titre assay (data not shown). To demonstrate the killing of *S. pneumoniae* by Pal and Cpl-1 without the interference of this activity, we repeated the assays with the isolated protein. The addition of 25 µg/mL of Pal or 20 µg/mL Cpl-1 reduced the bacterial titre of strain 16 NP3 by nearly 4 $\log_{10}$ units (Pal: 3.8 ± 0.5 and Cpl-1: 3.9 ± 0.2 $\Delta\log_{10}$ cfu/mL) in comparison with an untreated control. Treatment of *S. pneumoniae* D39 with 22.5 µg/mL of Pal reduced the cfu/mL by 3.6 ± 0.3 $\log_{10}$ units. A treatment with 45 µg/mL of Pal was sufficient to kill all *S. pneumoniae* cells in suspensions of D39 and 16 NP3, which was a reduction by more than 7 $\log_{10}$ units for both strains. This supports the TRA results that the recombinant Pal and Cpl-1 are able to effectively kill *S. pneumoniae* through cell lysis. Furthermore, it shows that the two endolysins retain antibacterial activity during the purification procedure.

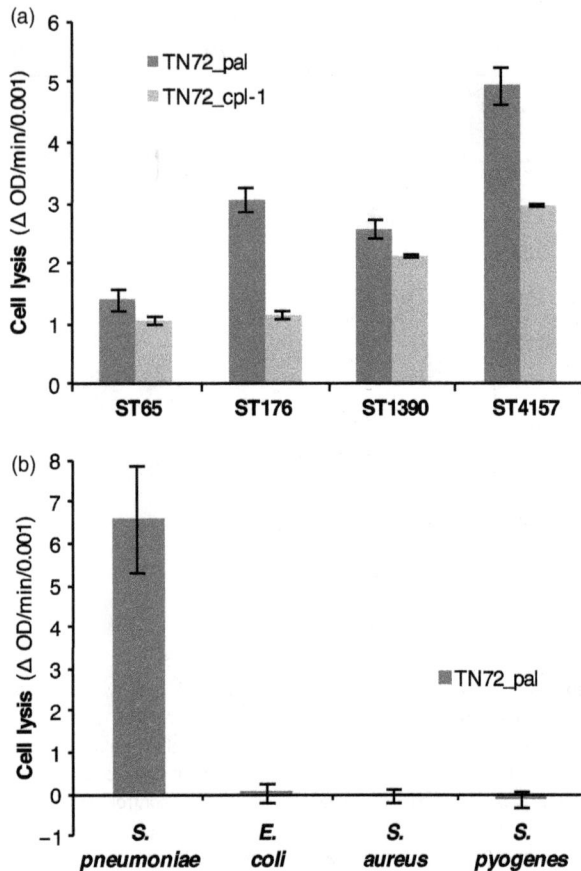

Figure 7 Lytic activity of Pal and Cpl-1 against clinical isolates of *S. pneumoniae* and specificity of Pal for *S. pneumoniae*. (a) Suspensions of clinical isolates with an initial OD$_{595}$ of 0.5, (0.15 for ST4157) were treated with crude extracts of equal concentration of TN72_pal, TN72_cpl-1 or TN72_control. (b) TN72_pal and TN72_control extracts were added to cultures of *S. pneumoniae* (16 NP3, serotype 19F), *Escherichia coli, Staphylococcus aureus* and *Streptococcus pyogenes* (initial OD$_{600}$ 0.8–1.0). The graphs show the decrease in OD per minute (divided by 0.001) in the linear range with changes in the OD of the corresponding control subtracted from the results. ST65 = serotype 6A, ST176 = serotype 6B, ST1390 = serotype 6C, ST4157 = serotype 6B, 35 NP1. The error bars show ± one standard deviation ((a) n = 3, (b) n = 2)).

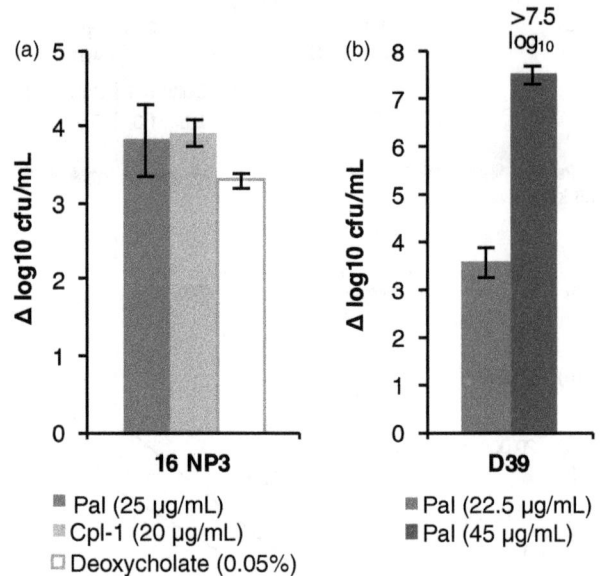

Figure 8 *In vitro* killing of *S. pneumoniae* by isolated Pal and Cpl-1. (a) *S. pneumoniae* 16 NP3 or (b) D39 was incubated with isolated Pal, Cpl-1 or deoxycholate as a positive control. In (b), 45 μg/mL of Pal killed all *S. pneumoniae* D39 cells and therefore reduced the cfu/mL by more than 7.5 log$_{10}$ units. The graph shows the decrease in bacterial titres in powers of 10 in comparison with an untreated control. The error bars show ± one standard deviation (n = 3).

Discussion

In this study, we investigated the synthesis of two bacteriophage endolysins specific to the human pathogen *S. pneumoniae* in the chloroplast of *C. reinhardtii*. We found that Pal and Cpl-1 accumulate as predominately full-length proteins in the transgenic lines. Furthermore, for Pal we demonstrated that this endolysin is highly stable despite the presence of a suite of proteases within the algal chloroplast (Adam *et al.*, 2006). The antibacterial activity of Pal and Cpl-1 against different serotypes of *S. pneumoniae*, including clinical isolates, was demonstrated using both cell lysates and the isolated proteins. Incubation with 25 μg/mL Pal or 20 μg/mL Cpl-1 resulted in a decrease in colony-forming units (cfu) for *S. pneumoniae* of nearly four log$_{10}$ units. These activities are similar to the antibacterial activities and minimal inhibitory concentrations (MIC) described for Pal and Cpl-1 synthesized in *E. coli* or tobacco (Loeffler and Fischetti, 2003;

Loeffler *et al.*, 2001; Oey *et al.*, 2009b; Rodríguez-Cerrato *et al.*, 2007), which suggests that Pal and Cpl-1 are produced in a predominately active form in the algal chloroplast.

The production of the endolysins did not have any negative impact on the viability of the alga suggesting that the production in *C. reinhardtii* could be performed using continuous cultivation systems. Both proteins account for approximately 1% of total soluble protein (TSP) during mixotrophic cultivation, which equates to ~1.3 mg of recombinant endolysin per gram of cell dry weight. These yields are similar to previous reports on the production of recombinant proteins in the *C. reinhardtii* chloroplast, which range from 0.1% to 5% of TSP (Dreesen *et al.*, 2010; He *et al.*, 2007; Manuell *et al.*, 2007; Rasala *et al.*, 2010; Sun *et al.*, 2003; Tran *et al.*, 2013), although significantly lower than yields reported for tobacco chloroplasts which are typically 10%–30% of the TSP in leaves, and in extreme cases are as much as 70% (Bock, 2015). The reason for this difference is not clear, but might reflect the much tighter anterograde regulation of chloroplast gene expression by the numerous nuclear-encoded factors in the single chloroplast of the algal cell (Douchi *et al.*, 2016; Lefebvre-Legendre *et al.*, 2015). Many of these factors act on the 5'UTR of specific gene transcripts and are required for RNA stability or translation initiation; they therefore also regulate transgenes driven from the same 5'UTR. Competition between endogenous and foreign transcripts with the same 5'UTR for low abundance factors might therefore limit the level of recombinant protein. Indeed, higher yields were achieved using the *psbA* promoter/5'UTR to drive transgene expression in a *psbA* deletion mutant (Manuell *et al.*, 2007; Rasala *et al.*, 2010). However, the generation of nonphotosynthetic transgenic lines negates the key value of the alga as a light-driven cell factory; we have therefore focused on maximizing endolysin production in a photosynthetic line. Further studies of the mechanisms of anterograde regulation in the *C. reinhardtii* chloroplast combined with the redesign of UTR elements to eliminate regulatory elements (Specht and Mayfield, 2013) should result in significant improvements in protein production levels (Rasala and Mayfield, 2015).

Microalgae are gaining increasing attention as alternative expression platforms due to the simple and cheap nutrient requirements, as well as easy and cost-effective scalability, which will potentially result in low production costs for recombinant proteins (Barrera and Mayfield, 2013; Gimpel *et al.*, 2014). Furthermore, the GRAS status of several green algal species including *C. reinhardtii*, which lack endotoxins and infectious agents, might decrease downstream processing costs and enables the use of dried algal biomass for the administration of therapeutics and vaccines (Dreesen *et al.*, 2010; Gregory *et al.*, 2012). However, *C. reinhardtii* cultures reach lower final cell densities and have lower recombinant protein yields than conventional expression systems such as mammalian cells, yeast or bacterial systems (Andersen and Krummen, 2002; Jørgensen *et al.*, 2014). Therefore, to determine whether the *C. reinhardtii* chloroplast has the potential to be cost-competitive with conventional expression platforms, an accurate costing for the production of recombinant proteins at scale in *C. reinhardtii* needs to be generated.

This study represents a first report and 'proof of concept' for the production of endolysins in the *C. reinhardtii* chloroplast. The use of endolysins as antibacterial agents is a relatively new, but potentially immense field. It is estimated that, in total, 10^{31} bacteriophages exist globally (Abedon, 2008), so there is a huge untapped resource of endolysins that could be investigated for their potential use as antimicrobials in a wide range of applications (Borysowski *et al.*, 2006). A recent study even describes an endolysin that can traverse epithelial membranes and eliminate intracellular *Streptococcus pyogenes* cells (Shen *et al.*, 2016). Additionally, several research groups have started to create designer endolysins by removing cell wall binding domains, adding additional catalytic domains or combining domains from different endolysins (Schmelcher *et al.*, 2012). Furthermore, endolysins combined with polycationic peptides have been created that are effective against Gram-negative bacteria. These enzymes are referred to as artilysins (Briers *et al.*, 2014). The *C. reinhardtii* chloroplast might offer advantages especially for the production of endolysins and artilysins specific to Gram-negative bacteria. These enzymes are less specific and can have adverse effects on bacterial production systems containing the peptidoglycan substrate (Briers *et al.*, 2014; Oey *et al.*, 2009b). The successful production in the *C. reinhardtii* chloroplast of a Gram-negative endolysin that cannot be produced in bacterial systems would demonstrate a clear advantage of this microalga as an expression platform for endolysins.

Overall it is acknowledged that the field of therapeutic protein production in the green microalga *C. reinhardtii* is in its infancy, and at present is not a viable alternative to more mature technologies. That being said, the use of an algal platform does hold a number of intrinsic advantages over other systems including cost of cultivation and sustainability. The path to greater adoption of this and similar platforms will likely start with niche applications specifically tailored to the unique characteristics exhibited by the algal host; the particular suitability of the *C. reinhardtii* chloroplast for recombinant endolysin expression can be seen as a first step on this road.

Experimental procedures

C. reinhardtii strains and growth conditions

The recipient strain used to create the transformant lines was TN72, a *psbH* deletion mutant of the cell wall-deficient *C. reinhardtii* strain cw15.mt+ (Wannathong *et al.*, 2016). TN72 was maintained on Tris-acetate-phosphate (TAP) medium at 20 °C and 5–10 µmol/m²/s light, as were transformant lines following selection on high salt minimal (HSM) medium (Harris, 2009). Liquid cultures were grown in Erlenmeyer flasks at 100–200 µmol/m²/s, 120 rpm shaking and 25 °C in either illuminated shaking incubators or Algem photobioreactors (Algenuity, Stewartby, UK) for growth studies. Growth was measured by recording the optical density at 750 nm with a Unicam UV/Vis Spectrometer (Thermo Electron Corporation, USA) or 740 nm in the Algem photobioreactor.

Bacterial strains and growth conditions

Activity assays were performed with the *S. pneumoniae* strain 16 NP3 (serotype 19F) or D39. Additional assays were with clinical isolates: ST65 = serotype 6A, H08212 0259; ST176 = serotype 6B, H08052 0052; ST1390 = serotype 6C, H05252 0075; ST4157 = serotype 6B, 35 NP1. The *S. pneumoniae* strains and *S. pyogenes* (ATCC 19615, *Streptococcus* group A) were grown on Columbia blood agar plates overnight at 35 °C under anaerobic or aerobic conditions, respectively. Liquid cultures were grown in trypticase soy broth (30 g/L trypticase soy broth, 3 g/L yeast extract) overnight at 35 °C without shaking. Anaerobic conditions were generated using Oxoid AGS CO_2Gen Compact gas packs and an anaerobic jar from Oxoid. *S. aureus* (ATCC 28213) liquid cultures were grown in ISO-Sensitest broth from Oxoid overnight at 35 °C with shaking. All target bacteria were obtained from the Royal Free Hospital (London, UK).

Plasmid construction

Endolysin coding sequences (including a C-terminal human influenza haemagglutinin (HA) tag sequence) were codon-optimized for the *C. reinhardtii* chloroplast to a Codon Adaption Index (CAI) of 0.8 using the Kazusa CAI table (Nakamura *et al.*, 2000) and synthesized *de novo* by GENEART (Regensburg, Germany). A 5′ SapI and a 3′ SphI restriction site were included in each synthetic gene to facilitate cloning into the chloroplast expression vectors pASapI (*atpA* promoter/5′UTR used to drive GOI expression) and pSRSapI (*psaA* exon 1 promoter/5′UTR) as described in Wannathong *et al.* (2016). Plasmids were prepared from *E. coli* DH5α by alkaline lysis (Sambrook and Russell, 2001) and checked by SphI digestion and by Sanger DNA sequencing (Wolfson Institute for Biomedical Research, University College London). For algal transformation, larger plasmid preparations were made using a QIAfilter Plasmid Midi kit (Qiagen).

Transformation of *C. reinhardtii*

The endolysin genes were introduced into the *psbH–trnE2* intergenic region of the chloroplast genome of TN72 by homologous recombination, restoring a functional *psbH* gene and the ability to grow photosynthetically. The transformation procedure was based on that described by Kindle (1990). Cells were grown to mid-log phase (1–2 × 10^6 cells/mL) in TAP medium, harvested by centrifugation (8000 **g**, 8 min) and resuspended in TAP medium to 2 × 10^8 cells/mL. The cell suspension (300 µL per transformation) was vortexed for 15 s together with 0.3 g of glass beads (Sigma-Aldrich, diameter 0.4–0.6 mm) and 2–20 µg of circular plasmid DNA. Molten top agar (3.5 mL HSM + 0.5% agar) at 42 °C was mixed with the cell suspensions and poured onto HSM plates. The plates were incubated for 1 h in the dark, followed by 3–4 weeks at 20 °C under 50 µmol/m²/s light. To confirm gene insertion and homoplasmicity of transformants, DNA was extracted using the Chelex 100 method as described by Werner and Mergenhagen (1998), and 2 µL of the DNA extract used in PCR reactions with the primers F1, R1 and R3 (Young and Purton, 2014).

Southern blot analysis

Genomic DNA (4 µg) of the recipient strain TN72, TN72_pal and TN72_cpl-1 was cut with the restriction enzymes SphI and EcoRI and separated on an agarose gel (1% w/v). The DNA was transferred to a Hybond N (Amersham Biosciences) membrane as described in Sambrook and Russell (2001). Subsequently, the blot was hybridized with DNA probes (produced either by PCR or the digest of a plasmid) that bind within the *pal* and *cpl-1* genes, as well as a probe that binds just before the insertion site of the expression cassette in all strains. DNA labelling and detection was performed with the DIG High Prime DNA Labelling and Detection Starter Kit II (Roche Diagnostics GmbH) according to the manufacturer's instructions.

SDS-PAGE and Western blot analysis

SDS-PAGE was carried out using the Mini-PROTEAN Tetra cell system (Bio-Rad) and gels based on the recipe by Laemmli (1970) containing 15% acrylamide. Samples were supplemented with sample loading buffer, boiled at 99 °C for 3 min and centrifuged at 21 000 g for 2 min. The gels were run at 120–150 V for 90–120 min. To visualize all proteins, the gels were stained with Coomassie Brilliant Blue R solution for 1 h and destained for at least 2 h. The stained gels were scanned using the Odyssey® imaging system from LI-COR.

For Western blot analyses, the proteins were transferred to Hybond-ECL nitrocellulose membranes (GE Healthcare), at 19 V for 1 h using a Trans-Blot SD semi-dry electrophoretic transfer cell (Bio-Rad). The membranes were blocked overnight in TBS-T (TBS + 0.1% Tween) with 0.5% milk and incubated with the primary antibody (α-HA antibody from rabbit (Sigma-Aldrich product H6908) diluted 1:2000 or anti-Pal antibodies) for 1–3 h followed by incubation with the secondary antibody for 1 h. Both antibodies were resuspended in TBS-T + 0.5% milk, and the membranes were washed after each incubation 3× in TBS-T for 5–10 min. When ECL IgG horseradish peroxidase-linked secondary antibodies (GE Healthcare, 1:10 000) were used, the membranes were incubated with SuperSignal® West Pico Chemiluminescence Substrate (Thermo Scientific) for 5 min and exposed to Hyperfilm ECL (GE Healthcare). For quantitative Western blot analyses, IRDye® secondary antibodies (Dylight™ 800, Thermo Scientific, 1:20 000) and the Odyssey® Infrared Imaging system (Li-COR Biosciences) were used. Custom made anti-Pal antibodies were raised by Eurogentec (Belgium) using two peptides of 17 amino acids (see Figure S2) attached to keyhole limpet haemocyanin carrier protein.

Preparation of *C. reinhardtii* extract

Chlamydomonas reinhardtii cultures were grown to an OD₇₅₀ of 2–3, harvested by centrifugation at 8000 g for 10 min and resuspended in 20 mM sodium phosphate (Na-Pi) buffer (pH 6.9) + protease inhibitor (Roche cOmplete, EDTA-free). Cells were either broken by three cycles of freezing and thawing (liquid nitrogen, 30 °C) or sonication with the Cup Horn (Qsonica, LLC), followed by centrifugation at 21 000 g for 5 min (=crude extract).

Endolysin isolation and quantification

After cell breakage, cell suspensions were centrifuged at 5000 g for 20 min, followed by ultracentrifugation at 100 000 g for 1 h. Ultracentrifugation extracts were applied to a diethylethanolamine (DEAE) cellulose column. The column was washed with six column volumes (CV) of Na–Pi, eight CV of Na–Pi + 1.5 M NaCl and four CV of Na–Pi + 0.1 M NaCl. Pal and Cpl-1 were eluted with 2 CV of Na–Pi + 0.1 M NaCl + 6.5% (w/v) choline. The elution fractions with the highest amount of Pal and Cpl-1 were determined in protein microarrays (2 µL spots on nitrocellulose membranes, followed by Western blot analysis protocol), combined and dialysed using Slide-A-Lyzer G2 dialysis cassettes (20 000 MWCO) (Thermo Scientific). After dialysis, the elution was concentrated 10-fold by ammonium sulphate (AS) precipitation: Pal was precipitated with 35% AS, and Cpl-1 with 50% AS. Alternatively, the chromatography as described above was performed with an ÄKTA pure system (GE Healthcare Life Sciences) and the fractions with the highest amount of endolysin were concentrated with centrifugal concentrators (5000 MWCO). Protein concentrations were determined using Bradford reagent (Sigma-Aldrich) and bovine serum albumin as standard (taking impurities into account).

Endolysin concentration within *C. reinhardtii* lines TN72_pal and TN72_cpl-1 was quantified using serial dilutions of the isolated Pal or Cpl-1 in Western blot analyses with anti-HA antibodies and IRDye® secondary antibodies. The IR fluorescence signals of the serial dilutions were plotted against the protein concentration (determined by Bradford assay, taking impurities into account), and the equations of the resulting trend lines were used to calculate the amount of Pal/Cpl-1 in cell lysates. The amount of total soluble protein (TSP) in the cell suspensions was determined by Bradford assay after cell breakage and centrifugation at 13 000 g for 15 min (Manuell et al., 2007). The cell dry weight was determined after lyophilization of the cell suspensions for 16 h.

Assays of endolysin activity

S. pneumoniae cultures were grown to mid-log phase, (*S. pyogenes*, *E. coli* and *S. aureus* overnight) harvested and resuspended in 20 mM Na–Pi. Most TRAs were performed at 37 °C in 96-well microtitre plates with 20 µL of crude extract in a final volume of 200 µL using a FLUOstar OPTIMA Microplate Reader (BMG Labtech Ltd) recording the OD₅₉₅ over times courses of 60–180 min. The assays in Figure 7b and S10 were performed in cuvettes with 50 µL of crude extract in a final volume of 1 mL at room temperature. For the preparation of the crude extracts used in the TRAs, cultures of TN72_pal, TN72_cpl-1 and TN72_control were grown to the same OD₇₅₀, harvested by centrifugation and resuspended in 1/100 of the culture volume. After cell breakage, the suspension was centrifuged at 21 000 g for 5 min and the supernatant was used as crude extract in the TRAs.

To determine the colony-forming units (cfu) that survived the endolysin treatment, samples from TRAs with the isolated endolysins were diluted in serial 10-fold dilutions and plated in triplicate onto Columbia blood agar plates. After incubation overnight, the cfu were counted and the decrease in cfu/mL calculated relative to an untreated control. For the cfu assays with

S. pneumoniae D39, frozen aliquots (−80 °C) from an overnight culture were thawed and incubated with isolated endolysin at 37 °C for up to 200 min. The decrease in cfu was determined as described above.

Acknowledgements

This work was funded by the UK's Biotechnology and Biological Sciences Research Council (awards BB/F016948/1 and BB/I007660/1). LS was supported by a PhD studentship funded jointly by UCL and Supreme Biotechnologies Ltd. We thank Rosie Young for critical reading of the manuscript, Stavros Panagiotou (University of Liverpool) for providing *S. pneumoniae* D39, Amandine Maréchal and Tom Warelow (UCL) for access to the ÄKTA pure system and advice. The authors declare no conflict of interests.

References

Abedon, S.T. (2008) *Bacteriophage Ecology: Population Growth, Evolution, and Impact of Bacterial Viruses.* Cambridge: Cambridge University Press.

Adam, Z., Rudella, A. and van Wijk, K.J. (2006) Recent advances in the study of Clp, FtsH and other proteases located in chloroplasts. *Curr. Opin. Plant Biol.* **9**, 234–240.

Almaraz-Delgado, A.L., Flores-Uribe, J., Pérez-España, V.H., Salgado-Manjarrez, E. and Badillo-Corona, J.A. (2014) Production of therapeutic proteins in the chloroplast of *Chlamydomonas reinhardtii. AMB Express,* **4**, 57.

Andersen, D.C. and Krummen, L. (2002) Recombinant protein expression for therapeutic applications. *Curr. Opin. Biotechnol.* **13**, 117–123.

Barrera, D.J. and Mayfield, S.P. (2013) High-value recombinant protein production in microalgae. In: *Handbook of Microalgal Culture* (Richmond, A. and Hu, Q. eds), pp. 532–544. New Jersey: John Wiley and Sons, Ltd.

Blaser, M. (2011) Antibiotic overuse: stop the killing of beneficial bacteria. *Nature,* **476**, 393–394.

Bock, R. (2015) Engineering plastid genomes: methods, tools, and applications in basic research and biotechnology. *Annu. Rev. Plant Biol.* **66**, 211–241.

Bock, R. and Warzecha, H. (2010) Solar-powered factories for new vaccines and antibiotics. *Trends Biotechnol.* **28**, 246–252.

Borysowski, J., Weber-Dąbrowska, B. and Górski, A. (2006) Bacteriophage endolysins as a novel class of antibacterial agents. *Exp. Biol. Med.* **231**, 366–377.

Briers, Y., Walmagh, M., Puyenbroeck, V.V., Cornelissen, A., Cenens, W., Aertsen, A., Oliveira, H. *et al.* (2014) Engineered endolysin-based "artilysins" to combat multidrug-resistant Gram-negative pathogens. *mBio,* **5**, e01379–14.

Chen, Y. and Blaser, M.J. (2007) Inverse associations of *Helicobacter pylori* with asthma and allergy. *Arch. Intern. Med.* **167**, 821–827.

Daniell, H., Lin, C.S., Yu, M. and Chang, W.J. (2016) Chloroplast genomes: diversity, evolution, and applications in genetic engineering. *Genome Biol.* **17**, 134.

Douchi, D., Qu, Y., Longoni, P., Legendre-Lefebvre, L., Johnson, X., Schmitz-Linneweber, C. and Goldschmidt-Clermont, M. (2016) A nucleus-encoded chloroplast phosphoprotein governs expression of the photosystem I subunit PsaC in *Chlamydomonas reinhardtii. Plant Cell,* **28**, 1182–1199.

Dove, A. (2002) Uncorking the biomanufacturing bottleneck. *Nat. Biotechnol.* **20**, 777–779.

Dreesen, I.A.J., Hamri, G.C.-E. and Fussenegger, M. (2010) Heat-stable oral alga-based vaccine protects mice from *Staphylococcus aureus* infection. *J. Biotechnol.* **145**, 273–280.

Economou, C., Wannathong, T., Szaub, J. and Purton, S. (2014) A simple, low-cost method for chloroplast transformation of the green alga *Chlamydomonas reinhardtii.* In *Chloroplast Biotechnology* (Maliga, P. ed), pp. 401–411. New York: Humana Press.

Entenza, J.M., Loeffler, J.M., Grandgirard, D., Fischetti, V.A. and Moreillon, P. (2005) Therapeutic effects of bacteriophage Cpl-1 lysin against *Streptococcus pneumoniae* endocarditis in rats. *Antimicrob. Agents Chemother.* **49**, 4789–4792.

Fischer, R., Schillberg, S., Hellwig, S., Twyman, R.M. and Drossard, J. (2012) GMP issues for recombinant plant-derived pharmaceutical proteins. *Biotechnol. Adv.* **30**, 434–439.

Fuhrmann, M., Oertel, W. and Hegemann, P. (1999) A synthetic gene coding for the green fluorescent protein (GFP) is a versatile reporter in *Chlamydomonas reinhardtii. Plant J.* **19**, 353–361.

Garcia, P., Garcia, E., Ronda, C., Lopez, R. and Tomasz, A. (1983) A phage-associated murein hydrolase in *Streptococcus pneumoniae* infected with bacteriophage Dp-1. *J. Gen. Microbiol.* **129**, 489–497.

Ghasemi, Y., Moradian, A., Mohagheghzadeh, A., Shokravi, S. and Morowvat, M.H. (2007) Antifungal and antibacterial activity of the microalgae collected from paddy fields of Iran: characterization of antimicrobial activity of *Chroococcus dispersus. J. Biol. Sci.* **7**, 904–910.

Gimpel, J.A., Hyun, J.S., Schoepp, N.G. and Mayfield, S.P. (2014) Production of recombinant proteins in microalgae at pilot greenhouse scale. *Biotechnol. Bioeng.* **112**, 339–345.

Gregory, J.A., Li, F., Tomosada, L.M., Cox, C.J., Topol, A.B., Vinetz, J.M. and Mayfield, S. (2012) Algae-produced Pfs25 elicits antibodies that inhibit malaria transmission. *PLoS ONE,* **7**, e37179.

Grandgirard, D., Loeffler, J.M., Fischetti, V.A. and Leib, S.L. (2008) Phage lytic enzyme Cpl-1 for antibacterial therapy in experimental pneumococcal meningitis. *J. Infect. Dis.* **197**, 1519–1522.

Harris, E.H. (2009) *The Chlamydomonas Sourcebook - Volume 1 - Introduction to Chlamydomonas and its Laboratory Use.* Amsterdam: Elsevier.

He, D.-M., Qian, K.-X., Shen, G.-F., Zhang, Z.-F., Li, Y.-N., Su, Z.-L. and Shao, H.-B. (2007) Recombination and expression of classical swine fever virus (CSFV) structural protein E2 gene in *Chlamydomonas reinhardtii* chloroplasts. *Colloids Surf. B Biointerfaces,* **55**, 26–30.

Infectious Diseases Society of America (IDSA) (2011) Combating antimicrobial resistance: policy recommendations to save lives. *Clin. Infect. Dis.* **52**, S397–S428.

Jado, I., Lopez, R., Garcia, E., Fenoll, A., Casal, J. and Garcia, P. (2003) Phage lytic enzymes as therapy for antibiotic-resistant *Streptococcus pneumoniae* infection in a murine sepsis model. *Antimicrob. Chemother.* **52**, 967–973.

Jørgensen, E.G. (1962) Antibiotic substances from cells and culture solutions of unicellular algae with special reference to some chlorophyll derivatives. *Physiol. Plant.* **15**, 530–545.

Jørgensen, C.M., Vrang, A. and Madsen, S.M. (2014) Recombinant protein expression in Lactococcus lactis using the P170 expression system. *FEMS Microbiol. Lett.* **351**, 170–178.

Kindle, K.L. (1990) High-frequency nuclear transformation of *Chlamydomonas reinhardtii. Proc. Natl Acad. Sci. USA,* **87**, 1228–1232.

Laemmli, U.K. (1970) Cleavage of structural proteins during the assembly of the head of bacteriophage T4. *Nature,* **227**, 680–685.

Lee, S.-B., Li, B., Jin, S. and Daniell, H. (2011) Expression and characterization of antimicrobial peptides Retrocyclin-101 and Protegrin-1 in chloroplasts to control viral and bacterial infections. *Plant Biotechnol. J.* **9**, 100–115.

Lefebvre-Legendre, L., Choquet, Y., Kuras, R., Loubéry, S., Douchi, D. and Goldschmidt-Clermont, M. (2015) A nucleus-encoded chloroplast protein regulated by iron availability governs expression of the photosystem I subunit PsaA in *Chlamydomonas reinhardtii. Plant Physiol.* **167**, 1527–1540.

Levy, S.B. (2005) Antibiotic resistance—the problem intensifies. *Adv. Drug Deliv. Rev.* **57**, 1446–1450.

Levy, S.B. and Marshall, B. (2004) Antibacterial resistance worldwide: causes, challenges and responses. *Nat. Med.* **10**, S122–S129.

Loeffler, J.M. and Fischetti, V.A. (2003) Synergistic lethal effect of a combination of phage lytic enzymes with different activities on penicillin-sensitive and -resistant *Streptococcus pneumoniae* strains. *Antimicrob. Agents Chemother.* **47**, 375–377.

Loeffler, J.M., Nelson, D. and Fischetti, V.A. (2001) Rapid killing of *Streptococcus pneumoniae* with a bacteriophage cell wall hydrolase. *Science,* **294**, 2170–2172.

Loessner, M. (2005) Bacteriophage endolysins — current state of research and applications. *Curr. Opin. Microbiol.* **8**, 480–487.

Manuell, A.L., Beligni, M.V., Elder, J.H., Siefker, D.T., Tran, M., Weber, A., McDonald, T.L. *et al.* (2007) Robust expression of a bioactive mammalian protein in *Chlamydomonas* chloroplast. *Plant Biotechnol. J.* **5**, 402–412.

Mellroth, P., Daniels, R., Eberhardt, A., Rönnlund, D., Blom, H., Widengren, J., Normark, S. et al. (2012) LytA, major autolysin of Streptococcus pneumoniae, requires access to nascent peptidoglycan. J. Biol. Chem. 287, 11018–11029.

Michelet, L., Lefebvre-Legendre, L., Burr, S.E., Rochaix, J.-D. and Goldschmidt-Clermont, M. (2011) Enhanced chloroplast transgene expression in a nuclear mutant of Chlamydomonas. Plant Biotechnol. J. 9, 565–574.

Nakamura, Y., Gojobori, T. and Ikemura, T. (2000) Codon usage tabulated from international DNA sequence databases: status for the year 2000. Nucleic Acids Res. 28, 292.

Oey, M., Lohse, M., Kreikemeyer, B. and Bock, R. (2009a) Exhaustion of the chloroplast protein synthesis capacity by massive expression of a highly stable protein antibiotic. Plant J. 57, 436–445.

Oey, M., Lohse, M., Scharff, L.B., Kreikemeyer, B. and Bock, R. (2009b) Plastid production of protein antibiotics against pneumonia via a new strategy for high-level expression of antimicrobial proteins. Proc. Natl Acad. Sci. USA, 106, 6579–6584.

Potvin, G. and Zhang, Z. (2010) Strategies for high-level recombinant protein expression in transgenic microalgae: a review. Biotechnol. Adv. 28, 910–918.

Preiss, S., Schrader, S. and Johanningmeier, U. (2001) Rapid, ATP-dependent degradation of a truncated D1 protein in the chloroplast. Eur. J. Biochem. 268, 4562–4569.

Rasala, B.A. and Mayfield, S.P. (2015) Photosynthetic biomanufacturing in green algae; production of recombinant proteins for industrial, nutritional, and medical uses. Photosynth. Res. 123, 227–239.

Rasala, B., Muto, M., Lee, P.A., Jager, M., Cardoso, R.M.F., Behnke, C.A., Kirk, P. et al. (2010) Production of therapeutic proteins in algae, analysis of expression of seven human proteins in the chloroplast of Chlamydomonas reinhardtii. Plant Biotechnol. J. 8, 719–733.

Rodríguez-Cerrato, V., García, P., Del Prado, G., García, E., Gracia, M., Huelves, L., Ponte, C. et al. (2007) In vitro interactions of LytA, the major pneumococcal autolysin, with two bacteriophage lytic enzymes (Cpl-1 and Pal), cefotaxime and moxifloxacin against antibiotic-susceptible and -resistant Streptococcus pneumoniae strains. J. Antimicrob. Chemother. 60, 1159–1162.

Sambrook, J. and Russell, D.W. (2001) Molecular Cloning. A Laboratory Manual. New York: Cold Spring Harbor Laboratory Press.

Sanz, J.M., Lopez, R. and Garcia, J.L. (1988) Structural requirements of choline derivatives for "conversion" of pneumococcal amidase A new single-step procedure for purification of this autolysin. FEBS Lett. 232, 308–312.

Schmelcher, M., Donovan, D.M. and Loessner, M.J. (2012) Bacteriophage endolysins as novel antimicrobials. Future Microbiol. 7, 1147–1171.

Schuch, R., Nelson, D. and Fischetti, V.A. (2002) A bacteriolytic agent that detects and kills Bacillus anthracis. Nature, 418, 884–889.

Shen, Y., Barros, M., Vennemann, T., Gallagher, D.T., Yin, Y., Linden, S.B. and Heselpoth, R.D. et al. (2016) A bacteriophage endolysin that eliminates intracellular streptococci. eLife, 5, e13152.

Specht, E.A. and Mayfield, S.P. (2013) Synthetic oligonucleotide libraries reveal novel regulatory elements in Chlamydomonas chloroplast mRNAs. ACS Synth. Biol. 2, 34–46.

Specht, E., Miyake-Stoner, S. and Mayfield, S. (2010) Micro-algae come of age as a platform for recombinant protein production. Biotechnol. Lett. 32, 1373–1383.

Spellberg, B., Guidos, R., Gilbert, D., Bradley, J., Boucher, H.W., Scheld, W.M., Bartlett, J.G. et al. (2008) The epidemic of antibiotic-resistant infections: a call to action for the medical community from the Infectious Diseases Society of America. Clin. Infect. Dis. 46, 155–164.

Sun, M., Qian, K., Su, N., Chang, H., Liu, J., Shen, G. and Chen, G. (2003) Foot-and-mouth disease virus VP1 protein fused with cholera toxin B subunit expressed in Chlamydomonas reinhardtii chloroplast. Biotechnol. Lett. 25, 1087–1092.

Tran, M., Henry, R.E., Siefker, D., Van, C., Newkirk, G., Kim, J., Bui, J. et al. (2013) Production of anti-cancer immunotoxins in algae: ribosome inactivating proteins as fusion partners. Biotechnol. Bioeng. 110, 2826–2835.

Wannathong, T., Waterhouse, J.C., Young, R.E.B., Economou, C.K. and Purton, S. (2016) New tools for chloroplast genetic engineering allow the synthesis of human growth hormone in the green alga Chlamydomonas reinhardtii. Appl. Microbiol. Biotechnol. 100, 5467–5477.

Werner, R. and Mergenhagen, D. (1998) Mating type determination of Chlamydomonas reinhardtii by PCR. Plant Mol. Biol. Rep. 16, 295–299.

WHO report (2014) WHO | Antimicrobial resistance: global report on surveillance 2014.

Wright, G.D. (2012) Antibiotics: a new hope. Chem. Biol. 19, 3–10.

Young, R.E.B. and Purton, S. (2014) Cytosine deaminase as a negative selectable marker for the microalgal chloroplast: a strategy for the isolation of nuclear mutations that affect chloroplast gene expression. Plant J. 80, 915–925.

Young, R., Wang, I.-N. and Roof, W.D. (2000) Phages will out: strategies of host cell lysis. Trends Microbiol. 8, 120–128.

10

Down-regulation of BnDA1, whose gene locus is associated with the seeds weight, improves the seeds weight and organ size in *Brassica napus*

Jie-Li Wang[1,†], Min-Qiang Tang[2,†], Sheng Chen[1,†], Xiang-Feng Zheng[1], Hui-Xian Mo[3], Sheng-Jun Li[3], Zheng Wang[1], Ke-Ming Zhu[1], Li-Na Ding[1], Sheng-Yi Liu[2], Yun-Hai Li[3,*] and Xiao-Li Tan[1,*]

[1]*Institute of Life Sciences, Jiangsu University, Zhenjiang, China*
[2]*The Oil Crops Research Institute (OCRI) of the Chinese Academy of Agricultural Sciences (CAAS), Wuhan, China*
[3]*State Key Laboratory of Plant Cell and Chromosome Engineering, Institute of Genetics and Developmental Biology (IGDB), Chinese Academy of Sciences (CAS), Beijing, China*

*Correspondence
e-mail xltan@ujs.edu.cn and Y.H. Li;
e-mail yhli@genetics.ac.cn.
†These authors contributed equally to this work.

Keywords: Association analysis, *B. napus*, *DA1*, Overexpression, Seed size.

Summary

Brassica napus L. is an important oil crop worldwide and is the main raw material for biofuel. Seed weight and seed size are the main contributors to seed yield. DA1 (DA means big in Chinese) is an ubiquitin receptor and negatively regulates seed size. Down-regulation of AtDA1 in *Arabidopsis* leads to larger seeds and organs by increasing cell proliferation in integuments. In this study, BnDA1 was down-regulated in *B. napus* by over expressed of *AtDA1*^{R358K}, which is a functional deficiency of DA1 with an arginine-to-lysine mutation at the 358th amino acid. The results showed that the biomass and size of the seeds, cotyledons, leaves, flowers and siliques of transgenic plants all increased significantly. In particular, the 1000 seed weight increased 21.23% and the seed yield per plant increased 13.22% in field condition. The transgenic plants had no negative traits related to yield. The candidate gene association analysis demonstrated that the *BnDA1* locus was contributed to the seeds weight. Therefore, our study showed that regulation of DA1 in *B. napus* can increase the seed yield and biomass, and DA1 is a promising target for crop improvement.

Introduction

Rapeseed (*Brassica napus* L.) is an important oil crop. Its contribution to global oilseed production is considerable, and approximately 71.0 million metric tonnes were produced worldwide in 2014 (data from FAOSTAT http://faostat3.fao.org/browse/Q/QC/E). Although the yield of rapeseed is high, the supply of rapeseed oil is insufficient globally. Furthermore, market demand for vegetable oil-derived biodiesel is increasing rapidly because the amount of available fossil fuels is decreasing dramatically (Sidibe *et al.*, 2010). Therefore, there is a need to maximize the productivity of vegetable oil to solve the problem of edible oil supply and relieve pressure on energy supply (Chauhan *et al.*, 2012). Rapeseed oil is a crucial source of vegetable oil, and currently, many studies have focused on enhancing the synthetic activity of oil to elevate the oil content (Li *et al.*, 2013b, Tan *et al.*, 2014). However, lipid synthesis efficiency is limited, and achieving oil content to 50%–55% in *B. napus* seems to be a natural limit (Li *et al.*, 2006). Increasing the biomass of rapeseed could be an alternative way of increasing the total oil when oil content remains constant.

Seed weight is a main yield component and also a key trait that influences seedling establishment and seed dispersal (Gegas *et al.*, 2010; Kesavan *et al.*, 2013; Zhang *et al.*, 2015). The seedlings of large-seeded plants are capable to adapt the stressful environment, while small-seeded plants are thought to produce large numbers of seeds (Moles *et al.*, 2005; Westoby *et al.*, 2002). The seed and organs size is regulated by both cell number and cell size, which are controlled by coordinating cell proliferation and cell expansion during organogenesis (Mizukami, 2002; Sugimoto-Shirasu and Roberts, 2003). The mechanism that regulated the seed size and weight was well studied in *Arabidopsis*. For example, phytohormone signalling pathway is involved in the seed size regulation, and cytokinin acts downstream of the IKU pathway to regulate seed size (Li *et al.*, 2013a), while *iku* mutations reduce seed size due to precocious cellularization of the endosperm (Garcia *et al.*, 2003; Luo *et al.*, 2005; Wang *et al.*, 2012). AUXIN RESPONSE FACTOR2 (ARF2) regulates seed cell proliferation in the integuments to affect seed size (Schruff *et al.*, 2006). Other factors also influence the seed size. KLUH/CYTOCHROME P450 78A5 (CYP78A5) affects seed size by promoting cell proliferation in the integuments (Adamski *et al.*, 2009). In other side, TRANSPARENT TESTA GLABRA2 (TTG2) promotes cell elongation in the integuments to increase the seed size (Garcia *et al.*, 2005). On the contrary, APETALA2 (AP2) represses cell elongation in the integuments to suppress the seed size (Jofuku *et al.*, 2005; Ohto *et al.*, 2005, 2009). Overexpression of CYP78A6 promotes seed size by both increasing the cell proliferation and cell elongation in the integuments (Fang *et al.*, 2012). Seed size is also influenced by zygotic tissues. SHORT HYPOCOTYL UNDER BLUE1 (SHB1) promotes endosperm proliferation to increase seed growth (Zhou *et al.*, 2009). In addition, the endosperm growth is also affected by epigenetic mechanisms (Xiao *et al.*, 2006).

The ubiquitin receptor DA1 restricts cell proliferation in the integuments to affect seed size (Li *et al.*, 2008; Xia *et al.*, 2013). Mutations in EOD1, which encodes the E3 ubiquitin ligase (Disch *et al.*, 2006; Li *et al.*, 2008), synergistically promote the seed size of *da1-1*, showing that DA1 acts synergistically with EOD1/BB to regulate seed size. On the whole, factors that regulate the seeds size have successfully characterized, and many works have been reported. However, there was only one paper published on seed size in *B. napus* (Liu *et al.*, 2015).

The network for controlling seed size has been well described in model plants like *Arabidopsis*. But the similar study has not been reported in rapeseeds. AtDA1 negatively regulates seed and organ size, and the phenotype of the *da1-1* mutant shows large seeds and organs in *A. thaliana* (Li *et al.*, 2008). Then, we overexpressed deficient AtDA1 (AtDA1^{R358K}) in rapeseed to investigate whether larger seed sizes and higher seed yields could be obtained and confirm the potential way to improve oil crop yields.

Candidate gene association analysis is based on polymorphism at the DNA level. It is used to discover alleles that make large contributions to the target traits from the natural population and is helpful to further validate gene function and dissect the site of the key role. Additionally, it can be used to analyse multiple effects for pleiotropic genes (Chen and Lubbersted, 2010). This method was successfully applied to discover the *Dwarf8* (Thornsberry *et al.*, 2001) and *Vgt1* (Salvi *et al.*, 2007) polymorphisms associated with variation in flowering time. In addition, association analysis of flowering time genes *Hd1*, *Hd3a* and *Ehd1* showed that the Hd1 protein type, *Hd3a* promoter and *Ehd1* expression level were major factors in rice flowering (Takahashi *et al.*, 2009). However, association analysis has not been used to validate gene function in *B. napus* to our knowledge. In this work, candidate gene association analysis was also used to verify the contribution of *BnDA1* to seed weight in a natural *B. napus* population.

Results

BnDA1 is highly homologous with AtDA1

BnDA1 (BnaC05g14930D) and AtDA1 contain 507 and 532 amino acids, respectively; they share 83.15% identity (Figure S1). BnDA1 contains the LIM-DA1 domain, corresponding to the '<LIM···LIM>' in Figure 1a, and Zn binding sites, which are located in the LIM-DA1 domain. BnDA1 has another domain (DUF3633 superfamily), which corresponds to the '(DUF3633···DUF3633)' in Figure 1a. This domain family is found in bacteria and eukaryotes. This functional domain is very conservative in BnDA1 and AtDA1. The mutation site for AtDA1^{R358K} is in the domain of the DUF3633 superfamily (The '*' shown in Figure 1a). Showing that the DUF3633 functional domain is related to the activity of DA1.

The phylogenetic tree analysis, based on the amino acid sequences, showed that BnDA1 and AtDA1 were in one clade (Figure 1b). DA1-like isoform proteins, BrDA1-like X1, BrDA1-like X2, BrDA1-like X3 and BrDA1-like X4, are selected from *B. rapa*. The proteins BnDA1 (BnaC05DA1), BnaAO6DA1, BnaC08DA1 and BnaA08DA1 are all from *B. napus*. BnDA1 has similar domains with AtDA1. BnDA1 and AtDA1 also have the highest similarity in amino acid levels (Figures 1a, S1), so we named it BnDA1. From the sequence analysis, we can show that BnDA1 and AtDA1 are closely related in terms of amino acid sequence

and evolutionary relationship. This means that they are also likely to have similar functions.

Overexpression of *BnDA1* can recover the *da1-1* phenotype in *Arabidopsis thaliana*

To further prove that the function of BnDA1 and AtDA1 is conserved, 35S promoter::*AtDA1* and a 35S promoter::*BnDA1* expression vectors were constructed and transformed into the *Arabidopsis* mutant *da1-1*. We found that the *da1-1* phenotype was recovered in leaves (Figure S2a), flowers (Figure S2b) and seeds (Figure S2c). The complementary assays revealed that BnDA1 and AtDA1 had similar functions in regulating seed and organ size in *A. thaliana*, and potentially in *B. napus*.

The analysis of expression level of *DA1* in *AtDA1*^{R358K} overexpression lines in *B. napus*

As overexpression of *AtDA1*^{R358K} resulted in large seed and organs (Li *et al.*, 2008; Weng *et al.*, 2008), we overexpressed the deficient *AtDA1*^{R358K} in rapeseed to verify the function of AtDA1 in *B. napus* and to obtain potentially larger seeds. The binary vector containing *35S::AtDA1*^{R358K} was transformed into wild-type (WT) rapeseed plants by floral dipping approach (Li *et al.*, 2010). The transgenic plants were identified by PCR (the primers were shown in Table S1 and the PCR product is shown in Figure S3). We further identified the expression levels of *AtDA1* by real-time quantitative PCR (qRT-PCR) and RT-PCR. Among *AtDA1* over expression lines, Line 6 showed an almost threefold higher expression, and Line 8 and Line 11 were more than fivefold higher than in the WT (Figure 2a). Therefore, these three lines were chosen for further phenotype analysis. RT-PCR analysis also produced identical results to qPCR (Figure 2b). Due to the high sequence similarity between *AtDA1* and *BnDA1*, the set of primers chosen could bind both the *AtDA1* and *BnDA1* sequences (Figure S4). Thus, the WT control (CK) had the background in qPCR and RT-PCR analysis. Through expression analysis, we confirmed that the higher expression level of *AtDA1*^{R358K} lines in *Brassica napus* was obtained.

Overexpression of *AtDA1*^{R358K} increase thousand seed weight (TSW) and organs size in rapeseed

The TWS of 2 years data (Figure 3a) showed that the seed size of the transgenic lines was significantly larger than the CK seed. The TWS of line 11 was 21.23% higher than CK (Figure 3a, b). The size of a seed is regulated by the coordinated growth of the embryo, endosperm and maternal tissue. We therefore examined the size of the embryo and hypocotyl and found that the sizes of the embryo and hypocotyl in transgenic lines had also increased compared to CK (Figure 3c). These results suggested that overexpression of *AtDA1*^{R358K} increased the seed weight and size in *B. napus*.

The seed size also affected the seedlings size. After the seeds germinated, the cotyledon weight and area of 3-d-old seedlings from the three selected transgenic lines were measured. The results showed that overexpression of *AtDA1*^{R368K} increased cotyledon size compared to CK. Cotyledons of the 3-d-old transgenic line were significantly larger than the wild-type cotyledons (Figure 3d and e). The biomass data for the seedlings showed that the seedling weight had also increased (Figure 3f). Therefore, the overexpression of *AtDA1*^{R358K} increased the seed weight and size and further enhanced the size of the cotyledons.

(a)

(b)

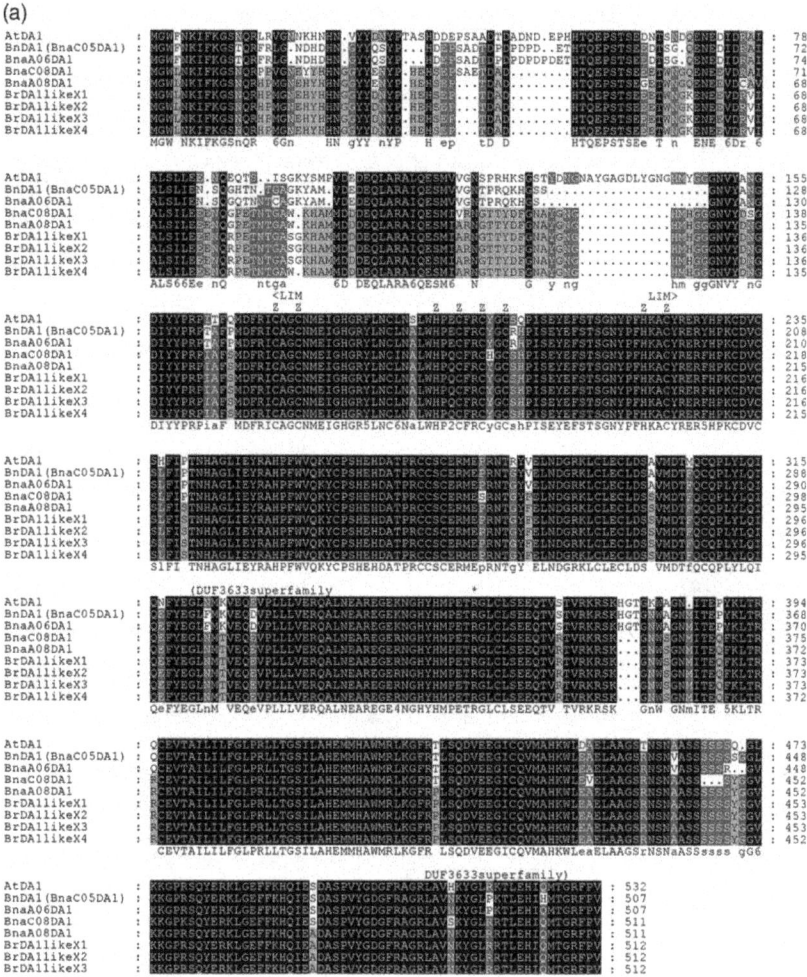

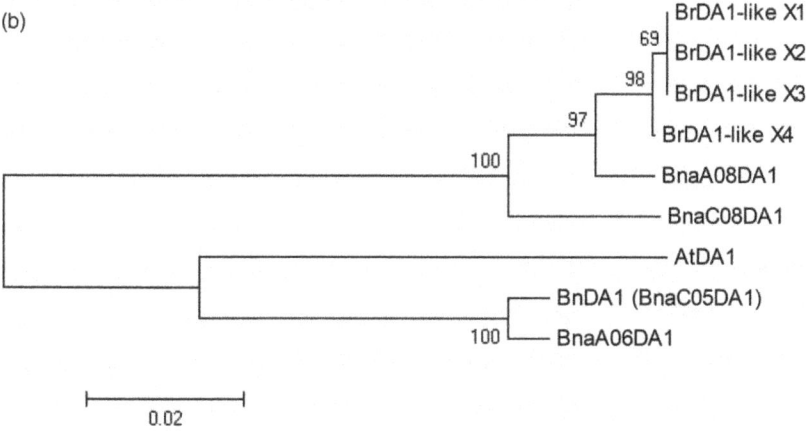

Figure 1 Analysis of the BnDA1 and AtDA1 amino acids sequences. (a) Multiple sequence alignments of the amino acid sequences. Proteins BrDA1-like X1 to BrDA1-like X4 come from *B. rapa*. They are DA1-like isoform proteins. The proteins BnaC08g18690D, BnaA08g22120D and BnDA1 (BnaC05g14930D) come from *B. napus*. They are DA1 related or have similar functions. AtDA1 is from *A. thaliana*. (b) Phylogenetic tree for BnDA1, AtDA1 and DA1-like protein from *B. rapa* and *B. napus*. The picture was constructed by the neighbour-joining method using IMAGE5.2.

We also collected the biomass data for leaves, which showed that overexpression of *AtDA1*^{R358K} produced large leaves compared to CK. The leaves of the transgenic lines were also more rounded than CK (Figure 4a), like those observed in *Arabidopsis da1-1* mutant. We further measured the length, width and area of 24-d-old leaves. The results demonstrated that they had all increased (Figure 4b, c and d). The leaf palisade cell sizes of CK and *AtDA1*^{R358K} overexpression plants were then measured to evaluate whether the leaf size increase in *AtDA1*^{R358K} overexpression plants was due to increased cell proliferation. The results showed

that the palisade cells in *AtDA1*^{R358K} overexpression plants were significantly smaller than in CK (Figure 4e). The average area of palisade cells in *AtDA1*^{R358K} overexpression plants was about 70% smaller than CK (Figure 4f). This implied that the larger leaf was caused by an increase in the number of cells in *AtDA1*^{R358K} overexpression plants.

The flower is one of the reproductive organs, and it is a major feature of higher plants. Flower pollination is an important process during sexual reproduction in flowering plants. As shown in Figure 5, the flowers were larger on the

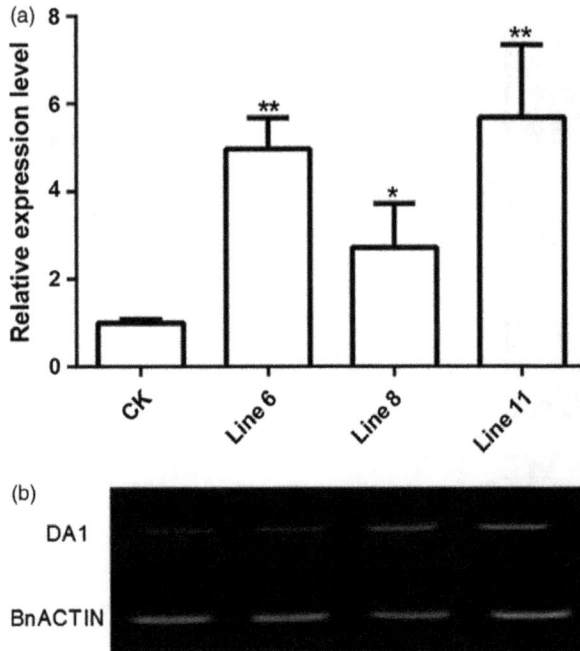

Figure 2 Relative expression levels of *AtDA1* in *AtDA1^{R358K}* transgenic plants. (a) Relative expression levels of *AtDA1* in homozygous *AtDA1^{R358K}* transgenic plants and CK were quantified by quantitative real-time PCR. The quantity of each transcript was measured using the $2^{-\Delta\Delta Ct}$ method. *BnACTIN2* was used as an internal control. Value represents mean ±the standard error from three independent rapeseed samples. (b) Semi-quantitative RT-PCR also showed that *AtDA1* expression levels significantly increased in Line 6, Line 8 and Line 11 compared to CK, *BnACTIN2* was also used as an internal control. Value represents mean ± standard error from three independent rapeseed samples. * means a significant difference at the $P < 0.05$ level, and ** represents a significant difference at the $P < 0.01$ level.

AtDA1^{R358K} overexpression plants compared to CK (Figure 5a). The area, length and width of the petals increased significantly (Figure 5b, c and d), indicating that overexpression of *AtDA1^{R358K}* had an important influence on flower development, but the timing, frequency and duration of flowering have no difference.

After flowering, the siliques developed. We compared the size of the siliques between transgenic plants and CK. As expected, plants overexpressing *AtDA1^{R358K}* formed wider siliques compared to CK (Figure 5e). The silique width of transgenic lines was noticeably wider than the width of CK (Figure 5f), but the number of seeds in the siliques did not differ.

Overexpression of *AtDA1^{R358K}* increases the seed yield per plant

The overexpression of *AtDAI^{R358K}* leads to larger seeds, seedlings, embryos, cotyledons, leaves, flowers and siliques compared to CK. Apart from these positive agronomic traits, the transgenic plants also had other superior agronomic traits. The whole plant seed weights increased significantly. Line 11 increased by 13.22% compared to CK (Figure 6a), and the plant average stem diameter and plant fresh weight were also significantly higher (Figure 6b and c). This may lead to the observed increased yield and biomass of the transgenic plants compared to CK. Both the seed weights and biomass increased together (Figure 6a and c), and the final

seed yield in the field could increase in a similar way. The over expression of *AtDA1^{R358k}* increased the seed yield of a single plant, whether it accompanied by other negative agronomic traits. Therefore, we measured primary branch number per plant, silique number per primary branches, silique number per rachis and seed number per pod. These agronomic traits are the major components of rapeseed yield. The data showed that all these agronomic traits did not significantly change (Figure 6d and e). Therefore, overexpression of *AtDA1^{R358K}* improved the seed yield of a single plant in filed without producing any negative agronomic traits.

BnDA1 homologous gene expression analysis and association analysis

As *B. napus* is a recent allotetraploid species derived from *B. rapa* and *B. oleracea* (Naganara, 1935), it has many highly homologous genes. To identify BnDA1 as a major functional gene, transcriptome analysis was carried out on the unfolded petal and ovule. Compared to expression level of three other homologous genes mentioned in Figure 1b, the FPKM (Reads Per Kilobase of exon model per Million mapped reads) of BnDA1 was significantly higher and more than twice as high in both tissues (Figure S5), this indicated that BnDA1 gene was a major functional gene in the two tissues. Furthermore, to validate the function of *BnDA1* for TSW in other *B. napus* accessions, we performed association analysis of a set of 224 accessions collected from different geographic position, TSW varied from 2.83 to 5.52 g with an average of 3.88 g, and 20 polymorphism SNPs were detected in BnDA1 by re-sequencing. The results displayed two significantly associated SNPs, BnDA1_8885118 and BnDA1_8885818, explaining 4.50% and 5.18% of TSW variation in this population, respectively (Figure 7a). For BnDA1_8885818, the corresponding line had the low TSW of 2.83 g, the SNP was in coding sequence of *BnDA1* and caused the wild-type form of BnDA1, which negatively regulated the seed size. Finally, led to the lower TSW. In the QQ plot, the observed value of the two SNPs significantly deviates from expected value (Figure 7b), indicating that they were associated with TSW. Therefore, BnDA1 was the main functional gene among it homologues in the unfolded petal and ovule, and the BnDA1 locus in natural population was also associated with the seeds TSW.

Discussion

Both *B. napus* and *A. thaliana* belong to the cruciferae family. The sequence of the functional genes is highly conserved (Jiang et al., 2015; Navabi et al., 2013). From the sequence analysis, we found that BnDA1 had a high sequence similarity with AtDA1 and that they have the same functional domain, suggesting that the BnDA1 and AtDA1 have conserved functions. The BnDA1 could functionally complement the *da1-1* phenotype, providing further evidence that BnDA1 and AtDA1 have the same function with regard to regulating seed and organs size. Being able to produce larger seeds and organs in rapeseed oil crops has enormous economic value. If DA1 negatively regulates the seed size, then overexpression of the function deficient *AtDA1R^{358K}* gene in rapeseed could compete with BnDa1, resulting in the down-regulation of BnDa1 and the production of larger seeds and organs. Similarly, the protein encoded by *da1-1* (*AtDA1^{R358K}*) has negative effects on DA1 and DA1-related proteins (Kesavan et al., 2013; Wang et al., 2012; Zhao et al., 2015). Our experiments demonstrated that overexpression of AtDA1^{R358K} in

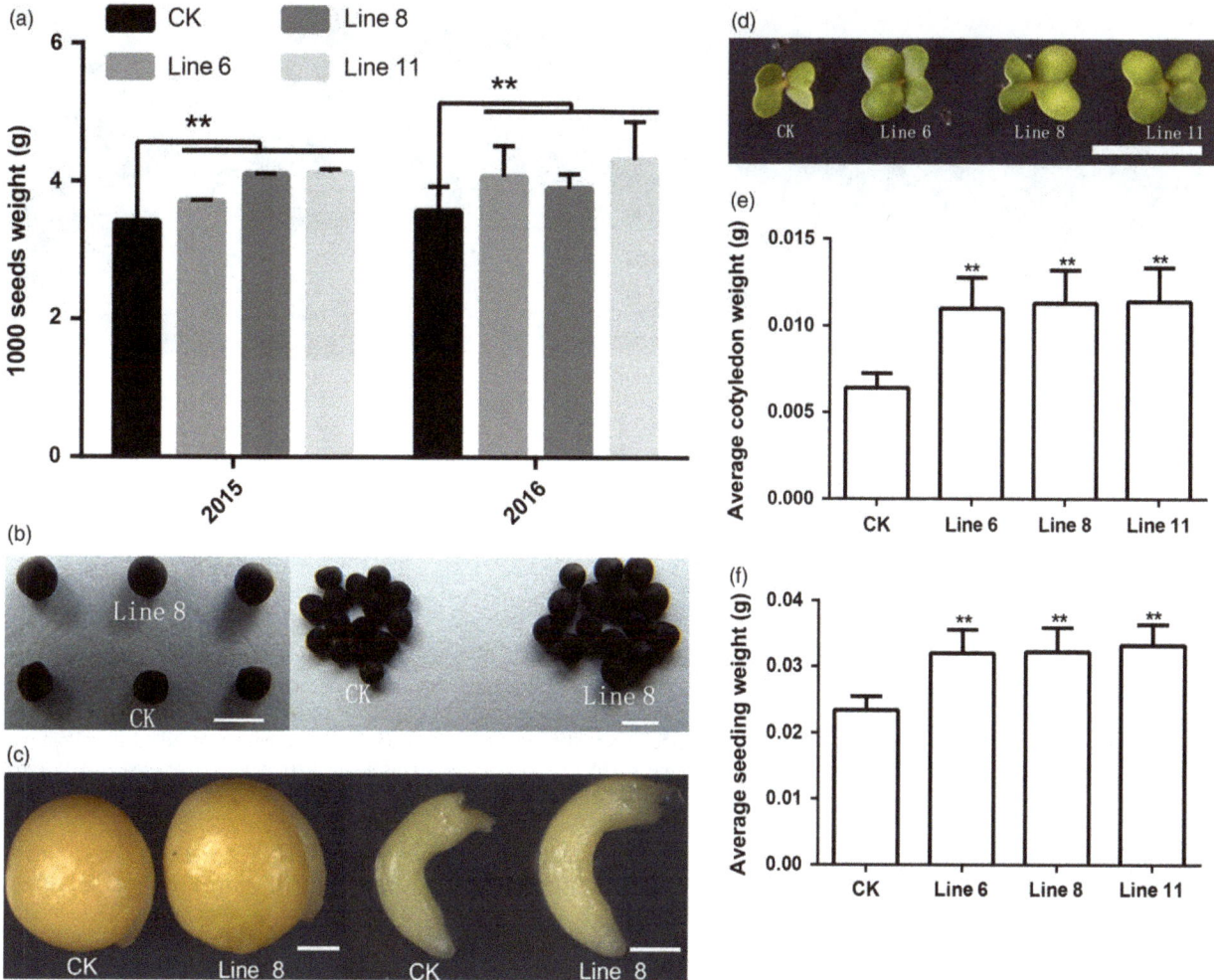

Figure 3 Overexpression of $AtDA1^{R358K}$ increases the weight and size of the seeds. (a) The 1000-seed average weights of CK and the transgenic plant lines 6, 8 and 11. Standard deviations are shown ($n = 5$). (b) The seeds from transgenic plant line 8 were compared to seeds from CK. Bar = 2 mm. (c) A comparison of the embryo and hypocotyl from CK and transgenic plant line 8. (d) The 3-d-old seedlings of CK and the transgenic plant lines 6, 8 and 11, Bar = 5 mm. (e) The average cotyledon weight of CK and the transgenic plant lines 6, 8 and 11. (f) The average seedling weight and the seedling average weight. Value represents mean ± standard error from three independent rapeseed samples. ** represents significant differences at the $P < 0.01$ level.

rapeseed comprehensively enhanced the seed, cotyledon, leaf, flower and silique size. These results showed that BnDA1 and AtDA1 are functionally conserved. In addition, the larger seed and organ sizes have potential economic value in oil crop improvement and bioenergy production.

Overexpression of $AtDA1^{R358K}$ in rapeseed negatively regulated BnDA1, which led to larger seed and organs sizes. In agriculture, seed size is a main components of seed yield. Overexpression of $AtDA1^{R358K}$ in B. napus produced a 21.23% TSW increase and 13.22% increase in seed yield per plant. In addition, larger flowers help attract insects and improve pollination. Higher pollination efficiency could improve seed setting rate, which can directly affect the yield of flowering plants. In short, overexpression of $AtDA1^{R358K}$ in B. napus could increase rapeseed yield. In addition, the vegetative organs, such as seedlings, leaves and fresh weight of the whole plant, also increased in size without any negative influences on major agronomic traits and oil content.

The increased biomass production suggested that the DA1 gene has a potential application in the bioenergy industry and in cash crop improvement. However, we used a transgenic approach to achieve this goal, and the public have concerns about genetically modified organisms. The newly developed genome edit approach, CRISPR/Cas9, could circumvent public apprehension (Shan et al., 2014; Xu et al., 2015). Therefore, DA1 is a promising target for editing the genome with CRISPR/Cas9 because one amino acid alteration may produce the desired positive trait.

With the constant evolution of polyploid species, the expression and biological functions of homologous genes are possibly subject to subfunctionalization, and even produce neofunctionalization (Liu and Adams, 2007). Different homologous genes could be expressed in different tissues and organs at specific time. The homologous genes TaWLHS1 (Shitsukawa et al., 2007) and TaMBD2 (Hu et al., 2011) have different expression pattern in different tissues in wheat. In the transcriptome analysis, compared with three other homologous genes, BnDA1 was shown as the major functional gene for unfolded petal and ovule, but whether it is also the major functional gene for other tissues requires further and analysis. Although the association analysis

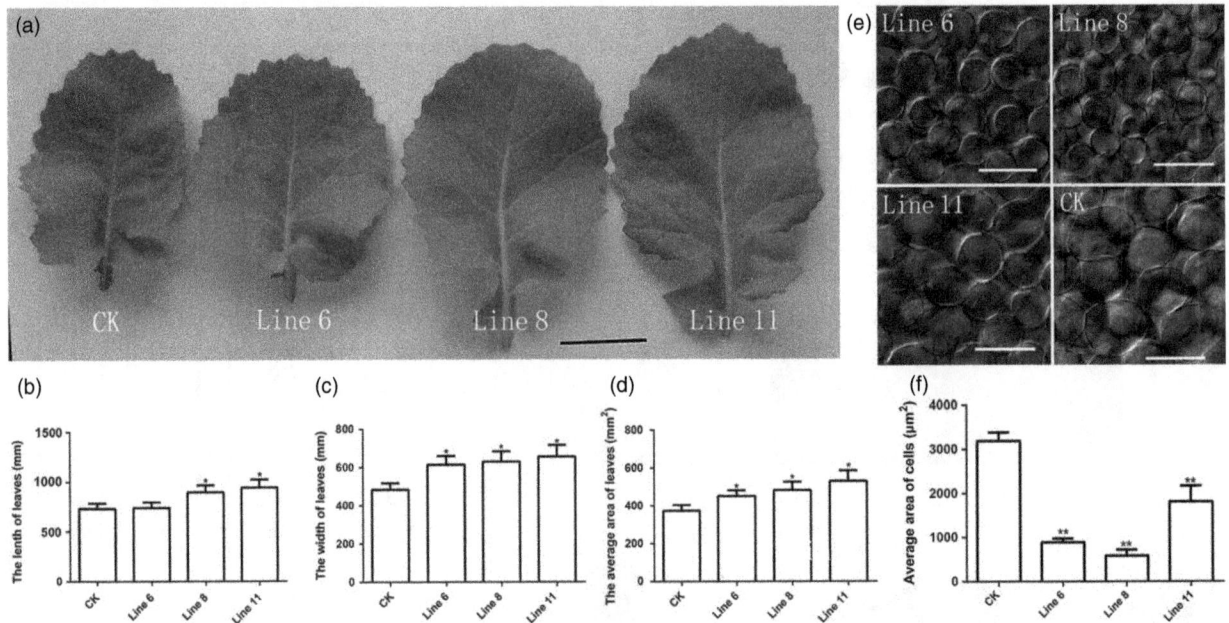

Figure 4 Overexpression of $AtDA1^{R358K}$ increases the size of leaf. (a) The leaves of CK and the transgenic plant lines 6, 8 and 11. Bar = 1 cm. (b) The leaf lengths of CK and transgenic plant lines 6, 8 and 11. (c) The leaf widths of CK and transgenic plant lines 6, 8 and 11. (d) The leaf average areas of CK and transgenic plant lines 6, 8 and 11. (e) The palisade cells of CK and the transgenic plant lines 6, 8 and 11, Bar = 50 μm. (f) The palisade cell average size of CK and the transgenic plant lines 6, 8 and 11. The leaves were collected from 24-d-old seedlings. Value represents mean ± standard error from three independent rapeseed samples.* means a significant difference at the $P < 0.05$ level, and ** represents a significant difference at the $P < 0.01$ level.

showed *BnDA1* was associated with TWS, the phenotypic contribution of the two significant SNPs detected by association analysis was lower than linkage mapping research (Fan *et al.*, 2010). The reason may be that TSW is a complex quantitative trait controlled by many QTLs (genes) and that phenotypic variation is not wide enough (range from 2.83 to 5.52 g) in this natural population.

In conclusion, we have demonstrated that the overexpression of $AtDA1^{R358K}$ with a single amino change in ubiquitin receptor DA1 caused larger seeds and organs in *B. napus*. The average TSW increased almost 21.23%, and seed production per plant increased by 13.22%. Other organs, including cotyledons, leaves, flowers and siliques, were also larger. These phenotype changes suggested that BnDA1 is a promising target for crop improvement and that it is feasible to edit BnDA1 with a CRISPR/Cas9 approach. Because the BnDA1 locus contributed to TSW of *B. napus* in the natural population, this locus could be developed as a functional molecular marker in marker assistant breeding for TSW improvement.

Experimental procedures

Plant materials and growth conditions

Rapeseed (*Brassica napus L*) cv Zhongshuang 9 (ZS9) was as the wild-type control (CK). The ZS9 plants were planted in the experimental fields at Jiangsu University, Zhenjiang, China. *A. thaliana Landsberg erecta* (Ler) was also used as a WT. The plants were grown under long-day conditions, which were 14-h light at 22 °C and 10-h dark at 20 °C.

Sequence analysis of *BnDA1*

The *Arabidopsis thaliana DA1* gene was used to do a blast search of the *BnDa1* sequences and other closed genes in the NCBI. We

identified a highly similar sequence in *Brassica napus L* named *Bnac05g14930D* (GenBank NO.), which we named BnDA1. Multiple sequence alignments of the amino acids sequences for AtDA1, BnDA1 and homologous proteins were performed using GeneDoc software. Phylogeny tests were accomplished using the bootstrap method with 1000 replications to reconstruct a neighbour-joining tree using MEGA5.2 software. Pairwise deletion of gaps/missing data was employed, and uniform rates among sites and similar patterns among lineages were selected for the neighbour-joining (NJ) trees.

Vector constructs and plant transformation

The 35S::*AtDA1* and 35S::*BnDA1* constructs were made using a PCR-based Gateway system (the PCR primers were shown in Table S1), according to the manufacture guide (Invitrogen, Carlsbad, CA). The *AtDA1* and *BnDA1* genes were subcloned into the Gateway binary vector pMDC32. The 35S::*AtDA1* and 35S:*BnDA1* plasmids were transformed with *Agrobacterium* GV3101, and the transformants were selected on the hygromycin-contained medium. The pMDC32-35S::$AtDA1^{R358K}$ vector was introduced into ZS9 according our previous report (Li *et al.*, 2010).

RNA isolation, RT-PCR and q PCR analysis

Total RNA was extracted using TRIzol (Invitrogen, Carlsbad, CA), and mRNA was reverse transcribed using Revert Aid first-strand cDNA Synthesis Kit (#K1622; Thermo Fisher Scientific, Dreieich, Germany). The samples of cDNA were standardized based on the amount of *BnACTIN* transcript using the primers *BnACTIN*-F and *BnACTIN*-R (Table S1). This pair of primers was also used in the RT-PCR and q PCR analyses as a reference gene. The q PCR analysis was performed with SYBR Green format and SYBR premix Ex Taq II (Takara Biotechnology, Dalian, China) using the

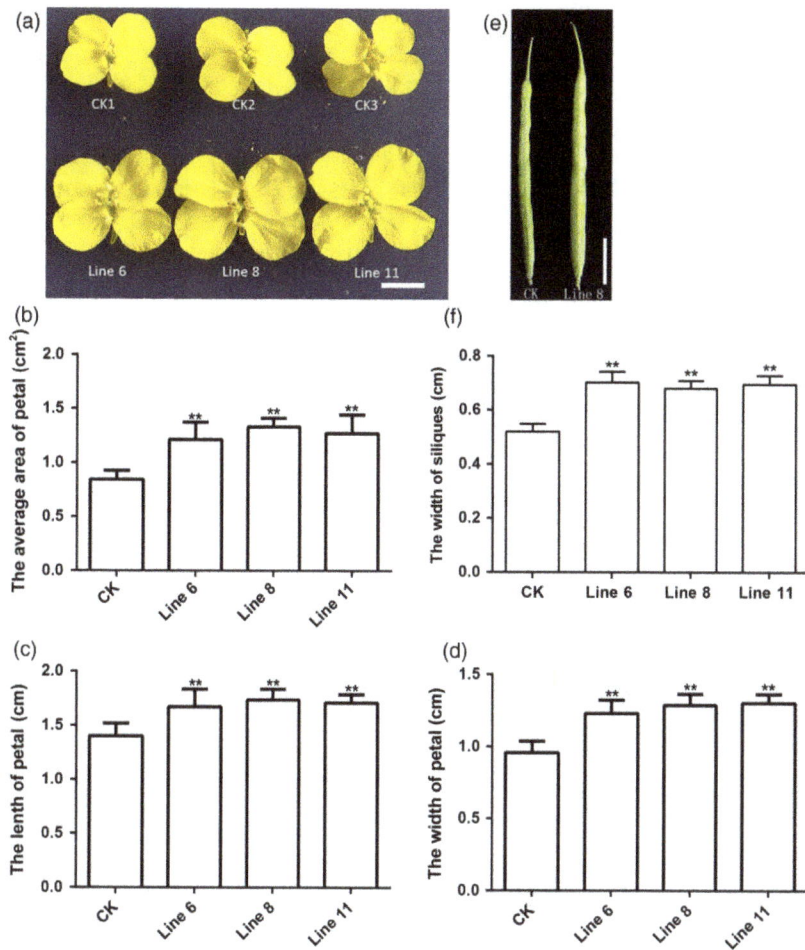

Figure 5 Overexpression of *AtDA1^R358K^* increases the size of the flowers and siliques. (a) The flowers of CK and the transgenic plant lines 6, 8 and 11. Bar = 1 cm. (b) The petal average areas of CK and the transgenic plant lines 6, 8 and 11. (c) The petal average lengths of CK and the transgenic plant lines 6, 8 and 11. (d) The petal average width of CK and the transgenic plant lines 6, 8 and 11. (e) The siliques of CK and the transgenic plant line 8. Bar = 2 cm. (f) The siliques average width of CK and the transgenic plant lines 6, 8 and 11. Standard deviations are shown ($n = 10$). Value represents mean ± standard error from three independent rapeseed samples. ** represents a significant difference at the $P < 0.01$ level.

Applied Biosystems 7300 Fast Real-Time PCR System (ABI, Carlsbad, CA). The primers used for RT-PCR and q PCR were are described in Table S1.

Morphological analysis

Average seed weight was weighted by an electronic analytical balance (BS223S; Sartorius, Gottingen, Germany) with mature dry seeds in batches of 1000. The seeds were photographed using a camera (COOLPIXP7000; Nikon, Tokyo, Japan), and then, seed size, petals area and leaf area were measured using Image J software. Mature plant biomass accumulation was measured by weighing the different organs. The seedlings for analysis were planted in an incubator. The samples for analysis of fresh weight, plant average stem diameter, siliques on each branch, seeds number in each silique and seed oil content were all collected from the mature plants.

Leaf and cell observations

Chloral hydrate was used to make the leaves transparent. Firstly, entire or small pieces of leaves were fixed in the same amount of absolute alcohol and glacial acetic acid mixture for 24 h. Secondly, the leaves were dipped in a saturated aqueous solution of chloral hydrate. Then, the leaves were washed carefully with pure water after the leaves were transparent. Finally, the leaves were floated on glycerol before observation. A metallurgical microscope (Axio Imager A1; ZEISS, Jena, Germany) was used to observe and record the cell size, and the data were analysed using Image J software.

Analysis of seed oil content

Oil content was analysed by NMR (PC120; BRUKER, Karlsruhe, Germany). The data were classified and analysed via one-way analysis of variance using the SAS statistical package (SAS Institute, Cary, NC). Comparisons between the treatment means were made using Duncan's multiple range test at the $P < 0.05$ level. Each line was measured three times.

Association analysis of *BnDA1*

A worldwide collection of 224 rapeseed accessions was used for targeted gene association analysis. The phenotypic data were collected from the field experiments over 3 years in two locations (Wuhan, Hubei province and Yangzhou, Jiangsu province). Field experiments were designed in a randomized complete block design with three replicates and plot size of 3 m². The open-pollinated seeds were harvested from 10 individual plants of each plot when they were mature and measured for TSW. An R script (www.eXtension.org/pages/61006) based on a linear model was used to obtain the best linear unbiased prediction of TSW as phenotypic values in each line.

Genomic DNA was extracted from juvenile leaves of 224 self-pollinated lines using the modified CTAB method. The polymorphic SNPs of *BnDA1* were genotyped by genomic DNA

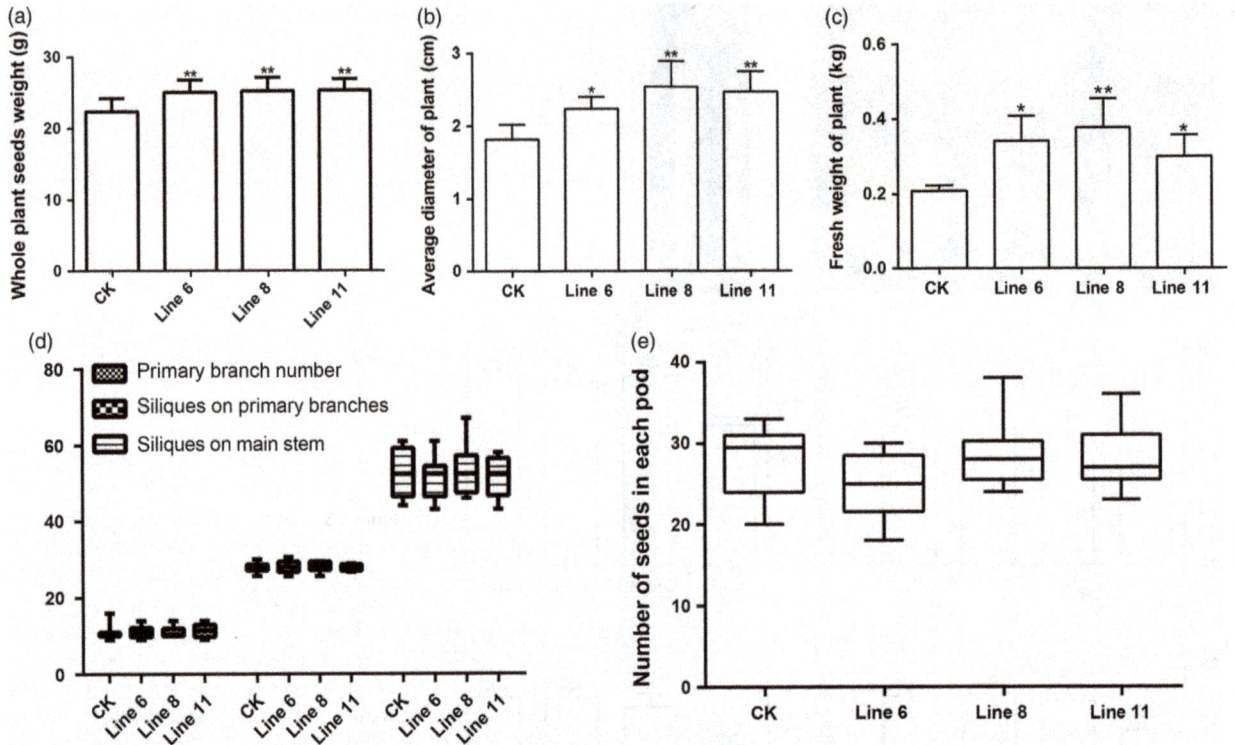

Figure 6 *AtDA1^R358K* overexpression plants have no negative agronomic traits. (a) Whole plant seed weights of CK and transgenic plant lines 6, 8 and 11. (b) The single plant fresh weights of CK and transgenic plant lines 6, 8 and 11. (c) The plant average stem diameters of CK and transgenic plant lines 6, 8 and 11. The whole plant seed weights, the single plant fresh weight and the plant average stem diameter of transgenic plants significantly increased compared to CK. (d) Primary branches per plant, siliques on primary branches (siliques per primary branch) and siliques on main stem (Siliques per main stem) of CK and the transgenic plant lines 6, 8 and 11. (e) Number of seeds in each pod (Seeds per siliques) of CK and the transgenic plant lines 6, 8 and 11. Value represents mean ± standard error from three independent rapeseed samples. ** represents a significant difference at the $P < 0.01$ level.

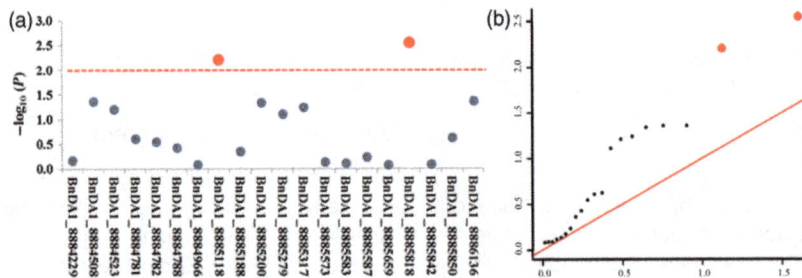

Figure 7 Association of SNP polymorphisms with TWS across the *BnDA1*. (a) The red-dotted line was the significant threshold $-\log_{10}(p) = -\log_{10}(0.01) = 2.0$; the red dot above the red-dotted line represents a significantly associated SNP. (b) The Q-Q plots for thousand seed weight (TSW) from association analysis. The red line was the unbiased estimates of the expected and observed value. Red dots were the significant SNPs associated with TSW based on threshold.

re-sequencing. Association analysis was performed using the software Tassel 5.0.

BnDA1 homologous gene expression analysis

The total RNA from unfolded petal and ovule of cv. Zhongshuang 11 were extracted and then sequenced. The sequenced reads were mapped to the reference genome, and expression quantity was calculated as FPKM (Trapnell *et al.*, 2010).

Acknowledgements

This work was supported by National Key Research and Development Program of China (2016YFD0100305 and 2016YFD0101904), the National Natural Science Foundation of China (31271760 and 31471527). The authors declare no conflict of interest.

References

Adamski, N.M., Anastasiou, E., Eriksson, S., O'Neill, C.M. and Lenhard, M. (2009) Local maternal control of seed size by *KLUH/CYP78A5*-dependent growth signaling. *Proc. Natl Acad. Sci. USA*, **106**, 20115–20120.

Chauhan, B.S., Kumar, N. and Cho, H.M. (2012) A study on the performance and emission of a diesel engine fueled with Jatropha biodiesel oil and its blends. *Fuel Energy Abstracts*, **37**, 616–622.

Chen, Y. and Lubberstedt, T. (2010) Molecular basis of trait correlations. *Trends Plant Sci.* **15**, 454–461.

Disch, S., Anastasiou, E., Sharma, V.K., Laux, T., Fletcher, J.C. and Lenhard, M. (2006) The E3 ubiquitin ligase BIG BROTHER controls *Arabidopsis* organ size in a dosage-dependent manner. *Curr. Biol. Cb.* **16**, 272–279.

Fan, C., Cai, G., Qin, J., Li, Q., Yang, M., Wu, J., Fu, T. *et al.* (2010) Mapping of quantitative trait loci and development of allele-specific markers for seed weight in *Brassica napus. Theor. Appl. Genet.* **121**, 1289–1301.

Fang, W., Wang, Z., Cui, R., Li, J. and Li, Y. (2012) Maternal control of seed size by *EOD3/CYP78A6* in *Arabidopsis thaliana. Plant J.* **70**, 929–939.

Garcia, D., Saingery, V., Chambrier, P., Mayer, U., Jurgens, G. and Berger, F. (2003) *Arabidopsis haiku* mutants reveal new controls of seed size by endosperm. *Plant Physiol.* **131**, 1661–1670.

Garcia, D., Fitz Gerald, J.N. and Berger, F. (2005) Maternal control of integument cell elongation and zygotic control of endosperm growth are coordinated to determine seed size in *Arabidopsis. Plant Cell*, **17**, 52–60.

Gegas, V.C., Nazari, A., Griffiths, S., Simmonds, J., Fish, L., Orford, S., Sayers, L. *et al.* (2010) A genetic framework for grain size and shape variation in wheat. *Plant Cell*, **22**, 1046–1056.

Hu, Z., Yu, Y., Wang, R., Yao, Y.Y., Peng, H.R., Ni, Z.F. and Sun, Q.X. (2011) Expression divergence of *TaMBD2* homoeologous genes encoding methyl CpG-binding domain proteins in wheat (*Triticum aestivum* L.). *Gene*, **471**, 13–18.

Jiang, J., Wang, Y., Zhu, B., Fang, T., Fang, Y. and Wang, Y. (2015) Digital gene expression analysis of gene expression differences within *Brassica* diploids and allopolyploids. *BMC Plant Biol.* **15**, 1–13.

Jofuku, K.D., Omidyar, P.K., Gee, Z. and Okamuro, J.K. (2005) Control of seed mass and seed yield by the floral homeotic gene *APETALA2. Proc. Natl Acad. Sci. USA*, **102**, 3117–3122.

Kesavan, M., Song, J.T. and Seo, H.S. (2013) Seed size: a priority trait in cereal crops. *Physiol. Plant.* **147**, 113–120.

Li, Y., Beisson, F., Pollard, M. and Ohlrogge, J. (2006) Oil content of *Arabidopsis* seeds: the influence of seed anatomy, light and plant-to-plant variation. *Phytochemistry*, **67**, 904–915.

Li, Y., Zheng, L., Corke, F., Smith, C. and Bevan, M.W. (2008) Control of final seed and organ size by the *DA1* gene family in *Arabidopsis thaliana. Genes Dev.* **22**, 1331–1336.

Li, J., Tan, X., Zhu, F. and Guo, J. (2010) A rapid and simple method for *Brassica Napus* floral-dip transformation and selection of transgenic plantlets. *Int. J. Biol.* **2**, 127–131.

Li, J., Nie, X., Tan, J.L. and Berger, F. (2013a) Integration of epigenetic and genetic controls of seed size by cytokinin in *Arabidopsis. Proc. Natl Acad. Sci. USA*, **110**, 15479–15484.

Li, M., Bahn, S.C., Fan, C., Li, J., Phan, T., Ortiz, M., Roth, M.R. *et al.* (2013b) Patatin-related phospholipase pPLAIIIδ increases seed oil content with long-chain fatty acids in *Arabidopsis. Plant Physiol.* **162**, 39–51.

Liu, Z. and Adams, K.L. (2007) Expression partitioning between genes duplicated by polyploidy under abiotic stress and during organ development. *Curr. Biol.* **17**, 1669–1674.

Liu, J., Hua, W., Hu, Z., Yang, H., Zhang, L., Li, R., Deng, L. *et al.* (2015) Natural variation in *ARF18* gene simultaneously affects seed weight and silique length in polyploid rapeseed. *Proc. Natl Acad. Sci. USA*, **112**, 5123–5132.

Luo, M., Dennis, E.S., Berger, F., Peacock, W.J. and Chaudhury, A. (2005) *MINISEED3 (MINI3)*, a WRKY family gene, and *HAIKU2 (IKU2)*, a leucine-rich repeat (*LRR*) *KINASE* gene, are regulators of seed size in *Arabidopsis. Proc. Natl Acad. Sci. USA*, **102**, 17531–17536.

Mizukami, Y. (2002) A matter of size: developmental control of organ size in plants. *Curr. Opin. Plant Biol.* **4**, 533–539.

Moles, A.T., Ackerly, D.D., Webb, C.O., Tweddle, J.C., Dickie, J.B. and Westoby, M. (2005) A brief history of seed size. *Science*, **307**, 576–580.

Naganara, U. (1935) Genomic analysis in *Brassica* with special reference to the experimental formation of *B. napus* and peculiar mode of fertilization. *Jpn J. Bot.* **7**, 389–452.

Navabi, Z.-K., Huebert, T., Sharpe, A.G., O'Neill, C.M., Bancroft, I. and Parkin, I.A. (2013) Conserved microstructure of the *Brassica B* Genome of *Brassica nigra* in relation to homologous regions of *Arabidopsis thaliana*, B. rapa and B. oleracea. *BMC Genom.* **14**, 54–64.

Ohto, M.A., Fischer, R.L., Goldberg, R.B., Nakamura, K. and Harada, J.J. (2005) Control of seed mass by *APETALA2. Proc. Natl Acad. Sci. USA*, **102**, 3123–3128.

Ohto, M.A., Floyd, S.K., Fischer, R.L., Goldberg, R.B. and Harada, J.J. (2009) Effects of APETALA2 on embryo, endosperm, and seed coat development determine seed size in *Arabidopsis. Sex. Plant Reprod.* **22**, 277–289.

Salvi, S., Sponza, G., Morgante, M., Tomes, D.T., Niu, X., Fengler, K.A. and Tuberosa, R. (2007) Conserved noncoding genomic sequences associated with a flowering-time quantitative trait locus in maize. *Proc. Natl Acad. Sci. USA*, **104**, 11376–11381.

Schruff, M.C., Spielman, M., Tiwari, S., Adams, S., Fenby, N. and Scott, R.J. (2006) The *AUXIN RESPONSE FACTOR 2* gene of *Arabidopsis* links auxin signalling, cell division, and the size of seeds and other organs. *Development*, **133**, 251–261.

Shan, Q., Wang, Y., Li, J. and Gao, C. (2014) Genome editing in rice and wheat using the CRISPR/Cas system. *Nat. Protoc.* **9**, 2395–2410.

Shitsukawa, N., Tahira, C., Kassai, K., Hirabayashi, C., Shimizu, T., Takumi, S., Mochida, K. *et al.* (2007) Genetic and epigenetic alteration among three homoeologous genes of a class E MADS box gene in hexaploid wheat. *Plant Cell*, **19**, 1723–1737.

Sidibe, S.S., Blin, J., Vaitilingom, G. and Azoumah, Y. (2010) Use of crude filtered vegetable oil as a fuel in diesel engines state of the art: Literature review. *Renew. Sustain. Energy Rev.* **14**, 2748–2759.

Sugimoto-Shirasu, K. and Roberts, K. (2003) "Big it up": endoreduplication and cell-size control in plants. *Curr. Opin. Plant Biol.* **6**, 544–553.

Takahashi, Y., Teshima, K.M., Yokoi, S., Innan, H. and Shimamoto, K. (2009) Variations in Hd1 proteins, Hd3a promoters, and Ehd1 expression levels contribute to diversity of flowering time in cultivated rice. *Proc. Natl Acad. Sci. USA*, **106**, 4555–4560.

Tan, X.L., Zheng, X.F., Zhang, Z.Y., Wang, Z., Xia, H.C., Lu, C. and Gu, S.L. (2014) Long chain acyl-coenzyme a synthetase 4 (*BnLACS4*) gene from *Brassica napus* enhances the yeast lipid contents. *J. Integrative Agric.* **13**, 54–62.

Thornsberry, J.M., Goodman, M.M., Doebley, J., Kresovich, S., Nielsen, D.M. and Buckler, E.S. (2001) Dwarf8 polymorphisms associate with variation in flowering time. *Nat. Genet.* **28**, 286–289.

Trapnell, C., Williams, B.A., Pertea, G., Mortazavi, A., Kwan, G., van Baren, M.J., Salzberg, S.L. *et al.* (2010) Transcript assembly and quantification by RNA-Seq reveals unannotated transcripts and isoform switching during cell differentiation. *Nat. Biotechnol.* **28**, 511–515.

Wang, X., Liu, B., Huang, C., Zhang, X., Luo, C., Cheng, X., Yu, R. *et al.* (2012) Over expression of *Zmda1-1* gene increases seed mass of corn. *Afr. J. Biotechnol.* **11**, 13387–13395.

Weng, J., Gu, S., Wan, X., Gao, H., Guo, T., Su, N., Lei, C. *et al.* (2008) Isolation and initial characterization of *GW5*, a major QTL associated with rice grain width and weight. *Cell Res.* **18**, 1199–1209.

Westoby, M., Falster, D.S., Moles, A.T., Vesk, P.A. and Wright, I.J. (2002) Plant ecological strategies: some leading dimensions of variation between species. *Annu. Rev. Ecol. Syst.* **33**, 125–159.

Xia, T., Li, N., Dumenil, J., Li, J., Kamenski, A., Bevan, M.W., Gao, F. *et al.* (2013) The ubiquitin receptor DA1 interacts with the E3 ubiquitin ligase DA2 to regulate seed and organ size in *Arabidopsis. Plant Cell*, **25**, 3347–3359.

Xiao, W., Brown, R.C., Lemmon, B.E., Harada, J.J., Goldberg, R.B. and Fischer, R.L. (2006) Regulation of seed size by hypomethylation of maternal and paternal genomes. *Plant Physiol.* **142**, 1160–1168.

Xu, R.F., Li, H., Qin, R.Y., Li, J., Qiu, C.H., Yang, Y.C., Ma, H. *et al.* (2015) Generation of inheritable and "transgene clean" targeted genome-modified rice in later generations using the CRISPR/Cas9 system. *Sci. Rep.* **5**, 1–10.

Zhang, Y., Du, L., Xu, R., Cui, R., Hao, J., Sun, C. and Li, Y. (2015) Transcription factors SOD7/NGAL2 and DPA4/NGAL3 act redundantly to regulate seed size by directly repressing *KLU* expression in *Arabidopsis thaliana. Plant Cell*, **27**, 620–632.

Zhao, M., Gu, Y., He, L., Chen, Q. and He, C. (2015) Sequence and expression variations suggest an adaptive role for the *DA1*-like gene family in the evolution of soybeans. *BMC Plant Biol.* **15**, 1–12.

Zhou, Y., Zhang, X., Kang, X., Zhao, X., Zhang, X. and Ni, M. (2009) SHORT HYPOCOTYL UNDER BLUE1 associates with *MINISEED3* and *HAIKU2* promoters in vivo to regulate *Arabidopsis* seed development. *Plant Cell*, **21**, 106–117.

11

Hypoxia-responsive *ERFs* involved in postdeastringency softening of persimmon fruit

Miao-miao Wang[1,2], Qing-gang Zhu[1], Chu-li Deng[1], Zheng-rong Luo[3], Ning-jing Sun[4], Donald Grierson[1,5], Xue-ren Yin[1,2,*] and Kun-song Chen[1,2]

[1]*Zhejiang Provincial Key Laboratory of Horticultural Plant Integrative Biology, Zhejiang University, Hangzhou, China*
[2]*The State Agriculture Ministry Laboratory of Horticultural Plant Growth, Development and Quality Improvement, Zhejiang University, Hangzhou, China*
[3]*Key Laboratory of Horticultural Plant Biology, Ministry of Education, Huazhong Agricultural University, Wuhan, China*
[4]*Department of Horticultural Sciences, College of Agriculture, Guangxi University, Nanning, China*
[5]*Plant & Crop Sciences Division, School of Biosciences, University of Nottingham, Loughborough, UK*

*Correspondence
xuerenyin@zju.edu.cn

Summary

Removal of astringency by endogenously formed acetaldehyde, achieved by postharvest anaerobic treatment, is of critical importance for many types of persimmon fruit. Although an anaerobic environment accelerates de-astringency, it also has the deleterious effect of promoting excessive softening, reducing shelf life and marketability. Some hypoxia-responsive ethylene response factors (*ERFs*) participate in anaerobic de-astringency, but their role in accelerated softening was unclear. Undesirable rapid softening induced by high CO_2 (95%) was ameliorated by adding the ethylene inhibitor 1-MCP (1 μL/L), resulting in reduced astringency while maintaining firmness, suggesting that CO_2-induced softening involves ethylene signalling. Among the hypoxia-responsive genes, expression of eight involved in fruit cell wall metabolism (*Dkβ-gal1/4, DkEGase1, DkPE1/2, DkPG1, DkXTH9/10*) and three ethylene response factor genes (*DkERF8/16/19*) showed significant correlations with postdeastringency fruit softening. Dual-luciferase assay indicated that *DkERF8/16/19* could trans-activate the *DkXTH9* promoter and this interaction was abolished by a mutation introduced into the C-repeat/dehydration-responsive element of the *DkXTH9* promoter, supporting the conclusion that these DkERFs bind directly to the *DkXTH9* promoter and regulate this gene, which encodes an important cell wall metabolism enzyme. Some hypoxia-responsive *ERF* genes are involved in deastringency and softening, and this linkage was uncoupled by 1-MCP. Fruit of the Japanese cultivar 'Tonewase' provide a model for altered anaerobic response, as they lost astringency yet maintained firmness after CO_2 treatment without 1-MCP and changes in cell wall enzymes and ERFs did not occur.

Keywords: Astringency removal, *ERF*, high CO_2, hypoxia, persimmon fruit, postharvest softening, transcriptional regulation.

Introduction

Plant responses to anoxia involve a range of metabolic and morphological changes on different timescales, including rapid induction of anaerobic metabolism (Kennedy *et al.*, 1992; Voesenek and Bailey-Serres, 2015). For persimmon (*Diospyros kaki*), the anaerobic metabolite acetaldehyde, which accumulates under high CO_2 treatment (95%), participates in fruit postharvest deastringency by converting the soluble tannins to insoluble products (Min *et al.*, 2012; Pesis and Ben-Arie, 1984; Taira *et al.*, 1992, 2001). However, deastringency is usually also accompanied by rapid fruit softening (Arnal and Del Río, 2004; Yin *et al.*, 2012). Thus, although useful for taste improvement, anaerobic treatment has very adverse effects on persimmon fruit storage life.

Softening in most fruit occurs naturally at the commencement of ripening and continues after harvest. It is due primarily to partial cell wall degradation and a reduction in intercellular adhesion (Li *et al.*, 2010) catalysed by a battery of enzymes including pectin methylesterase (PME, EC 3.1.1.11), polygalacturonase (PG, endo-type, EC 3.2.1.15; exo-type, EC 3.2.1.67) and β-galactosidase (β-gal, EC 3.2.1.23) (Brummell and Harpster, 2001; Payasi *et al.*, 2009; Vicente *et al.*, 2007). Understanding the roles of individual genes and enzymes in changing fruit texture has continued to advance since the first experiments on *PG* antisense transgenic tomato (Smith *et al.*, 1988). Some reports indicated that modulation of individual genes effectively influenced fruit texture/softening, such as *PG* (Carrington *et al.*, 1993; Kramer *et al.*, 1992), *TBG4* (a β-gal gene, Smith *et al.*, 2002), *Pmeu1* (a *pectinesterase* gene, Phan *et al.*, 2007) and *PL* (pectate lyase, Silvia *et al.*, 2002). Additional research has indicated that multiple genes contribute to fruit texture, and in tomato, *Fir*[s.p.] QTL2.5 (containing three *PME* genes) is tightly correlated with fruit firmness (Chapman *et al.*, 2012) and double-suppression of *LePG* and *LeEXP1* resulted in increased firmness compared to single gene repression (Powell *et al.*, 2003). Recently, a tomato pectate lyase has been implicated in playing a major role in reducing cell adhesion and firmness (Uluisik et al., 2016). Many of these investigations have been conducted in tomato and strawberry. Limited studies on cell wall-related genes in persimmon have highlighted the importance for fruit softening, of genes such as DkXTH1 and DkXTH2, encoding xyloglucan endotrans-glycosylase/hydrolases (Nakatsuka et al., 2011; Zhu et al., 2013) and DkExp3, which encodes an expansin (Zhang et al., 2012).

Plants respond to hypoxia by the N-end rule pathway which post-translationally regulates levels of group VII ERFs (Gibbs et al., 2011; Licausi et al., 2011). Among other responses, this induces alcohol dehydrogenase (ADH) and pyruvate decarboxy-lase (PDC) genes (Hinz et al., 2010; Licausi et al., 2010; Papdi et al., 2015; Yang et al., 2011) which generate the acetaldehyde which in persimmon fruit precipitates soluble tannins. Several transcription factors have been implicated in regulating this process, including ethylene response factors (ERFs), and some may also be involved in the softening that accompanies deastringency. Some AP2/ERF transcription factors have also been associated with fruit cell degradation and softening in various other fruit (Xie et al., 2016), such as kiwifruit AdERF9, which is a transcriptional repressor acting on the AdXET5 promoter (Yin et al., 2010), and MdCBF (an AP2/ERF member), which activates MdPG1 (Tacken et al., 2010). Using introgression lines, SlERF2.2 was shown to underlie the firmness QTL, Fir$^{s.p.}$ QTL2.2, and its expression is tightly correlated with fruit texture (Chapman et al., 2012). Another AP2/ERF gene, SlAP2a, is also associated with retarding fruit softening, as SlAP2a antisense tomatoes are softer than wild-type fruit (Chung et al., 2010), and LeERF1 antisense transgenic fruit had significant longer shelf life (Li et al., 2007). However, the hypoxia-responsive ERFs have mainly been investigated for their regulation of anaerobic related genes, and their possible relationship with other target genes is largely unknown.

Twenty-two hypoxia-responsive DkERF genes have been iso-lated from persimmon (Min et al., 2012, 2014; Yin et al., 2012) and DkERF9/10/19/22 were shown to be involved directly in transcriptional regulation of anaerobic metabolism genes involved in persimmon deastringency (Min et al., 2012, 2014), but a possible role in postdeastringency fruit softening was not investigated. In the present research, a combination of high CO_2 storage and 1-MCP treatment (together with CO_2 treatment) was shown to maintain fruit firmness while removing astringency, implicating some ERFs in the excessive postdeastringency soften-ing. The roles of DkERFs in controlling cell wall metabolism-related genes were analysed during these treatments and in a Japanese cultivar 'Tonewase', which appears to have an altered anaerobic response, and lacks the ethylene and softening response to high CO_2.

Results

Effects of CO_2 and 1-MCP on 'Mopanshi' persimmon deastringency and softening

Mature 'Mopanshi' persimmon fruit were astringent at harvest and the soluble tannin content was maintained at approximately 1.1%, during 4-day storage (Figure 1a). CO_2 treatment (95%, 1 day) caused a decline in soluble tannins to 0.65% after 1 day and 0.47% after 2 day (Figure 1a). This was accompanied by a rapid decrease in firmness in CO_2-treated fruit, to 33.7 N at 3 day and 22.6 N at 4 day, compared to control fruit firmness of 48.7 N at 3 day and 47.9 N at 4 day (Figure 1b).

The effects of adding the ethylene action antagonist 1-MCP (1 µL/L) to the CO_2 (95%) treatment were investigated in order to test whether this could alleviate the rapid softening that occurred during astringency removal. The results indicated that 1-MCP-treated fruit had higher firmness than the control fruit in CO_2 alone, with 42.4 N at 3 days and 42.5 N at 4 days, slightly lower than control fruit in air (Figure 1b). CO_2 + 1-MCP also enhanced 'Mopanshi' persimmon astringency removal, as indicated by the decrease in soluble tannin, to 0.59% after 1 day and 0.46% after 2 day, values which were similar to those in CO_2-treated fruit without 1-MCP (Figure 1a). Similar effects were confirmed in a subsequent replication in a different year with 'Mopanshi' fruit (data not shown).

Isolation and analysis of deastringency-responsive cell wall degradation-related and DkERF genes associated with 'Mopanshi' persimmon softening

Using the previously generated RNA-seq data (Min et al., 2014), deastringency-responsive cell wall-related unigenes were obtained. After RACE experiments, 35 genes were cloned encoding the following cell wall degrading enzymes: α-L-arabinofuranosidase (DkAraf1-2, KX259530–KX259531), endoglucanase (DkEGase1-2, KX259532–KX259533), β-galacto-sidase (Dkβ-gal1-8, KX259534–KX259541), Mannan endo-1, 4-beta-mannosidase (DkMAN1, KX259542), polygalacturonase (DkPG1, Jiang et al., 2010; DkPG2-5, KX259551–KX259554), pectinesterase (DkPE1-8, KX259543–KX259550), pectate lyase (DkPL1, KX259555) and xyloglucan endotransglycosylase/hydro-lase (DkXTH1-2, DkXTH4, Han et al., 2015; DkXTH8-9, KF318888-9; DkXTH10-12, KX259556-KX259558).

These genes were all expressed in fruit but responded differentially to CO_2 treatment, with only DkEGase1, Dkβ-gal1, Dkβ-gal4, DkPG1, DkPG4, DkPE1, DkPE2, DkXTH1, DkXTH4, DkXTH9-11 showing a strong increase, while the others were nonresponsive or were repressed by CO_2 treatment (Figures 2, S1). Among the CO_2 treatment-induced genes, DkEGase1 and

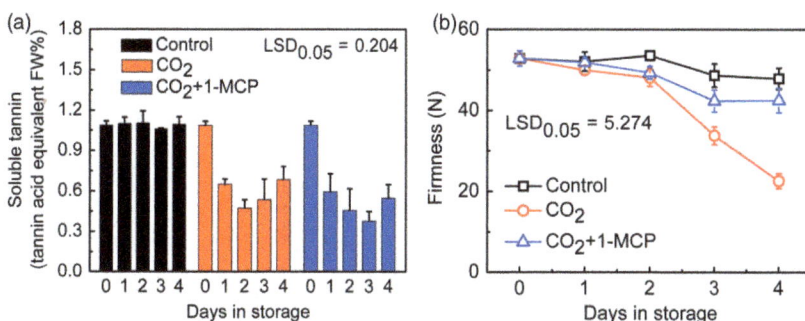

Figure 1 Effects of CO_2 and CO_2 + 1-MCP treatments on soluble tannins (a) and firmness (b) in 'Mopanshi' persimmon fruit at 20 °C. The persimmon fruit were treated with CO_2 (95%) and CO_2 + 1-MCP (95% CO_2 and 1 µL/L 1-MCP) for 1 day, while control fruit were sealed in airtight containers. All treatments and subsequent storage was at 20 °C. Error bars indicate SEs from 3 (for soluble tannins) or 10 (for firmness) replicates.

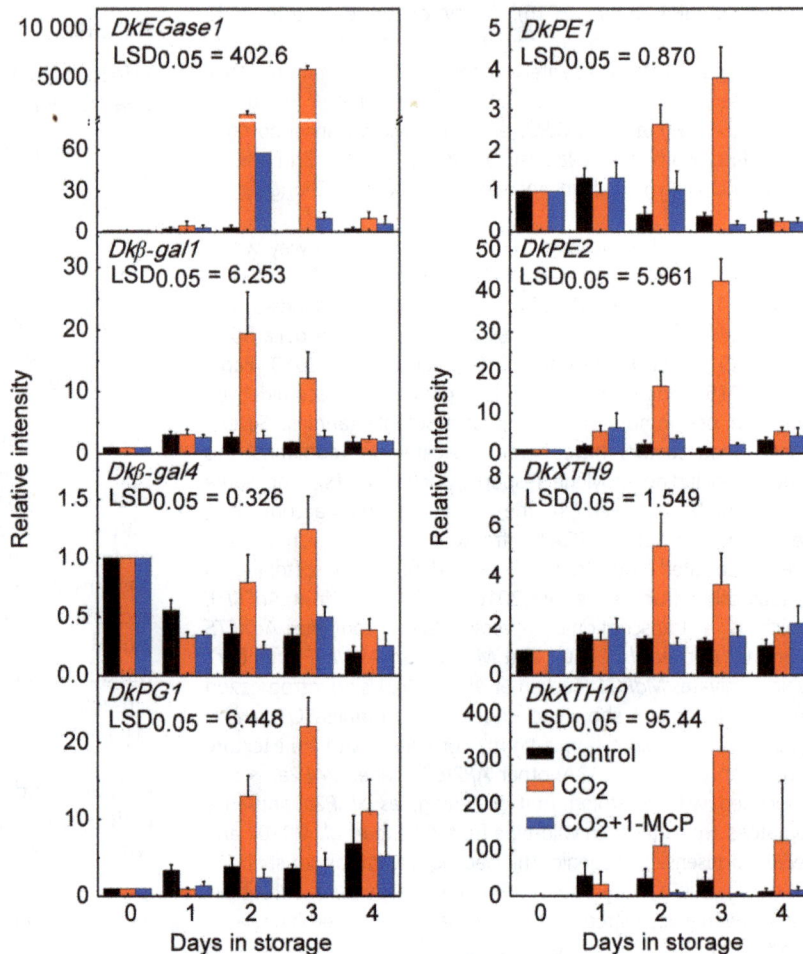

Figure 2 Accumulation of mRNAs from eight cell wall-related genes in response to CO_2 and CO_2 + 1-MCP treatments in 'Mopanshi' persimmon fruit at 20 °C. Gene expression was analysed by real-time PCR. Error bars indicate SEs from three replications. *DkEGase1*: endoglucanase; *Dkβ-gal1/4*: β-galactosidase; *DkPG1*: polygalacturonase; *DkPE1/2*: pectinesterase; *DkXTH9/10*: xyloglucan endotransglycosylase/hydrolase.

DkXTH10 were highly responsive, and at 2 days, their mRNA abundance increased approximately 1233- and 110-fold, respectively (Figure 2). Furthermore, increases in mRNA from most of the CO_2-induced cell wall degrading genes were reduced in CO_2 + 1-MCP-treated fruit, with the exception of *DkPG4*, *DkXTH1*, *DkXTH4* and *DkXTH11*. For instance, *DkXTH1* mRNA was enhanced 6.36-fold at 1 day by CO_2 treatment, and a similar increase was found in CO_2 + 1-MCP-treated fruit (5.71-fold at 1 day; Figure S1). Thus, of the 35 cell wall degradation-related genes, *DkEGase1*, *Dkβ-gal1*, *Dkβ-gal4*, *DkPG1*, *DkPE1*, *DkPE2*, *DkXTH9* and *DkXTH10* were the most likely to be involved in major persimmon fruit softening during and after CO_2 treatment (Figures 2 and S1).

mRNAs from twenty-two *DkERF* genes were found to increase in abundance during deastringency treatment (high CO_2); however, only four, *DkERF9/10/19/22*, have been shown previously to be involved in transcriptional regulation of anoxia-related genes during persimmon fruit deastringency (Min *et al.*, 2012, 2014). The possibility that other *DkERF* genes might also participate in fruit softening during astringency removal was investigated. Using the 'Mopanshi' persimmon, expression of twenty-two *DkERF* genes was analysed and mRNAs for all of these were up-regulated by CO_2 treatment (Figure 3). Adding 1-MCP blocked the enrichment of four *DkERF* genes transcripts, *DkERF7*, *DkERF8*, *DkERF16* and *DkERF19*, suggesting they could play a role in fruit softening (Figures 1b, 3). One, *DkERF7*, contained an EAR motif within the coding region and

is a putative transcriptional repressor (Kagale and Rozwadowski, 2011), while the accumulation of mRNAs from the other three *DkERF* genes was positively correlated with fruit softening.

Effect of CO_2 and CO_2 + 1-MCP treatments on persimmon fruit deastringency and softening in various cultivars

In order to confirm the effect of CO_2 + 1-MCP treatment on persimmon fruit deastringency and softening, another Chinese astringent-type cultivar, 'Jingmianshi', was studied. CO_2 (95%, 1 day) treatment effectively accelerated deastringency as indicated by the darker staining from tannin printing of control fruit, compared with treated fruit (Figure 4). CO_2 + 1-MCP treatment had similar effects on deastringency as CO_2 alone (Figure 4), but there were major differences in fruit softening (Figure 4). Whereas the firmness of 'Jingmianshi' fruit in CO_2 decreased from 21.48 N at 0 day to 1.19 N at 6 day, fruit in CO_2 + 1-MCP retained a firmness of 16.26 N at 6 day (Figure 4).

A comparison was made with a Japanese astringent cultivar, 'Tonewase', which responded similarly to deastringency in CO_2 (95%, 1 day) but did not exhibit the rapid softening observed in the Chinese cultivars, even in the absence of 1-MCP (Figure 4). During the 10-day storage, both CO_2-treated and control fruit firmness decreased only slightly from 17.2 N at 0 day to 15.86 N and 15.15 N at 10 days for control and CO_2-treated fruit, respectively.

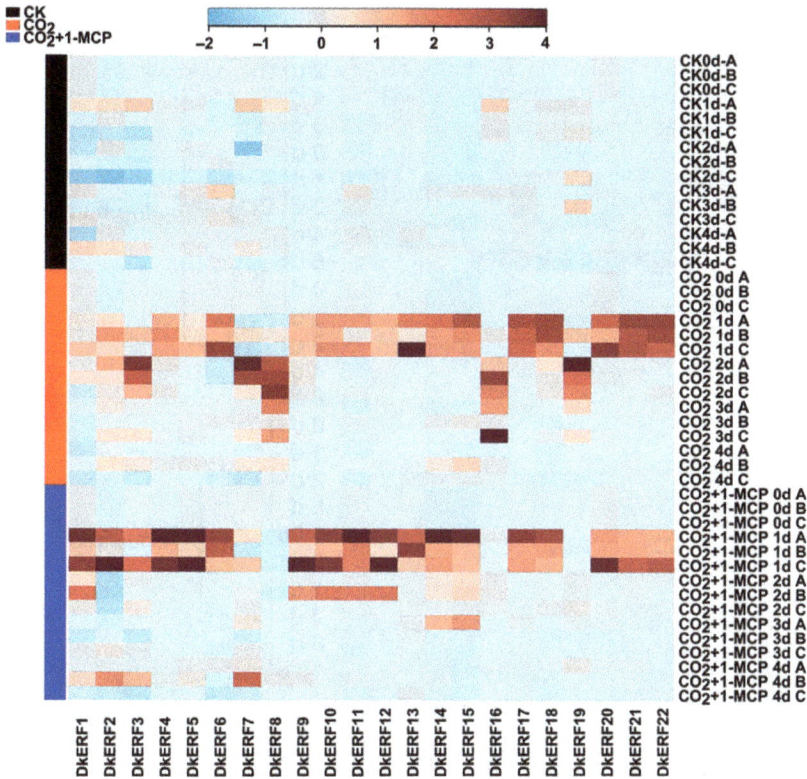

Figure 3 Accumulation of mRNA from hypoxia-responsive *DkERF* genes in response to CO_2 and CO_2 + 1-MCP treatments in 'Mopanshi' persimmon fruit at 20 °C. Gene expression was analysed by real-time PCR. The heatmap was constructed by MetaboAnalyst 3.0 and indicates the mRNA abundance. A, B and C represent three biological replicates. *DkERF1-22*: ethylene response factors.

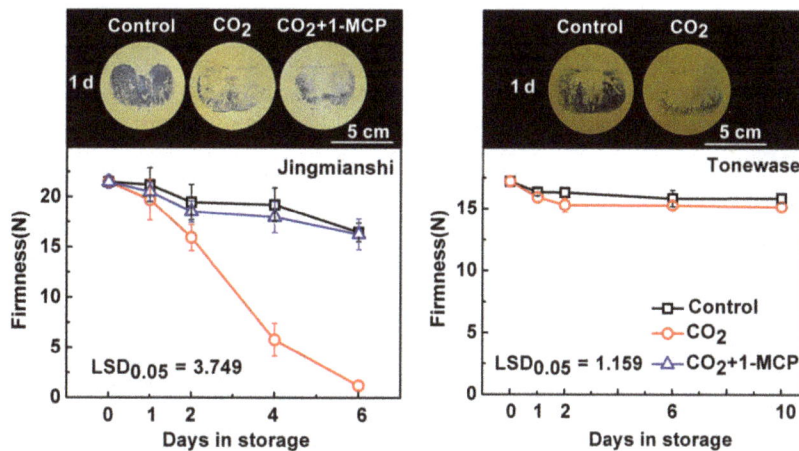

Figure 4 Changes in soluble tannins and firmness in two astringent cultivars, 'Jingmianshi' and 'Tonewase', in response to postharvest treatments. The Chinese cultivar 'Jinmianshi' fruit were treated with CO_2 (95%) or CO_2 + 1-MCP (95% CO_2 and 1 µL/L 1-MCP) for 1 day, separately. Japanese astringent cultivar 'Tonewase' fruit were only treated with CO_2 (95%) for 1 day. Control fruit was sealed in airtight containers. All treatments and subsequent storage were at 20 °C. Astringency was indicated by soluble tannin content, using the tannin printing method. The black colour indicates soluble tannins and the intensity of black reflects the soluble tannin content. Error bars indicate SEs from eight replicates.

Accumulation of mRNAs from fruit softening-related genes in various cultivars

Using 'Jingmianshi' and 'Tonewase', the involvement of the selected eight cell wall-related genes and three *DkERF* genes in fruit softening was assessed. The heatmap (Figure 5) shows that these eleven genes were all highly expressed in CO_2-treated fruit of the 'Jingmianshi' cultivar, which undergoes both deastringency and rapid softening, and were significantly inhibited in the CO_2 + 1-MCP treatment, where fruit underwent deastringency but remained firm. In the astringent Japanese cultivar 'Tonewase', however, mRNAs for these eleven genes remained at basal levels in response to CO_2 treatment alone, resulting in deastringency but maintaining firmness (Figure 5). Taking the results from the three cultivars 'Mopanshi', 'Jingmianshi' and 'Tonewase' together, accumulation of mRNAs for these eleven genes, encoding five cell wall enzymes (*DkEGase1*, *Dkβ-gal1*, *Dkβ-gal4*, *DkPG1*, *DkPE1*, *DkPE2*, *DkXTH9*, *DkXTH10*) and three ERF transcription factors (*DkERF8*, *DkERF16* and *DkERF19*) is highly correlated with CO_2-triggered softening associated with deastringency in Chinese cultivars of persimmon.

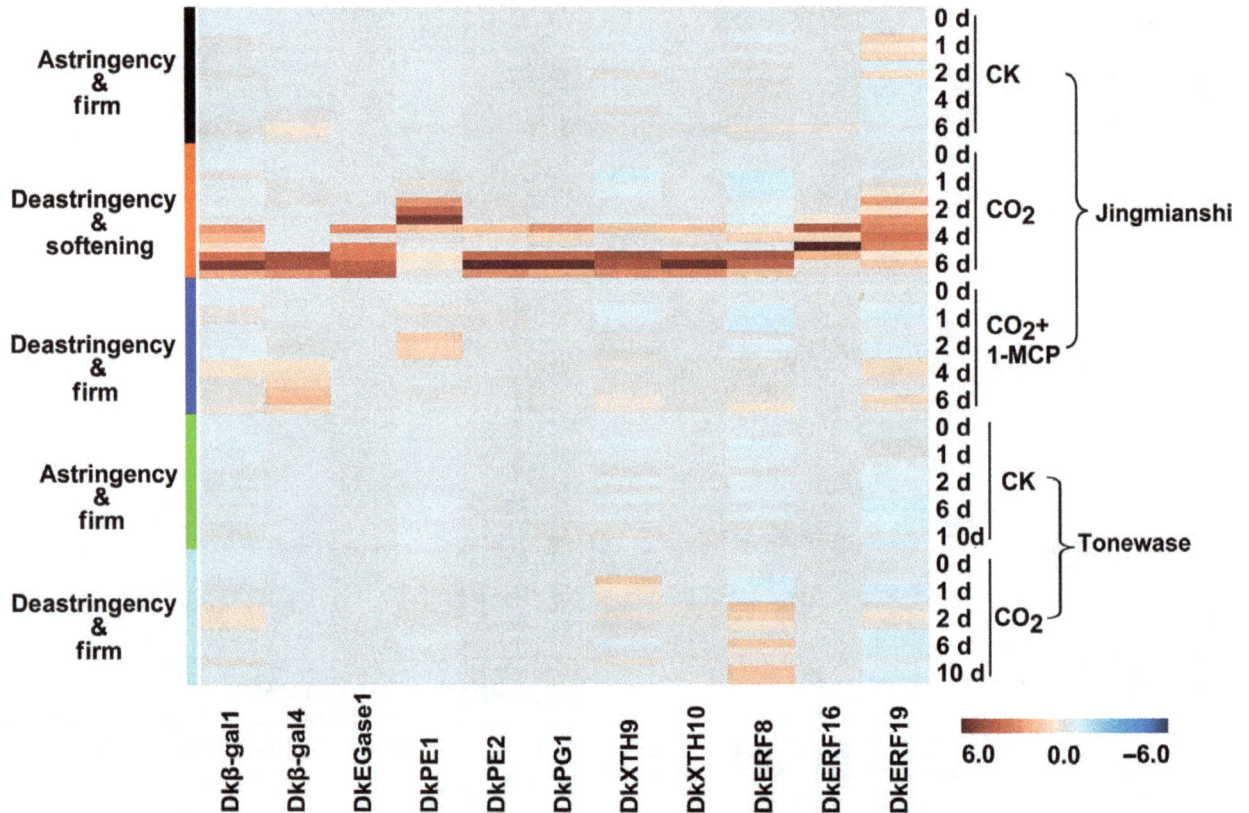

Figure 5 Relationship between accumulation of mRNAs from cell wall-related genes, *DkERF* and persimmon fruit softening in different cultivars. The astringent Chinese cultivar 'Jingmianshi' was treated with CO_2 and CO_2 + 1-MCP, while the Japanese astringent cultivar 'Tonewase' was only treated with CO_2. The concentration of CO_2 and 1-MCP was 95% and 1 μL/L. All treatments and post-treatment storage were conducted at 20 °C. Gene expression was analysed by real-time PCR. The heatmap was constructed by MetaboAnalyst 3.0 and indicates the mRNA abundance. *DkEGase1*: endoglucanase; *Dkβ-gal1/4*: β-galactosidase; *DkPG1*: polygalacturonase; *DkPE1/2*: pectinesterase; *DkXTH9/10*: xyloglucan endotransglycosylase/hydrolase; *DkERF8/16/19*: ethylene response factors.

Roles of softening-related *DkERFs* in controlling expression of genes encoding cell wall degrading enzymes

In order to investigate the possible regulatory linkage between *DkERF* genes and transcription of cell wall-related genes, the promoters of *DkEGase1*, *Dkβ-gal1*, *Dkβ-gal4*, *DkPG1*, *DkPE1*, *DkPE2*, *DkXTH9* and *DkXTH10* were isolated (Table S5), using genome walking, due to the lack of persimmon genome information. Analysis of *cis*-elements indicated that the GCC box and C-repeat/dehydration-responsive element (DRE), recognized by *AP2/ERF* transcription factors, were only distributed in some promoters, such as *Dkβ-gal4*, *DkPE1* and *DkXTH9* (Figure 6a). Dual-luciferase assays indicated that DkERF8, DkERF16 and DkERF19 could trans-activate *DkXTH9* promoters, with 2.13-, 2.13- and 1.74-fold enhancement, respectively (Figure 6b). However, these *DkERF* genes had limited effects on the promoters of the other seven putative softening-related genes, although some *in vivo* regulations had the statistical differences.

DkERF interaction with the *DkXTH9* promoter

DkERF trans-activated the *DkXTH9* promoter, but in a yeast one-hybrid system, the *DkXTH9* promoter exhibited auto-activation (Figure S2), so an alternative approach was used to test the

interaction. When the two DRE elements (CCGAC) within the *DkXTH9* promoter were mutated to CTGAG (*mDkXTH9*, Figure 7a), dual-luciferase assay indicated that all three DkERFs had lower trans-activation activity on the mutated *DkXTH9* promoter, compared to *DkXTH9* promoter (Figure 7b).

Discussion

Regulation of anaerobic treatment on persimmon fruit deastringency and softening

For persimmon, postharvest softening not only depends on ripening stage, but also occurs rapidly during astringency removal (Arnal and Del Río, 2004; Nakatsuka *et al.*, 2011; Ortiz *et al.*, 2005; Taira *et al.*, 1992), which is essential because the main persimmon cultivars are of the astringent type (Luo *et al.*, 2014; Yamada *et al.*, 2002). Thus, although deastringency treatments (including the most widely used high CO_2 treatment, and also ethylene treatment (Min *et al.*, 2012)) remove soluble tannins successfully, they significantly shorten persimmon fruit shelf life. Here, fruit of two Chinese persimmon cultivars, 'Mopanshi' and 'Jingmianshi', of the astringent type exhibited rapid softening after high CO_2 (95%) treatment, which is similar to the results from other persimmon cultivars, such as 'Rojo Brillante' (Arnal and Del Río, 2004) and 'Saijo' (Xu *et al.*, 2003). 'Tonewase', a Japanese persimmon cultivar, maintained firmness after

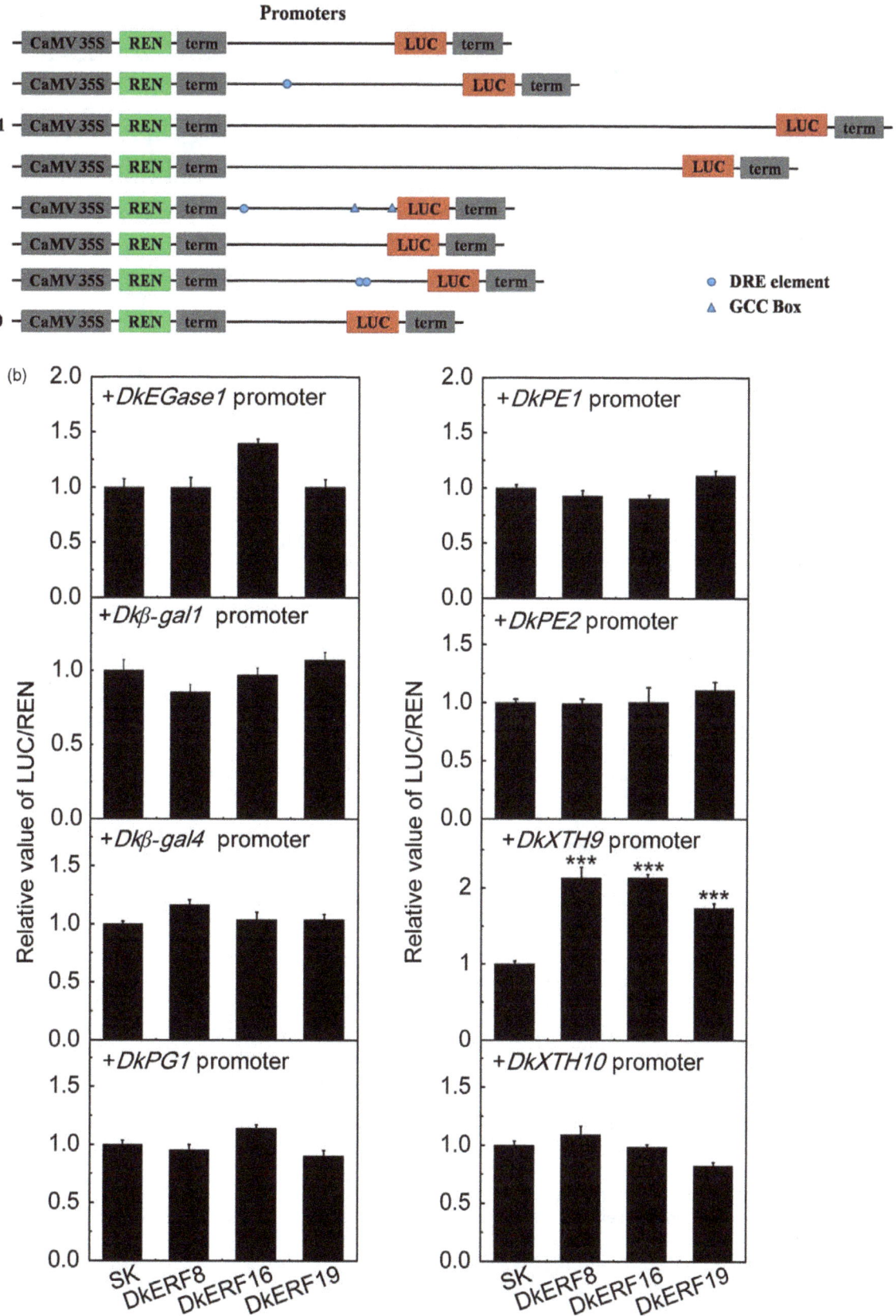

Figure 6 Effect of DkERF on transcription from the promoters of cell wall metabolism-related genes. (a) Schematics of promoters: lines indicate promoter length, triangles show GCC box, circles represent DRE motifs, and triangles indicate GCC boxes. (b) *In vivo* interactions between DkERF and promoters were measured by dual-luciferase assay. Error bars indicate SE from five replicates (***$P < 0.001$).

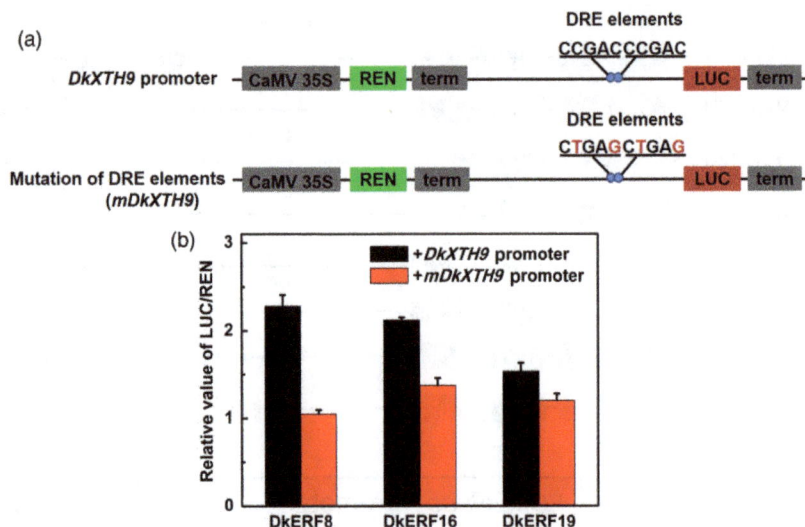

Figure 7 Effect of mutation of DRE elements in the *DkXTH9* promoter on transcription by DkERF. (a) Mutation of DRE elements. (b) Activity of DkERF on transcription from normal and mutated *DkXTH9* promoters. Error bars indicate SE from five replicates.

CO_2-driven deastringency (Figure 4). It should note that 'Tonewase' obtained from different orchards (Itamura *et al.*, 1995) or environments (e.g. humidity, Nakano *et al.*, 2002) has been reported to have varied softening rates. The responses of 'Tonewase', the Japanese cultivar, to high CO_2 treatment (same treatment as for the two Chinese cultivars) show that in this cultivar, deastringency and the accompanying softening are effectively uncoupled.

In the Chinese cultivars 'Mopanshi' and 'Jingmianshi', the softening that accompanies the anoxia-induced deastringency could be prevented by application of the ethylene antagonist 1-MCP and softening was retarded while astringency was reduced or removed (Figures 1 and 4), indicating that ethylene signalling was probably responsible for the softening process during deastringency without 1-MCP. While 1-MCP had little effect on the removal of astringency, it was extremely effective at preventing fruit softening in the Chinese cultivars, which is to be expected for an ethylene response.

Characterization of genes involved in persimmon fruit softening during deastringency

Fruit texture is generally considered as a quantitative trait, which is regulated by multiple genes (Li *et al.*, 2010), such as *PG* (Atkinson *et al.*, 2002; Smith *et al.*, 1988), *XTH* (Miedes *et al.*, 2010), β-*Gal* (Kitagawa *et al.*, 1995; Nakamura *et al.*, 2003; Smith *et al.*, 2002) and pectate lyase (Uluisik *et al.*, 2016). Here, eight cell wall-related genes (*Dkβ-gal1/4*, *DkEGase1*, *DkPE1/2*, *DkPG1*, *DkXTH9/10*) were identified after a preliminary search and their expression correlated with fruit firmness in CO_2- and CO_2 + 1-MCP-treated 'Mopanshi' and 'Jingmianshi' fruit and CO_2-treated 'Tonewase' fruit. The expression of *DkEGase1* in particular was significantly correlated with fruit firmness (Figure 2), indicating the possible involvement of cellulose metabolism during persimmon fruit softening. Furthermore, mRNAs for five other genes (*Dkβ-gal1/4*, *DkPE1/2* and *DkPG1*) related to modifications to pectin, one of the main cell wall components, accumulated during softening (Figure 2), which is consistent with previous demonstrations of the large decrease in pectin content during persimmon fruit softening (Cutillas-Iturralde *et al.*, 1993; Luo, 2007; Taira *et al.*, 1997). None of these genes, however, has been previously reported in persimmon fruit.

XTH is involved in hemicellulose metabolism and is considered one of the most important enzymes that contribute to persimmon fruit softening (Cutillas-Iturralde *et al.*, 1994). Previously, *DkXTH1* and *DkXTH2* have been shown to be associated with persimmon softening (Nakatsuka *et al.*, 2011; Zhu *et al.*, 2013), but these two *DkXTH* genes were different genes with same names. Based on the similarity of nucleotide acid sequences, the present *DkXTH1* and *DkXTH2* were similar to *DkXTH1* and *DkXTH2* from 'Fupingjianshi' (Zhu *et al.*, 2013), while the present *DkXTH8* and *DkXTH9* were similar to *DkXTH1* and *DkXTH2* from 'Saijo' (Nakatsuka *et al.*, 2011). Here, two *DkXTH* genes (*DkXTH9* and *DkXTH10*) were characterized, which indicated potential overlap between natural postharvest softening and anoxia-induced softening on *DkXTH9* (which was named as *DkXTH2* in 'Saijo', Nakatsuka *et al.*, 2011).

Involvement of hypoxia-responsive *ERFs* in regulating postdeastringency softening-related genes

Twenty-two *DkERF* genes were previously characterized as responsive to high CO_2 treatment (Figure 3; Min *et al.*, 2012, 2014), but only four (*DkERF9/10/19/22*) are transcriptional regulators of persimmon fruit astringency removal by regulating genes encoding PDC and ADH (Min *et al.*, 2012, 2014). Thus, only a minority of *DkERF* genes are involved in deastringency, and others, including three *DkERF* genes (*DkERF8/16/19*) were identified as candidates for regulation of postdeastringency softening.

DkERF8/16/19 were shown to trans-activate the *DkXTH9* promoter, but not the promoters of the other seven cell wall-related genes that show increased mRNA accumulation in response to anoxia. In apple, *MdCBF2* can activate the *MdPG* promoter (Tacken *et al.*, 2010), and in kiwi fruit, AdERF9 represses the *AdXET5* (Yin *et al.*, 2010). In most reported cases, however (including *SlERF.B3*, *ERF2.2*, *MdCBF2* and *AdERF9*), the binding of these transcription factors to target promoters was not addressed. Our results indicate a direct *in vivo* activation by DkERF8/16/19 of the *DkXTH9* promoter, as the effect was greatly reduced by targeted mutations on DRE (a known binding motif for *ERF*) in the *DkXTH9* promoter. Direct binding of transcription factors on targets promoters could be analysed by various methods, such as EMSA. Here, dual-luciferase assay and motif mutagenesis were also widely used methods to indicate the potential direct binding. Moreover, motif mutagenesis also provided the self-explanations for *in vivo* effects of DkERF8/16/19 were through the recognization on DRE motif, but not the side effects of genes themselves or infiltrations. These results indicated hypoxia-responsive *DkERF8/16/19* contributed to the softening that accompanied deastringency. Previously, four hypoxia-responsive ERFs, *DkERF9/10/19/22*, were shown to be regulators of *ADH* and *PDC* and contribute to fruit deastringency (Min *et al.*, 2012, 2014). Here, except for *DkERF19*, the other three deastringency-related *DkERF9/10/22* were also response to deastringency treatment in 'Tonewase' persimmon (Figure S3), which were similar to previous findings. A model incorporating these findings is shown in Figure 8. The action of some ERFs could explain the linkage between deastringency and softening, processes which are generally regarded as quite distinct (Figure 8). Furthermore, *DkERF19* has regulatory roles in both deastringency (*DkPDC2*, Min *et al.*, 2014) and softening (*DkXTH9*; Figures 6 and 7), and could genuinely be considered to have dual-functions (Figure 8). Our previous findings (Min *et al.*, 2012, 2014) and the present research indicated that *DkERF* genes appear to be involved in both astringency removal and softening of persimmon fruit. The difficulties in generating transgenic persimmon fruit can be circumvented by population screening and other approaches, to identify variants in expression of *DkERF* genes which can be further developed as markers, enabling breeding of persimmon varieties that rapidly lose astringency while maintaining firmness. Meanwhile, ectopic overexpression of the *DkERF* to tomato (a model fruit) is another alternative, but different genetic background might give the different results. The mechanism for this difference in the anaerobic response between the Chinese and Japanese cultivars requires further investigation. Furthermore, although eight cell wall-related genes were shown to be closely associated with persimmon fruit softening during deastringency, only *DkXTH9* was responsive to the *DkERF* transcription factors and it seems likely that further investigation will reveal other factors involving different signalling pathways, which also participate in softening regulation.

Experimental procedures

Plant material and treatments

Three astringent-type persimmon (*Diospyros kaki*) fruit were selected for this study, including two Chinese cultivars 'Mopanshi' (previously named 'Mopan', Min *et al.*, 2012, 2014) and 'Jingmianshi' and one Japanese cultivar 'Tonewase'.

'Mopanshi' persimmon fruit were harvested from a commercial orchard at Fangshan (Beijing, China) in 2012. Fruit without disease or mechanical wounding were selected. Three different treatments were conducted: (i) the first batch of fruit was treated with 95% CO_2 for 1 day to remove astringency and the post-treatment fruit exhibited rapid softening, (ii) the second batch was treated with a combination of 95% CO_2 and 1 μL/L 1-MCP for 1 day (CO_2 and 1-MCP treatments were performed at the same time), which removed the astringency and maintained fruit firmness, and (iii) the third batch of fruit was sealed in containers similar to those in the above treatments for 1 day, as control. The fruit after treatment were transferred to storage in air. All treatments and post-treatment storage were at 20 °C.

In order to verify the combined effects of 95% CO_2 and 1 μL/L 1-MCP treatment (CO_2 + 1-MCP), 'Mopanshi' and another astringent-type cultivar ('Jingmianshi') were collected from a commercial orchard at Qingdao (Shandong, China) in 2014. Treatments and the conditions were the same as in 2012 and the effect of CO_2 + 1-MCP on fruit deastringency and softening was similar to that in 2012 (data not shown).

In the 2014 season, an additional sample collection was conducted with a Japanese astringent-type cultivar, 'Tonewase'. 'Tonewase' fruit maintain firmness after astringency removal; thus, only CO_2 treatment and control treatment were performed. The fruit were obtained from a commercial orchard at Qingdao (Shandong, China).

All of the above treatments were designed with three biological replicates (150 fruit in each). At each sampling time, flesh samples (three replicates, three fruit in each) were bulked

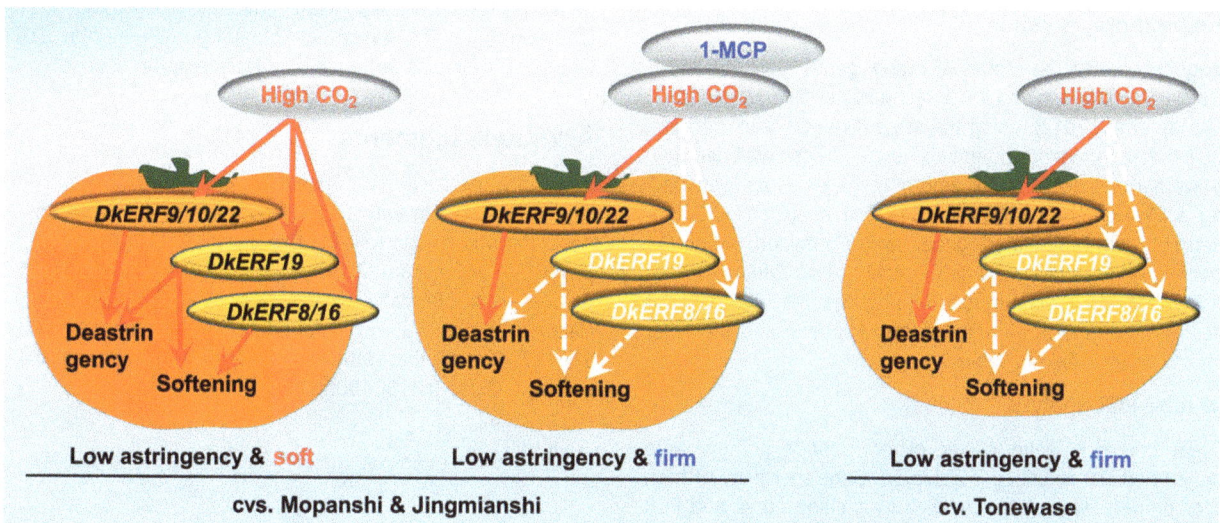

Figure 8 Proposed regulatory model for ethylene response factors in persimmon fruit deastringency and softening. High CO_2 treatment is an effective treatment for postharvest deastringency and widely used for industry. Six *DkERF* genes were characterized to be responsive to high CO_2 treatment: DkERF9/10/22 trans-activate the DkADH/DkPDC and are involved in deastringency; DkERF8/16 are involved in regulation of cell wall metabolism-related *DkXTH9* and associated with fruit softening; DkERF19 has a dual regulatory role in deastringency and softening. Red arrows indicated significant activations; white arrows indicated the absence of activations.

and frozen in liquid nitrogen and stored at −80 °C until further use (soluble tannin measurements and RNA extraction).

Fruit firmness

Firmness was measured with a TA-XT2i texture analyser (Stable Micro Systems, UK), and the penetration indices were calculated according to Yin et al. (2012). Over two different seasons, the firmness was measured using two different texture analyser probes, a 7.5-mm-diameter probe for 'Mopanshi' fruit in 2012 and a 5-mm-diameter probe used for 'Jingmianshi' and 'Tonewase' fruit in 2014. For each fruit, firmness was measured twice at the equator region at 90° intervals, after removal of 1-mm peel. Fruit firmness was measured with 10 fruit replicates.

Soluble tannin content

Soluble tannins are the main source of persimmon fruit astringency. Here, measurements of soluble tannin content were made using two different methods. For 'Mopanshi' fruit, soluble tannins were measured with the Folin-Ciocalteu reagent, using frozen samples, according to the method described by Yin et al. (2012). The results were calculated using the standard curve of tannin acids equivalents g^{-1} fresh weight. Soluble condensed tannin content was measured with three biological replicates. For 'Jingmianshi' and 'Tonewase', soluble tannins were visualized by the tannin printing method, according to Min et al. (2015). Fruit after treatment were immediately cut lengthwise and then printed onto 5% $FeCl_2$-soaked filter paper for 5 s. After removal of fruit, the soluble tannin contents were observed by the intensity of black colour and the filter paper was photographed.

RNA extraction and cDNA synthesis

Total RNA extractions were conducted by the methods described by Yin et al. (2012). Each extraction was conducted with 2.0 g frozen persimmon fruit flesh. The total RNA was treated with TURBO DNAse (Ambion) to remove the contaminated gDNA, and then used for cDNA synthesis, using an iScript™ cDNA Synthesis Kit (Bio-Rad). Three biological replicates were used for RNA extraction.

Gene/promoter isolation

Twenty-two high CO_2-responsive DkERF genes were previously isolated using RNA-seq and RACE technologies (Min et al., 2012, 2014; Yin et al., 2012). Using the same RNA-seq data, differentially expressed unigenes potentially related to cell wall degradation were obtained. The full-length unigenes were obtained, using a SMART RACE cDNA amplification Kit (Clontech). Promoters of cell wall-related genes were obtained using the Genome Walker Universal Kit (Clontech). Two DRE motifs of DkXTH9 promoter were mutated using the Fast Mutagenesis System Kit (Transgen). All primers used for gene and promoter isolation are described in Table S1.

Real-time PCR analysis

For real-time PCR, gene-specific oligonucleotide primers were designed and are described in Table S2. Gene specificity of each pair of primers was double-checked by melting curve and PCR product re-sequencing. The DkACT was chosen as a housekeeping gene to monitor the abundance of mRNA (Min et al., 2014).

Real-time PCR reactions were performed on a CFX96 instrument (Bio-Rad). The PCR program comprised an initial step at 95 °C 3 min, 45 cycles of 95 °C 10 s and 60 °C 30 s, ending with a melting curve analysis programme. The PCR mixture (20 μL total volume) consisted of 10 μL Ssofast EvaGreen Supermix (Bio-Rad), 1 μL of each primer (10 μM), 2 μL of 10-fold diluted cDNA and 6 μL H_2O. No-template controls and melting curve analysis were included for each gene during each run. $2^{-\Delta\Delta Ct}$ method was used to calculate the relative expression levels of genes (Livak and Schmitten, 2001).

The heatmap was used to present the genes expression results and was constructed by MetaboAnalyst 3.0. The different colours indicated up (red) or down (blue) regulations. mRNA abundance was indicated with the intensity of colours. The A, B and C represent three biological replicates.

Dual-luciferase assay

Dual-luciferase assay was used as an efficient and rapid method to detect in vivo trans-activation or trans-repression effects of transcription factors (Yin et al., 2010; Zeng et al., 2015). Full-length DkERF genes were fused to pGreen II 0029 62-SK vector (SK) by Min et al. (2012, 2014). Due to the lack of availability of a persimmon genome, promoters of Dkβ-gal1/4, DkEGase1, DkPE1/2, DkPG1 and DkXTH9/10 were obtained with GenomeWalker™ Universal Kit (Clontech), using the primers described in Table S3. The promoters were amplified with the primers described in Table S4 and were cloned to pGreen II 0800-LUC vector (LUC). Details of vector information are described in Hellens et al. (2005).

All of the constructs were electroporated into Agrobacterium tumefaciens GV3101. The dual-luciferase assays were performed with Nicotiana benthamiana leaves. Agrobacterium cultures were prepared with infiltration buffer (10 mM MES, 10 mM $MgCl_2$, 150 μM acetosyringone, pH 5.6) to an OD_{600} from 0.7 to 1.0. Agrobacterium culture mixtures of transcription factor and promoter (v/v, 10 : 1) were infiltrated into tobacco leaves using needleless syringes. Tobacco plants were grown in a growth chamber, with light : dark cycles of 16 : 8 h. Three days after infiltration, firefly luciferase and renilla luciferase were assayed using the dual-luciferase assay reagents (Promega). The results were calculated from at least five replicates.

Statistical analysis

Statistical significance of differences was calculated using Student's t-test or least significant difference (LSD) using DPS7.05 (Zhejiang University, Hangzhou, China). Figures were drawn using Origin 8.0 (Microcal Software Inc., Northampton, MA).

Acknowledgements

We wish to thank Dr. Shaolan Yang (Qingdao Agriculture University) for help with the collection of persimmon fruit. This research was supported by the National Key Research and Development Program (2016YFD0400100), the Special Fund for Agro-scientific Research in the Public Interest (201203047), National Natural Science Foundation of China (31372114; 31672204) and the Natural Science Foundation of Zhejiang Province, China (LR16C150001).

References

Arnal, L. and Del Río, M.A. (2004) Effect of cold storage and removal astringency on quality of persimmon fruit (Diospyros kaki, L.) cv. Rojo Brillante. Food Sci. Technol. Int. **10**, 179–185.

Atkinson, R.G., Schroder, R., Hallett, I.C., Cohen, D. and MacRae, E.A. (2002) Overexpression of *POLYGALACTURONASE* in transgenic apple trees leads to a range of novel phenotypes involving changes in cell adhesion. *Plant Physiol.* **129**, 122–133.

Brummell, D.A. and Harpster, M.H. (2001) Cell wall metabolism in fruit softening and quality and its manipulation in transgenic plants. *Plant Mol. Biol.* **47**, 311–340.

Carrington, C.M.S., Greve, L.C. and Labavitch, J.M. (1993) Cell wall metabolism in ripening fruit. VI. Effect of the antisense polygalacturonase gene on cell wall changes accompanying ripening in transgenic tomatoes. *Plant Physiol.* **103**, 429–434.

Chapman, N.H., Bonnet, J., Grivet, L., Lynn, J., Graham, N., Smith, R., Sun, G.P. *et al.* (2012) High-resolution mapping of a fruit firmness-related quantitative trait locus in tomato reveals epistatic interactions associated with a complex combinatorial locus. *Plant Physiol.* **159**, 1644–1657.

Chung, M.Y., Vrebalov, J., Alba, R., Lee, J., McQuinn, R., Chung, J.D., Klein, P. *et al.* (2010) A tomato (*Solanum lycopersicum*) *APETALA2/ERF* gene, *SlAP2a*, is a negative regulator of fruit ripening. *Plant J.* **64**, 936–947.

Cutillas-Iturralde, A., Zarra, I. and Lorences, E.P. (1993) Metabolism of cell wall polysaccharides from persimmon fruit. Pectin solubilization during fruit ripening occurs in apparent absence of polygalacturonase activity. *Physiol. Plantarum*, **89**, 369–375.

Cutillas-Iturralde, A., Zarra, I., Fry, S.C. and Lorences, E.P. (1994) Implication of persimmon fruit hemicellulose metabolism in the softening process. Importance of xyloglucan endotransglycosylase. *Physiol. Plantarum*, **91**, 169–176.

Gibbs, D.J., Lee, S.C., Isa, N.M., Gramuglia, S., Fukao, T., Bassel, G.W., Correia, C.S. *et al.* (2011) Homeostatic response to hypoxia is regulated by the N-end rule pathway in plants. *Nature*, **479**, 415–418.

Han, Y., Zhu, Q.G., Zhang, Z.K., Meng, K., Hou, Y.L., Ban, Q.Y., Suo, J.T. *et al.* (2015) Analysis of xyloglucan endotransglycosylase/hydrolase (*XTH*) genes and diverse roles of isoenzymes during persimmon fruit development and postharvest softening. *PLoS ONE*, **10**, e0123668.

Hellens, R.P., Allan, A.C., Friel, E.N., Bolitho, K., Grafton, K., Templeton, M.D., Karunairetnam, S. *et al.* (2005) Transient expression vectors for functional genomics, quantification of promoter activity and RNA silencing in plants. *Plant Methods*, **1**, e13.

Hinz, M., Wilson, I.W., Yang, J., Buerstenbinder, K., Llewellyn, D., Dennis, E.S., Sauter, M. *et al.* (2010) Arabidopsis *RAP2.2*: an ethylene response transcription factor that is important for hypoxia survival. *Plant Physiol.* **153**, 757–772.

Itamura, H., Tanigawa, T. and Yamamura, H. (1995) Composition of cell-wall polysaccharides during fruit softening in 'Tonewase' Japanese persimmon. *Acta Hortic.* **398**, 131–138.

Jiang, N.N., Rao, J.P., Fu, R.S. and Suo, J.T. (2010) Effects of propylene and 1-methylcyclopropene on PG activities and expression of *DkPG1* gene during persimmon softening process. *Acta Hortic. Sin.* **37**, 1507–1512. (In Chinese, with English abstract)

Kagale, S. and Rozwadowski, K. (2011) EAR motif-mediated transcriptional repression in plants: an underlying mechanism for epigenetic regulation of gene expression. *Epigenetics*, **6**, 141–146.

Kennedy, R.A., Rumpho, M.E. and Fox, T.C. (1992) Anaerobic metabolism in plants. *Plant Physiol.* **100**, 1–6.

Kitagawa, Y., Kanayama, Y. and Yamaki, S. (1995) Isolation of β-galactosidase fractions from Japanese pear: activity against native cell wall polysaccharides. *Physiol. Plantarum*, **93**, 545–550.

Kramer, M., Sanders, R., Bolkan, H., Waters, C., Sheeny, R.E. and Hiatt, W.R. (1992) Postharvest evaluation of transgenic tomatoes with reduced levels of polygalacturonase: processing, firmness and disease resistance. *Postharvest Biol. Technol.* **1**, 241–255.

Li, Y.C., Zhu, B.Z., Xu, W.T., Zhu, H.L., Chen, A.J., Xie, Y.H., Shao, Y. *et al.* (2007) *LeERF1* positively modulated ethylene triple response on etiolated seedling, plant development and fruit ripening and softening in tomato. *Plant Cell Rep.* **26**, 1999–2008.

Li, X., Xu, C.J., Korban, S.S. and Chen, K.S. (2010) Regulatory mechanisms of textural changes in ripening fruits. *Crit. Rev. Plant Sci.* **29**, 222–243.

Licausi, F., van Dongen, J.T., Giuntoli, B., Novi, G., Santaniello, A., Geigenberger, P. and Perata, P. (2010) *HRE1 and HRE2*, two hypoxia-inducible ethylene response factors, affect anaerobic responses in *Arabidopsis thaliana*. *Plant J.* **62**, 302–315.

Licausi, F., Kosmacz, M., Weits, D.A., Giuntoli, B., Giorgi, F.M., Voesenek, L.A.C.J., Perata, P. *et al.* (2011) Oxygen sensing in plants is mediated by an N-end rule pathway for protein destabilization. *Nature*, **479**, 419–422.

Livak, J. and Schmitten, T.D. (2001) Analysis of relative gene expression data using real-time quantitative PCR and the $2^{-\Delta\Delta CT}$ method. *Methods*, **25**, 402–408.

Luo, Z.S. (2007) Effect of 1-methylcyclopropene on ripening of postharvest persimmon (*Diospyros kaki* L.) fruit. *LWT Food Sci. Technol.* **40**, 285–291.

Luo, C., Zhang, Q.L. and Luo, Z.R. (2014) Genome-wide transcriptome analysis of Chinese pollination-constant nonastringent persimmon fruit treated with ethanol. *BMC Genomics*. **15**, e112.

Miedes, E., Herbers, K., Sonnewald, U. and Lorences, E.P. (2010) Overexpression of a cell wall enzyme reduces xyloglucan depolymerization and softening of transgenic tomato fruits. *J. Agric. Food Chem.* **58**, 5708–5713.

Min, T., Yin, X.R., Shi, Y.N., Luo, Z.R., Yao, Y.C., Grierson, D., Ferguson, I.B. *et al.* (2012) Ethylene-responsive transcription factors interact with promoters of *ADH* and *PDC* involved in persimmon (*Diospyros kaki*) fruit de-astringency. *J. Exp. Bot.* **63**, 6393–6405.

Min, T., Fang, F., Ge, H., Shi, Y.N., Luo, Z.R., Yao, Y.C., Grierson, D. *et al.* (2014) Two novel anoxia-induced ethylene response factors that interact with promoters of deastringency-related genes from persimmon. *PLoS ONE*, **9**, e97043.

Min, T., Wang, M.M., Wang, H.X., Liu, X.F., Fang, F., Grierson, D., Yin, X.R. *et al.* (2015) Isolation and expression of *NAC* genes during persimmon fruit postharvest astringency removal. *Int. J. Mol. Sci.* **16**, 1894–1906.

Nakamura, A., Maeda, H., Mizuno, M., Koshi, Y. and Nagamatsu, Y. (2003) β-Galactosidase and its significance in ripening of "Saijyo" Japanese persimmon fruit. *Biosci. Biotechnol. Biochem.* **67**, 68–76.

Nakano, R., Inoue, S., Kubo, Y. and Inaba, A. (2002) Water stress-induced ethylene in the calyx triggers autocatalytic ethylene production and fruit softening in 'Tonewase' persimmon grown in a heated plastic-house. *Postharvest Biol. Technol.* **25**, 293–300.

Nakatsuka, A., Maruo, T., Ishibashi, C., Ueda, Y., Kobayashi, N., Yamagishi, M. and Itamura, H. (2011) Expression of genes encoding xyloglucan endotransglycosylase/hydrolase in 'Saijo' persimmon fruit during softening after deastringency treatment. *Postharvest Biol. Technol.* **62**, 89–92.

Ortiz, G.I., Sugaya, S., Sekozawa, Y., Ito, H., Wada, K. and Gemma, H. (2005) Efficacy of 1-Methylcyclopropene (1-MCP) in prolonging the shelf-life of 'Rendaiji' persimmon fruits previously subjected to astringency removal treatment. *J. Jpn. Soc. Hortic. Sci.* **74**, 248–254.

Papdi, C., Pérez-Salamó, I., Joseph, M.P., Giuntoli, B., Bögre, L., Koncz, C. and Szabados, L. (2015) The low oxygen, oxidative and osmotic stress responses synergistically act through the ethylene response factor VII genes *RAP2.12*, *RAP2.2* and *RAP2.3*. *Plant J.* **82**, 772–784.

Payasi, A., Mishra, N.N., Chaves, A.L.S. and Singh, R. (2009) Biochemistry of fruit softening: an overview. *Physiol. Mol. Biol. Plants*, **15**, 103–113.

Pesis, E. and Ben-Arie, R. (1984) Involvement of acetaldehyde and ethanol accumulation during induced deastringency of persimmon fruits. *J. Food Sci.* **49**, 896–899.

Phan, T.D., Bo, W., West, G., Lycett, G.W. and Tucker, G.A. (2007) Silencing of the major salt-dependent isoform of pectinesterase in tomato alters fruit softening. *Plant Physiol.* **144**, 1960–1967.

Powell, A.L.T., Kalamaki, M.S., Kurien, P.A., Gurrieri, S. and Bennett, A.B. (2003) Simultaneous transgenic suppression of LePG and LeExp1 influences fruit texture and juice viscosity in a fresh market tomato variety. *J. Agric. Food Chem.* **51**, 7450–7455.

Silvia, J.B., José, R.N., Juan, M.B., Caballero, J.L., López-Aranda, J.M., Victoriano, V., Pliego-Alfaro, F. *et al.* (2002) Manipulation of strawberry fruit softening by antisense expression of a pectate lyase gene. *Plant Physiol.* **128**, 751–759.

Smith, C.J.S., Watson, C.F., Ray, J., Bird, C.R., Morris, P.C., Schuch, W. and Grierson, D. (1988) Antisense RNA inhibition of polygalacturonase gene expression in transgenic tomatoes. *Nature*, **334**, 724–726.

Smith, D.L., Abbott, J.A. and Gross, K.C. (2002) Down-regulation of tomato β-galactosidase 4 results in decreased fruit softening. *Plant Physiol.* **129**, 1755–1762.

Tacken, E., Ireland, H., Gunaseelan, K., Karunairetnam, S., Wang, D., Schultz, K., Bowen, J. *et al.* (2010) The role of ethylene and cold temperature in the regulation of the apple *POLYGALACTURONASE1* gene and fruit softening. *Plant Physiol.* **153**, 294–305.

Taira, S., Oba, S. and Watanabe, S. (1992) Removal of astringency from 'Hiratanenashi' persimmon fruit with a mixture of ethanol and carbon dioxide. *J. Jpn. Soc. Hortic. Sci.* **61**, 437–443.

Taira, S., Ono, M. and Matsumoto, N. (1997) Reduction of persimmon astringency by complex formation between pectin and tannins. *Postharvest Biol. Technol.* **12**, 265–271.

Taira, S., Ikeda, K. and Ohkawa, K. (2001) Comparison of insolubility of tannins induced by acetaldehyde vapor in fruit of three types of astringent persimmon. *J. Jpn. Soc. Hortic. Sci.* **48**, 684–687.

Uluisik, S., Chapman, N., Smith, R., Poole, M., Adams, G., Gillis, R.B., Besong, T.M.D. *et al.* (2016) Genetic improvement of tomato by targeted control of fruit softening. *Nat. Biotechnol.* **34**, 950–952.

Vicente, A.R., Saladié, M., Rose, J.K. and Labavitch, J.M. (2007) The linkage between cell wall metabolism and fruit softening: looking to the future. *J. Sci. Food Agric.* **87**, 1435–1448.

Voesenek, L.A.C.J. and Bailey-Serres, J. (2015) Flood adaptive traits and processes: an overview. *New Phytol.* **206**, 57–73.

Xie, X.L., Yin, X.R. and Chen, K.S. (2016) Roles of APETALA2/ethylene responsive factors in regulation of fruit quality. *Crit. Rev. Plant Sci.* **35**, 120–130.

Xu, C., Nakatani, Y., Nakatsuka, A. and Itamura, H. (2003) Effects of different methods of deastringency and storage on the shelf life of 'Saijo' persimmon fruit. *Food Preserv. Sci.* **29**, 191–196.

Yamada, M., Taira, S., Ohtsuki, M., Sato, A., Iwanami, H., Yakushiji, H., Wang, R.Z. *et al.* (2002) Varietal differences in the ease of astringency removal by carbon dioxide gas and ethanol vapor treatments among Oriental astringent persimmons of Japanese and Chinese origin. *Sci. Hortic.* **94**, 63–72.

Yang, C.Y., Hsu, F.C., Li, J.P., Wang, N.N. and Shih, M.C. (2011) The AP2/ERF transcription factor AtERF73/HRE1 modulates ethylene responses during hypoxia in Arabidopsis. *Plant Physiol.* **156**, 202–212.

Yin, X.R., Allan, A.C., Chen, K.S. and Ferguson, I.B. (2010) Kiwifruit *EIL* and *ERF* genes involved in regulating fruit ripening. *Plant Physiol.* **153**, 1280–1292.

Yin, X.R., Shi, Y.N., Min, T., Luo, Z.R., Yao, Y.C., Xu, Q., Ferguson, I. *et al.* (2012) Expression of ethylene response genes during persimmon fruit astringency removal. *Planta*, **235**, 895–906.

Zeng, J.K., Li, X., Xu, Q., Chen, J.Y., Yin, X.R., Ferguson, I.B. and Chen, K.S. (2015) *EjAP2-1*, an *AP2/ERF* gene, is a novel regulator of fruit lignification induced by chilling injury, via interaction with *EjMYB* transcription factors. *Plant Biotechnol. J.* **13**, 1325–1334.

Zhang, Z.K., Fu, R.S., Huber, D.J., Rao, J.P., Chang, X.X., Hu, M.J., Zhang, Y. *et al.* (2012) Expression of expansin gene (*CDK-Exp3*) and its modulation by exogenous gibberellic acid during ripening and softening of persimmon fruit. *HortScience*, **47**, 378–381.

Zhu, Q.G., Zhang, Z.K., Rao, J.P., Huber, D.J., Lv, J.Y., Hou, Y.L. and Song, K.H. (2013) Identification of xyloglucan endotransglucosylase/hydrolase genes (*XTHs*) and their expression in persimmon fruit as influenced by 1-methylcyclopropene and gibberellic acid during storage at ambient temperature. *Food Chem.* **138**, 471–477.

12

Genomic regions, cellular components and gene regulatory basis underlying pod length variations in cowpea (*V. unguiculata* L. Walp)

Pei Xu[1,2,*], Xinyi Wu[1], María Muñoz-Amatriaín[3], Baogen Wang[1], Xiaohua Wu[1], Yaowen Hu[1], Bao-Lam Huynh[4], Timothy J. Close[3], Philip A. Roberts[4], Wen Zhou[1], Zhongfu Lu[1] and Guojing Li[1,2,*]

[1]Institute of Vegetables, Zhejiang Academy of Agricultural Sciences, Hangzhou, China
[2]State Key Lab Breeding Base for Sustainable Control of Plant Pest and Disease, Zhejiang Academy of Agricultural Sciences, Hangzhou, China
[3]Department of Botany and Plant Sciences, University of California-Riverside, Riverside, CA, USA
[4]Department of Nematology, University of California-Riverside, Riverside, CA, USA

*Correspondence
email peixu@mail.zaas.ac.cn (Pei Xu) and
Ligj@mail.zaas.ac.cn
(Guojing Li)

Keywords: Cowpea, Domestication, GWAS, Pod length, Selection, Transcriptome.

Summary

Cowpea (*V. unguiculata* L. Walp) is a climate resilient legume crop important for food security. Cultivated cowpea (*V. unguiculata* L) generally comprises the bushy, short-podded grain cowpea dominant in Africa and the climbing, long-podded vegetable cowpea popular in Asia. How selection has contributed to the diversification of the two types of cowpea remains largely unknown. In the current study, a novel genotyping assay for over 50 000 SNPs was employed to delineate genomic regions governing pod length. Major, minor and epistatic QTLs were identified through QTL mapping. Seventy-two SNPs associated with pod length were detected by genome-wide association studies (GWAS). Population stratification analysis revealed subdivision among a cowpea germplasm collection consisting of 299 accessions, which is consistent with pod length groups. Genomic scan for selective signals suggested that domestication of vegetable cowpea was accompanied by selection of multiple traits including pod length, while the further improvement process was featured by selection of pod length primarily. Pod growth kinetics assay demonstrated that more durable cell proliferation rather than cell elongation or enlargement was the main reason for longer pods. Transcriptomic analysis suggested the involvement of sugar, gibberellin and nutritional signalling in regulation of pod length. This study establishes the basis for map-based cloning of pod length genes in cowpea and for marker-assisted selection of this trait in breeding programmes.

Introduction

Cowpea (*V. unguiculata* L. Walp., 2n = 2x = 22), native to Africa, is a worldwide important legume used as a grain, fodder or vegetable crop. Sub-Saharan Africa and Brazil are the major producers of cowpea grain, while East/South-East Asia is the main vegetable-type cowpea producer (Rachie, 1985; Timko *et al.*, 2007). Grain-type cowpea, commonly known as African cowpea or common cowpea (*V. unguiculata* L. Walp. ssp. *unguiculata*), provides a major source of dietary protein for millions of people in developing countries (Singh, 2002). The vegetable cowpea, also known as asparagus bean or 'yardlong' bean (*V. unguiculata* L. Walp. ssp. *sesquipedalis*), is characterized by its long tender pods that are harvested when immature, narrow kidney-shaped seeds, and strong trailing and climbing growth habit that is rare in ssp. *unguiculata* germplasm. The asparagus bean pods can either be snapped and cooked in stew or stir-fried, or preserved with salt and chile. They provide a good source of proteins, vitamins and minerals (Timko *et al.*, 2007). Besides the nutritional value, asparagus bean is among the top ten Asian cultivated vegetables due to its high tolerance to heat and drought (National Research Council, 2006).

Although vegetable cowpea is predominant in East/Southeast Asia, cowpea for grain use is also grown in some parts of these regions. Many of the latter are not typical ssp. *unguiculata* varieties but are landraces of ssp. *sesquipedalis* that can be classified into the 'nonstandard' vegetable type based on genetic compositions (Xu *et al.*, 2012). In some regions of Africa, Europe and America, vegetable cowpea is also cultivated, although these forms usually exhibit a 'bush' phenotype and develop relatively shorter pods than asparagus beans (National Research Council, 2006). To date, it is still unknown how the vegetable and grain cowpea types have diverged. It has been postulated that ssp. *sesquipedialis* was derived from domesticated ssp. *unguiculata* after it was brought to parts of Asia through intense selection for pod characteristics favourable for vegetable use (Fang *et al.*, 2007; Kongjaimun *et al.*, 2012a,b). In this regard, asparagus bean would have gone through double domestication bottleneck in that only a small portion of the genetic variation present in the progenitor African cowpea germplasm would have constituted the foundation germplasm of ssp. *sesquipedialis* (Fang *et al.*, 2007; Timko *et al.*, 2007). This theory has been partially validated by the lower genetic diversity in Chinese asparagus bean germplasm compared to African common cowpea through SSR and SNP marker analyses (Xu *et al.*, 2010,

2012). However, evidence that selection for pod characteristics, particularly pod length, has contributed essentially to the formation of present-day asparagus bean is lacking. Also, the relationships between pod length QTLs and other domestication genes/QTLs are not well understood. To our knowledge, the only QTL analysis of pod length in asparagus bean was reported by Kongjaimun et al. (2012a) using SSR markers.

The recently developed Cowpea iSelect Consortium Array (Illumina, Inc.) provides an opportunity to further dissect pod length and other domestication traits. This array contains 51 128 SNPs that derive from WGS sequencing of 37 diverse cowpea accessions (33 grain cowpeas and four asparagus bean) (Muñoz-Amatriaín et al., 2016). Using this assay, a cowpea consensus genetic map was constructed with SNP data from five mapping populations (Muñoz-Amatriaín et al., 2016). In the current study, we describe the application of this assay to dissect pod length QTLs in cowpea. In addition, histological and gene expression analyses were performed to uncover the cellular components and gene regulatory basis underlying pod length variations.

Results

SNP genotyping and data curation

We SNP genotyped 432 DNA samples including 132 recombinant inbred lines (RILs) and a diversity panel of 299 landraces, cultivars and breeding lines (Figure 1a, b; Table S1). A total of 49 194 SNPs produced successful calls. After the removal of samples showing high heterozygosity, missing data or nonparental alleles (for the RIL population), 119 RILs and all germplasm accessions were used for further analysis. After eliminating monomorphic SNPs and those with excessive number of missing and/or heterozygous calls and very low minor allele frequencies, 7988 high-quality SNPs were retained for the RIL population and 30 211 for the diversity panel.

Phenotypic analysis of pod length

Zhijiang282, the male parent of the RIL population, exhibited longer pods (mean = 47.0 cm) than the female parent ZN016 (mean = 31.9 cm) in all experiments. Among the RILs, pod length displayed a continuous distribution with the population means falling between the parental values (Figure 1c). Transgressive segregation was observed as some RILs exhibited a pod length outside the parental value range, suggesting the existence of intragenic (e.g. incomplete dominance or codominance) or intergenic interactions. High correlation (0.7–0.82, $P < 0.001$) was observed for pod length between different trials. The estimated broad-sense heritability was 70.9% over the 2 years. A wider range of variation (12.3–74.5 cm) and a continuous distribution based on multiyear pod length data were observed for the diversity panel. Correlation coefficients between experiments ranged from 0.87 to 0.93 ($P < 0.001$) for this population.

iSelect genetic map construction and QTL mapping

At a LOD score of 10 for marker grouping, 7964 of the 7988 SNPs (99.7%) were mapped to 697 bins in 11 linkage groups (LGs) (Figure S1). This map, hereafter referred to as 'ZZ map v.2' in consistency with its earlier version (Xu et al., 2011a), covered 803.4 cM with an average distance between marker bins of 1.15 cM and an average density of 11.4 SNPs per bin. The length of individual LGs varied from 45.1 cM to 124 cM, and the

average marker distance per LG ranged from 0.05 cM to 0.16 cM (Table S2). The map contained two gaps of >10 cM, located on LG1 and LG5, respectively.

Inclusive composite interval mapping identified two major QTLs for pod length under the additive model (ICIM-ADD) (Figure 2a, Table 1). The larger effect QTL, designated as Qpl.zaas-3, was mapped within a 1.3-cM interval on LG3 between the markers 2_04960 and 2_02274 in all four experiments. The LOD scores in different experiments varied from 14.7 to 18.7, and the average phenotypic variance explained was 44%. A significant interaction of Qpl.zaas-3 with environment was also detected (Table 1), but its effect on phenotypic variance was small (8%). Another QTL spanning a 9.5-cM interval on LG5 (Qpl.zaas-5 hereafter) was detected at a LOD score above 3 in the 2010HN trial, and above LOD = 6 in the 2009HN trial. It explained 11% of the phenotypic variance on average. Genome-wide scan of digenic interactions identified four pairs of epistatic interactions, three interchromosomal and one intrachromosomal (Figure 2b). The phenotypic variance explained by the individual pairs of epistatic interactions ranged from 6.4% to 18.2%. No common epistatic interactions were detected across all four environments, and none of the loci involved in those epistatic interactions coincided with Qpl.zaas-3 or Qpl.zaas-5.

Genetic diversity and population structure of the germplasm panel

The overall heterozygosity rate of the germplasm panel estimated with 30 211 high-quality SNPs was 2.23%, fitting the expectation for inbred plant materials. Pairwise genetic distances between the 299 accessions ranged from 0.0003 to 0.5556, with a mean of 0.253. A model-based clustering method identified two major subpopulations in the germplasm panel (Figure 3a, b, Table S1). Subpopulation 1 consisted of 79 accessions, including all nine American accessions, three of the four Filipino accessions and 67 Chinese accessions. The mean and median pod length in this subpopulation were 26.1 cm and 25.6 cm, respectively. All members of this subgroup except for seven Chinese accessions and two American accessions are grain types, and all the Chinese accessions are landraces. Subpopulation 2, consisting of 99 accessions with 98 from China and one from Thailand, had a mean and median pod length of 53.1 cm and 53.4 cm, respectively. All accessions from this subpopulation were typical vegetable cowpea accessions. The rest of the accessions with the probability of belonging to a subpopulation lower than 0.7 were grouped as 'admixed'. Principal component analysis (PCA) supported the subgroup assignment of model-based method in that the accessions assigned to subpopulations 1, 2 and the admixed group, in particular subpopulation 1, were well distinguishable along the first PC (Y-axis, Figure 3c). The PCA result also revealed that the two subpopulations were mainly differentiated by pod length rather than by geographic origin or improvement status (Figure 3d, e, f), reinforcing the critical role of pod length in the diversification of cowpea. An unrooted neighbour-joining tree constructed for subpopulation 1 and subpopulation 2 accessions showed a clear (albeit not perfect) branching between the short-podded and long-podded types (Figure 3g, Figure S2).

Genome-wide association studies (GWAS) for pod length

Prior to GWAS, the extent of LD was analysed for the whole set of accessions and for each of the subpopulations. The LD decay

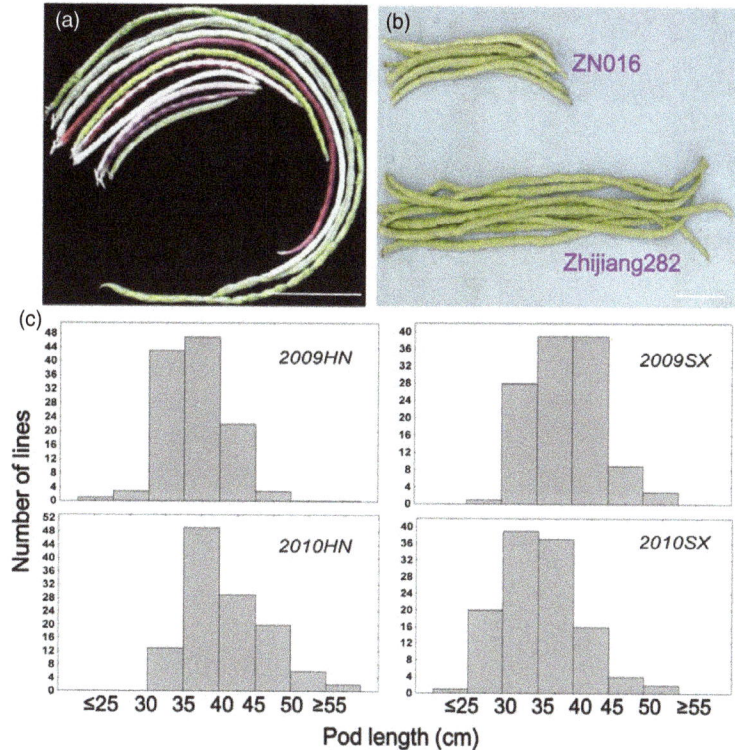

Figure 1 Pod length phenotypes. (a) Variation of pod length among selected germplasm lines. (b) Pod morphology of ZN016 and Zhijiang282, the parents of the RIL population. (c) Distribution of pod length in the ZN016× Zhijiang282 RIL population. Number of RILs = 119. Scale bar = 10 cm.

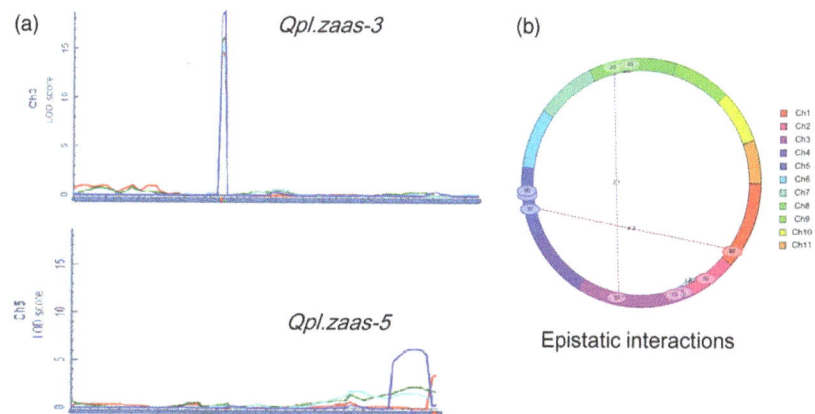

Figure 2 QTLs detected under the additive (a) and epistatic interaction (b) models implemented in QTL iCIMapping software. In (a), different colours represent results obtained from different trials. Dark blue: 2009HN; Red: 2010HN; Green: 2010SX; Light blue: 2009SX. Number of RILs for mapping = 119.

(r^2 = 0.2) with genetic distance occurred at ~2 cM across the whole panel. LD decayed (cM values) faster in subpopulation 1 than in subpopulation 2 where more breeding lines were included (Figure 3h). GWAS was performed for pod length using a mixed-linear model (MLM) method correcting for population structure and kinship. We identified 72 SNP loci associated with pod length, of which 55 had known map positions (Figure 4, Table S3). The 55 SNPs were distributed among nine of the 11 LGs, representing 16 SNP clusters. Observed phenotypic variation explained by a single SNP varied from 4.6% to 7.1%. The two significant SNP makers on LG3 coincided with *Qpl.zaas-3*, which spans the region 51.87– 54.99 cM on the cowpea consensus map (Muñoz-Amatriaín *et al.*, 2016).

Population genome-wide scan for selection signals

A genome-wide scan of population differentiation index (F_{ST}) and nucleotide diversity (π) was carried out first between the

two subgene pools and then between the landraces and cultivars/breeding lines from subpopulation 2. Overall F_{ST} between subpopulation 1 and subpopulation 2 was 0.262 (Table S4), indicating a moderate population divergence. The mean π value in subpopulation 1 was 0.314, more than twice that of in subpopulation 2 (0.13, Table S4), suggesting a severe loss of genetic diversity in the vegetable cowpea subgene pool. A fine-scale inspection of π and F_{ST} in a window sizing 0.15 cM (roughly 100 Kb of physical distance) with a step size of 0.03 cM across the genome revealed variations among and fluctuations across LGs, as expected. A total of 40.5 cM of differentiated genomic regions as indicated by F_{ST} outliers were determined between the two subgroups ($P \leq 0.05$, Figure 5a, Table S5), accounting for 4.6% of the total length of the genetic map. Selective sweeps as suggested by nucleotide diversity ratio ($\pi_{subpopulation1/subpopulation2}$) outliers were detected in genomic regions totalling 15.33 cM in length, representing ~1.8% of the genome ($P \leq 0.05$,

Table 1 QTLs detected with a LOD score ≥3 under the ICIM-ADD model

QTL	LG	Position (cM)	Marker interval	Single environment					LOD.[c]A×E
				Environment	LOD	[a]PVE%	[b]Additive		
Qpl.zaas-3	3	44.8-46.1	2_04960-2_02274	2009HN	18.7	45.7	−3.8547		4.4
				2009SX	15.4	45	−3.398		
				2010HN	14.7	39.1	−2.6831		
				2010SX	16.1	46	−3.6866		
Qpl.zaas-5	5	66-75.5	2_12600-1_1130	2009HN	6.3	13.7	−2.053		
				2010HN	3.6	7.8	−1.1896		

[a]PVE, phenotypic variation explained. Average proportion of variation explained.

[b]The negative additive effects indicated that ZN016 contributed the allele to a decrease in pod length.

[c]A by E effect, the additive and dominance × environment effect.

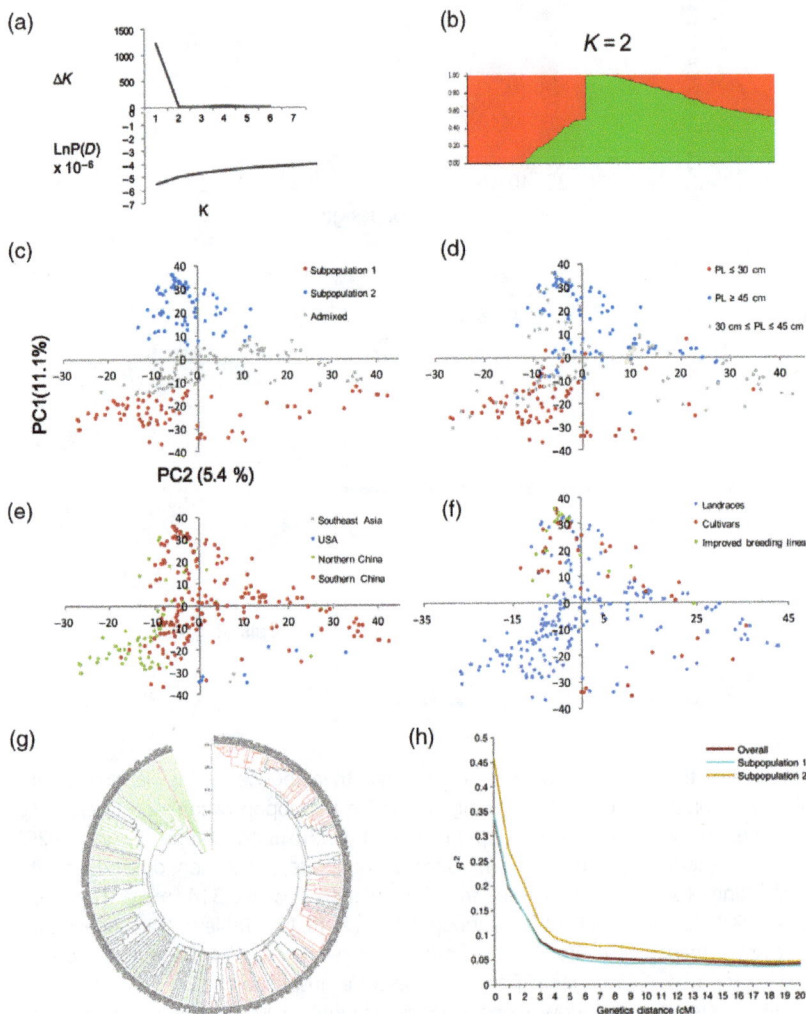

Figure 3 Subpopulation inference, principal component analysis (PCA) and dendrogram of the germplasm collection (299 accessions). (a) A plot of LnP(D) and delta K against K. Note that the scales for Y-axis are not proportional above and under zero; (b) estimated population structure of the germplasm collection inferred at K = 2; c–f, display of PCA results with the accessions coloured by population subgrouping (c), pod length group (d), geographic origin (e) or breeding status (f); (g) an unrooted phylogenetic tree showing the dendrogram of all samples. Accessions with pod length shorter than 30 cm are marked in red, longer than 45 cm in green and between 30 and 45 cm in black; h, decay of linkage disequilibrium (LD) in all samples, subpopulation 1 and subpopulation 2. The strength of LD was measured by r^2.

Figure 5a, Table S5). When comparing the locations of these regions with 42 reported SNP-tagged qualitative genes/QTLs governing domestication/resistance traits (Table S6) and 55 pod length-associated SNPs identified in this study, many of them were found to be overlapping (Figure 5a). Given the very small portion of the genome tagged as diverged and selected regions, the coincidences are unlikely to occur by chance alone. Among these overlapping regions, ten were related to pod length. The remaining seven regions harbour genes/QTLs governing many other traits such as seed size, seed coat colour, flower colour, leaf shape and root-knot nematode resistance. As the two subgene pools under

Figure 4 Manhattan plot displaying significant associations between SNP markers and pod length. The position of the major QTL *Qpl.zaas-3* detected in single-family QTL analysis in this study and the position of the major QTL *Pdl7.1* reported in Kongjaimun et al. (2012a) are indicated with black arrows. The horizontal axes indicate the consensus map position of each SNP while the vertical axes indicate the -log10 of the *P*-values. The dash line indicates the threshold of 3. The false discovery rate (FDR) for significant SNPs is listed in Table S3.

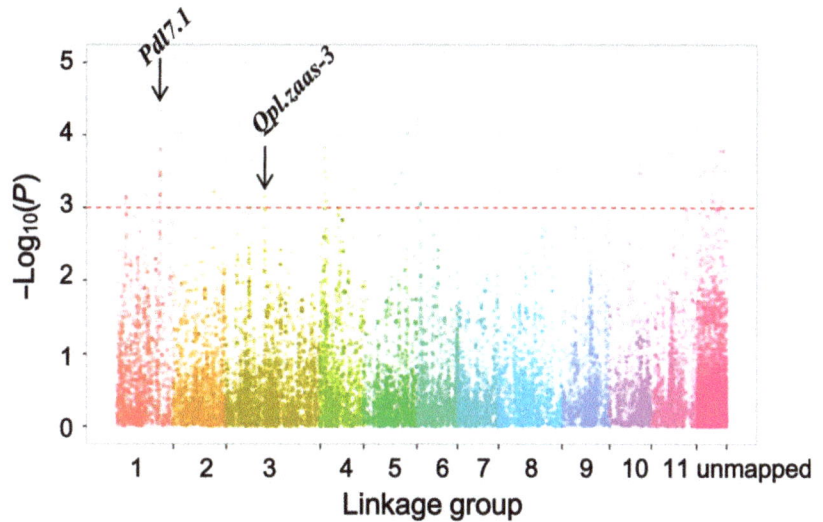

investigation are predominantly comprised of grain and vegetable cowpeas, respectively, our results may suggest that domestication of the vegetable cowpea was accompanied by selection for both pod length and other domestication/resistance traits.

F_{ST} and π ratio outliers between landraces and cultivars are considered to be indicative of selective signals during crop improvement after domestication (Shi and Lai, 2015). Our F_{ST} outliers analysis between landraces and cultivars/breeding lines in subpopulation 2 revealed a total of 16.32 cM of differentiated genome regions ($P \leq 0.05$, Figure 5b). Selective sweeps ($\pi_{cultivars}/\pi_{landraces}$ outliers) were detected in genome regions totalling 13.74 cM in length. Seven of these regions were found to be coincident with pod length QTLs, and only three with QTLs controlling other traits. These results suggest that pod length appears to be the primary selection target during improvement of vegetable cowpea after its domestication. Among the seven regions overlapping with pod length QTLs, two were specific to the improvement process.

Growth kinetics and histological analysis of the long and short pods

The genotypic differences of pod length might result either from different cell size or from cell number. To elucidate this, cell diameters were measured for sectioned pods among three genotypes representing the long- (Zhijiang282), medium- (ZN016) and short-podded (G314) genotypes, respectively. Their growth kinetics after anthesis are shown in Figure 6a. We found that Zhijiang282 reached a stable pod length much later than G314, with ZN016 being intermediate. This indicates more durable pod growth activity in Zhijiang282. At the early stage of postanthesis (1 dpa) during pod elongation, the cell diameters were similar among genotypes (Figure 6b, c). At the later stages (5 dpa and 10 dpa), Zhijiang282 did not show more elongated or enlarged cells, but rather had a similar or even smaller cell sizes than the shorter-pod lines. This phenomenon was explained by the slower growth rate of pods in Zhijiang282 and suggested that the eventually long-pod phenotype in Zhijiang282 was due primarily to more durable cell proliferation rather than to cell elongation/enlargement.

Comparative transcriptomic analysis between the long- and short-pod pools

Custom microarray hybridizations of a long-pod and a short-pod pool were performed to reveal the gene regulatory basis underlying pod length differences. Three technical replicates were included and showed a high consistency between each other (pairwise correlation coefficients ranged from 0.931 to 0.997, Figure S3). A total of 873 and 1228 genes were found to be more abundantly expressed in long-pod and short-pod pools, respectively (fold change ≥ 2, FDR ≤ 0.05, Table S7). *SWEETIE*, encoding a glycosyl transferase known to involve in the regulation of sugar flux (Veyres et al., 2008), was found among the top differentially expressed genes (DEGs). It suggests that sugar signalling is involved in pod length regulation. Gene ontology (GO) enrichment analysis of the DEGs revealed that GO terms including 'mitotic cell cycle process', 'cytokinesis by cell plate formation' and 'cyclin-dependent protein serine/threonine kinase regulator activity' were significantly enriched (Table S8). This conformed to the results of histological analysis and suggests a role of cell division in pod length regulation. Additionally, the enrichment of 'gibberellin mediated signalling pathway', 'nutrient reservoir activity' and 'response to carbon dioxide' indicates that hormonal and nutritional signalling are also important mechanisms related to pod length control.

Discussion

The past decade has witnessed significant advances in genome research for cowpea including asparagus bean (e.g. Kongjaimun et al., 2012a,b; Muchero et al., 2009; Xu et al., 2012). Here, we used the Cowpea iSelect Consortium Array to genotype a RIL population and a large association panel for asparagus bean to map QTLs controlling pod length. Our results, in line with previous studies (Kongjaimun et al., 2012a; Vidya et al., 2002), show a high heritability of pod length in asparagus bean, suggesting that the use of marker-assisted selection (MAS) in the improvement of this trait in early generations is feasible. Consistent also with previous studies (Hazra et al., 2007; Rashwan, 2010), we found that additive effects serve as the major

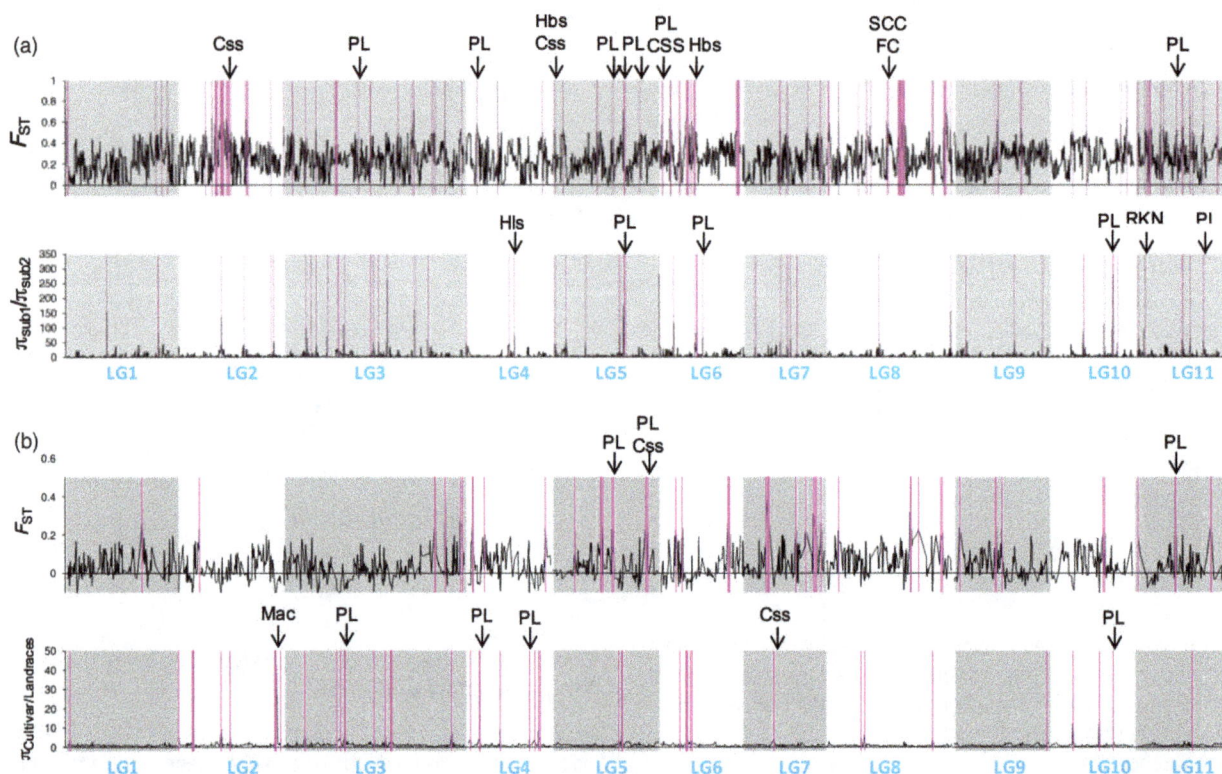

Figure 5 Genome scan for F_{ST} and π ratio. For (a), the parameters were calculated between the two subgene pools and for (b) between the cultivars/breeding lines and landraces of subpopulation 2. Both parameters were first calculated at each SNP site and then were averaged and plotted via a kernel-smoothing moving method that took 0.15 cM sliding windows with 0.03 cM steps to generate genome-wide distributions. A bootstrap resampling technique was applied for assigning significance threshold values. One million replicates were run for each statistic. Putative selective signal regions (outliers of F_{ST} and π ratio, $P \leq 0.05$) are highlighted in purple. Qualitative genes/QTLs overlapping with the selective signals are marked with black arrows. PL: pod length; Hbs: heat-induced brown discoloration of seed coats; RKN: resistance to root-knot nematodes; SSC: seed coat colour; FC: flower colour; Css: seed size; Hls: hastate leaf shape; Mac: resistance to *Macrophomina phaseolina*.

genetic basis for pod length. In particular, the 'one major QTL + minor QTLs' mode of pod length determination as discovered in our intervarietal population is similar to the case in an asparagus bean × grain cowpea population (Kongjaimun *et al.*, 2012a). It is interesting to note that in some other related *Vigna* species such as azuki bean (Isemura *et al.*, 2007), rice bean (Isemura *et al.*, 2010) and mung bean, a similar mode of pod length inheritance was reported. Nevertheless, the extremely broad variability of pod length has been only observed in cowpea. This implies the occurrence of species-specific mutations in pod length genes in *V. unguiculata*. By incorporating a broader range of genetic variations available in the germplasm collection, the GWAS analysis uncovered additional QTLs associated with pod length. *Qpl.zaas-3*, the major QTL detected in the biparental RIL mapping study, was also detected by GWAS. Due to a shared SSR marker (cp07863) between the earlier version of the ZZ map (Xu *et al.*, 2011b) and the Kongjaimun *et al.* (2012a) genetic map, we found that the major QTL *Pdl7.1* reported in Kongjaimun *et al.* (2012a) was coincident with a cluster of significant SNPs on LG1 in our study (Figure 4). Some other highly significant SNP clusters (e.g. on LG4, 5, 10 and 11) may represent novel pod length QTLs. Besides additive gene effects, epistasis as part of the genetic components affecting pod length was also disclosed, suggesting the feasibility of utilizing heterosis in breeding programmes targeting pod length.

It is a common phenomenon that QTLs governing different domestication traits are linked or are involved in pleiotropic

effects. In cowpea, tightly linked or pleiotropic QTLs were reported for pod length, plant height, seed size, flowering time, seed coat colour, flower colour and other traits (Peksen, 2004; Ubi *et al.*, 2000; Xu *et al.*, 2011b, 2013). Among the QTLs detected in our population, *Qpl.zaas-3* overlaps a previously mapped major QTL for pod number per plant (*Qpn.zaas-3*, Xu *et al.*, 2013), which could explain the long-standing findings of strong correlation between the two traits (Aggarwal *et al.*, 1982; Bapna *et al.*, 1972). Recently, extensive co-localizations of QTLs controlling pod tenderness and pod length were also reported (Kongjaimun *et al.*, 2013), pointing to the co-domestication of agricultural traits in vegetable cowpea.

Population structure analyses suggested a general division of the grain cowpea and vegetable cowpea subgene pools among the investigated diversity panel. Based on comparative analysis of known QTLs and putative selective signals during vegetable cowpea domestication and improvement, we were able to tentatively propose a model for the diversification of grain and vegetable cowpea (Figure 7). We assume that founder germplasm of vegetable cowpea, after being introduced into the place of their domestication in Asia, was naturally or artificially selected for both pod length and many other domestication/resistance traits important to its adaptation to the local target agroecosystem. The long-lasting wet season in Asia must have favoured vegetative growth, and thus late maturing and pest/disease-resistant cowpea became more adapted. The domesticated landrace vegetable cowpea may have then undergone

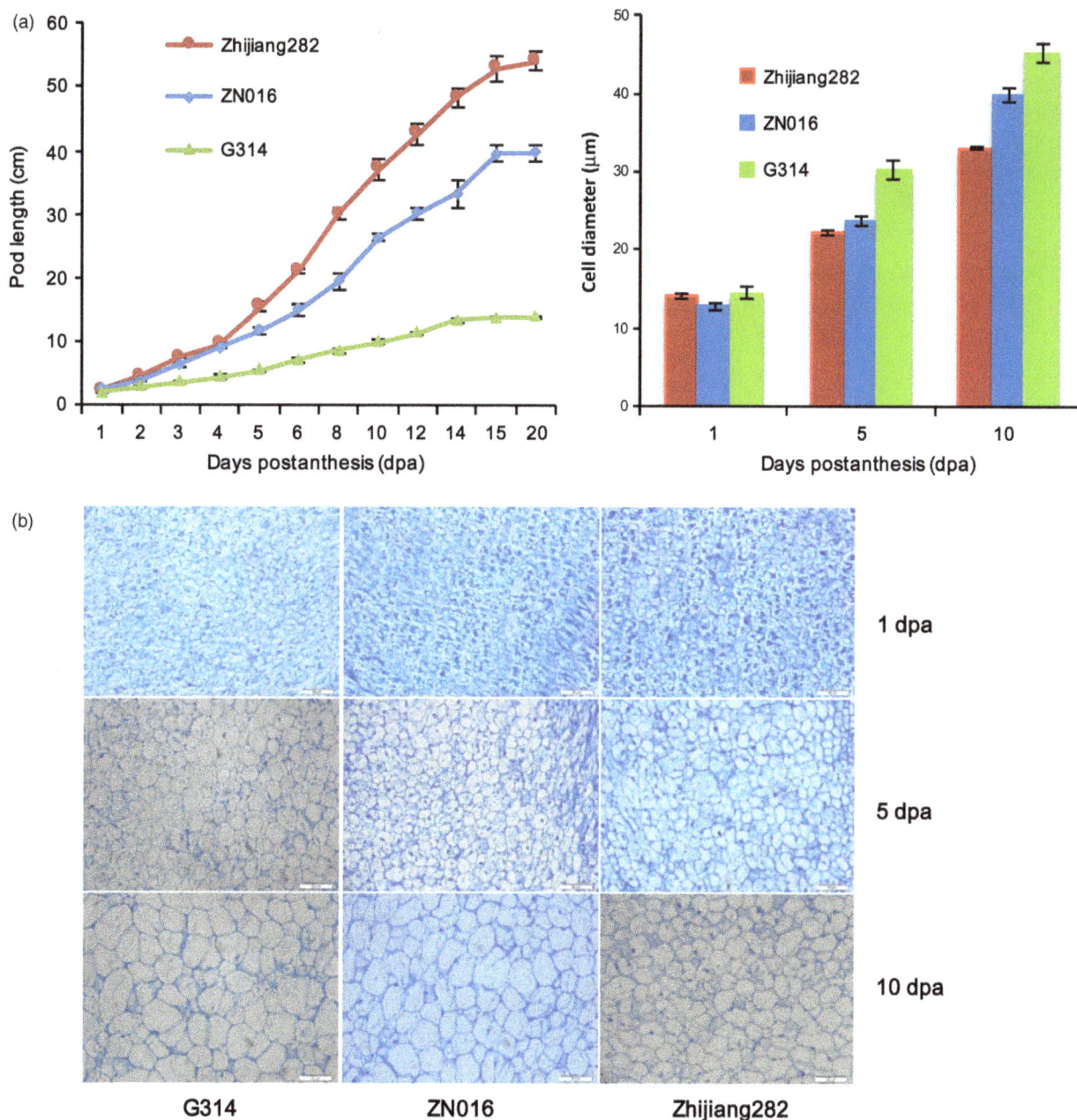

Figure 6 Growth kinetics and cellular morphology of sectioned pods in three genotypes. (a) kinetics of pod elongation after anthesis (left panel) and cell diameters measured at different dpa, $n = 20$ (right panel); (b) longitudinally sectioned pod cells at different dpa. Error bars indicate SE. Scale bar = 50 μm.

further intensive selections towards increased pod length, leading to the present-day asparagus bean. This may partly be owing to the fact that cowpea varieties with longer soft pods are more attractive to local consumers and growers due to cooking convenience and higher yield contribution (Umaharan et al., 1997). During this process, severe loss of genetic diversity would have occurred, resulting in the low genetic diversity in asparagus bean. Hence, restoring specific beneficial or favourable alleles in the current asparagus bean varieties and breeding lines may require introgression from landrace asparagus bean lines or grain-type cowpeas.

Our study provides novel insights into the diversification of cowpea selected for vegetable and grain uses. Nevertheless, several caveats should be mentioned regarding the interpretation of our results due to technical limitations. First, we used a fixed SNP chip

(albeit high genome coverage and density) for genotyping which may suffer from ascertainment bias (Albrechtsen et al., 2010). However, asparagus beans were included in the SNP discovery panel (Muñoz-Amatriaín et al., 2016). Also, the impact of a possible ascertainment biases on our results may be low because one of the key statistics, F_{ST}, used in this study was reported to be 'tolerant' to ascertainment bias (Albrechtsen et al., 2010). Second, the uneven distribution of the markers on the genetic map and co-segregation of physically distant loci may confound the results. Therefore, drawing a more conclusive picture of the domestication history of vegetable cowpea will benefit from whole genome (re-) sequencing studies. In the near future, a complete cowpea reference genome is expected to enable identification of causal genes from the QTL regions. The analysis of pod development and its transcriptional regulations conducted in this study will be of

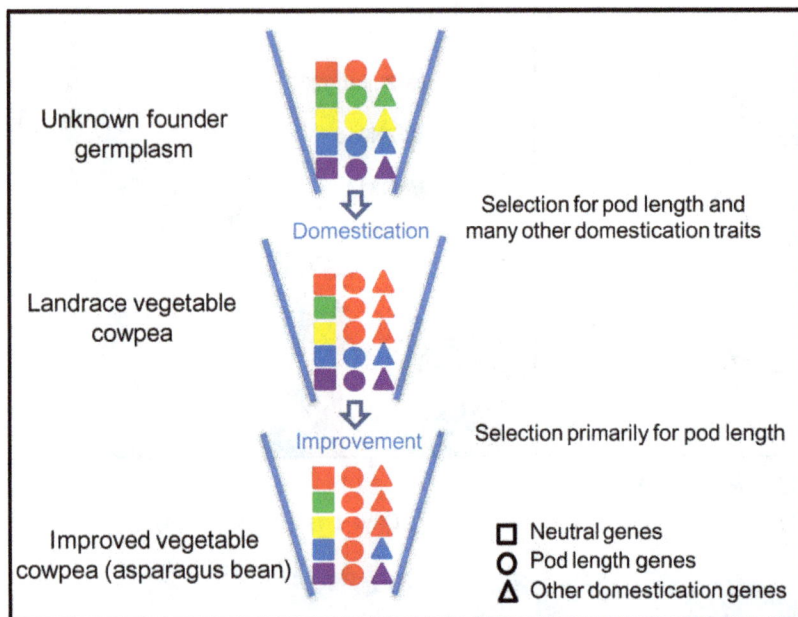

Figure 7 A graphical model depicting the signatures of natural or artificial selection driving the diversification of grain and vegetable cowpea. Different colours represent genetic diversity. We assume that the founder germplasm of vegetable cowpea was naturally or artificially selected for many domestication and resistance traits including pod length; the domesticated vegetable cowpea may have then undergone further selections towards pod length primarily, leading to the present-day asparagus bean.

particular use for such task as it provides biological information complementary to genetic analyses. Another interesting future work is to measure the dynamic behaviour of pod length for all individuals from the mapping population. These data, when integrated into a functional mapping framework (Wu and Lin, 2006), will provide new insights into the developmental genetic mechanisms of pod length.

Experimental procedures

Plant materials and growth conditions

The population used in genetic mapping included 132 recombinant inbred lines ($F_{6:8}$, 'ZZ' population) developed by single-seed descent from the cross 'ZN016' × 'Zhijiang282'. 'ZN016' is a landrace vegetable cowpea accession originating from southern China with medium-long pods (~38 cm) while 'Zhijiang282' is a typical asparagus bean cultivar with long pods (~55 cm, Figure 1a). Field experiments were carried out in two consecutive years (2009 and 2010), each in two locations, that is Haining County (HN, 30°32′N, 120°41′E) and Shaoxing County (SX, 29°43′N, 120°14′E). Each experiment had two replicates, except for SX2009 with one replicate. Eight to ten seeds per line were sown every 28 cm in 25-m-long plots on rows 75 cm apart, but only four uniform seedlings were retained per line after seedling emergence due to the size and trailing habit of the adult plants. The plots were spaced by 50 cm to avoid border effect. A set of 299 cowpea accessions, either grain or vegetable type, was used as the diversity panel (Table S1, Figure 1b). The accession 'G93' was sampled twice in order to provide an internal control of data quality, leading to a total sample number of 300. The plants were grown in HN in 2014 and 2015 for pod length measurement, with two replicates in each experiment. Growth conditions and field management were similar to those for the RIL population.

DNA extraction, SNP genotyping and raw data processing

Genomic DNA was extracted from leaves of field-grown plants 1 month after sowing using DNeasy Plant DNA miniprep kits (Qiagen, Hilden, Germany). Standardized DNA for the samples,

the majority at a concentration of 50 ng/μL and a small fraction 10–30 ng/μL were hybridized to the Cowpea iSelect Consortium Array according to the standard protocol. Single-base extension was performed and the chips were scanned using the Illumina iScan. Image files were saved for cluster file analysis. The clustering algorithm of GenomeStudio Genotyping Module (V 1.9.4, Illumina, Inc.) was used for SNP calling. To increase the accuracy of the clustering algorithm, additional samples including synthetic heterozygous were added to the workspace. Polymorphic SNPs were identified based on Illumina GenTrain score (proximity of clusters) and call frequencies across samples. All markers were manually inspected and curated based on manufacturer's best-practice instructions. All marker sequences can be retrieved from the HarvEST database (http://harvest.ucr.edu) and detailed information from Muñoz-Amatriaín et al. (2016).

Linkage mapping

Linkage mapping of the SNP markers was performed with MSTmap (Wu et al., 2008). Before computing the marker orders and distances, SNPs were filtered using the following criteria: missing call rate ≤20% (equivalent to a minimum effective population size of 106), heterozygous call rate ≤10% and minor allele frequency (MAF) ≥0.25. Critical parameters for mapping were as follows: population type = RIL at generation 6; no. of mapping size threshold = 2; no. of mapping distance threshold = 15 cM; no. of mapping missing threshold = 25%; genetic mapping function = Kosambi; try to detect genotyping errors = no.

Phenotyping

Pod length was phenotyped by measuring the distance from the peduncle connection point to the apex of the pod. Ten representative pods were measured per line, and the data were averaged.

QTL mapping

QTL IciMapping V4.0 (Wang et al., 2014) was used to perform inclusive composite interval mapping (ICIM-ADD) for pod length.

Critical mapping parameters were as follows: step size = 1 cM, PIN = 0.001. A LOD threshold of three was used to determine major QTLs. The ICIM-EPI function implemented in the software was used to scan for QTL interactions with a LOD threshold of three. QTLs × environment interactions were detected under the 'multi-environmental trials' mode with a LOD score threshold of three. QTLs detected in different environments within same, adjacent or overlapping marker intervals were designated as the same.

Diversity, population structure and kinship analyses

Pairwise genetic distances were calculated using TASSEL 5.1 (Bradbury et al., 2007) based on 30 211 SNPs, which was a subset of the 49 194 technically successful SNPs after filtration with the following criteria: missing call rate ≤20%, heterozygous call rate ≤20% and MAF ≥ 0.01. Population structure was inferred using a two-step procedure. Firstly, we ran the software STRUCTURE 2.3.4 (Pritchard et al., 2000) under the 'admixture model' with a burn-in period of 100 000 followed by 100 000 replications of Markov chain Monte Carlo with 2320 SNPs each with a discrete map position on the latest cowpea consensus map (Muñoz-Amatriaín et al., 2016). Five independent runs each were performed with the number of clusters (K) varying from 1 to 8. The Evanno method (Evanno et al., 2005) was used to determine the optimal K for subgrouping. Next, the software FastStructure (Raj et al., 2014) that enables time-efficient handling of a large number of markers was used to run all the 30 211 SNPs at the optimal K to give the whole genome ancestry estimates (Q-matrix). Lines with a probability of membership ≥70% were assigned to a subgroup (otherwise, 'admixed'). Relative kinship matrix was generated using the same set of 30 211 SNP markers with TASSEL 5.1. The neighbour-joining tree was generated by MEGA5 based on genetic distance data (Tamura et al., 2011). LD was measured by calculating the square value of correlation coefficient (r^2) between each SNP pair.

Genome-wide trait–marker association study (GWAS)

A trait–marker association analysis was performed using Tassel 5.1 under a mixed-linear model (MLM) that corrects for both population structure (Q) and relative kinship (K). The phenotypic data used were a combined set of field data from multiple experiments. Significant SNPs were defined if showing a minus log10-transformed $P \geq 3$. The FDR multiple test correction was performed using the software QVALUE (Storey, 2002). SNPs with a genetic distance less than 2 cM were considered to be in a LD extension block and belong to the same SNP cluster.

Population genomic parameters calculation and scanning across the genome

Nucleotide diversity (π) was calculated using the formula

$$\pi = 1 - \sum_i \binom{n_i}{2} / \binom{n}{2},$$ where n_i denotes the count of allele i and $n = \Sigma \, n_i$. Population differentiation index (F_{ST}) was estimated

adapting the formula $F_{ST} = 1 - \dfrac{\sum_j \binom{n_j}{2} \pi_j}{\pi \cdot \sum_j \binom{n_j}{2}}$ that accounts for

unbalanced population sizes according to Nielsen et al. (2009). These parameters were first calculated at each of the 25 873 SNP site mapped onto the cowpea consensus map and then were averaged and plotted via a kernel-smoothing moving method

that took 0.15 cM sliding windows with 0.03 cM steps to generate genome-wide distributions. With the estimated cowpea genome size of 630 Mb and the consensus genetic map length of 837.11 cM, the window and the sliding step size roughly equalled 100 Kb and 20 Kb of physical distances, respectively. Wherever the π value as denominator was zero, the corresponding F_{ST} value was replaced with the genomic mean. The method for assigning significance threshold values was based on a bootstrap resampling technique as described in Hohenlohe et al. (2010). One million replicates were run for each statistic.

Comparative analysis of known genes/QTLs and putative selective signals

Map position information of known qualitative genes or QTLs governing domestication and resistance traits in cowpea was assembled from a literature search. Only genes/QTLs tagged by SNP markers that are comparable to our marker system were used in the analysis. SNP markers reported to be tightly linked to qualitative genes or located around the peak regions of QTLs were extracted. A coincidence between a known gene/QTL and a selective signal was declared if the SNP markers fell into the outlier window regions of F_{ST} or π ratio. If the known QTLs (or single associated SNPs) were detected by GWAS, their coincidence with a selective signal was determined by the overlap of the LD block where the SNP resides (2 cM downstream and upstream of the significant SNP) and the window region of the selective signal.

Pod growth kinetics assay, paraffin sectioning and histological analysis

Pod samples collected at 1, 5 and 10 dpa were fixed in FAA (5 mL 38% formalin: 5 mL glacialacetic acid: 90 mL 70% alcohol) for 24 h and then dehydrated in a graded ethanol series, substituted with 1-butanol and embedded in Paraplast Plus. The samples were sectioned longitudinally at 4 μm thick using a rotary microtome (RM2135,LEICA). The sections were stained with methylene blue and observed under a light microscope (BH2, OLYMPUS). For cell diameter comparison, ten cells from each sample were measured and the data averaged.

Custom cDNA microarray construction, hybridization and data processing

A custom Agilent microarray targeting 29 471 cowpea unigenes was transferred from a previous Roche NimbleGen cowpea microarray (Xu et al., 2015). Hybridizations were performed for long-pod pools and short-pod pools, each consisting of RNA from seven independent genotypes, in three replicates. Details of the 14 genotypes used for pooling can be found in Table S1. Methods for RNA extraction, purification and quantification were as described in Xu et al. (2015). All cDNA were labelled with the fluorescent dye Cy5-dCTP using cRNA amplification and labelling kit (CapitalBio, Beijing, China). Array hybridization, washing and image scanning were conducted according to the manufacturer's instructions. The raw data were summarized and normalized using the GeneSpring software V12.0 (Agilent). Differentially expressed genes (DEGs) were defined by an expression fold change ≥2 between the two pools and a corrected P-value (Benjamini-Hochberg FDR) ≤0.05. Hierarchical clustering was performed with Cluster 3.0 using the average linkage method, and the results were visualized using TreeView (Eisen et al., 1998).

Gene ontology (GO) enrichment analyses

Gene ontology (GO) enrichment analyses were performed using GOrilla (Eden et al., 2009) under a P-value threshold of $10E^{-3}$ for statistical significance. The GObase used was the version updated on 5 December 2015. Prior to running GOrilla, the cowpea unigenes were BLASTx searched against the Arabidopsis protein sequences database under a P-value cut-off of $10E^{-5}$ to generate legible sequence IDs for recognition in the program. The entire list of unigenes on the chip was inputted as the background for calculation.

Acknowledgements

This work was supported by the National Key Research & Development Program of China (2016YFD0100204-32); the National Natural Science Foundation of China (31572135); the Natural Science Foundation of Zhejiang Province (LY15C150002); the Major Science and Technology Project of Zhejiang Province (2012C12903); and the National Ten-Thousand Talent Program of China (to P. Xu). Development of the cowpea iSelect was supported by the Feed the Future Innovation Lab for Climate Resilient Cowpea (USAID Cooperative Agreement AID-OAA-A-13-00070) and the Illumina Agricultural Greater Good Initiative. Partial support was also provided by Hatch Project CA-R-BPS-5306-H. We thank Mitchell Lucas and Yi-Ning Guo (UC-Riverside), and Ye Tao and Liang Zeng (Biozeron Biotech, Shanghai) for technical assistances. The authors declare no conflict of interest.

References

Aggarwal, V.D., Natare, R.B. and Smithson, J.B. (1982) The relationship among yield and other characters in vegetable cowpea and the effect of different trellis management on pod yield. Trop. Grain Leg. Bull. **25**, 8–14.

Albrechtsen, A., Nielsen, F.C. and Nielsen, R. (2010) Ascertainment biases in SNP Chips affect measures of population divergence. Mol. Biol. Evol. **27**, 2534–2547.

Bapna, C.S., Joshi, S.N. and Kabria, M.M. (1972) Correlation studies on yield and agronomic characters in cowpea. Indian J. Agron. **17**, 321–324.

Bradbury, P.J., Zhang, Z.W., Kroon, D.E., Casstevens, T.M., Ramdoss, Y. and Buckler, E.S. (2007) TASSEL: software for association mapping of complex traits in diverse samples. Bioinformatics, **23**, 2633–2635.

Eden, E., Navon, R., Steinfeld, I., Lipson, D. and Yakhini, Z. (2009) Gorilla: a tool for discovery and visualization of enriched GO terms in ranked gene lists. BMC Bioinformatics, **10**, 48.

Eisen, M.B., Spellman, P.T., Brown, P.O. and Botstein, D. (1998) Cluster analysis and display of genome-wide expression patterns. Proc. Natl Acad. Sci. USA, **95**, 14863–14868.

Evanno, G., Regnaut, S. and Goudet, J. (2005) Detecting the number of clusters of individuals using the software STRUCTURE: a simulation study. Mol. Ecol. **14**, 2611–2620.

Fang, J.G., Chao, C.C.T., Roberts, P.A. and Ehlers, J.D. (1996) Genetic diversity of cowpea [Vigna unguiculata. L. Walp.] in four West African and USA breeding programs as determined by AFLP analysis. Genet. Resour. Crop Ev. **54**, 1197–1209.

Hazra, P., Chattopadhaya, A., Dasgupta, T., Kar, N., Das, P.K. and Som, M.G. (2007) Breeding strategy for improving plant type, pod yield and protein content in vegetable cowpea (Vigna unguiculata). Acta Hortic. **752**, 725–780.

Hohenlohe, P.A., Bassham, S., Etter, P.D., Stiffler, N., Johnson, E.A. and Cresko, W.A. (2010) Population genomics of parallel adaptation in threespine stickleback using sequenced RAD tags. PLoS Genet. **6**, e1000862.

Isemura, T., Kaga, A., Konishi, S., Ando, T., Tomooka, N., Han, O.K. and Vaugha, D.A. (2007) Genome dissection of traits related to domestication in azuki bean (Vigna angularis) and comparison with other warm-season legumes. Ann. Bot. **100**, 1053–1071.

Isemura, T., Kaga, A., Tomooka, N., Shimizu, T. and Vaughan, D.A. (2010) The genetics of domestication of rice bean, Vigna umbellata. Ann. Bot. **106**, 927–944.

Kongjaimun, A., Kaga, A., Tomooka, N., Somta, P., Vaughan, D.A. and Srinives, P. (2012a) An SSR-based linkage map of yardlong bean (Vigna unguiculata (L.) Walp. subsp. unguiculata Sesquipedalis Group) and QTL analysis of pod length. Genome, **55**, 81–92.

Kongjaimun, A., Kaga, A., Tomooka, N., Somta, P., Vaughan, D.A. and Srinives, P. (2012b) The genetics of domestication of yardlong bean, Vigna unguiculata (L.) Walp. ssp. unguiculata cv. -gr. sesquipedalis. Ann. Bot. **109**, 1185–2000.

Kongjaimun, A., Somta, P., Tomooka, N., Kaga, A., Vaughan, D.A. and Srinives, P. (2013) QTL mapping of pod tenderness and total soluble solid in yardlong bean [Vigna unguiculata (L.) Walp. subsp. unguiculata cv.-gr. sesquipedalis]. Euphytica, **189**, 217–223.

Muchero, W., Diop, N.N., Bhat, P.R., Fenton, R.D., Wanamaker, S., Pottorff, M., Hearne, S. et al. (2009) A consensus genetic map of cowpea [Vigna unguiculata L. Walp] and synteny based on EST derived SNPs. Proc. Natl Acad. Sci. USA, **106**, 18159–18164.

Muñoz-Amatriaín, M., Mirebrahim, H., Xu, P., Wanamaker, S., Luo, M.C., Alhakami, H., Alpert, M. et al. (2016) Genome resources for climate-resilient cowpea, an essential crop for food security. bioRxiv, 059261; doi: http://dx.doi.org/10.1101/059261.

National Research Council. (2006) Lost Crops of Africa: Volume II: Vegetables. Washington, DC: The National Academies Press.

Nielsen, R., Hubisz, M.J., Hellmann, I., Torgerson, D., Andrés, A.M., Albrechtsen, A., Gutenkunst, R. et al. (2009) Darwinian and demographic forces affecting human protein coding genes. Genome Res. **19**, 838–849.

Peksen, A. (2004) Fresh pod yield and some pod characteristics of cowpea (Vigna unguiculata L. Walp.) genotypes from Turkey. Asia J. Plant Sci. **3**, 269–273.

Pritchard, J.K., Stephens, M. and Donnelly, P. (2000) Inference of population structure from multilocus genotype data. Genetics, **155**, 945–959.

Rachie, K.O. (1985) Introduction: Cowpea research, Production and Utilization. Edited by Singh and Rachie. New York, NY: John Wiley and Sons Ltd p. 320.

Raj, A, Stephens, M., Jonathan, K. and Pritchard, J.K. (2014) fastSTRUCTURE: variational inference of population structure in large SNP data sets. Genetics, **197**, 573–589.

Rashwan, A.M.A. (2010) Estimation of some genetic parameters using six populations of two cowpea hybrids. Asia. J. Crop Sci. **2**, 261–267.

Shi, J. and Lai, J. (2015) Patterns of genomic changes with crop domestication and breeding. Curr. Opin. Plant Biol. **24**, 47–53.

Singh, B.B. (2002) Recent genetic studies in cowpea. In Challenges and Opportunities for Enhancing Sustainable Cowpea Production(Fatokun, C.A., Tarawali, S.A., Singh, B.B., Kormawa, P.M. and Tamo, M., eds), pp. 3–13. Ibadan, Nigeria: International Institute of Tropical Agriculture.

Storey, J.D. (2002) A direct approach to false discovery rates. J. R. Stat. Soc. B, **64**, 479–498.

Tamura, K., Peterson, D., Peterson, N., Stecher, G., Nei, M. and Kumar, S. (2011) MEGA5: molecular evolutionary genetics analysis using maximum likelihood, Evolutionary Distance, and maximum parsimony methods. Mol. Biol. Evol. **28**, 2731–2739.

Timko, M.P., Ehlers, J.D. and Roberts, P.A. (2007) Cowpea. In Pulses, Sugar and Tuber Crops, Genome Mapping and Molecular Breeding in Plants, vol. **3** (Kole, C., ed), pp. 49–67. Berlin Heidelberg: Springer-Verlag.

Ubi, B.E., Mignouna, H. and Thottappilly, G. (2000) Construction of a genetic linkage map and QTL analysis using a recombinant inbred population derived from an intersubspecific cross of cowpea (Vigna unguiculata (L.) Walp.). Breed. Sci. **50**, 161–172.

Umaharan, P., Ariyanayagam, R.P. and Haque, S.Q. (1997) Genetic analysis of yield and its components in vegetable cowpea (Vigna unguiculata L. Walp). Euphytica, **96**, 207–213.

Veyres, N., Danon, A., Aono, M., Galliot, S., Karibasappa, Y.B., Diet, A., Grandmottet, F. *et al.* (2008) The Arabidopsis *sweetie* mutant is affected in carbohydrate metabolism and defective in the control of growth, development and senescence. *Plant J.* **55**, 665–686.

Vidya, C., Sunny, K., Oommen, S.K. and Kumar, V. (2002) Genetic variability and heritability of yield and related characters in yard-long bean. *J. Trop. Agri.* **40**, 11–13.

Wang, J., Li, H., Zhang, L. and Meng, L. (2014) *Users' Manual of QTL IciMapping*. The Quantitative Genetics Group, Institute of Crop Science, Chinese Academy of Agricultural Sciences (CAAS), Beijing 100081, China, and Genetic Resources Program, International Maize and Wheat Improvement Center (CIMMYT), Apdo. Postal 6-641, 06600 Mexico, D.F., Mexico

Wu, R. and Lin, M. (2006) Functional mapping – How to map and study the genetic architecture of dynamic complex traits. *Nature Rev. Genet.* **7**, 229–737.

Wu, Y., Bhat, P.R., Close, T.J. and Lonardi, S. (2008) Efficient and accurate construction of genetic linkage maps from the minimum spanning tree of a graph. *PLoS Genet.* **4**, e1000212.

Xu, P., Wu, X.H., Wang, B.G., Liu, Y.H., Qin, D.H., Ehlers, J.F., Close, T.J. *et al.* (2010) Development and polymorphism of *Vigna unguiculata* ssp. *unguiculata* microsatellite markers used for phylogenetic analysis in asparagus bean (*Vigna unguiculata* ssp. *sesquipedialis*. L. Verdc.). *Mol. Breed.* **25**, 675–684.

Xu, P., Hu, T., Yang, Y., Wu, X., Wang, B., Liu, Y., Qin, D. *et al.* (2011a) Mapping genes governing flower and seed coat color in asparagus bean (*Vigna unguiculata* ssp *sesquipedalis*) based on SNP and SSR markers. *HortScience*, **46**, 1102–1104.

Xu, P., Wu, X.H., Wang, B.G., Liu, Y.H., Ehlers, J.D. and Close, T.J. (2011b) A SNP and SSR based genetic map of asparagus bean (*Vigna. unguiculata* ssp. *sesquipedialis*) and comparison with the broader species. *PLoS ONE*, **6**, 1–8.

Xu, P., Wu, X.H., Wang, B.G., Luo, J., Liu, Y.H., Ehlers, J.D., Close, T.J. *et al.* (2012) Genome wide linkage disequilibrium in Chinese asparagus bean (*Vigna. unguiculata* ssp. *sesquipedialis*) germplasm: implications for domestication history and genome wide association studies. *Heredity*, **109**, 34–40.

Xu, P., Wu, X., Wang, B., Hu, T., Lu, Z., Liu, Y., Qin, D. *et al.* (2013) QTL mapping and epistatic interaction analysis in asparagus bean for several characterized and novel horticulturally important traits. *BMC Genet.* **14**, 4.

Xu, P., Moshelion, M., Wu, X., Halperin, O., Wang, B., Luo, J., Wallach, R. *et al.* (2015) Natural variation and gene regulatory basis for the responses of asparagus beans to soil drought. *Front. Plant Sci.* **6**, 891.

13

Engineering the production of conjugated fatty acids in *Arabidopsis thaliana* leaves

Olga Yurchenko[1], Jay M. Shockey[2], Satinder K. Gidda[3], Maxwell I. Silver[3], Kent D. Chapman[4], Robert T. Mullen[3] and John M. Dyer[1],*

[1]*USDA-ARS, US Arid-Land Agricultural Research Center, Maricopa, AZ, USA*

[2]*USDA-ARS, Southern Regional Research Center, New Orleans, LA, USA*

[3]*Department of Molecular and Cellular Biology, University of Guelph, Guelph, ON, Canada*

[4]*Department of Biological Sciences, University of North Texas, Denton, TX, USA*

Correspondence
email john.dyer@ars.usda.gov

Keywords: *Arabidopsis thaliana*, conjugated fatty acids, bio-based feedstocks, biofuels, oil in leaves, *Vernicia fordii*.

Summary

The seeds of many nondomesticated plant species synthesize oils containing high amounts of a single unusual fatty acid, many of which have potential usage in industry. Despite the identification of enzymes for unusual oxidized fatty acid synthesis, the production of these fatty acids in engineered seeds remains low and is often hampered by their inefficient exclusion from phospholipids. Recent studies have established the feasibility of increasing triacylglycerol content in plant leaves, which provides a novel approach for increasing energy density of biomass crops. Here, we determined whether the fatty acid composition of leaf oil could be engineered to accumulate unusual fatty acids. Eleostearic acid (ESA) is a conjugated fatty acid produced in seeds of the tung tree (*Vernicia fordii*) and has both industrial and nutritional end-uses. *Arabidopsis thaliana* lines with elevated leaf oil were first generated by transforming wild-type, *cgi-58* or *pxa1* mutants (the latter two of which contain mutations disrupting fatty acid breakdown) with the diacylglycerol acyltransferases (*DGAT1* or *DGAT2*) and/or oleosin genes from tung. High-leaf-oil plant lines were then transformed with tung *FADX*, which encodes the fatty acid desaturase/conjugase responsible for ESA synthesis. Analysis of lipids in leaves revealed that ESA was efficiently excluded from phospholipids, and co-expression of tung *FADX* and *DGAT2* promoted a synergistic increase in leaf oil content and ESA accumulation. Taken together, these results provide a new approach for increasing leaf oil content that is coupled with accumulation of unusual fatty acids. Implications for production of biofuels, bioproducts, and plant–pest interactions are discussed.

Introduction

Modern society is heavily reliant on fossil oil as a source of fuel and chemical feedstocks, and with continued growth of world population, it is expected that this need will only continue to increase. Given the finite nature of fossil oils, as well as environmental concerns associated with its usage, there is a clear and pressing need to develop more sustainable, environmentally friendly alternatives to petroleum. The fatty acid components of plant oils are chemically similar to the long-chain hydrocarbons of fossil oil and thus represent outstanding renewable sources of raw materials (Biermann *et al.*, 2011; Carlsson *et al.*, 2011; Dyer *et al.*, 2008; Horn and Benning, 2016). Indeed, a significant proportion of oilseed crops is already diverted for usage as feedstocks for biodiesel production (Durrett *et al.*, 2008), and government mandates for increased usage of renewable fuels have put additional pressure on agricultural production systems (Robbins, 2011; Zilberman *et al.*, 2013). Given that oilseed crops also serve as important sources of food and feed, there is significant interest in developing novel approaches for producing high amounts of energy-dense oils in dedicated, nonfood bioenergy crops and algae.

One approach that shows great potential for increasing oil production in plants is the elevation of neutral lipid content in vegetative biomass, such as leaves and stems (Chapman *et al.*, 2013). While plant oils are typically derived from seeds, vegetative cell types also have the capacity to synthesize triacylglycerol (TAG), the major component of plant oil. In seeds, TAG accumulates to high levels (~35%–45% dry weight) and serves as an important carbon and energy reserve to fuel postgerminative growth, prior to photosynthetic establishment. In leaves, the TAG pool is much smaller (generally ≪1% dry weight) and more dynamic in nature, acting as a buffer against excess lipids and serving as a transient depot for fatty acids involved in membrane remodelling, lipid signalling, and/or fatty acid turnover (Chapman *et al.*, 2012; Xu and Shanklin, 2016). However, recent research has shown that the TAG pool can be dramatically enhanced in vegetative cells using various engineering strategies that 'push' more carbon into the fatty acid biosynthetic pathway, 'pull' more fatty acids towards TAG synthesis, and 'protect' the TAG pool from turnover and/or fatty acid degradation (Vanhercke *et al.*, 2013a; Weselake, 2016; Xu and Shanklin, 2016). While many of these studies have been conducted using the model plant *Arabidopsis thaliana*, combinatorial approaches have been used to increase oil content of tobacco leaves up to 30% dry weight (Vanhercke *et al.*, 2016) and up to 4.7% in sugarcane (Zale *et al.*, 2016), suggesting that commercial high oil biomass crops are just on the horizon.

The energy obtained from fossil oils can be derived from a number of alternative sources including wind, solar, nuclear, and hydropower. The petrochemical industry, however, requires carbon-based feedstocks, and plant oils show great potential for fulfilling this need. Indeed, approximately 10% of plant oil is already used in various industrial applications (Biermann et al., 2011), but the fatty acid composition of vegetable oils is typically limited to just five basic fatty acid structures. There are hundreds of structurally diverse fatty acids synthesized in nature, and in many plant species, their seed oil is enriched in a single unusual fatty acid that can accumulate up to 90% of fatty acid composition (Badami and Patil, 1981; Smith, 1971). Many of the plants that produce these valuable oils, however, have poor agronomic traits or limited geographical growing areas. Therefore, a major goal of the plant biotechnology community has been to identify enzymes for unusual fatty acid synthesis and express them in higher yielding platform crops (Napier et al., 2014; Vanhercke et al., 2013b). Results to date have been mixed, however. For instance, engineering changes in fatty acid chain lengths or production of wax esters has been particularly successful (Lardizabal et al., 2000; Vanhercke et al., 2013b; Voelker et al., 1992), but production of unusual oxidized fatty acids has remained a challenge (Bates, 2016; Cahoon et al., 2007).

It is now widely recognized that many of the unusual fatty acids in plants are synthesized by divergent forms of fatty acid desaturase 2 (FAD2), an endoplasmic reticulum (ER) membrane-bound enzyme that typically acts upon phosphatidylcholine (PC)-linked oleate to produce linoleic acid (Okuley et al., 1994; Shanklin and Cahoon, 1998). Subtle changes in the polypeptide sequence of duplicated and diverged FAD2 enzymes alter their active site chemistry, allowing for production of a variety of oxidized products including hydroxy, epoxy, conjugated, and acetylenic fatty acids. Expression of diverged FAD2 enzymes in transgenic plants typically results in much lower accumulation of unusual fatty acids in seeds in comparison with seed oil from the plant in which the gene was sourced (Vanhercke et al., 2013b), and metabolic labelling studies have revealed inefficient removal of the unusual fatty acid from PC (Bates, 2016; Bates and Browse, 2011; Bates et al., 2014). This is particularly problematic for production of conjugated fatty acids, which can account for up to 25% of fatty acids in phospholipids of engineered seeds (Cahoon et al., 2006). Given that these fatty acids are likely disruptive to membrane structure, their accumulation in phospholipids is likely to contribute to the reported negative effects on embryo development and reduced germination potential (Cahoon et al., 1999, 2006).

Based on the recent success of increasing neutral lipid content in plant leaves, we asked whether this TAG pool might be engineered for accumulation of industrially important fatty acids, thereby bypassing some of the problems encountered with seeds. Towards that end, we focused on production of conjugated fatty acids, as these fatty acids have potential usage as industrial 'drying oils' in formulations of paints, inks, dyes, and resins (Sonntag, 1979), and they also have lipid-lowering and possibly anticancer effects in animals (Lee et al., 2002; Thiel-Cooper et al., 2001; Yuan et al., 2014). The source of genes for our study was the tung (Vernicia fordii) tree, which accumulates up to 80% eleostearic acid (ESA) in seed oil (Smith, 1971). The tung fatty acid conjugase (a diverged FAD2 termed FADX) has also previously been described, as have the tung diacylglycerol acyltransferases (DGAT1 and DGAT2) and oleosin genes (Cao et al., 2014; Dyer et al., 2002; Shockey et al., 2006). Notably, prior studies revealed

that tung DGAT2 likely plays a more important role in channelling of ESA into tung oil than DGAT1 (van Erp et al., 2015; Shockey et al., 2006).

Here, we developed a two-step approach for producing high amounts of ESA in leaf tissues. The first step was to engineer elevated leaf oil content, in general, using strategies that were also likely to be important for accumulation of unusual fatty acids in leaves. Prior studies on ectopic expression of hydroxylases and conjugases in plants using constitutive gene promoters resulted in the accumulation of unusual fatty acids in plant seeds, but not in leaves (Iwabuchi et al., 2003; van de Loo et al., 1995). This suggested that leaf tissues contained robust mechanisms for exclusion of the unusual fatty acid from membranes, likely resulting in their degradation via peroxisomal β-oxidation. To increase the possibility of channelling ESA to oil rather than turnover, we explored the usage of Arabidopsis pxa1 mutant plants. PXA1 (peroxisomal ABC-transporter 1) is a peroxisomal membrane protein that transports fatty acids into peroxisomes, and its disruption results in reduced fatty acid turnover and increase in leaf TAG (Kunz et al., 2009; Slocombe et al., 2009; Zolman et al., 2001). Given that disruption of PXA1 blocks fatty acid breakdown, and accumulation of high amounts of ESA in leaf tissues might be toxic, we also explored the usage of Arabidopsis cgi-58 mutant plants. CGI-58 (comparative gene identification-58) is thought to act by stimulating the transport activity of PXA1, and loss of CGI-58 also results in an elevation in leaf TAG (James et al., 2010; Park et al., 2013). To help redirect ESA from the fatty acyl-CoA pool to TAG, we also explored the ectopic expression of the DGAT enzymes (tung DGAT1 and DGAT2) in leaf tissues, which has previously been shown to be an effective strategy for increasing oil content of plant leaves (Andrianov et al., 2010; Bouvier-Navé et al., 2000; Vanhercke et al., 2014). We further combined the expression of tung DGAT1 or DGAT2 with expression of tung OLEOSIN (Cao et al., 2014), with the premise that oleosin can stabilize TAG by coating leaf lipid droplets and reducing accessibility to enzymes that might otherwise promote TAG turnover (Vanhercke et al., 2014; Winichayakul et al., 2013).

Our results show that constitutive expression of tung FADX in leaves of Arabidopsis resulted in low accumulation of ESA in phospholipids (<1% of fatty acids), revealing that leaf tissues do indeed contain robust mechanisms for exclusion of conjugated fatty acids from cellular membranes. The plant lines expressing FADX in leaves also displayed poor plant growth, however, indicative of possible cytotoxicity in leaves. By contrast, co-expression of FADX and tung DGAT2 resulted in a strong synergistic increase in total Arabidopsis leaf oil content, significantly improved channelling of ESA into TAG, and suppressed the poor growth phenotype observed with FADX alone. Furthermore, there were no observed negative effects on seed development or germination. Taken together, these results open a new avenue for producing high amounts of oil in plant leaves that is coupled with accumulation of unusual, oxidized fatty acids. The results should serve as a useful guide for production of other types of high-value oils in the leaves of plants.

Results

Generation of high-leaf-oil plant lines expressing tung FADX

High-leaf-oil plant lines were generated by transforming wild-type (Col-0) (WT), cgi-58, pxa1, or cgi-58/pxa1 double-mutant

Arabidopsis plants with either an empty binary plasmid or the same plasmid expressing (via constitutive promoters) tung *DGAT1*, *DGAT2*, *DGAT1/OLEOSIN* (*OLEO*), or *DGAT2/OLEO*. Seeds were selected on Basta, and then, resistant plants were advanced to the T_2 stage to identify single copy (or closely spaced multicopy) insertions. Plants of homozygous T_3 lines were then analysed for elevated leaf oil (TAG) content by thin-layer chromatography (TLC), transgene expression determined by RT-PCR, and leaf cytosolic lipid droplets visualized using confocal microscopy (Figure S1). Notably, some of the genotype/transgene combinations expressing tung *DGAT1* did not show elevated TAG content (Figure S1a), and thus, only a subset of lines were used for subsequent transformation with either a second empty binary plasmid or the same plasmid constitutively expressing tung *FADX* (Table S1). Transgenic seeds were selected on hygromycin, followed by progeny analysis and isolation of homozygous T_3 lines (Table S1). Analysis of leaf lipids using gas chromatography with flame ionization detector (GC/FID) revealed a wide distribution in the number of plants showing at least traces of ESA in leaf lipids (Table S1), but for several genotype/transgene combinations, particularly the *cgi-58/pxa1* double mutants and those expressing tung *DGAT1*, no lines with ESA were recovered. As such, we reduced the number of lines analysed to a subset that would allow for more direct comparisons of effects of genetic background and transgene combination on ESA production and accumulation. These included WT, *cgi-58,* and *pxa1* mutant backgrounds transformed with either empty plasmid, *FADX* alone, *DGAT2* alone, or a combination of *FADX* and *DGAT2*. Notably, no transgenic lines were generated with *FADX* alone in the *pxa1* mutant background, and thus, this combination was not included in the analysis.

Co-expression of tung *FADX* and *DGAT2* results in a synergistic increase in total leaf oil content and ESA accumulation

Analysis of lipids in 15-day-old *Arabidopsis* seedlings and in mature, fully expanded leaves of 42-day-old plants revealed that total neutral lipid content was moderately, yet significantly increased in WT plants expressing *FADX* or *DGAT2* alone, but co-expression of *FADX* and *DGAT2* together resulted in a substantial, synergistic increase in neutral lipid content (Figure 1a). The increase in neutral lipids was not likely due to differences in transgene expression, as *DGAT2* expression was fairly consistent between all of the plant lines examined and *FADX* expression showed no correlation with neutral lipid content (Figure 2). Analysis of fatty acid composition (Figure 1b) further revealed that co-expression of *FADX* and *DGAT2* also increased the percentage of ESA in neutral lipids in both 15-day-old seedlings and leaves of 42-day-old plants (Figure 1b), accounting for approximately 2% in lines expressing *FADX* alone, and 12% in lines co-expressing *FADX* and *DGAT2* together. Notably, the percentage of oleic acid (18 : 1) was also increased in lines expressing *FADX* (Figure 1b), which is often observed in plants engineered for expression of divergent FAD2 enzymes (Broun and Somerville, 1997; Cahoon et al., 1999; Singh et al., 2001). It is generally believed that the diverged FAD2 inhibits endogenous FAD2 activity through direct protein–protein interaction and/or product-mediated inhibition (Lou et al., 2014). A different mechanism might also be involved here, however, as lines with the highest percentage of 18 : 1 in leaves were reduced primarily in linolenic acid (18 : 3), rather than linoleic acid (18 : 2) (Figure 1b), the product of FAD2. The increases in total neutral

lipid content of engineered lines were further supported by confocal microscopy, which revealed an increase in lipid droplet abundance in *DGAT2*-expressing plants that was further enhanced by co-expression with *FADX* (Figure 1c).

Similar increases in neutral lipid and ESA contents were observed when *FADX* and *DGAT2* were co-expressed in *pxa1* and *cgi-58* mutant backgrounds (Figure 3 and S2). Results with *cgi-58* lines were not, however, as pronounced as in WT or *pxa1* transgenics (Figure S2), likely due to the relatively low level of transgene expression (Figure 2). The neutral lipid content of 15-day-old *pxa1* seedlings was already elevated in comparison with 15-day-old WT plants (compare Figure 3a and 1a), which is consistent with an inability of *pxa1* mutant plants to degrade fatty acids (Kunz et al., 2009; Slocombe et al., 2009; Zolman et al., 2001). Analysis of fatty acid composition further revealed the presence of very-long-chain fatty acids (e.g. 20 : 1, 20 : 2, and 22 : 0) in neutral lipids of 15-day-old seedlings (Figure 3b), likely due to the persistence of seed storage oils in these tissues. Co-expression of *FADX* and *DGAT2*, however, led to an increase in neutral lipid content above and beyond that observed in either *pxa1* seedlings or those expressing *DGAT2* (Figure 3a). Notably, the total neutral lipid content of *pxa1/DGAT2/FADX* 15-day-old seedlings was similar to that of WT/*DGAT2/FADX* seedlings (i.e. ~0.6 µg/mg FW). In mature, fully expanded leaves of 42-day-old plants, however, the neutral lipid content of *pxa1/DGAT2/FADX* plants was nearly doubled that of WT/*DGAT2/FADX* plants (i.e. ~1.5 µg/mg FW and ~0.7 µg/mg FW, respectively) (compare Figures 3a and 1a), indicating that loss of PXA1 function leads to a further increase in steady-state accumulation of neutral lipids as the plants age. Like WT transgenic lines, analysis of fatty acid composition showed an increase in percentage of 18 : 1 in *pxa1* lines expressing *FADX* and *DGAT2* (Figure 3b), and confocal microscopy revealed comparable increases in lipid droplet abundance (Figure 3c).

Engineering ESA production in leaves has no effect on total polar lipid content, but causes changes in fatty acid composition

We next investigated whether there were any changes in polar lipid content and composition of leaves engineered for production of ESA. As shown in Figure 4a for 15-day-old seedlings, and Figure S3a for mature, fully expanded leaves of 42-day-old plants, there were no significant changes in total polar lipid content in any of the WT plant lines investigated. Inspection of fatty acid composition of total polar lipids, however, revealed trace amounts of ESA (<1% of total fatty acids) in WT lines expressing *FADX*, with or without co-expression of *DGAT2* (Figures 4b and S3b). Notably, the majority of WT lines expressing *FADX*, with or without *DGAT2*, also showed elevated 18 : 1 content and reduced 18 : 3 content, which is consistent with results observed for neutral lipids (Figure 1b). Similar results were observed for analysis of polar lipids in *pxa1* and *cgi-58* mutant plant lines (Figures S4 and S5, respectively).

To determine whether the changes in fatty acid composition were present primarily in phospholipids or galactolipids, the total polar lipid fraction from 15-day-old WT transgenic seedlings was separated by TLC then major lipid classes were isolated and analysed by GC-FID. As shown in Figure 4, the high oleate (18 : 1) and reduced linolenate (18 : 3) phenotype was more pronounced in phospholipids (Figure 4c) than galactolipids (Figure 4d). Similar results were observed for phospholipids and galactolipids of *pxa1* mutant plant lines (Figure S4). These data

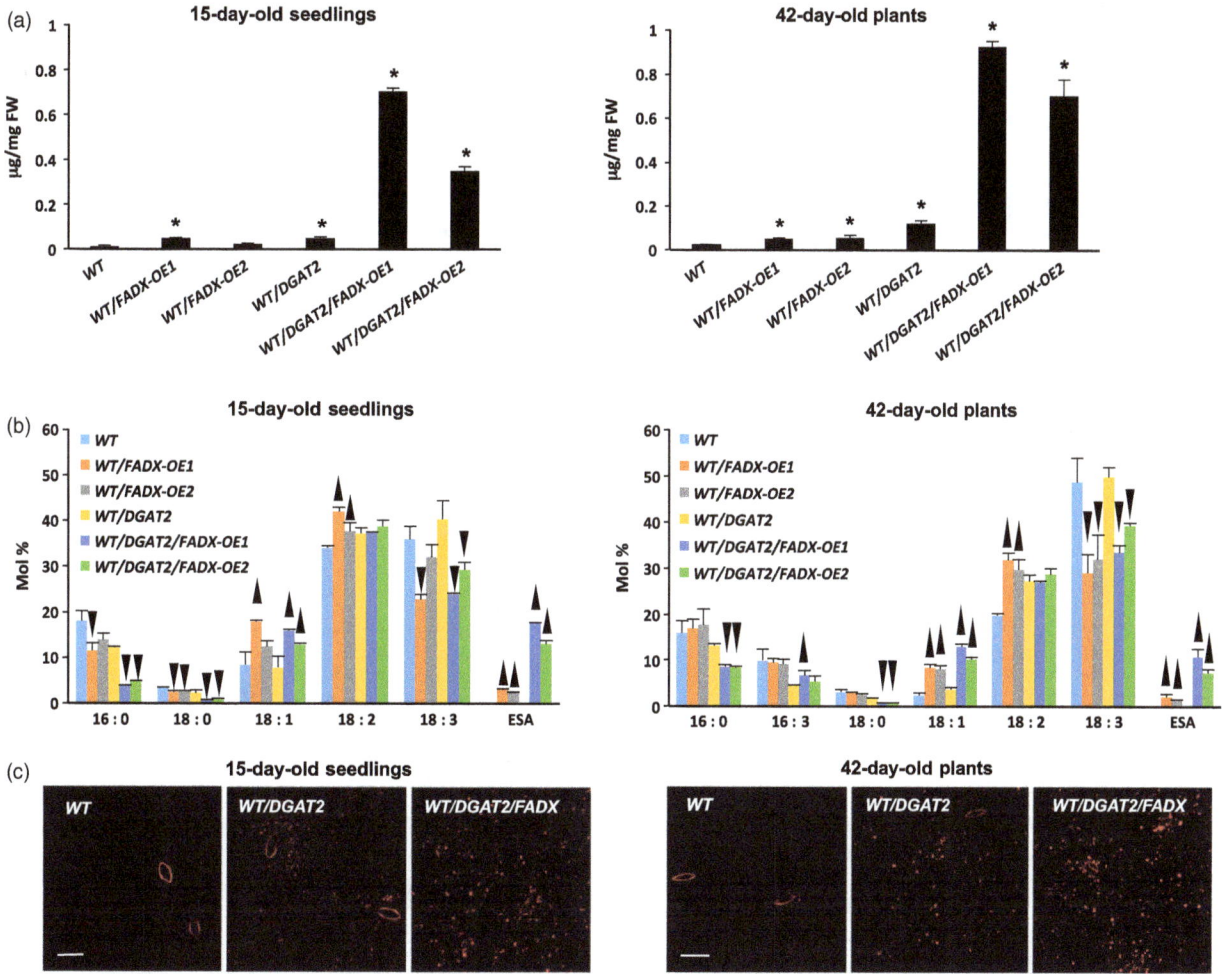

Figure 1 Analysis of neutral lipids and lipid droplets in *Arabidopsis* WT plant lines. (a) Content of neutral lipids in 15-day-old seedlings and in mature, fully expanded leaves of 42-day-old, soil-grown plants (mean ± SD, $n = 3$; asterisks denote significant difference from respective empty-vector control at $P = 0.05$). (b) Fatty acid composition of neutral lipids in 15-day-old seedlings and in mature leaves of 42-day-old plants (mean ± SD, $n = 3$; up and down arrowheads denote values significantly higher or lower, respectively, compared to the respective empty-vector control at $P = 0.05$). (c) Confocal fluorescence micrographs of Nile red-stained lipid droplets in 15-day-old seedlings and mature leaves of 42-day-old plants. Scale bar = 20 μm.

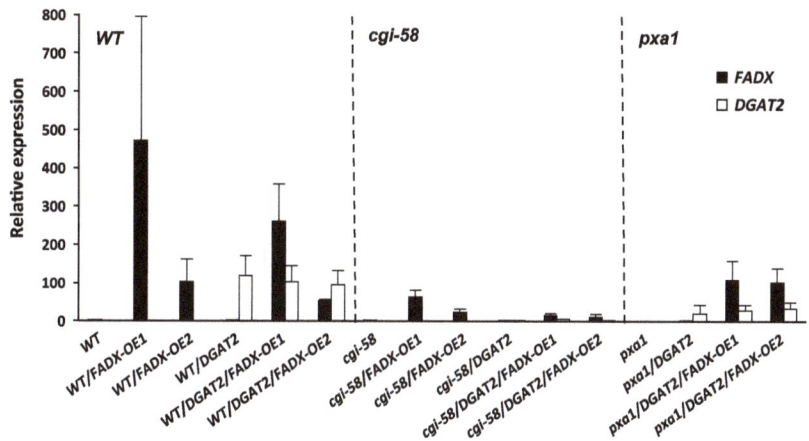

Figure 2 qRT-PCR analysis of ectopically expressed tung *FADX* and *DGAT2* transcripts relative to endogenously expressed *Arabidopsis* *ACTIN8* in 15-day-old seedlings (mean ± SD, $n = 3$).

are consistent with the known localization of FADX in the ER (Dyer *et al.*, 2002) and activity of conjugases towards PC-linked substrates (Liu *et al.*, 1997); the changes in galactolipid composition likely reflect the extensive exchanges of glycerolipids known

to occur between ER and chloroplast membranes (Benning *et al.*, 2006; Browse *et al.*, 1986). The consistency of high 18 : 1 and reduced 18 : 3 in both phospholipid and neutral lipid fractions of engineered lines further supports a metabolic relationship

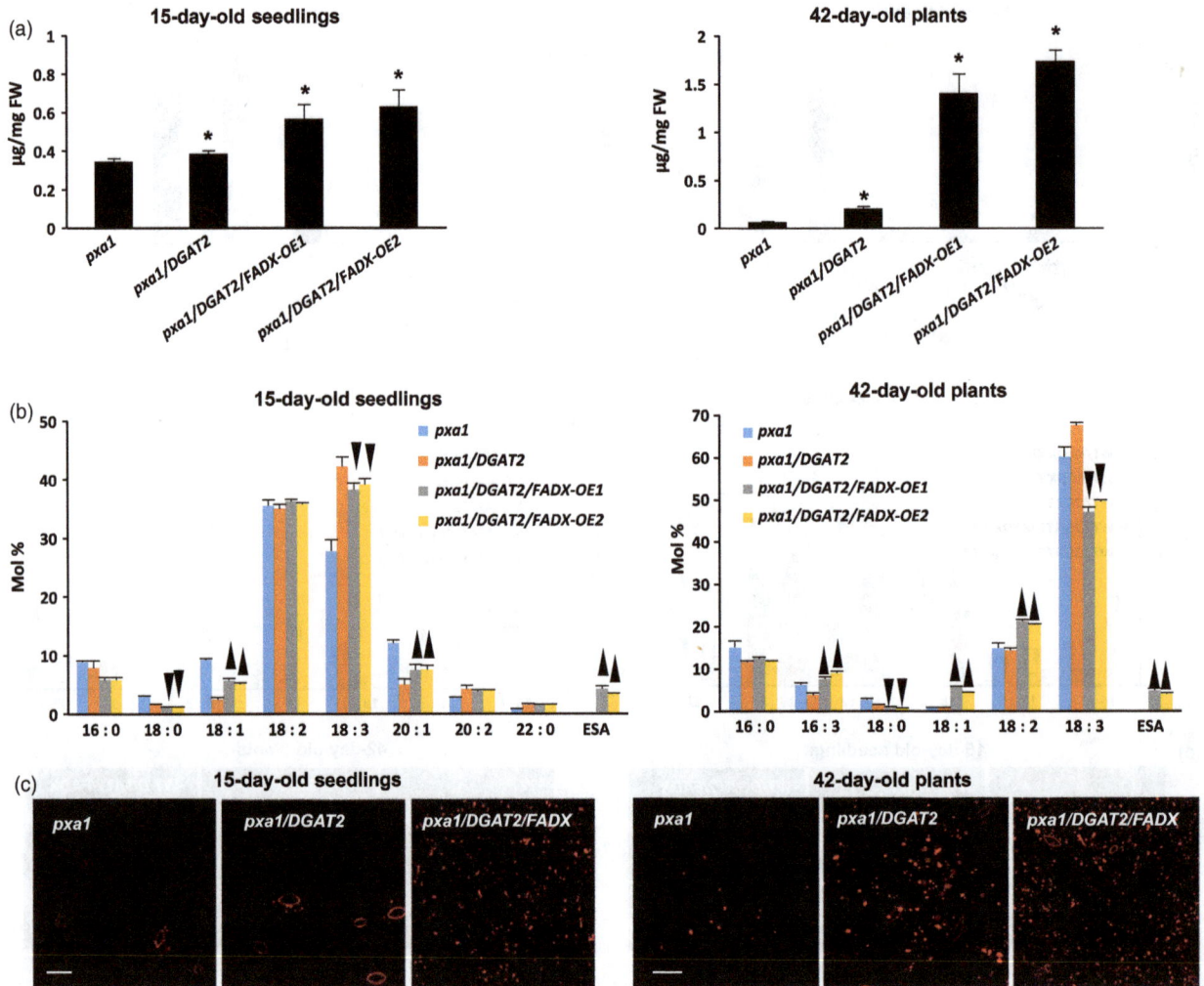

Figure 3 Analysis of neutral lipids and lipid droplets in *Arabidopsis pxa1* mutant plant lines. (a) Content of neutral lipids in 15-day-old seedlings and in mature, fully expanded leaves of 42-day-old, soil-grown plants (mean ± SD, *n* = 3; asterisks denote significant difference from respective empty-vector control at *P* = 0.05). (b) Fatty acid composition of neutral lipids in 15-day-old seedlings and in mature leaves of 42-day-old plants (mean ± SD, *n* = 3; up and down arrowheads denote values significantly higher or lower, respectively, compared to the respective empty-vector control at *P* = 0.05). (c) Confocal fluorescence micrographs of Nile red-stained lipid droplets in 15-day-old seedlings and in mature leaves of 42-day-old plants. Scale bar = 20 μm.

between these two lipid classes for production of neutral lipids containing ESA.

Production of ESA in plant leaves is detrimental to plant growth, but is improved by co-expression of *FADX* and *DGAT2*

WT plants expressing *FADX* alone often exhibited reduced plant growth and yellowing of leaves in comparison with empty plasmid controls, with obvious differences apparent by 35 days after germination (Figure 5a–c). Co-expression of *FADX* and *DGAT2* in the WT background, however, suppressed the phenotype, resulting in more normal-sized plants and less leaf discoloration (Figure 5a–c). Further, measurement of ion leakage, an indicator of cell death (Kawai-Yamada *et al.*, 2004), showed higher relative conductivity in leaves expressing *FADX* alone, and the phenotype was partially suppressed by co-expressing *DGAT2* (Figure 5d). Measurement of endogenous *ACYL-CoA OXIDASE 4* (*ACX4*) and *LYSOPHOSPHATIDYLCHOLINE ACYLTRANSFERASE 1* (*LPCAT1*) expression, which is induced during plant senescence

(Troncoso-Ponce *et al.*, 2013), showed no obvious changes in any of the lines examined (Figure 5e). These data are generally consistent with a model whereby ESA is synthesized in the phospholipids of the ER, and then removed from membranes as a free fatty acid or acyl-CoA, which promotes cytotoxicity, and DGAT2 reduces cytotoxic effects by more effectively capturing ESA and sequestering it in TAG.

Consistent with this premise, the cytotoxic effects were more pronounced in *pxa1* mutant plant lines (Figure 6). Plants harbouring mutations in *PXA1* are known to be more sensitive to the cytotoxic effects of free fatty acids, due in large part to their inability to degrade fatty acids (Kunz *et al.*, 2009). In *pxa1* lines expressing *FADX* and *DGAT2* together, plant size was similar to *DGAT2*-only controls (Figure 6a and b), but the leaves of *pxa1/DGAT2/FADX* plants showed more obvious yellowing and necrosis in comparison with WT/*DGAT2*/*FADX* lines (compare Figures 6 and 5). Furthermore, ion leakage assays revealed that relative conductivity remained high in *pxa1/DGAT2/FADX* lines (Figure 6d), and *ACX4* and *LPCAT1* were induced in comparison

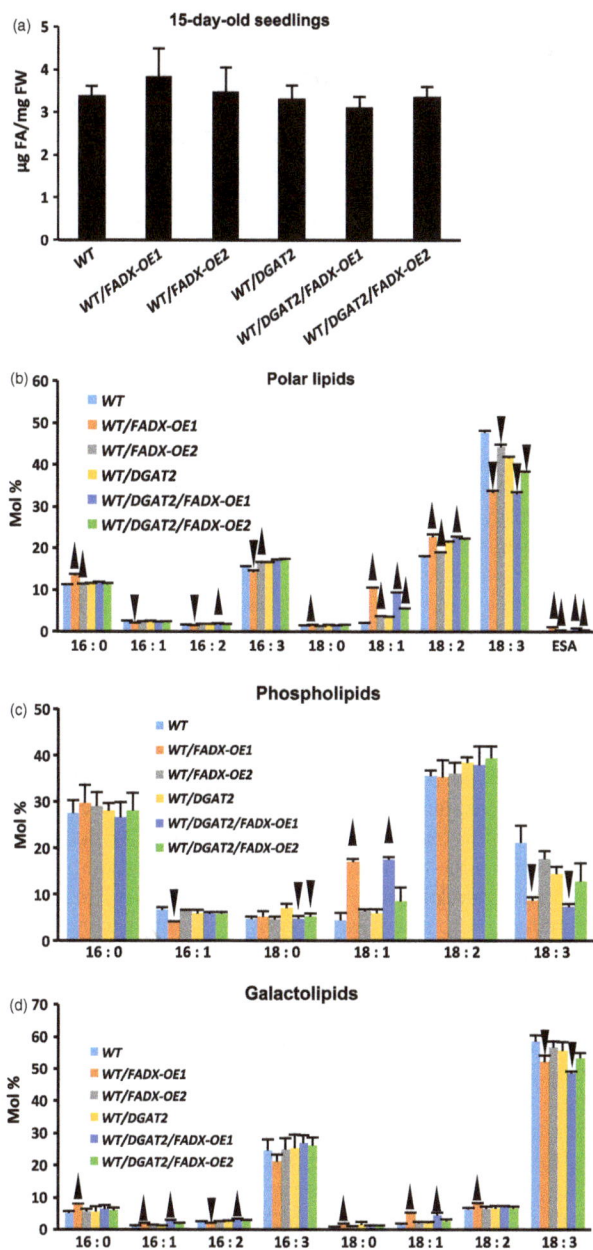

Figure 4 Analysis of polar lipids in 15-day-old seedlings of *Arabidopsis* WT plant lines. Total polar lipid content (a) and fatty acid composition of total polar lipids (b), phospholipids (c), and galactolipids (d), in 15-day-old seedlings of WT plant lines (mean ± SD, $n = 3$; up and down arrowheads denote values significantly higher or lower, respectively, compared to the respective empty-vector control at $P = 0.05$).

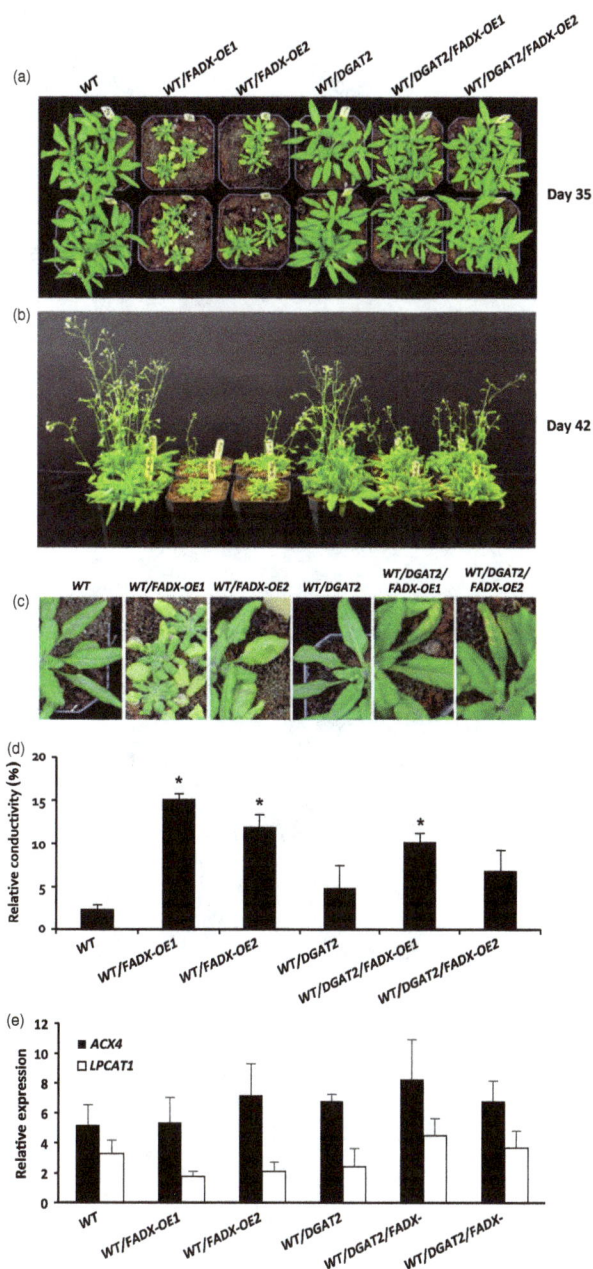

Figure 5 Phenotypes of *Arabidopsis* WT plant lines expressing tung *FADX* and/or *DGAT2*. Images of 35-day-old (a) and 42-day-old (b) soil-grown plants, and mature, fully expanded leaves of 35-day-old plants (c). Electrolyte leakage assay (d) of mature leaves from 35-day-old plants (mean ± SD, $n = 3$; asterisks denote significant difference from respective empty-vector control at $P = 0.05$). (e) qRT-PCR analysis of *Arabidopsis* *LPCAT1* and *ACX4* transcripts in leaves of 43-day-old plants relative to *Arabidopsis* *ACTIN8* (mean ± SD, $n = 3$).

with controls (Figure 6e). Taken together, these data indicate that *pxa1* mutants are more sensitive than WT to ESA production and accumulation. By contrast, *cgi-58* transgenic plants showed essentially normal plant growth and development (Figure S6), again likely due to the relatively low level of *FADX* and *DGAT2* expression in this mutant background (Figure 2).

Engineering the production of ESA in plant leaves has essentially no effect on seeds

One of the major challenges for producing high amounts of ESA in transgenic seeds is that the seeds are often shrunken and

wrinkled, and have poor germination rates (Cahoon *et al.*, 1999, 2006). To determine whether the engineering of ESA production in leaves had any effects on seeds, we examined seed morphology, germination potential, and oil content and composition. As shown in Figure 7a, seeds of WT transgenic lines showed similar overall morphology, size and pigmentation in comparison with the empty plasmid control seeds. The engineered seeds also showed normal germination rates, which was nearly 100%

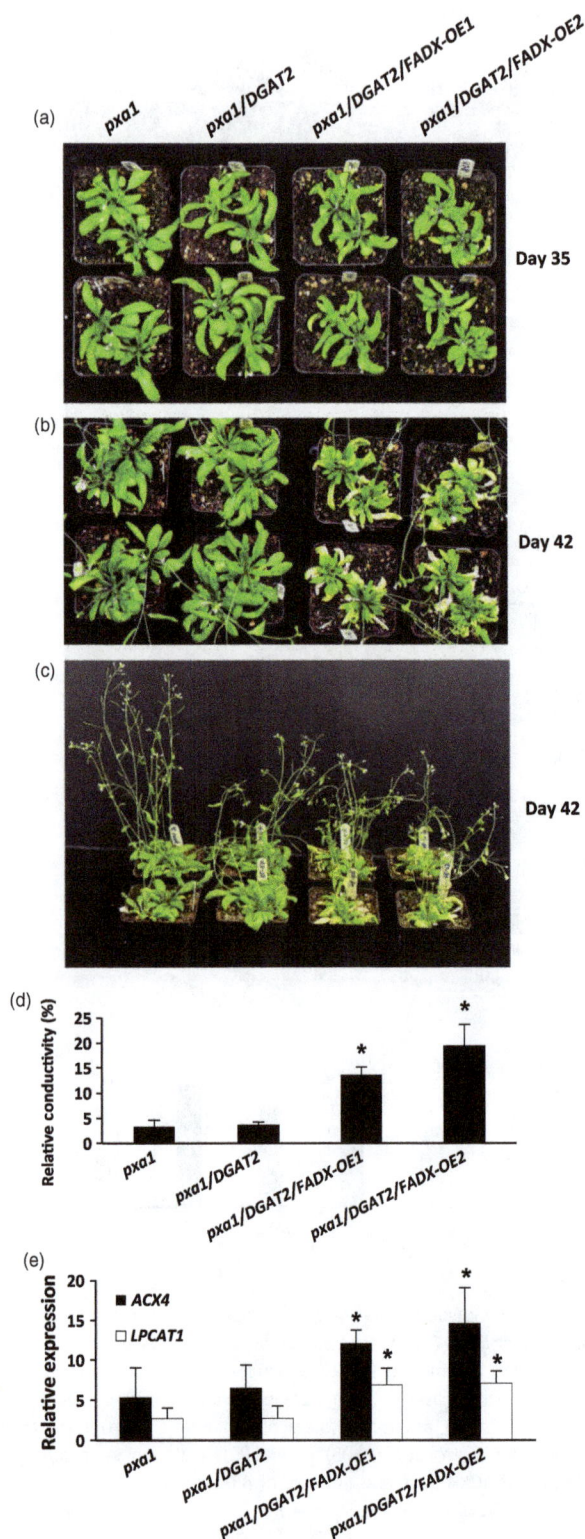

Figure 6 Phenotypes of *Arabidopsis pxa1* mutant plant lines expressing tung *FADX* and/or *DGAT2*. Images of 35-day-old (a) and 42-day-old (b and c) soil-grown plants. Electrolyte leakage assay (d) of mature, fully expanded leaves from 37-day-old plants. (e) qRT-PCR analysis of *Arabidopsis LPCAT1* and *ACX4* transcripts in mature leaves of 43-day-old plants relative to *Arabidopsis ACTIN8*. Values in (d) and (e) represent mean ± SD, *n* = 3; asterisk denotes significant difference from respective empty-vector control at *P* = 0.05.

germination frequency for all plant lines (Figure 7b). Examination of total oil content revealed that some of the plant lines had reduced seed oil (Figure 7c), which might be due to the smaller leaf sizes and/or necrosis and senescence in the associated transgenic plants (Figure 5), but no ESA, or only trace amounts, was detected in seed lipids (Figure 7d). Similar results were observed for seeds derived from both *pxa1* and *cgi-58* mutant plant lines (Figure S7 and S8, respectively).

Discussion

Production of industrially important fatty acids in engineered seeds is often limited by their inefficient exclusion from phospholipid membranes, resulting in reduced flux of the fatty acid to storage oil and, particularly in the case of conjugated fatty acids, negative effects on seed development and germination (Cahoon et al., 2006; Napier et al., 2014). Here, we examined the possibility of coupling the production of unusual fatty acids in plant leaves with accumulation of high amounts of TAG, thereby developing a novel strategy for producing high-value oils in plants. The engineering approach was designed, in part, based on prior studies that examined constitutive expression of fatty acid hydroxylases and conjugases in transgenic plants, which showed efficient exclusion of unusual fatty acids from phospholipids of leaves (Iwabuchi et al., 2003; van de Loo et al., 1995). Expression of tung *FADX* in WT leaves resulted in low accumulation of ESA in polar lipids (<1% weight of total fatty acids; Figure 4), consistent with these prior observations. While it is possible that the low amounts of ESA in leaf lipids are due to low FADX activity, the negative effects of *FADX* expression on plant growth and development (Figure 5a–c), coupled with the observation that co-expression with *DGAT2* significantly enhances TAG and ESA accumulation (Figure 1a and b), would more likely suggest that ESA is synthesized in phospholipids, as expected, and then rapidly excluded from membranes, possibly by phospholipase A2 (PLA2) or the reverse reaction catalysed by LPCAT (Bates, 2016). Indeed, LPCATs are known to show preferential activity towards oxidized fatty acids in PC (Lager et al., 2013), which might suggest a role in exclusion of 'unusual' fatty acids that are otherwise disruptive to cellular membranes. This activity is likely to be particularly important in photosynthetic tissues, where oxidative stress often contributes to lipid peroxidation (Triantaphylidès et al., 2008). Regardless, the identification of the enzyme(s) responsible for exclusion of ESA from phospholipids in leaves would be a useful tool for increasing flux of ESA in engineered plants.

The low accumulation of ESA in neutral lipids of plant leaves expressing tung *FADX* alone (Figure 1b) suggests also that the endogenous acyltransferases catalysing TAG synthesis in leaves do not effectively metabolize substrates containing ESA. Both phospholipid:diacylglycerol acyltransferase1 (PDAT1) and DGAT1 are known to play a role in production of TAG in leaves (Zhang et al., 2009), and studies employing *tgd1* mutants of *Arabidopsis* (Fan et al., 2013, 2014) have shown that PDAT1 is required for synthesis of TAGs that serve as a buffer against excess lipids and cytotoxic free fatty acids. Given that only trace amounts of ESA were observed in the neutral lipids of leaves expressing *FADX* alone (Figure 1b), it is unlikely that ESA was excluded from phospholipids by the activity of endogenous PDAT. Rather, ESA is probably removed first by a PLA2 and/or the reverse reaction of LPCAT, and then inefficiently used by endogenous DGAT for

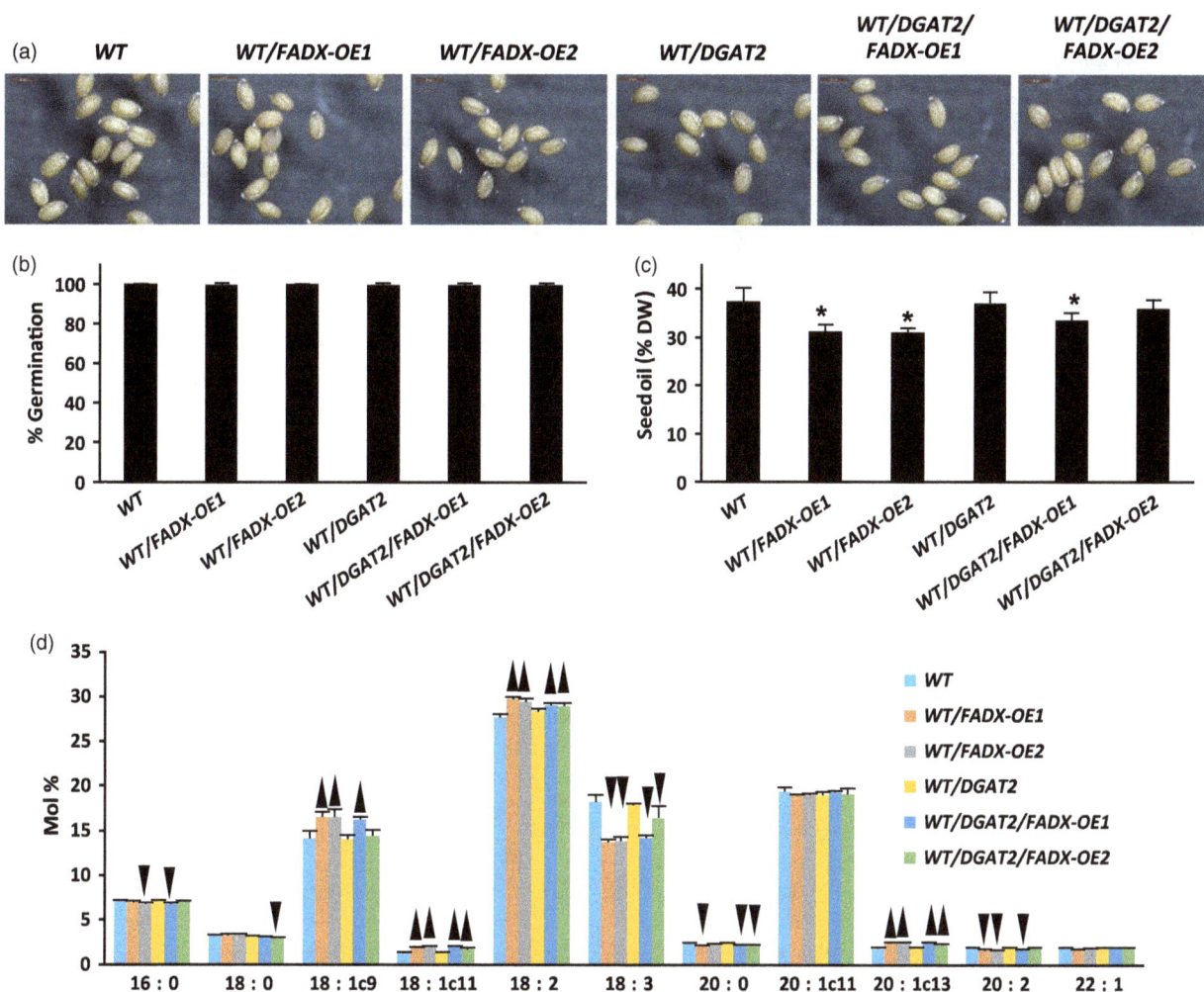

Figure 7 Properties of seeds derived from *Arabidopsis* WT plant lines expressing tung *FADX* and/or *DGAT2*. (a) Images of mature, dry seeds (at 3.5×magnification). (b) Percentage of seed germination. (c) Seed oil content determined by NMR (mean ± SD, *n* = 5; asterisks denote values significantly different from respective empty-vector control at *P* = 0.05). (d) Fatty acid composition of seed oil (mean ± SD, *n* = 5; up and down arrowheads denote values significantly higher or lower, respectively, compared to the respective empty-vector control at *P* = 0.05).

formation of TAG. Notably, the stunted size and appearance of white, necrotic spots and leaf yellowing in lines expressing tung *FADX* alone (Figure 5c) is similar to the phenotype of *tgd1-1/pdat1-2* mutant plants (Fan *et al.*, 2013), which also contained elevated amounts of diacylglycerol (DAG) and free fatty acids. While we did not measure ESA content in the free fatty acid fraction, due to the very labile nature of ESA as a free acid (Yang *et al.*, 2009), the relative increase in cytotoxic effects of *FADX* when expressed in the *pxa1* mutant background (Figure 6) would further support a model involving fatty acid cytotoxicity (Kunz *et al.*, 2009).

Co-expression of tung *FADX* and *DGAT2* improves plant growth, increases leaf oil content, and greatly enhances channelling of ESA into neutral lipids

Expression of tung *DGAT2* alone in plant leaves resulted in at least a doubling in neutral lipid content (Figure 1a), consistent with prior studies that employed DGATs from other source species for increasing oil content in leaves (Andrianov *et al.*, 2010; Bouvier-Navé *et al.*, 2000; Vanhercke *et al.*, 2014). Co-expression of tung *FADX* and *DGAT2*, however, resulted in a synergistic increase in

neutral lipid content, increase in ESA accumulation, and improved plant growth and development (Figures 1 and 5). These results support a model whereby ESA is first excluded from phospholipids, and then effectively captured and channelled into TAG by tung DGAT2. By plotting the amount of ESA in lines expressing *FADX* alone versus *FADX* and *DGAT2*, as a percentage of total ESA present in polar or neutral lipids (Figures 8 and S9), it becomes readily apparent that tung DGAT2 helps to partition ESA from polar lipids into the neutral lipid fraction. This partitioning likely contributes to improved plant growth and development through sequestration of ESA metabolites that would otherwise be toxic.

Blocking fatty acid degradation by disrupting PXA1 elevates steady-state accumulation of neutral lipids and ESA, but also has undesirable effects on plant growth and development

Prior research indicated that unusual fatty acids are often degraded in engineered seeds, resulting in a futile cycle of synthesis and turnover that limits their steady-state accumulation (Eccleston and Ohlrogge, 1998). Therefore, we expected that

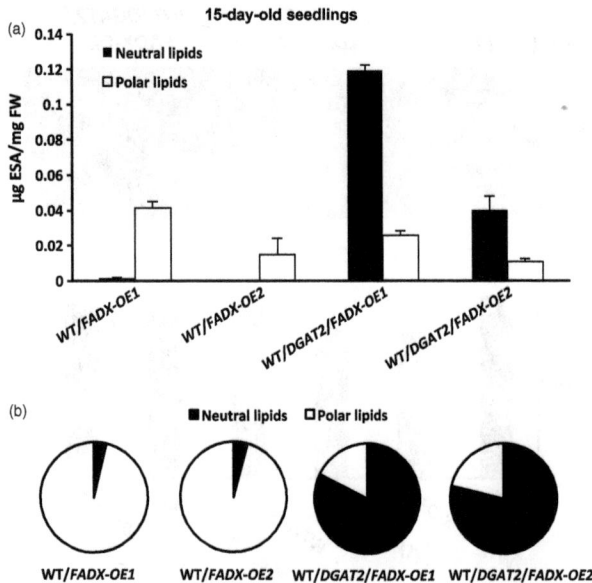

Figure 8 Distribution of ESA in lipids of 15-day-old *Arabidopsis* WT plant lines. (a) Content of ESA in neutral and polar lipids, plotted based on total mass amounts (mean ± SD, *n* = 3). (b) Percentage of ESA in neutral and polar lipids, plotted by setting the total mass amounts of ESA for each plant line in (a) to 100%.

disruption of fatty acid breakdown would be an important strategy for elevating unusual fatty acid content in leaves. Somewhat surprisingly, this was not entirely the case. Comparisons of 15-day-old WT and *pxa1* lines co-expressing tung *FADX* and *DGAT2* revealed similar increases in total neutral lipid and ESA content (Figures 1a and 3a). Comparison of lipids in 42-day-old plants, however, revealed additional increases in total neutral lipid content in *pxa1* transgenic lines relative to WT lines, although trends in polar lipids remained the same (Figures 1a and 3a; Figures 4 and S4). These data indicate that *pxa1* mutants can indeed be used to increase neutral lipid content above and beyond that observed in WT lines, but the plants are significantly less healthy at this stage of development (compare Figures 5 and 6). Furthermore, in addition to a role in fatty acid breakdown, PXA1 is known to function in the transport of lipid hormone precursors into peroxisomes for their subsequent β-oxidation to produce indole-3-acetic acid and jasmonic acid, and thus, *pxa1* mutants are compromised in some aspects of plant growth, development, and stress response (Dave *et al.*, 2011; Theodoulou *et al.*, 2005; Zolman *et al.*, 2001). And finally, *pxa1* mutant plants are also unable to degrade seed storage oil and thus require an exogenous source of sugar for seedling establishment (Footitt *et al.*, 2002; Hayashi *et al.*, 2002). As such, any engineering strategy that employs disruption of PXA1 must account for the difficulties with seed germination and establishment. One potential mechanism to circumvent these difficulties is to use inducible RNAi methods to silence *PXA1* expression at particular stages of development, and/or only in certain tissues (Kim *et al.*, 2015).

Future directions

To further increase oil and ESA contents in leaves, obvious next steps include an increase in fatty acids available for TAG synthesis, which can be accomplished using various 'push'-related

strategies, including ectopic expression of the transcription factor WRINKLED1, which activates multiple genes involved in fatty acid synthesis (Vanhercke *et al.*, 2013a), suppression of ADP-glucose pyrophosphorylase, which alters flux of carbon from starch into fatty acid production (Sanjaya *et al.*, 2011), or over-expression of FAX1 (fatty acid export 1), which enhances transport of fatty acids out of chloroplasts (Li *et al.*, 2015). Other enzymes for increased channelling of ESA into TAG would also likely include tung homologs of glycerol-3-phosphate acyltransferase 9 (Shockey *et al.*, 2016; Singer *et al.*, 2016) and lysophosphatidyl acyltransferase 2 (Chen *et al.*, 2016). Additional enzymes, such as phosphatidylcholine diacylglycerol cholinephosphotransferase, might also be required to increase flux of ESA-containing metabolites through PC (Bates and Browse, 2011; Bates *et al.*, 2012; Hu *et al.*, 2012).

While the production of elevated oil in plant leaves effectively increases the caloric content of biomass crops, which is desirable for biofuel or animal feed production (Horn and Benning, 2016), there is also potential for altering plant/pest interactions. For instance, feeding studies have demonstrated increased caterpillar weights when insects were fed a diet of leaves containing elevated oil content (Sanjaya *et al.*, 2013), and thus, strategies for mitigating pest predation should be considered. Unusual fatty acids, including hydroxy or acetylenic, have potential antiphysiological effects (Cahoon *et al.*, 2003; Tunaru *et al.*, 2012). As such, the production of unusual fatty acids in leaf oil might serve a dual purpose in elevating oil content in leaves, while at the same time discouraging plant–pest interactions.

Overall, the demonstration provided here, showing a synergistic relationship between an enzyme for unusual fatty acid synthesis (FADX) and an enzyme for selective channelling into TAG (DGAT2), should serve as a useful guide for production of other industrially important fatty acids in plants. There are many other divergent FAD2 enzymes responsible for synthesis of a wide array of fatty acid structures, including epoxy, hydroxy, and acetylenic fatty acids (Lee *et al.*, 1998; van de Loo *et al.*, 1995; Shanklin and Cahoon, 1998), and DGAT2 enzymes are known to be important for their accumulation in engineered seeds (Burgal *et al.*, 2008; van Erp *et al.*, 2011; Li *et al.*, 2010a). These observations suggest that other, structurally diverse fatty acids can be produced in plant leaves by matching the fatty acid-modifying enzyme with the acyl-CoA-dependent DGAT from the same source plant species.

Experimental procedures

Gene cloning and construction of tung gene binary plasmids

The contents and available restriction sites for all cloning and binary plasmids described below are described in Shockey *et al.* (2015), except where noted. The full-length open reading frames (ORFs) for myc-epitope-tagged tung DGAT1 and DGAT2 were initially cloned into the dual CaMV 35S promoter/terminator shuttle plasmid K34. The two promoter:tung *DGAT1/2*:terminator *Asc*I cassettes were then ligated separately into the *Asc*I site of the binary plasmid B9. The ORF for tung *OLEOSINII*, which was identified in the tung seed cDNA 454 sequencing project (Pastor *et al.*, 2013), was cloned into the nos promoter-nos terminator cassette in cloning plasmid K33. Thereafter, the *Asc*I promoter: *OLEOSINII*:terminator cassette from this plasmid was ligated into the *Mlu*I site of each of the tung *DGAT*-B9 binary plasmids. In this study, B9 is also referred to as EV1 (empty vector 1) because it served as the negative control plasmid used in the first round of plant transformations. The haemagglutinin (HA)-epitope-tagged tung *FADX* ORF was cloned into the binary plasmid pMDC32 (Curtis and Grossniklaus, 2003) using the *AscI* and *SacI* sites located between the dual CaMV 35S promoter and terminator. Empty vector 2 (EV2), which was used in the second round of plant transformations, was constructed by removal of the attR1-ccdB-attR2 cassette from pMDC32 as an *AscI-SacI* fragment, followed by restoration of blunt ends by Klenow fill-in, and self-ligation.

Plant material, growth conditions, transformation, and seed germination

Arabidopsis lines used in this study were WT Columbia-0 ecotype and derivatives thereof, including the T-DNA insertional mutant line, *cgi-58* [SALK_136871] (James *et al.*, 2010), *pxa1*, an ethyl methanesulfonate-generated, splice variant mutant (Zolman *et al.*, 2001), and *cgi-58/pxa1*, a double mutant generated by crossing *pxa1* and *cgi-58* plants (Park *et al.*, 2013). Plants were grown on soil in a growth chamber set for 16-h light/8-h dark cycle at 22 °C, 40% RH, and 50 μE/m^2/s. Seeds were surface-sterilized and plated on ½ MS media (Murashige and Skoog, 1962) solidified with 0.8% Gelzan (Sigma-Aldrich) with (for *pxa1* mutants) or without 1% sucrose, and antibiotics, as specified. After 3 days of stratification in the dark at 4 °C, plates with seeds were moved to the growth chamber with growth conditions as described above. Seedlings were transferred to soil at 10 days after stratification or harvested for analysis at 15 days after stratification.

Agrobacterium-mediated transformation of *Arabidopsis* plants was performed using the floral dip method of Clough and Bent (1998), using *A. tumefaciens* strain GV3101, as described previously (Cai *et al.*, 2015). Bulk T$_1$ seeds were collected from mature transformed T$_0$ plants and sown on soil wetted with 0.15% final concentration Basta (ChemService Inc.). At least 24 individual T$_1$ plants were chosen at random from each *Agrobacterium*/plant genotype transformation and transplanted to individual pots and grown to maturity. Segregating T$_2$ seed samples were used to further select homozygous T$_3$ lines for further analysis. High-performing lines (as evidenced by TLC, qRT-PCR, and confocal microscopy) were re-transformed with *Agrobacterium* strains carrying either hygromycin-resistant binary plasmid pMDC32 with or without the HA-tagged tung *FADX* gene. Additional rounds of selection for hygromycin-resistant double transgenic lines were carried out as described in Harrison *et al.* (2006).

Seed germination was evaluated by first sterilizing, plating, and stratifying seeds as described above, and then, plates were moved to a growth chamber with the same conditions as before. The assay was performed in triplicate, and germination was scored by radical emergence after 3 days.

Lipid analysis

For semi-quantitative analysis of neutral lipids from vegetative tissues by TLC, total lipids were extracted from 50 mg FW of mature (fully expanded, but not senescing) leaves of soil-grown 42-day-old plants using a hexane/isopropanol method (Hara and Radin, 1978; Gidda *et al.*, 2016). Total lipids in chloroform were separated on a silica TLC plate (Merck) using hexane: diethyl ether: acetic acid (70:30:1, v/v/v), stained with 0.05% primuline in 80% acetone and visualized under UV light. C17:0 TAG (Sigma-Aldrich) was used as an external standard.

For analysis of content and fatty acid composition of neutral and polar lipids from vegetative tissues, total lipids were extracted from 500 mg FW of 15-day-old seedlings grown on ½ MS medium (and 1% sucrose for *pxa1* mutants) or from 500 mg of mature, fully expanded leaves from 42-day-old soil-grown plants, with addition of C17:0 TAG (Sigma-Aldrich) and C15:0 PC (Avanti Polar Lipid, Inc.) as internal standards. Total lipid extracts in hexane were separated into neutral and polar lipids on solid-phase extraction cartridges (Supelco Discovery DSC-Si 6 mL), as described (Gidda *et al.*, 2016). To prepare fatty acid methyl esters (FAMEs), 0.5 mL of 0.5 N sodium methoxide solution in methanol was added to neutral or polar lipid extracts, and samples were incubated at room temperature in the dark for 25 min. The reaction was quenched with 1 mL of saturated NaCl solution in water, and FAMEs were extracted with 1 mL of hexane. FAME samples were analysed on an Agilent HP 6890 series GC system equipped with a 7683 series injector and autosampler and a BPX70 (SGE Analytical Science) capillary column (10 m × 0.1 mm × 0.2 μm) with a constant pressure of 25 PSI, as described in Gidda *et al.* (2016). Compounds were identified by comparing with the GLC-10 FAME standard mix (Sigma-Aldrich) and FAMEs prepared from tung oil. Two-tailed Student's *t*-tests were used for comparisons to the respective empty-vector control; $P = 0.05$.

For analysis of the fatty acid composition of galactolipids and phospholipids, a portion of the polar lipid fraction from 15-day-old seedlings was applied on a silica TLC plate and developed in a solvent system of acetone:toluene:water (91: 30: 7.5, v/v/v), as described (Wang and Benning, 2011). Lipids were stained with 0.05% primuline in 80% acetone and visualized under UV light. Silica spots corresponding to position of galactolipids (MGDG and DGDG) and phospholipids (PC, PE, PI) were scraped and used for direct transmethylation with 1.25 M HCl in methanol at 85°C for 2 h. FAMEs were extracted with 1 mL of hexane and analysed by GC as described above.

Seed oil content was determined on 50 mg samples of dry seeds using an mq20 NMR Analyzer (Bruker). For analysis of fatty acid composition of seed oil, samples of 50 seeds were homogenized in 0.5 mL hexane on GenoGrinder (SPEX SamplePrep) for 3 min, 0.5 mL of 0.5N sodium methoxide solution in methanol was added to the homogenate, and then, samples were incubated at room temperature in the dark under the nitrogen for 25 min. Extraction and analysis of FAMEs was performed as described above.

qRT-PCR

For analysis of expression levels of tung transgenes, total RNA was extracted from ~100 mg of the 15-day-old *Arabidopsis* seedlings using the RNeasy Plant Mini Kit (Qiagen) and treated with DNase (Qiagen). Complementary DNA (cDNA) was synthesized from 1 μg of total RNA using iScript Reverse Transcription Supermix for RT-qPCR (Bio-RAD) following the manufacturer's protocol. Quantitative PCR of tung *DGAT1*, *DGAT2*, *OLEOSINII*, and *FADX*, and *Arabidopsis* *ACTIN8* or *18S* (as reference transcripts) was performed using iTaq Universal SYBR Green Supermix (Bio-RAD) using gene-specific forward and reverse primers (Table S2) on Bio-RAD CFX96 Real-Time System with C1000 Thermal cycler. All samples were also subjected to a melt curve analysis between 65 and 95°C with 0.5°C increment. For analysis of the transcript levels of *Arabidopsis* *ACX4*, *LPCAT1*, and *ACTIN8*, the same protocol was used except that cDNA was prepared from leaves of 43-day-old plants. Data were quantified using the Delta CT method, and two-tailed Student's *t*-tests were used for comparisons to the empty-vector control ($P = 0.05$).

Microscopy

WT and transgenic 15-day-old *Arabidopsis* seedlings, or mature, fully expanded leaves of 28- or 42-day-old plants, were collected from the growth chamber at the end of the night cycle, when the abundance of LDs is at its highest (Gidda *et al.*, 2016), and then processed for confocal microscopy, including staining with BODIPY 493/503 (Invitrogen) or Nile red (Sigma-Aldrich) as described previously (Cai *et al.*, 2015; Park *et al.*, 2013). Microscopic images were acquired using a Leica DM RBE microscope with a Leica 63× Plan Apochromat oil-immersion objective, a Leica TCS SP2 scanning head, and the Leica TCS NT software package. Nile red was excited using a 543-nm laser and BODIPY and chlorophyll autofluorescence were excited with a 488-nm laser. All fluorophore emissions were collected sequentially as 15 µm Z-series of the adaxial surface of true leaves, and all images of cells shown are representative of at least three separate experiments.

Electrolyte leakage assay

Relative conductivity was measured essentially as described (Wan *et al.*, 2014). Briefly, three detached leaves from 35-day-old *Arabidopsis* plants were rinsed with deionized water and immersed in 5 mL of deionized water. Samples were gently agitated (120 rpm) for 16 h. Conductivity was measured using an Accumet Excel XL50 conductivity meter, and then, leaf samples were incubated at 95 °C for 15 min, allowed to cool to room temperature for 1 h, and the second conductivity measurement was taken. Relative conductivity was calculated as a percentage of the final conductivity measurement: $RC = (Cond_{16 h}/Cond_{95C}) \times 100$.

Acknowledgements

The authors thank Ashley Ganceres, Amanda Smith, Reavelyn Pray, Marina Mehling, Jaime Adame, and Catherine Mason for excellent technical assistance in generation, screening, and analysis of transgenic plant lines. This work was supported initially by a grant from the U.S. Department of Energy (DOE), BER Division, DE-FG02-09ER64812, and completed with funds from the U.S. DOE, Office of Science, BES-Physical Biosciences program under DE-SC0016536. Funding for a portion of this work was also provided by the Natural Sciences and Engineering Research Council of Canada and by a University of Guelph CBS Summer Research Assistantship (to M. Silver). Mention of trade names or commercial products in this article is solely for the purpose of providing specific information and does not imply recommendation or endorsement by the US Department of Agriculture. USDA is an equal opportunity provider and employer. The authors declare no conflict of interest.

References

Andrianov, V., Borisjuk, N., Pogrebnyak, N., Brinker, A., Dixon, J., Spitsin, S., Flynn, J. *et al.* (2010) Tobacco as a production platform for biofuel: overexpression of *Arabidopsis* DGAT and LEC2 genes increases accumulation and shifts the composition of lipids in green biomass. *Plant Biotechnol. J.* **8**, 277–287.

Badami, R.C. and Patil, K.B. (1981) Structure and occurrence of unusual fatty acids in minor seed oils. *Prog. Lipid Res.* **19**, 119–153.

Bates, P.D. (2016) Understanding the control of acyl flux through the lipid metabolic network of plant oil biosynthesis. *Biochim. Biophys. Acta*, **1861**, 1214–1225.

Bates, P.D. and Browse, J. (2011) The pathway of triacylglycerol synthesis through phosphatidylcholine in Arabidopsis produces a bottleneck for the accumulation of unusual fatty acids in transgenic seeds. *Plant J.* **68**, 387–399.

Bates, P.D., Fatihi, A., Snapp, A.R., Carlsson, A.S., Browse, J. and Lu, C. (2012) Acyl editing and headgroup exchange are the major mechanisms that direct polyunsaturated fatty acid flux into triacylglycerols. *Plant Physiol.* **160**, 1530–1539.

Bates, P.D., Johnson, S.R., Cao, X., Li, J., Nam, J.W., Jaworski, J.G., Ohlrogge, J.B. *et al.* (2014) Fatty acid synthesis is inhibited by inefficient utilization of unusual fatty acids for glycerolipid assembly. *Proc. Natl Acad. Sci. USA*, **111**, 1204–1209.

Benning, C., Xu, C. and Awai, K. (2006) Non-vesicular and vesicular lipid trafficking involving plastids. *Curr. Opin. Plant Biol.* **9**, 241–247.

Biermann, U., Bornscheuer, U., Meier, M.A., Metzger, J.O. and Schäfer, H.J. (2011) Oils and fats as renewable raw materials in chemistry. *Angew. Chem. Int. Ed. Engl.* **50**, 3854–3871.

Bouvier-Navé, P., Benveniste, P., Oelkers, P., Sturley, S.L. and Schaller, H. (2000) Expression in yeast and tobacco of plant cDNAs encoding acyl CoA:diacylglycerol acyltransferase. *Eur. J. Biochem.* **267**, 85–96.

Broun, P. and Somerville, C. (1997) Accumulation of ricinoleic, lesquerolic, and densipolic acids in seeds of transgenic Arabidopsis plants that express a fatty acyl hydroxylase cDNA from castor bean. *Plant Physiol.* **113**, 933–942.

Browse, J., Warwick, N., Somerville, C.R. and Slack, C.R. (1986) Fluxes through the prokaryotic and eukaryotic pathways of lipid synthesis in the '16:3' plant *Arabidopsis thaliana*. *Biochem. J.* **235**, 25–31.

Burgal, J., Shockey, J., Lu, C., Dyer, J., Larson, T., Graham, I. and Browse, J. (2008) Metabolic engineering of hydroxy fatty acid production in plants: RcDGAT2 drives dramatic increases in ricinoleate levels in seed oil. *Plant Biotechnol. J.* **6**, 819–831.

Cahoon, E.B., Carlson, T.J., Ripp, K.G., Schweiger, B.J., Cook, G.A., Hall, S.E. and Kinney, A.J. (1999) Biosynthetic origin of conjugated double bonds: production of fatty acid components of high-value drying oils in transgenic soybean embryos. *Proc. Natl Acad. Sci. USA*, **96**, 12935–12940.

Cahoon, E.B., Schnurr, J.A., Huffman, E.A. and Minto, R.E. (2003) Fungal responsive fatty acid acetylenases occur widely in evolutionarily distant plant families. *Plant J.* **34**, 671–683.

Cahoon, E.B., Dietrich, C.R., Meyer, K., Damude, H.G., Dyer, J.M. and Kinney, A.J. (2006) Conjugated fatty acids accumulate to high levels in phospholipids of metabolically engineered soybean and Arabidopsis seeds. *Phytochemistry*, **67**, 1166–1176.

Cahoon, E.B., Shockey, J.M., Dietrich, C.R., Gidda, S.K., Mullen, R.T. and Dyer, J.M. (2007) Engineering oilseeds for sustainable production of industrial and nutritional feedstocks: solving bottlenecks in fatty acid flux. *Curr. Opin. Plant Biol.* **10**, 236–244.

Cai, Y., Goodman, J.M., Pyc, M., Mullen, R.T., Dyer, J.M. and Chapman, K.D. (2015) Arabidopsis SEIPIN proteins modulate triacylglycerol accumulation and influence lipid droplet proliferation. *Plant Cell*, **27**, 2616–2636.

Cao, H., Zhang, L., Tan, X., Long, H. and Shockey, J.M. (2014) Identification, classification and differential expression of oleosin genes in tung tree (*Vernicia fordii*). *PLoS ONE*, **9**, e88409.

Carlsson, A.S., Yilmaz, J.L., Green, A.G., Stymne, S. and Hofvander, P. (2011) Replacing fossil oil with fresh oil - with what and for what? *Eur. J. Lipid Sci. Technol.* **113**, 812–831.

Chapman, K.D., Dyer, J.M. and Mullen, R.T. (2012) Biogenesis and functions of lipid droplets in plants: thematic review series: lipid droplet synthesis and metabolism: from yeast to man. *J. Lipid Res.* **53**, 215–226.

Chapman, K.D., Dyer, J.M. and Mullen, R.T. (2013) Commentary: why don't plant leaves get fat? *Plant Sci.* **207**, 128–134.

Chen, G.Q., van Erp, H., Martin-Moreno, J., Johnson, K., Morales, E., Browse, J., Eastmond, P.J. *et al.* (2016) Expression of castor LPAT2 enhances ricinoleic acid content at the *sn*-2 position of triacylglycerols in Lesquerella seed. *Int. J. Mol. Sci.* **17**, 507.

Clough, S.J. and Bent, A.F. (1998) Floral dip: a simplified method for Agrobacterium-mediated transformation of *Arabidopsis thaliana*. *Plant J.* **16**, 735–743.

Curtis, M.D. and Grossniklaus, U. (2003) A gateway cloning vector set for high-throughput functional analysis of genes in planta. Plant Physiol. **133**, 462–469.

Dave, A., Hernández, M.L., He, Z., Andriotis, V.M., Vaistij, F.E., Larson, T.R. and Graham, I.A. (2011) 12-Oxo-phytodienoic acid accumulation during seed development represses seed germination in Arabidopsis. Plant Cell, **23**, 583–599.

Durrett, T.P., Benning, C. and Ohlrogge, J. (2008) Plant triacylglycerols as feedstocks for the production of biofuels. Plant J. **54**, 593–607.

Dyer, J.M., Chapital, D.C., Kuan, J.C., Mullen, R.T., Turner, C., Mckeon, T.A. and Pepperman, A.B. (2002) Molecular analysis of a bifunctional fatty acid conjugase/desaturase from tung. Implications for the evolution of plant fatty acid diversity. Plant Physiol. **130**, 2027–2038.

Dyer, J.M., Stymne, S., Green, A.G. and Carlsson, A.S. (2008) High-value oils from plants. Plant J. **54**, 640–655.

Eccleston, V.S. and Ohlrogge, J.B. (1998) Expression of lauroyl-acyl carrier protein thioesterase in Brassica napus seeds induces pathways for both fatty acid oxidation and biosynthesis and implies a set point for triacylglycerol accumulation. Plant Cell, **10**, 613–622.

van Erp, H., Bates, P.D., Burgal, J., Shockey, J. and Browse, J. (2011) Castor phospholipid:diacylglycerol acyltransferase facilitates efficient metabolism of hydroxy fatty acids in transgenic Arabidopsis. Plant Physiol. **155**, 683–693.

van Erp, H., Shockey, J., Zhang, M., Adhikari, N.D. and Browse, J. (2015) Reducing isozyme competition increases target fatty acid accumulation in seed triacylglycerols of transgenic Arabidopsis. Plant Physiol. **168**, 36–46.

Fan, J., Yan, C. and Xu, C. (2013) Phospholipid:diacylglycerol acyltransferase-mediated triacylglycerol biosynthesis is crucial for protection against fatty acid-induced cell death in growing tissues of Arabidopsis. Plant J. **76**, 930–942.

Fan, J., Yan, C., Roston, R., Shanklin, J. and Xu, C. (2014) Arabidopsis lipins, PDAT1 acyltransferase, and SDP1 triacylglycerol lipase synergistically direct fatty acids toward beta-oxidation, thereby maintaining membrane lipid homeostasis. Plant Cell, **26**, 4119–4134.

Footitt, S., Slocombe, S.P., Larner, V., Kurup, S., Wu, Y., Larson, T., Graham, I. et al. (2002) Control of germination and lipid mobilization by COMATOSE, the Arabidopsis homologue of human ALDP. EMBO J. **21**, 2912–2922.

Gidda, S.K., Park, S., Pyc, M., Yurchenko, O., Cai, Y., Wu, P., Andrews, D.W. et al. (2016) Lipid droplet-associated proteins (LDAPs) are required for the dynamic regulation of neutral lipid compartmentation in plant cells. Plant Physiol. **170**, 2052–2071.

Hara, A. and Radin, N.S. (1978) Lipid extraction of tissues with a low-toxicity solvent. Anal. Biochem. **90**, 420–426.

Harrison, S.J., Mott, E.K., Parsley, K., Aspinall, S., Gray, J.C. and Cottage, A. (2006) A rapid and robust method of identifying transformed Arabidopsis thaliana seedlings following floral dip transformation. Plant Methods, **2**, 19.

Hayashi, M., Nito, K., Takei-Hoshi, R., Yagi, M., Kondo, M., Suenaga, A., Yamaya, T. et al. (2002) Ped3p is a peroxisomal ATP-binding cassette transporter that might supply substrates for fatty acid β-oxidation. Plant Cell Physiol. **43**, 1–11.

Horn, P.J. and Benning, C. (2016) The plant lipidome in human and environmental health. Science, **353**, 1228–1232.

Hu, Z., Ren, Z. and Lu, C. (2012) The phosphatidylcholine diacylglycerol cholinephosphotransferase is required for efficient hydroxy fatty acid accumulation in transgenic Arabidopsis. Plant Physiol. **158**, 1944–1954.

Iwabuchi, M., Kohno-Murase, J. and Imamura, J. (2003) Delta 12-oleate desaturase-related enzymes associated with formation of conjugated trans-delta 11, cis-delta 13 double bonds. J. Biol. Chem. **278**, 4603–4610.

James, C.N., Horn, P.J., Case, C.R., Gidda, S.K., Zhang, D., Mullen, R.T., Dyer, J.M. et al. (2010) Disruption of the Arabidopsis CGI-58 homologue produces Chanarin-Dorfman-like lipid droplet accumulation in plants. Proc. Natl Acad. Sci. USA, **107**, 17833–17838.

Kawai-Yamada, M., Ohori, Y. and Uchimiya, H. (2004) Dissection of Arabidopsis Bax inhibitor-1 suppressing Bax-, hydrogen peroxide-, and salicylic acid-induced cell death. Plant Cell, **16**, 21–32.

Kim, H.U., Lee, K.R., Jung, S.J., Shin, H.A., Go, Y.S., Suh, M.C. and Kim, J.B. (2015) Senescence-inducible LEC2 enhances triacylglycerol accumulation in leaves without negatively affecting plant growth. Plant Biotechnol. J. **13**, 1346–1359.

Kunz, H.H., Scharnewski, M., Feussner, K., Feussner, I., Flügge, U.I., Fulda, M. and Gierth, M. (2009) The ABC transporter PXA1 and peroxisomal beta-oxidation are vital for metabolism in mature leaves of Arabidopsis during extended darkness. Plant Cell, **21**, 2733–2749.

Lager, I., Yilmaz, J.L., Zhou, X.R., Jasieniecka, K., Kazachkov, M., Wang, P., Zou, J. et al. (2013) Plant acyl-CoA:lysophosphatidylcholine acyltransferases (LPCATs) have different specificities in their forward and reverse reactions. J. Biol. Chem. **288**, 36902–36914.

Lardizabal, K.D., Metz, J.G., Sakamoto, T., Hutton, W.C., Pollard, M.R. and Lassner, M.W. (2000) Purification of a jojoba embryo wax synthase, cloning of its cDNA, and production of high levels of wax in seeds of transgenic Arabidopsis. Plant Physiol. **122**, 645–655.

Lee, M., Lenman, M., Banas, A., Bafor, M., Singh, S., Schweizer, M., Nilsson, R. et al. (1998) Identification of non-heme diiron proteins that catalyze triple bond and epoxy group formation. Science, **280**, 915–918.

Lee, J.S., Takai, J., Takashi, K., Endo, Y., Fujimoto, K., Koike, S. and Matsumoto, W. (2002) Effect of dietary tung oil on the growth and lipid metabolism of laying hens. J. Nutr. Sci. Vitaminol. **48**, 142–148.

Li, R., Yu, K., Hatanaka, T. and Hildebrand, D.F. (2010a) Vernonia DGATs increase accumulation of epoxy fatty acids in oil. Plant Biotechnol. J. **8**, 184–195.

Li, N., Gügel, I.L., Giavalisco, P., Zeisler, V., Schreiber, L., Soll, J. and Philippar, K. (2015) FAX1, a novel membrane protein mediating plastid fatty acid export. PLoS Biol. **13**, e1002053.

Liu, L., Hammond, E.G. and Nikolau, B.J. (1997) In vivo Studies of the biosynthesis of α-eleostearic acid in the seed of Momordica charantia L. Plant Physiol. **113**, 1343–1349.

van de Loo, F.J., Broun, P., Turner, S. and Somerville, C. (1995) An oleate 12-hydroxylase from Ricinus communis L. is a fatty acyl desaturase homolog. Proc. Natl Acad. Sci. USA, **92**, 6743–6747.

Lou, Y., Schwender, J. and Shanklin, J. (2014) FAD2 and FAD3 desaturases form heterodimers that facilitate metabolic channeling in vivo. J. Biol. Chem. **289**, 17996–18007.

Murashige, T. and Skoog, F. (1962) A revised medium for rapid growth and bio assays with tobacco tissue cultures. Plant Physiol. **15**, 473–497.

Napier, J.A., Haslam, R.P., Beaudoin, F. and Cahoon, E.B. (2014) Understanding and manipulating plant lipid composition: metabolic engineering leads the way. Curr. Opin. Plant Biol. **19**, 68–75.

Okuley, J., Lightner, J., Feldmann, K., Yadav, N., Lark, E. and Browse, J. (1994) Arabidopsis FAD2 gene encodes the enzyme that is essential for polyunsaturated lipid synthesis. Plant Cell, **6**, 147–158.

Park, S., Gidda, S.K., James, C.N., Horn, P.J., Khuu, N., Seay, D.C., Keereetaweep, J. et al. (2013) The alpha/beta hydrolase CGI-58 and peroxisomal transport protein PXA1 coregulate lipid homeostasis and signaling in Arabidopsis. Plant Cell, **25**, 1726–1739.

Pastor, S., Sethumadhavan, K., Ullah, A.H., Gidda, S., Cao, H., Mason, C., Chapital, D. et al. (2013) Molecular properties of the class III subfamily of acyl-coenzyme A binding proteins from tung tree (Vernicia fordii). Plant Sci. **203–204**, 79–88.

Robbins, M. (2011) Policy: fuelling politics. Nature, **474**, S22–S24.

Sanjaya, Durrett, T.P., Weise, S.E. and Benning, C. (2011) Increasing the energy density of vegetative tissues by diverting carbon from starch to oil biosynthesis in transgenic Arabidopsis. Plant Biotechnol. J. **9**, 874–883.

Sanjaya, Miller, R., Durrett, T.P., Kosma, D.K., Lydic, T.A., Muthan, B., Koo, A.J. et al. (2013) Altered lipid composition and enhanced nutritional value of Arabidopsis leaves following introduction of an algal diacylglycerol acyltransferase 2. Plant Cell, **25**, 677–693.

Shanklin, J. and Cahoon, E.B. (1998) Desaturation and related modifications of fatty acids. Annu. Rev. Plant Physiol. Plant Mol. Biol. **49**, 611–641.

Shockey, J.M., Gidda, S.K., Chapital, D.C., Kuan, J.C., Dhanoa, P.K., Bland, J.M., Rothstein, S.J. et al. (2006) Tung tree DGAT1 and DGAT2 have nonredundant functions in triacylglycerol biosynthesis and are localized to different subdomains of the endoplasmic reticulum. Plant Cell, **18**, 2294–2313.

Shockey, J., Mason, C., Gilbert, M., Cao, H., Li, X., Cahoon, E. and Dyer, J. (2015) Development and analysis of a highly flexible multi-gene expression system for metabolic engineering in Arabidopsis seeds and other plant tissues. Plant Mol. Biol. **89**, 113–126.

Shockey, J., Regmi, A., Cotton, K., Adhikari, N., Browse, J. and Bates, P.D. (2016) Identification of Arabidopsis GPAT9 (At5 g60620) as an essential gene involved in triacylglycerol biosynthesis. *Plant Physiol.* **170**, 163–179.

Singer, S.D., Chen, G., Mietkiewska, E., Tomasi, P., Jayawardhane, K., Dyer, J.M. and Weselake, R.J. (2016) Arabidopsis GPAT9 contributes to synthesis of intracellular glycerolipids but not surface lipids. *J. Exp. Bot.* **67**, 4627–4638.

Singh, S., Thomaeus, S., Lee, M., Stymne, S. and Green, A. (2001) Transgenic expression of a delta 12-epoxygenase gene in *Arabidopsis* seeds inhibits accumulation of linoleic acid. *Planta,* **212**, 872–879.

Slocombe, S.P., Cornah, J., Pinfield-Wells, H., Soady, K., Zhang, Q., Gilday, A., Dyer, J.M. *et al.* (2009) Oil accumulation in leaves directed by modification of fatty acid breakdown and lipid synthesis pathways. *Plant Biotechnol. J.* **7**, 694–703.

Smith, C.R. Jr. (1971) Occurrence of unusual fatty acids in plants. *Progr. Chem. Fats Lipids* **11**, 137–177.

Sonntag, N.O.V. (1979) Composition and characteristics of individual fats and oils. In *Bailey's Industrial Oil and Fat Products*(Swern, D., ed), pp. 289–477. New York: John Wiley & Sons.

Theodoulou, F.L., Job, K., Slocombe, S.P., Footitt, S., Holdsworth, M., Baker, A., Larson, T.R. *et al.* (2005) Jasmonic acid levels are reduced in COMATOSE ATP-binding cassette transporter mutants. Implications for transport of jasmonate precursors into peroxisomes. *Plant Physiol.* **137**, 835–840.

Thiel-Cooper, R.L., Parrish, F.C. Jr, Sparks, J.C., Wiegand, B.R. and Ewan, R.C. (2001) Conjugated linoleic acid changes swine performance and carcass composition. *J. Anim. Sci.* **79**, 1821–1828.

Triantaphylidès, C., Krischke, M., Hoeberichts, F.A., Ksas, B., Gresser, G., Havaux, M., Van Breusegem, F. *et al.* (2008) Singlet oxygen is the major reactive oxygen species involved in photooxidative damage to plants. *Plant Physiol.* **148**, 960–968.

Troncoso-Ponce, M.A., Cao, X., Yang, Z. and Ohlrogge, J.B. (2013) Lipid turnover during senescence. *Plant Sci.* **205–206**, 13–19.

Tunaru, S., Althoff, T.F., Nüsing, R.M., Diener, M. and Offermanns, S. (2012) Castor oil induces laxation and uterus contraction via ricinoleic acid activating prostaglandin EP3 receptors. *Proc. Natl Acad. Sci. USA,* **109**, 9179–9184.

Vanhercke, T., Petrie, J.D. and Singh, S. (2013a) Energy densification in vegetative biomass through metabolic engineering. *Biocatal. Agric. Biotechnol.* **1**, 75–80.

Vanhercke, T., Wood, C.C., Stymne, S., Singh, S.P. and Green, A.G. (2013b) Metabolic engineering of plant oils and waxes for use as industrial feedstocks. *Plant Biotechnol. J.* **11**, 197–210.

Vanhercke, T., El Tahchy, A., Liu, Q., Zhou, X.R., Shrestha, P., Divi, U.K., Ral, J.P. *et al.* (2014) Metabolic engineering of biomass for high energy density: oilseed-like triacylglycerol yields from plant leaves. *Plant Biotechnol. J.* **12**, 231–239.

Vanhercke, T., Divi, U.K., El Tahchy, A., Liu, Q., Mitchell, M., Taylor, M.C., Eastmond, P.J. *et al.* (2016) Step changes in leaf oil accumulation via iterative metabolic engineering. *Metab. Eng.* **39**, 237–246.

Voelker, T.A., Worrell, A.C., Anderson, L., Bleibaum, J., Fan, C., Hawkins, D.J., Radke, S.E. *et al.* (1992) Fatty acid biosynthesis redirected to medium chains in transgenic oilseed plants. *Science,* **257**, 72–74.

Wan, F., Pan, Y., Li, J., Chen, X., Pan, Y., Wang, Y., Tian, S. *et al.* (2014) Heterologous expression of *Arabidopsis C-repeat binding factor 3* (*AtCBF3*) and *cold-regulated 15A* (*AtCOR15A*) enhanced chilling tolerance in transgenic eggplant (*Solanum melongena* L.). *Plant Cell Rep.* **33**, 1951–1961.

Wang, Z. and Benning, C. (2011) *Arabidopsis thaliana* polar glycerolipid profiling by thin layer chromatography (TLC) coupled with gas-liquid chromatography (GLC). *J. Vis. Exp.* **49**, e2518.

Weselake, R.J. (2016) Engineering oil accumulation in vegetative tissue. In *Industrial Oil Crops* (McKeon, T.A., Hayes, D.G., Hildebrand, D.F. and Weselake, R.J., eds), pp. 413–434. New York: Elsevier Inc..

Winichayakul, S., Scott, R.W., Roldan, M., Hatier, J.H., Livingston, S., Cookson, R., Curran, A.C. *et al.* (2013) In vivo packaging of triacylglycerols enhances Arabidopsis leaf biomass and energy density. *Plant Physiol.* **162**, 626–639.

Xu, C. and Shanklin, J. (2016) Triacylglycerol metabolism, function, and accumulation in plant vegetative tissues. *Annu. Rev. Plant Biol.* **67**, 179–206.

Yang, L., Cao, Y., Chen, J.-N. and Chen, Z.-Y. (2009) Oxidative stability of conjugated linolenic acids. *J. Agric. Food Chem.* **57**, 4212–4217.

Yuan, G.F., Chen, X.E. and Li, D. (2014) Conjugated linolenic acids and their bioactivities: a review. *Food Funct.* **5**, 1360–1368.

Zale, J., Jung, J.H., Kim, J.Y., Pathak, B., Karan, R., Liu, H., Chen, X. *et al.* (2016) Metabolic engineering of sugarcane to accumulate energy-dense triacylglycerols in vegetative biomass. *Plant Biotechnol. J.* **14**, 661–669.

Zhang, M., Fan, J.L., Taylor, D.C. and Ohlrogge, J.B. (2009) DGAT1 and PDAT1 acyltransferases have overlapping functions in Arabidopsis triacylglycerol biosynthesis and are essential for normal pollen and seed development. *Plant Cell,* **21**, 3885–3901.

Zilberman, D., Hochman, G., Rajagopal, D., Sexton, S. and Timilsina, G. (2013) The impact of biofuels on commodity food prices: assessment of findings. *Am. J. Agr. Econ.* **95**, 275–281.

Zolman, B.K., Silva, I.D. and Bartel, B. (2001) The Arabidopsis *pxa1* mutant is defective in an ATP-binding cassette transporter-like protein required for peroxisomal fatty acid β-oxidation. *Plant Physiol.* **127**, 1266–1278.

14

Significant enhancement of fatty acid composition in seeds of the allohexaploid, *Camelina sativa*, using CRISPR/Cas9 gene editing

Wen Zhi Jiang[1], Isabelle M. Henry[2], Peter G. Lynagh[2], Luca Comai[2], Edgar B. Cahoon[1] and Donald P. Weeks[1,*]

[1]*Department of Biochemistry and Center for Plant Science Innovation, University of Nebraska, Lincoln, NE, USA*
[2]*Department of Plant Biology and UC Davis Genome Center, University of California, Davis, CA, USA*

*Correspondence
email dweeks1@unl.edu

Summary

The CRISPR/Cas9 nuclease system is a powerful and flexible tool for genome editing, and novel applications of this system are being developed rapidly. Here, we used CRISPR/Cas9 to target the *FAD2* gene in *Arabidopsis thaliana* and in the closely related emerging oil seed plant, *Camelina sativa*, with the goal of improving seed oil composition. We successfully obtained Camelina seeds in which oleic acid content was increased from 16% to over 50% of the fatty acid composition. These increases were associated with significant decreases in the less desirable polyunsaturated fatty acids, linoleic acid (i.e. a decrease from ~16% to <4%) and linolenic acid (a decrease from ~35% to <10%). These changes result in oils that are superior on multiple levels: they are healthier, more oxidatively stable and better suited for production of certain commercial chemicals, including biofuels. As expected, *A. thaliana* T_2 and T_3 generation seeds exhibiting these types of altered fatty acid profiles were homozygous for disrupted *FAD2* alleles. In the allohexaploid, Camelina, guide RNAs were designed that simultaneously targeted all three homoeologous *FAD2* genes. This strategy that significantly enhanced oil composition in T_3 and T_4 generation Camelina seeds was associated with a combination of germ-line mutations and somatic cell mutations in *FAD2* genes in each of the three Camelina subgenomes.

Keywords: *Camelina sativa*, gene editing, CRISPR/Cas9, allohexaploid, oleic acid, fatty acid composition.

Introduction

Camelina sativa (hereafter, Camelina) is an oil seed crop of the Brassicaceae family that has attracted considerable attention because of its short growing season, and its productivity in geographic regions with limited rainfall and soil fertility (Iskandarov *et al.*, 2014; Pilgeram *et al.*, 2007; Zubr, 1997). Despite the increasing commercial interest in Camelina, a limitation to the wider use of its seed oil in biofuels, lubricants and food applications is its high content of polyunsaturated fatty acids, particularly linolenic acid (18:3), which accounts for 30%–40% of seed oil from most Camelina cultivars (Iskandarov *et al.*, 2014). The high polyunsaturated content makes Camelina oil more susceptible to oxidation and food products derived from this oil more prone to rancidity (Frolich and Rice, 2005). To address this deficiency in Camelina oil quality, efforts have been directed at increasing the content of the more oxidatively stable oleic acid by suppression of *FAD2* genes for the Δ12 oleic acid desaturase that converts oleic acid to linoleic acid (18:2) and linolenic acid (18:3) (Hutcheon *et al.*, 2010; Kang *et al.*, 2011; Nguyen *et al.*, 2013). The result of this genetic modification is an increase in oleic acid content and corresponding decreases in polyunsaturated fatty acid (18:2 and 18:3) content of seed oils (Nguyen *et al.*, 2013). Methods for *FAD2* suppression in Camelina and other crops have included RNA interference (RNAi; Clemente and Cahoon, 2009; Graef *et al.*, 2009; Jung *et al.*, 2011; Nguyen *et al.*, 2013), microRNAs (Belide *et al.*, 2012), TALENs (Haun *et al.*, 2014) and standard mutagenesis and selection (e.g. Kang *et al.*, 2011; Pham *et al.*, 2012; Thambugala *et al.*, 2013; Wells *et al.*, 2014).

The recent advent of the highly efficient and facile CRISPR/Cas9 system for gene editing (Cong *et al.*, 2013; Jinek *et al.*, 2012;

Mali *et al.*, 2013) in animals (Petersen and Niemann, 2015; Proudfoot *et al.*, 2015) and plants (Weeks *et al.*, 2015) offers the opportunity to determine whether the oil composition of Camelina seeds could be favourably altered by knocking out the activities of a few or all of the six fatty acid desaturase 2 (*FAD2*) genes present in the genome of this allohexaploid plant (Hutcheon *et al.*, 2010; Kang *et al.*, 2011). If successful, this strategy would increase oleic acid content and lower the content of linoleic acid, linolenic acid and other long-chain polyunsaturated fatty acids. Because the genomes of Arabidopsis and Camelina—and the *FAD2* genes, in particular (Hutcheon *et al.*, 2010; Kang *et al.*, 2011; Nguyen *et al.*, 2013)—share strong homology with each other, we designed sgRNA constructs for use in Camelina but tested them first in Arabidopsis to determine their efficacy in creating Cas9/sgRNA-mediated gene mutations and changing the fatty acid composition of seeds. This strategy assumed that Arabidopsis plants homozygous for *FAD2* gene mutations could be obtained within two or three generations, whereas three or more generations would be required to inactivate most or all of the *FAD2* genes in allohexaploid Camelina. The nuclear genome of Camelina, as well as that of other allohexaploids such as bread wheat, contains three separate subgenomes that, because they avoid intergenomic (homoeologous) recombination, behave as three distinct and separate diploid genomes (Comai, 2005; Feldman and Levy, 2012; Madlung and Wendel, 2013). While homoeologous recombination can be manipulated genetically in certain species, there is effectively no opportunity to replace a functional *FAD2* gene with a defective allele from another subgenome through classical breeding techniques. In other words, homozygous or biallelic knockouts of *FAD2* genes must be achieved independently for all three subgenomes if complete depletion of FAD2 enzyme activity is to be achieved.

In this report, we present evidence for efficient, multigenerational, knockout of *FAD2* genes by the Cas9/sgRNA gene editing system in somatic and germ-line cells of Arabidopsis and Camelina leaves and seeds. We demonstrate in these initial experiments that such alterations lead to increases in oleic acid composition from ~16% to >50% and total monounsaturated fatty acid (18:1, 20:1, 22:1) from ~32% to >70% with concurrent decreases in linoleic and linolenic fatty acid content.

Results and discussion

FAD2 as a target for Cas9/sgRNA modification

The results presented here are the first report from a long-term project to significantly change the oil composition of Camelina seeds using the Cas9/sgRNA gene editing system to knockout the activity of *FAD2*, a key gene involved in the synthesis of polyunsaturated fatty acids. All three pairs of *FAD2* genes in the allohexaploid, *C. sativa*, have all been shown to be active in developing seeds (Hutcheon *et al.*, 2010; Kang *et al.*, 2011; Nguyen *et al.*, 2013). Thus, knocking out one or more of these genes should decrease the conversion of oleic acid to linoleic acid and linolenic acid (Figure S1). Because Camelina *FAD2* genes share extremely high homology with their counterparts in the much easier and quicker to manipulate diploid plant, *Arabidopsis thaliana* (hereafter Arabidopsis) (Hutcheon *et al.*, 2010; Kang *et al.*, 2011; Nguyen *et al.*, 2013), our strategy (discussed in detail below) has been to design Cas9 and sgRNA genes that often, but not always, target the same 20-nucleotide (nt) DNA sequence [23 nt, if the NGG protospacer adjacent motif (PAM) region is included] present in both Arabidopsis and Camelina

genes. As detailed below, seed fatty acid profiles in T_2, T_3 and T_4 generations of transgenic Camelina were followed along with DNA sequence modifications in both leaf tissues and seeds for each generation.

For both Arabidopsis and Camelina, three 20-nt target sites for Cas9/sgRNA-mediated DNA cleavage of the *FAD2* gene were selected. These 20-nt target sequences were oriented within each gene either in the same 5′ to 3′ direction as the reading frame of the gene (the 'forward' direction) or in the 5′ to 3′ direction on the opposite (reverse) strand of DNA. The location of each of these sites in the *FAD2* genes of Camelina and Arabidopsis is shown in Figure 1a. Each of these sites was deliberately chosen to be located in the 5′ portion of each gene to ensure that gene disruptions altering the gene's reading frame would produce a translation product lacking enzymatic activity. The sites were also chosen to contain a restriction enzyme cut site at the site of Cas9/sgRNA-directed DNA cleavage three base pairs upstream of the PAM site. As described below, if Cas9/sgRNA activity results in DNA cleavage at this site and if the double-stranded DNA break (DSB) is repaired by the error-prone nonhomologous end-joining (NHEJ) DNA repair mechanism, the restriction enzyme site often will be destroyed. Thus, if the region containing this site is amplified using the appropriate PCR primers, amplicons containing nonmodified restriction enzyme sites will be cleaved, whereas amplicons containing a destroyed restriction site will remain intact. Size analysis of the restriction enzyme-digested amplicons thus allows a simple qualitative and rapid means for detecting leaf or seed samples that contain (or do not contain) genes whose DNA sequence has been altered by the Cas9/sgRNA gene editing complex.

Drawings depicting the structures of the Cas9 gene and sgRNA gene constructs designed to recognize and cleave each of the

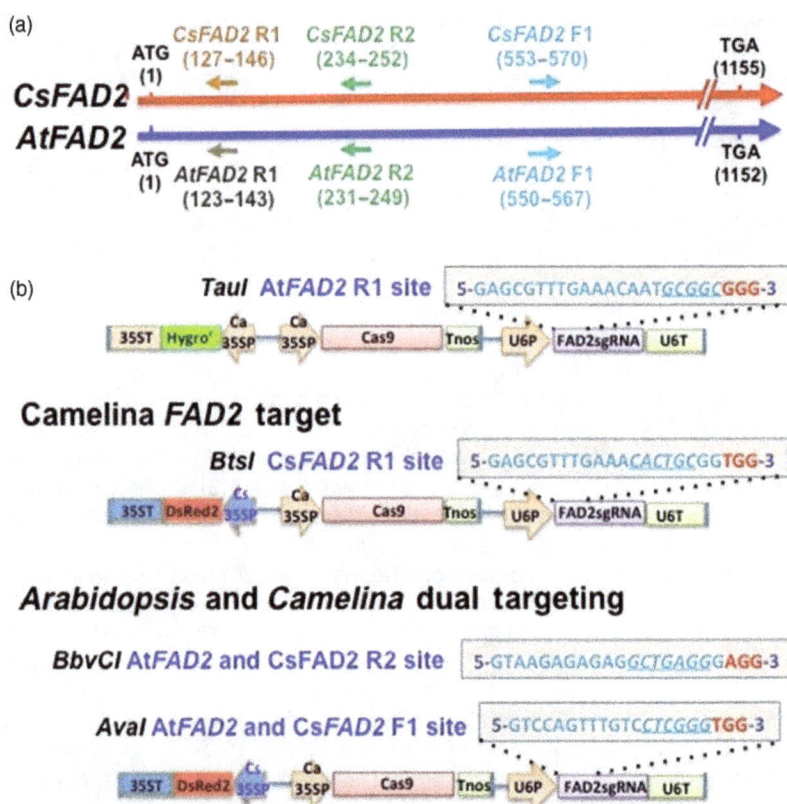

Figure 1 Targeting of the Camelina and Arabidopsis *FAD2* genes for knockout by the Cas9/sgRNA gene editing system. (a) Cas9/sgRNA targeting sites in the *FAD2* genes of Camelina (top line) and Arabidopsis (bottom line). ATG, the *FAD2* gene initiation codon starting at nucleotide 1; TGA, the *FAD2* gene termination codon ending at nucleotide 1155 (Camelina) or 1152 (Arabidopsis). (b) Cas9/sgRNA gene constructs for targeting the *FAD2* genes of Camelina and Arabidopsis. Boxed and lettered in blue are 20-nucleotide target sequences (plus 3-nucleotide PAM sequence—in red) contained in sgRNA genes for the Arabidopsis *FAD2*R1 target site (containing a *Tau*I restriction enzyme cut site at the predicted point of Cas9/sgRNA-mediated DNA cleavage), the Camelina RI site (with a *Bts*I site), Arabidopsis and Camelina R2 sites (with a *Bbv*CI site) and Arabidopsis and Camelina F1 sites (with an *Ava*I site). Transcription of the sgRNA gene is controlled by the Arabidopsis U6 gene promoter (U6P) and termination (U6T) regions, the Cas9 gene by Cauliflower mosaic virus (Ca) *35S* promoter and termination regions and the DsRed2 red fluorescence gene by cassava mosaic virus (Cs) *35S* promoter and termination regions.

three *FAD2* gene target sites (Figure 1a) are provided in Figure 1b. The complete DNA sequence for each construct is provided in Supporting Information and Experimental procedures. In each construct, the Cas9 gene is driven by the cauliflower mosaic virus *35S* gene promoter and terminated with the *Agrobacterium tumefaciens Tnos* gene termination sequence as described previously (Jiang *et al.*, 2013, 2014). The sgRNA genes are driven by the Arabidopsis U6 promoter and followed by the termination region from the same gene. The construct targeting the Arabidopsis *FAD2*R1 site contained a hygromycin resistance gene to allow for rapid selection of transgenic seeds (Figure 1b). The three other constructs each contained a DsRed2 gene that allowed the use of a hand-held green fluorescent flashlight to easily detect and select red fluorescing transgenic T_1 seeds (Lu and Kang, 2008; Nguyen *et al.*, 2013) obtained from T_0 plants produced by transformation of Camelina or Arabidopsis using the floral dip transformation technique. Importantly, as part of our strategy, only red fluorescent seeds were selected for further evaluation at each generation. Thus, presuming normal Mendelian inheritance of the T-DNA region containing the DsRed gene and the accompanying Cas9 and sgRNA genes, each plant analysed in this study had the potential to actively express the Cas9/sgRNA complex in most or all of its tissues. Important implications of this strategy are discussed below.

Extensive Cas9/sgRNA-mediated gene editing in leaf and seed tissues

Over 200 T_1 Arabidopsis plants and 300 T_1 Camelina plants that carried various Cas9 and sgRNA genes targeting *FAD2* genes were generated and analysed during this study (Data Set S1). From these T_1 plants, we produced large numbers of T_2 and T_3 plant progeny and seeds (Data Set S1) and limited samples of T_4 seeds. Each plant generated in this study was given a unique name (e.g. AtFAD2R1 T1-10-8-7 is the 7th T_3 progeny plant from the 8th T_2 progeny plant from the 10th T_1 Arabidopsis parent plant targeted for gene disruption at the R1 site of the *FAD2* gene).

To speed the selection of plants carrying mutations in a targeted gene and to discard plants lacking Cas9/sgRNA-directed gene mutations, DNA was extracted from leaf tissue of transgenic T_1 plants and used for PCR amplification of sgRNA target sites of interest using sets of primers listed in Table S1. As described above, each selected target site contained a restriction enzyme recognition sequence overlapping the expected Cas9/sgRNA cleavage site three base pairs upstream of the NGG PAM site, for easy detection of mutation events. As a result, if the Cas9/sgRNA complex created a DSB at the expected site and if an error occurred during DNA repair by the NHEJ mechanism, the restriction enzyme recognition site would be destroyed. In such cases, subjecting isolated DNA to digestion with the restriction enzyme appropriate to match the target site prior to PCR amplification of the target region would produce a nondigested, full-length, PCR product. Conversely, if no Cas9/sgRNA-directed mutation occurred at the site, the isolated DNA would be cleaved by the restriction enzyme and no PCR product would be visible on a standard ethidium bromide-stained agarose gel. An example of such a PCR/restriction enzyme (PCR/RE) analysis is shown in Figure S2 in which DNA from four different transgenic Arabidopsis plants displayed full-length, 322-bp PCR products, while DNA from a nontransgenic control plant produced the expected 244- and 78-bp restriction enzyme digestion fragments. Several DNA samples testing positive in the PCR/RE analyses were subjected to

Sanger DNA sequencing of PCR-amplified target sites to determine the exact nature of the mutation. In most experiments, a few DNA samples tested negative in the PCR/RE analyses were also sequenced as controls and to confirm that they, indeed, were true negatives.

The expected ability of the Cas9/sgRNA system to cause gene mutations in the *FAD2* genes of Arabidopsis and Camelina plants was confirmed by DNA sequencing. Sequence analyses of DNA from leaf and seed samples (Data Set S2) confirmed multiple mutations over multiple generations at each of the three target sites in each of the three different *FAD2* gene types present, respectively, in the A, B and C subgenomes of the allohexaploid genome. In the overview presented in Data Set S2, DNA sequences obtained by Sanger sequencing are provided for 328 *FAD2* gene target sites, 258 of which contained the kinds of short-length nucleotide insertions and deletions typical of NHEJ DNA repair that follows creation of DSBs by the Cas9/sgRNA complex. Interestingly, over half of these mutations were the result of a single-nucleotide insertion that was nearly always located at the predicted Cas9/sgRNA cut site 3 bp upstream of the PAM site. Among these single-nucleotide insertions, there was a marked preference for T (50%) and A (33%) nucleotide inserts over G (12%) and C (6%) nucleotide inserts. The summary at the bottom of Data Set S2 shows that approximately 19% of mutations involved nucleotide deletions over 3 bp in size and approximately 5% involved nucleotide insertions of 3 bp or more. Because all single-nucleotide insertions and most other insertions and deletions change the reading frame of the gene and because all target sites were deliberately chosen to be located in the 5′ region of the gene (Figure 1a), the vast majority of mutations created in the *FAD2* genes are predicted to lead to gene knockout. Because of single-nucleotide polymorphisms (SNPs) between the three types of *FAD2* genes found in the A, B and C subgenomes of Camelina (Hutcheon *et al.*, 2010), we were able to determine that all of the *FAD2* genes in each of the subgenomes were efficiently targeted by the Cas9/sgRNA complex.

To assess whether off-target mutations occurred, three sites in the genome most similar to the target *FAD2* site were PCR-amplified and sequenced by Sanger chemistry using *C. sativa* DNA from *C. sativa* plants with a high mutation rate at the *FAD2*R1 target site. These three best off-target candidate loci contain a PAM and complete homology to the 10-bp 'seed region' that is at the 3′ end of the *FAD2* protospacer (Jiang *et al.*, 2015). In the 5′ ends, they contain five or six mismatches to the *FAD2* protospacer (Figure S6). The resulting DNA sequence chromatograms from the Cas9-positive plants were indistinguishable from the WT chromatograms, suggesting that off-target mutations are not prevalent when using the *FAD2* protospacer of interest (Figure S6).

Examination of the data in Data Set S2 also revealed that approximately 8% of the DNA sequencing reads contained a combination of single-nucleotide polymorphisms found in two (or, in one case, three) of the homoeologous *FAD2* genes. Such combinations could be due to unexpected recombination of homoeologous chromosome fragments following Cas9/sgRNA-generated chromosome breaks or, alternatively, to generation of incomplete DNA transcripts during polymerase chain reaction (PCR) DNA amplification and the subsequent utilization of these incomplete strands as templates during ensuing rounds of amplification. To distinguish between these two possibilities, the same DNA samples used for PCR amplification and Sanger

DNA sequencing shown above were subjected to PCR amplification and DNA sequencing using Illumina Amplicon-Seq techniques in which thousands of DNA strands were sequenced from DNA isolated from each plant under study.

PCR artefacts are generated during amplification of the three highly homologous copies of the *FAD2* gene in Camelina

The allohexaploid Camelina contains three distinct yet closely related homoeologous subgenomes, called A, B and C. Two subgenomes consist of seven chromosomes and the third of six chromosomes adding to the somatic chromosome content of 20 pairs (Hutcheon *et al.*, 2010). Individual single-copy genes, such as the *FAD2* gene, thus come in three allelic pairs that can be distinguished based on single-nucleotide polymorphisms (SNPs) specific to each subgenome (Hutcheon *et al.*, 2010; Kang *et al.*, 2011; Nguyen *et al.*, 2013) (Data Set S2 and Figures 2 and 3). Our Sanger sequence data (Data Set S2) indicated that approximately

8% of the DNA sequencing reads contained a combination of single-nucleotide polymorphisms found in two (or, in one case, three) of the homoeologous *FAD2* genes. Specifically, of the 259 Cas9/sgRNA-mediated mutations observed in our Sanger sequencing of cloned FAD2 gene DNA, 21 contained apparent chimeras between homoeologous chromosomes (Data Set S2). All possible combinations of homoeologous chromosome exchanges (A + C, B + C, etc.) were observed. No chimeric sequences were observed in the limited number of DNA sequences (i.e. 30) obtained from Sanger DNA sequencing of wild-type Camelina DNA samples.

Because all homoeologs were targeted by the same guide RNA, it is possible that these sequences originated from recombination between homoeologous chromosomes following Cas9/sgRNA-generated chromosome breaks. Alternatively, as all sequences were amplified using the same primers, these sequences could have originated from chimeric PCR amplicons. To distinguish between these two possibilities, 30 transgenic plants expressing the Cas9 and sgRNA and two wild-type plants were selected from

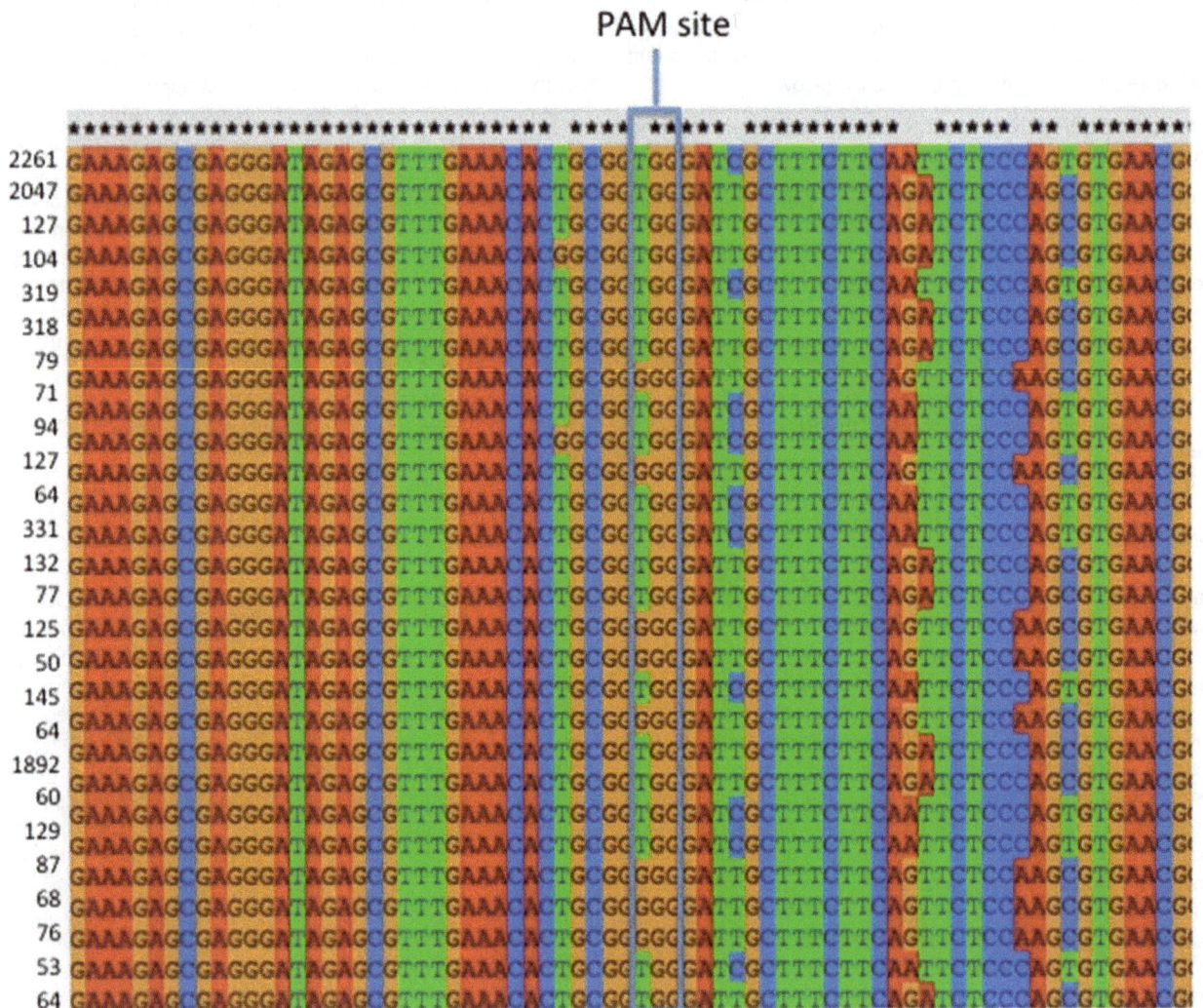

Figure 2 Clustal alignment of the sequences of the Cs*FAD2*R1 target site in *Camelina sativa* leaf wild-type sample. Illumina Amplicon-Seq was used for DNA sequencing. Read frequency counts for each sample are provided to the left of the sequences. Only the sequences found at least 50 times are depicted. The three most common sequences represent the three WT homoeologs sequences. Asterisks denote nucleotides conserved in all three subgenome homoeologs (A–C). Lack of an asterisk denotes a nucleotide position at which there is a single-nucleotide polymorphism in one of the three homoeologs.

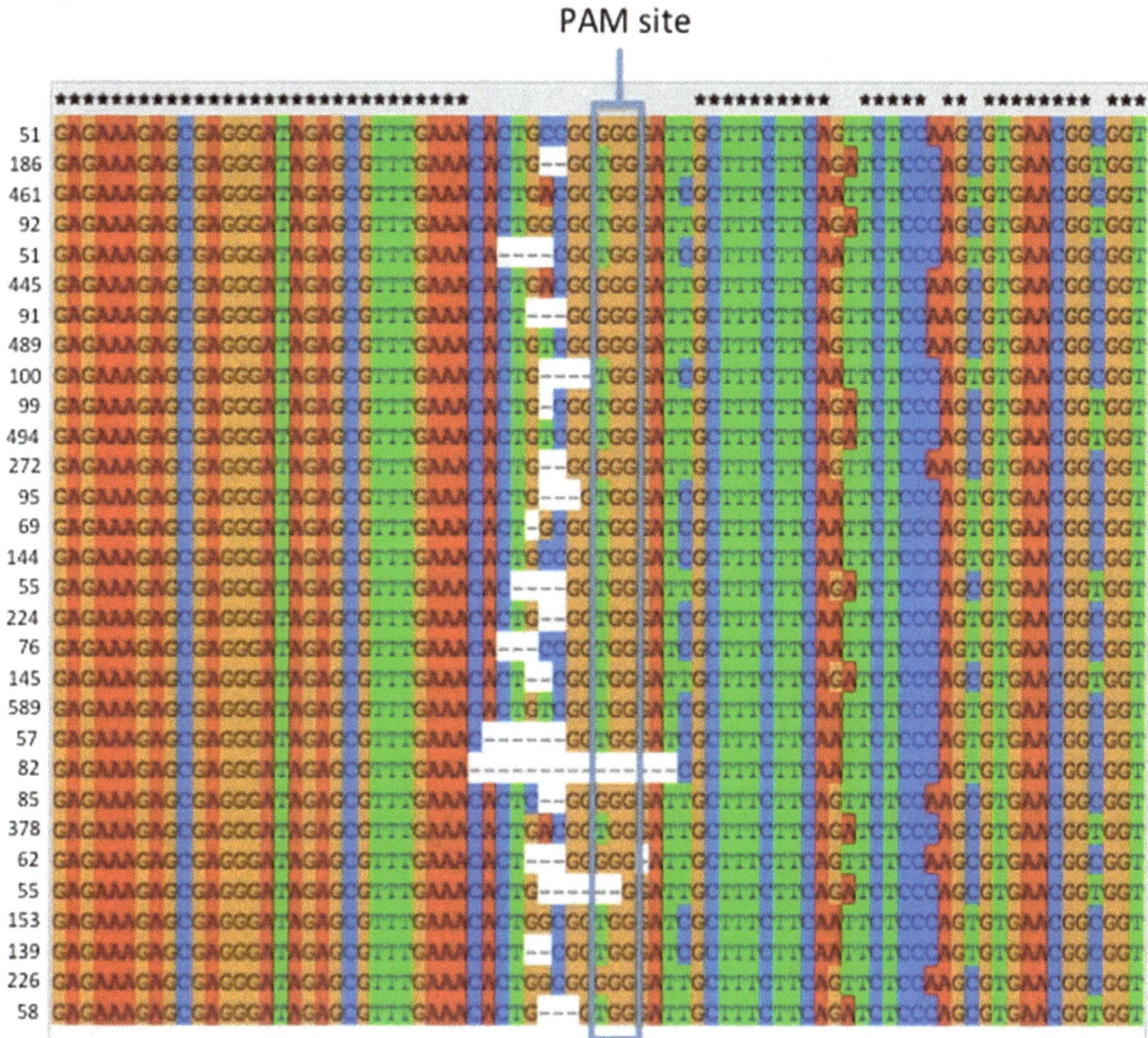

Figure 3 Clustal alignment of the sequences of the CsFAD2R1 target site in *Camelina sativa* leaf sample T1-1-3 exhibiting Cas9/sgRNA-mediated gene editing. Illumina Amplicon-Seq was used for DNA sequencing. Read frequency counts for each sample are provided to the left of the sequences. Only the most common sequences found at least 50 times are depicted. Asterisks denote nucleotides conserved in all three subgenome homoeologs (A, B and C). Lack of an asterisk denotes a nucleotide position at which there is a single-nucleotide polymorphism in one of the three homoeologs.

the samples analysed above and subjected to PCR amplification and DNA sequencing using Illumina Amplicon-Seq, generating thousands of sequences from each sample (Data Set S3 and S4). Analysed sequences were 350 bp long and amplified using nonhomoeologous specific primers, that is primers that perfectly matched all three homoeologous sequences. Paired-ended reads were obtained and processed using custom scripts (see Experimental procedures for details) to assign sequences to specific homoeologs, based on the SNPs distinguishing the three WT sequences. Chimeric sequences were observed in both transgenic and wild-type data sets (Data Set S3), and the percentage of chimeric sequence was at least as high in the control as in the transgenic samples (Table S2—summarized in Figure S3). These results thus provide no evidence for recombination between homoeologous DNA strands following Cas9/sgRNA-generated cleavage of the DNAs, but, rather, point to artificial recombination between homoeologous *FAD2* gene strands during PCR

amplification as the source of the observed chimeras. Reports in the recent literature (e.g. Liu *et al.*, 2014 and references therein) document that PCR amplification of DNA from distinctly different, but highly homologous, DNA molecules can lead to the generation of 'recombinant' DNA strands due to a small number of incomplete PCR amplicons being used as primer sequences. These studies coupled with the present study point to the need for extreme care in interpreting data from experiments using PCR amplification of DNA from polyploid species.

CRISPR/Cas9 action results in a variety of mutant alleles

Sequencing the targeted sites allowed for an in-depth analysis of the patterns of mutations as, in total, 329 Sanger sequences and 410 490 Illumina sequences were analysed. Analyses of the initial Sanger data and the Illumina Amplicon-Seq data provided useful comparisons. First, the Amplicon-Seq data confirmed the presence of many different types of insertions and deletions at the

CS*FAD2*R1 target site (average of 71% indels) and the lack of insertions and deletions at the same site in DNA samples obtained from wild-type plant tissues (average 0.6% indels) (Figures 3 and 4 and Data Set S3). Next, it confirmed that all three homoeologs were efficiently targeted although the percentage of mutant sequence was higher in homoeologs B sequences (84.4%) than in homoeologs A (60.7%) or C (67.3%). Averaged across all mutated samples and homoeologs, Sanger reads indicated 21.2% WT, 78.2% indel-containing and 0.6% SNP-containing sequencing. These numbers were similar to those obtained from the Illumina reads, which indicated 27.5%, 71.1% and 1.4% of WT, indel-containing and SNP-containing reads, respectively. There were nearly twice as many Illumina reads containing insertions as compared to deletions. Within the Illumina reads, >99% of the insertions were single-nucleotide insertions with the following distribution: 49.1% T, 35.3% A, 12.9% G and 2.8% C. This distribution was consistent with observation from the Sanger data (50% T, 33% A, 12% G and 6% C), demonstrating strong insertion preferences. Deletions were more variable in both data sets. Of the 64 368 Illumina reads that contained deletions, half were 1-bp deletions, a quarter were 2-bp deletions, 11% were 3-bp deletions, and the remaining 12% were >3-bp insertions. For WT1, the Illumina reads were 92.2% WT, 0% indel-containing and 7.1% SNP-containing, and for WT2, they were 91.9%, 1.3% and 6.8%, respectively. The SNPs and indels observed in WT1 and WT2 are assumed to originate primarily from PCR-derived artefacts documented above and provide a base estimate of false positives.

Significant Cas9/sgRNA-mediated changes in seed oil composition

Analyses of fatty acid composition of seeds from both Arabidopsis and Camelina plants containing Cas9/sgRNA-mediated mutations in their *FAD2* genes as well as from control plants lacking such modifications demonstrated a range of oleic acid content from the usual ~16% to greater than 50% (Data Set S5). One of the best-performing Arabidopsis lines (FAD2R2 T1-25-4 and its T_4 seed progeny) showed an oleic oil concentration of ~50% in the T_3 generation. This increased to a range between ~55% and ~60% in the T_4 generation (Figure 4). As expected from results of earlier studies with Cas9/sgRNA-disrupted genes of Arabidopsis (Feng *et al.*, 2014; Jiang *et al.*, 2014), DNA sequencing of the FAD2 genes in T_2 and T_3 plants

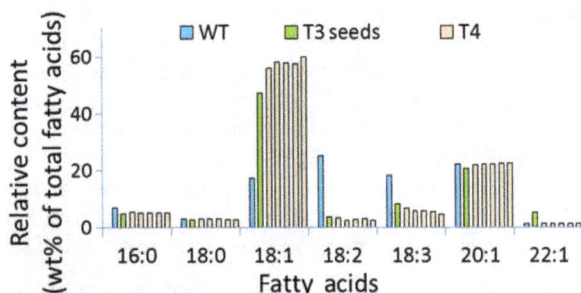

Figure 4 Seed oil profile of seeds from Arabidopsis plants transformed using Cas9/sgRNA targeting the R2 site of the AtFAD2 gene. Blue: wild-type control; green: T3 seeds (from T2 plant T1-25-4); light red: T4 seeds (from T3 plant, in order, T1-25-4-1, T1-25-4-2, T1-25-4-3, T1-25-4-4, T1-25-4-5) derived from T1-25-4. Methods for extraction and measurement of Arabidopsis seed fatty acids are detailed in Experimental procedures.

producing these high oleic acid seeds showed them to be homozygous for disruptions of the *FAD2* genes (Figure S4). Also, as expected, the number of generations needed for oleic acid seed content to increase in the allohexaploid, Camelina, was more than in diploid Arabidopsis—although a few T_4 Camelina seeds contained oleic acid levels of 50% or greater (Data Set S5 and Figure S5a). In each case in which there was a rise in oleic acid content, there was the expected proportional decrease in linoleic and linolenic acid content, leading to seeds with fatty acid composition more favourable for production of certain specialty oils and biodiesel. Targeting each of the three selected sites within the *FAD2* gene (F1, R1 and R2) for Cas9/sgRNA attack was effective in causing significant increases in production of oleic acid and decreases in linoleic acid and linolenic acid (Figure S5a) in the best-performing Camelina lines. Production of total monounsaturated fatty acids (18:1, 20:1, 22:1) increased from ~30% in wild-type seeds to as high as ~74% (Figure S5b). These results demonstrate that, as with the use of RNAi, gene silencing and other gene editing technologies, Cas9/sgRNA system can be used to cause significant and favourable changes in seed fatty acid composition in both Arabidopsis and the emerging oil seed crop, Camelina. Consistent with our molecular off-target analysis, no evidence of overt off-target effects was observed in the phenotypes of plants or seeds containing Cas9/sgRNA-mediated gene mutations.

Changes in seed oil composition originate from the combined contributions of Cas9/sgRNA-mediated mutations in somatic cells of Camelina seeds and germ-line mutations

While expression of the Cas9 and sgRNA genes is clearly responsible for the changes in seed oil composition documented in Data Set S3, the data presented in Data Set S2 and S4 raise questions with regard to the source of this phenotype and the probability of its stable inheritance—at least in the present set of Camelina plants that contain and appear to maintain an active set of Cas9 and sgRNA genes. Careful analysis of the DNA sequences in the mutant genes shown in Data Set S2–S4 suggest, as detailed below, that while some mutant *FAD2* genes observed in the present study may be inherited from one generation to the next, the large number of different mutations observed in DNA from a single sample of leaves or seeds from an individual plant points to a contribution of somatic cell mutations to the pool of DNA sequence variants observed. Specifically, cataloguing all types of mutations observed from the Amplicon-Seq data resulted in, on average, 59 different mutation types per sample, evenly distributed between the three homoeologous sequences (Data Set S4). We have also noted that the proportion of indel sequences was consistently higher in leaf samples (71.6%) compared to seed samples (58.5%), suggesting CRISPR/Cas9 activity is higher in somatic tissues than in germinal tissues (Data Set S3 and summarized in Figure S7). This is in agreement with our earlier observations and conclusions regarding Cas9/sgRNA-derived genetic mosaicism in somatic cells (Jiang *et al.*, 2014) and similar conclusions by others (e.g. Feng *et al.*, 2014). Thus, the changes in fatty acid composition observed in individual seeds (Data Set S5) could be due to the effects of germ-line mutations combined with *FAD2* mutations occurring in somatic cells during seed development. The latter would give rise to a mosaic of overlapping patches of somatic cells with varying combinations of mutations in the six copies of the *FAD2* genes. This conclusion is further strengthened by our observations that when fatty acid composition is determined for individual T_3 seeds from a single T_2 plant or individual T_4 seeds from a single T_3 plant (Figures 4, 5 and Data Set S5), there can be noticeable variability in oleic acid levels from seed to seed (e.g. 39%, 40%, 43%, 45% and 55% for seeds from plant T1-5-1-1)—a potential reflection of a mosaic of somatic cells in individual seeds that carry variable numbers of inactivated *FAD2* genes.

Germ-line mutations may be detected by the Mendelian inheritance of unique mutations in Cas9-positive plants or by the inheritance of Cas9-induced mutation in Cas9-negative plants. In the Amplicon-Seq data set from leaves of 27 individual Cas9-positive *C. sativa* plants, there are many cases in which a unique Cas9-induced mutation occurs in over 25% of the reads from a homologous pair (Data Set S4), suggesting that the mutation may have been inherited from one of the parents. On the other hand, many of these mutations are of a very common type and the high percentage could result from multiple somatic mutations as well. To better quantify the presence of germ-line mutations, we analysed Cas9-negative plants that were derived from a Cas9-positive *C. sativa* plant, T1-5 (Figure S8). Lack of the Cas9 gene in 20 nonfluorescent (i.e. 'black') seed progeny in each of three different T1-5 lines (i.e. T1-5-1-7B, T1-5-1-8B, T1-5-1-9B) was verified by PCR analyses (Figure S9). Sanger DNA sequence chromatograms indicate that some progeny from T1-5-1-8B contain a mutation in the B homoeolog and some progeny from T1-5-1-9B contain a mutation in the A homoeolog (Figure S10b). No mutation was detected in the other pools or in the C homoeolog (Figure S10c). From the pools that contained mutations, individual plants were sequenced and could easily be classified as 'WT', 'homozygous A insertion' or 'heterozygous'

(Figure S10a). From these allelic ratios, we inferred that two of the 18 FAD2 homoeologous copies, or 11.1%, carried a mutation in the three Cas9-negative parents. Additionally, as demonstrated by this germ-line inheritance, we have generated Cas9-negative *C. sativa* plants that contain homozygous FAD2 knockouts in homologous pairs (Figure S10b).

While the presence of *FAD2* gene knockouts in the germ-line may explain a substantial portion of the increase in oleic acid content observed, the high abundance of somatic cell mutations in developing seeds [documented by the large number of different Cas9/sgRNA-generated mutations in every line tested (Data Set S4)] suggests that *FAD2* knockout mutations in somatic cells also may contribute to increases in oleic acid concentrations.

Cas9/sgRNA activity may contribute to changes in oil composition via mechanisms other than reduced expression. This is consistent with previous observations that *FAD2* functions as a homodimeric enzyme (Lou et al., 2014) and that expression of nonfunctional mutants of *FAD2* in cotton (Chapman et al., 2001, 2008) or closely related desaturases, epoxygenases, and hydroxylases in a variety of other plant species (Broun and Somerville, 1997; Broun et al., 1998; Cahoon et al., 1999; Singh et al., 2001) can cause inhibition of *FAD2* desaturase activity—presumably through formation of nonfunctional heterodimers (i.e. 'subunit poisoning'). In these species, such dominant negative inhibition of *FAD2* activity results in striking increases in oleic acid and decreases in linoleic and linolenic acids. (Broun and Somerville, 1997; Broun et al., 1998; Cahoon et al., 1999; Chapman et al., 2001, 2008; Lou et al., 2014; Singh et al., 2001). Thus, because the Cas9/sgRNA constructs used in the present studies primarily cause frame-shift mutations in the 5' region of the targeted *FAD2*

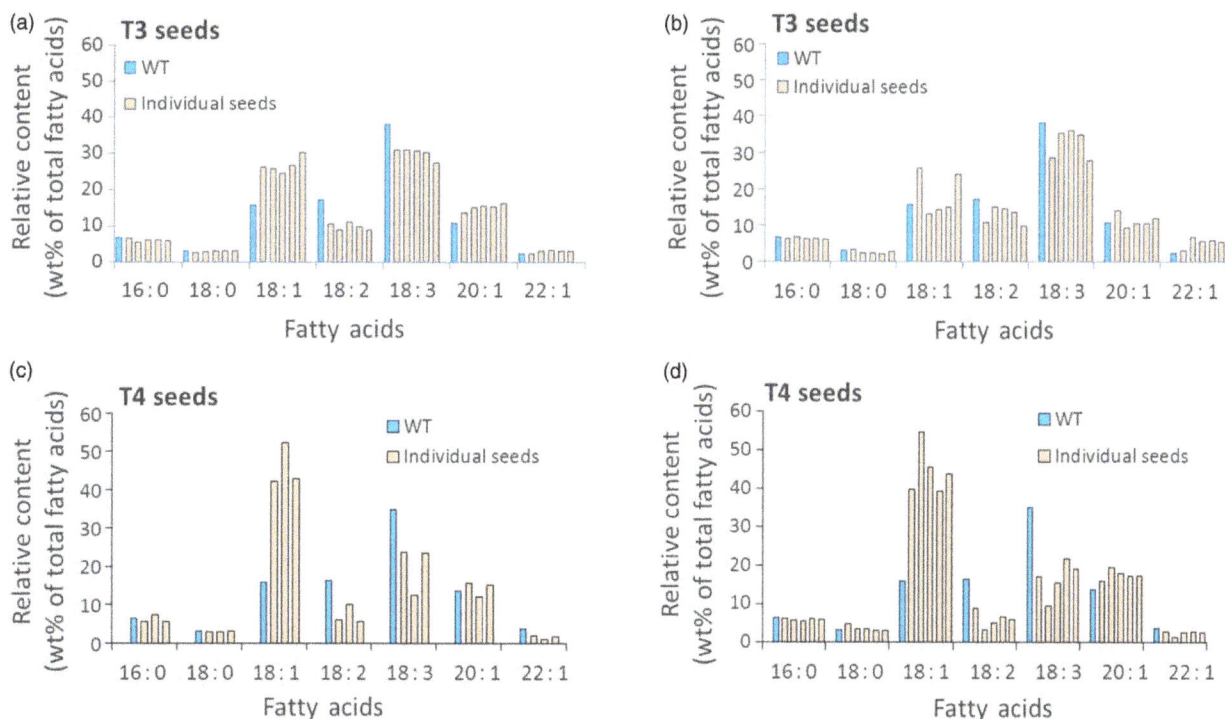

Figure 5 Oil profile in seeds of Camelina plants transformed using Cas9/sgRNA targeting the R1 site of the *FAD2* genes. Variation of T3 seed oil profiles of individual seeds from T_2 plant T1-1-3 (a) and T_2 plant T1-5-1 (b); variation of seed oil profiles of individual T_4 seeds from T_3 plant T1-3-2-3 (c) and T_3 plant T1-5-1-1 (d). Blue: wild-type control; light red: individual seed tested. Methods for extraction and measurement of Camelina seed fatty acids are detailed in Experimental procedures.

genes, it is likely that polypeptides are produced that contain authentic N-terminal *FAD2* domains. It is possible that such molecules can bind with functional *FAD2* monomers and partially inhibit desaturase activity. Future experimentation to test this hypothesis and, more importantly, future analyses of new generations of mutant Camelina plants containing homozygous or biallelic knockouts of *FAD2* genes will be needed to provide a better understanding of the high oleic acid phenotypes presently observed.

A potentially significant advantage of using gene editing techniques such as ZFNs, TALNs and Cas9/sgRNAs is that once the desired gene editing event(s) have been obtained, the editing genes can be eliminated by simple cross-breeding techniques and the desirable phenotype stabilized by production of homozygous progeny (e.g. Li *et al.*, 2012; reviewed in Weeks *et al.*, 2015 and Petolino and Kumar, 2015). A lesson from the present study is that obtaining the desired phenotypic stability in some allo-hexaploid crops, such as Camelina, likely will require several generations in which the Cas9/sgRNA genes remain present and active. The rapid screening of seeds carrying active Cas9 and sgRNA genes by virtue of their easily detectable DsRed fluorescence is an attractive tool for achieving this goal. One potential future approach to gaining a better, but not complete, picture in regard to this uncertainty will be to analyse DNA from pollen samples from individual plants (or, if technically possible, from single pollen grains) to better determine the complement of wild-type and mutant *FAD2* genes and the ratio between them. Alternatively, it is possible that germ-line mutations will be achieved more efficiently if the editing genes are driven by a different promoter, with documented expression in meristematic tissues. For example, homozygous mutants were recently obtained in a single generation in *A. thaliana* using a variety of promoters including an egg cell-specific promoter (Wang *et al.*, 2015), the *YAO* promoter that is active in meristematic cells (Yan *et al.*, 2015), the anther-specific *DD45* gene promoter (Mao *et al.*, 2016) and the *INCURVATA2* gene promoter (Hyun *et al.*, 2015).

Conclusions

Several important conclusions can be drawn from the present study. The Cas9/sgRNA system for gene editing is active in Camelina over multiple generations. Mutations in the *FAD2* genes of Camelina can result in significant increases in oleic acid content and concomitant large decreases in linoleic and linolenic acid content. Based on results with other oil seed crops, future Camelina lines emanating from this continuing long-term project that are homozygous for inactivated *FAD2* genes likely will exhibit even larger decreases in levels of long-chain polyunsaturated fatty acids and improved oleic acid content. Future projects aimed at producing plants that also contain Cas9/sgRNA-mediated knockouts of the fatty acid elongase, *FAE1*, genes should lead to additional boosts in oleic acid content of Camelina seeds.

Experimental procedures

Creation and analyses of transgenic Arabidopsis and Camelina plants and seeds: *A. thaliana* ecotype (Col-0) and *C. sativa*, cv. Suneson, were transformed using the floral dip method with vectors containing Cas9/sgRNA constructs targeting the respective Arabidopsis and Camelina *FAD2* genes. Transgenic plants were self-pollinated, and leaves and seeds were collected from sequential generations of plants to determine by PCR/restriction enzyme analyses and DNA sequencing if gene editing events had occurred at the gene target sites. The fatty acid compositions of seeds were determined using gas chromatography. Mutant FAD2 gene DNA sequences (somatic and germ-line) in Camelina lines containing (Cas9-positive) and lacking (Cas9-negative) functional Cas9 and sgRNA genes were obtained by various combinations of Sanger DNA sequencing and Illumina Amplicon-MiSeq DNA sequencing techniques. A search for potential off-target gene modifications was conducted using Sanger DNA sequencing of three DNA sites most closely related to the target sites within the Camelina *FAD2* genes. Details of each of these procedures are provided in Supporting Information.

Acknowledgements

We thank Aaron Duthoy and Andrew Blazek for technical assistance, Tara Nazarenus for assistance with fatty acid analyses and Dr. Avrahim Levy for his comments on an initial draft of the manuscript. This work was supported by the U.S. National Science Foundation (EPSCoR-1004094 to DPW and EBC), U.S. Department of Energy, Office of Science, OBER (DOE-BER SC0012459 to EBC and LC) and NSF Plant Genome Program (award 1444612 to LC and 13-39385 to EBC) and the University of Nebraska Foundation (Account #01133480 to DPW). The authors declare they have no conflicts of interest.

Author contributions

WJ developed methodology and performed or guided the experimentation. WJ and DPW designed the experiments. PL performed the Amplicon-Seq experiments. PL and IMH analysed the Amplicon-Seq data. WJ, DPW, EBC, LC and IMH analysed and interpreted the data and wrote the manuscript. DPW, EBC and LC are principal investigators on grants supporting this research.

References

Belide, S., Petrie, J.R., Shrestha, P. and Singh, S.P. (2012) Modification of seed oil composition in Arabidopsis by artificial microRNA-mediated gene silencing. *Front. Plant Sci.* **3**, 168–178.

Broun, P. and Somerville, C. (1997) Accumulation of ricinoleic, lesquerolic, and densipolic acids in seeds of transgenic Arabidopsis plants that express a fatty acyl hydroxylase cDNA from castor bean. *Plant Physiol.* **113**, 933–942.

Broun, P., Boddupalli, S. and Somerville, C. (1998) A bifunctional oleate 12-hydroxylase: desaturase from *Lesquerella fendleri*. *Plant J.* **13**, 201–210.

Cahoon, E.B., Carlson, T.J., Ripp, K.G., Schweiger, B.J., Cook, G.A., Hall, S.E. and Kinney, A.J. (1999) Biosynthetic origin of conjugated double bonds: production of fatty acid components of high-value drying oils in transgenic soybean embryos. *Proc. Natl Acad. Sci. USA*, **96**, 12935–12940.

Chapman, K.D., Austin-Brown, S., Sparace, S.A., Kinney, A.J., Ripp, K.G., Pirtle, I.L. and Pritle, R.M. (2001) Transgenic cotton plants with increased seed oleic acid content. *J. Am. Oil Chem. Soc.* **78**, 941–947.

Chapman, K.D., Neogi, P.B., Hake, K.D., Stawska, A.A., Speed, T.R., Cotte, M.Q., Garrett, D.C. *et al.* (2008) Reduced oil accumulation in cottonseeds transformed with a *Brassica* nonfunctional allele of a delta-12 fatty acid desaturase (FAD2). *Crop Sci.* **48**, 1470–1481.

Clemente, T.E. and Cahoon, E.B. (2009) Soybean oil: genetic approaches for modification of functionality and total content. *Plant Physiol.* **151**, 1030–1040.

Comai, L. (2005) The advantages and disadvantages of being polyploid. *Nat. Rev. Genet.* **6**, 836–846.

Cong, L., Ran, F.A., Cox, D., Lin, S., Barretto, R., Habib, N., Hsu, P.D. *et al.* (2013) Multiplex genome engineering using CRISPR/Cas systems. *Science*, **339**, 819–823.

Feldman, M. and Levy, A.A. (2012) Genome evolution due to allopolyploidization in wheat. *Genetics*, **192**, 763–774.

Feng, Z., Mao, Y., Xu, N., Zhang, B., Wei, P., Yang, D.L., Wang, Z. *et al.* (2014) Multigeneration analysis reveals the inheritance, specificity, and patterns of CRISPR/Cas-induced gene modifications in Arabidopsis. *Proc. Natl Acad. Sci. USA*, **111**, 4632–4637.

Frolich, A. and Rice, B. (2005) Evaluation of *Camelina sativa* oil as a feedstock for biodiesel production. *Ind. Crops Prod.* **21**, 25–31.

Graef, G., LaVallee, B.J., Tenopir, P., Tat, M., Schweiger, B., Kinney, A.J., Van Gerpen, J.H. *et al.* (2009) A high-oleic-acid and low-palmitic-acid soybean: agronomic performance and evaluation as a feedstock for biodiesel. *Plant Biotechnol. J.* **7**, 411–421.

Haun, W., Coffman, A., Clasen, B.M., Demorest, Z.L., Lowy, A., Ray, E., Retterath, A. *et al.* (2014) Improved soybean oil quality by targeted mutagenesis of the fatty acid desaturase 2 gene family. *Plant Biotechnol. J.* **12**, 934–940.

Hutcheon, C., Ditt, R.F., Beilstein, M., Comai, L., Schroeder, J., Goldstein, E., Shewmaker, C.K. *et al.* (2010) Polyploid genome of *Camelina sativa* revealed by isolation of fatty acid synthesis genes. *BMC Plant Biol.* **10**, 233–248.

Hyun, Y., Kim, J., Cho, S.W., Choi, Y., Kim, J.S. and Coupland, G. (2015) Site-directed mutagenesis in *Arabidopsis thaliana* using dividing tissue-targeted RGEN of the CRISPR/Cas9 system to generate heritable null alleles. *Planta*, **241**, 271–284.

Iskandarov, U., Kim, H.J. and Cahoon, E.B. (2014) Chapter 8: Camelina: an emerging oilseed platform for improved biofuels and bio-based materials. In *Plants and BioEnergy* (Carpita, N., McCann, M. and Buckeridge, M.S., eds), pp. 131–140. New York: Springer.

Jiang, W., Zhou, H., Bi, H., Fromm, M., Yang, B. and Weeks, D.P. (2013) Demonstration of CRISPR/Cas9/sgRNA-mediated targeted gene modification in Arabidopsis, tobacco, sorghum and rice. *Nucleic Acids Res.* **41**, e188.

Jiang, W., Yang, B. and Weeks, D.P. (2014) Efficient CRISPR/Cas9-mediated gene editing in *Arabidopsis thaliana* and inheritance of modified genes in the T2 and T3 generations. *PLoS One*, **9**, e99225.

Jiang, F., Zhou, K., Ma, L., Gressel, S. and Doudna, J.A. (2015) A Cas9-guide RNA complex preorganized for target DNA recognition. *Science*, **348**, 1477–1481.

Jinek, M., Chylinski, K., Fonfara, I., Hauer, M., Doudna, J.A. and Charpentier, E. (2012) A programmable dual-RNA-guided DNA endonuclease in adaptive bacterial immunity. *Science*, **337**, 816–821.

Jung, J.H., Kim, H., Go, Y.S., Lee, S.B., Hur, C.G., Kim, H.U. and Suh, M.C. (2011) Identification of functional BrFAD2-1 gene encoding microsomal delta-12 fatty acid desaturase from *Brassica rapa* and development of *Brassica napus* containing high oleic acid contents. *Plant Cell Rep.* **30**, 1881–1892.

Kang, J., Snapp, A.R. and Lu, C. (2011) Identification of three genes encoding microsomal oleate desaturases (*FAD2*) from the oilseed crop *Camelina sativa*. *Plant Physiol. Biochem.* **49**, 223–229.

Li, T., Liu, B., Spalding, M.H., Weeks, D.P. and Yang, B. (2012) High-efficiency TALEN-based gene editing produces disease-resistant rice. *Nat. Biotechnol.* **30**, 390–392.

Liu, J., Song, H., Liu, D., Zuo, T., Lu, F., Zhuang, H. and Gao, F. (2014) Extensive recombination due to heteroduplexes generates large amounts of artificial gene fragments during PCR. *PLoS One*, **9**, e106658.

Lou, Y., Schwender, J. and Shanklin, J. (2014) FAD2 and FAD3 desaturases form heterodimers that facilitate metabolic channeling in vivo. *J. Biol. Chem.* **289**, 17996–18007.

Lu, C. and Kang, J. (2008) Generation of transgenic plants of a potential oilseed crop *Camelina sativa* by *Agrobacterium*-mediated transformation. *Plant Cell Rep.* **27**, 273–278.

Madlung, A. and Wendel, J.F. (2013) Genetic and epigenetic aspects of polyploid evolution in plants. *Cytogenet. Genome Res.* **140**, 270–285.

Mali, P., Yang, L., Esvelt, K.M., Aach, J., Guell, M., DiCarlo, J.E., Norville, J.E. *et al.* (2013) RNA-guided human genome engineering via Cas9. *Science*, **339**, 823–826.

Mao, Y., Zhang, Z., Feng, Z., Wei, P., Zhang, H., Botella, J.R. and Zhu, J.K. (2016) Development of germ-line-specific CRISPR-Cas9 systems to improve the production of heritable gene modifications in Arabidopsis. *Plant Biotechnol. J.* **14**, 519–532.

Nguyen, H.T., Silva, J.E., Podicheti, R., Macrander, J., Yang, W., Nazarenus, T.J., Nam, J.W. *et al.* (2013) Camelina seed transcriptome: a tool for meal and oil improvement and translational research. *Plant Biotechnol. J.* **11**, 759–769.

Petersen, B. and Niemann, H. (2015) Molecular scissors and their application in genetically modified farm animals. *Transgenic Res.* **24**, 381–396.

Petolino, J.F. and Kumar, S. (2015) Transgenic trait deployment using designed nucleases. *Plant Biotechnol. J.* **14**, 503–509.

Pham, A.T., Shannon, J.G. and Bilyeu, K.D. (2012) Combinations of mutant *FAD2* and *FAD3* genes to produce high oleic acid and low linolenic acid soybean oil. *Theor. Appl. Genet.* **125**, 503–515.

Pilgeram, A.L., Smith, D.C., Boss, D., Dale, N., Wichman, S., Lamb, P., Lu, C. *et al.* (2007) *Camelina sativa*: a Montana omega-3 and fuel crop. In *Issues in New Crops and New Uses* (Janik, J. and Whipkey, A., eds), pp. 129–131. Alexandra, VA: ASHS Press.

Proudfoot, C., Carlson, D.F., Huddart, R., Long, C.R., Pryor, J.H., King, T.J., Lillico, S.G. *et al.* (2015) Genome edited sheep and cattle. *Transgenic Res.* **24**, 147–153.

Singh, S., Thomaeus, S., Lee, M., Stymne, S. and Green, A. (2001) Transgenic expression of a delta 12-epoxygenase gene in Arabidopsis seeds inhibits accumulation of linoleic acid. *Planta*, **212**, 872–879.

Thambugala, D., Duguid, S., Loewen, E., Rowland, G., Booker, H., You, F.M. and Cloutier, S. (2013) Genetic variation of six desaturase genes in flax and their impact on fatty acid composition. *Theor. Appl. Genet.* **126**, 2627–2641.

Wang, Z.P., Xing, H.L., Dong, L., Zhang, H.Y., Han, C.Y., Wang, X.C. and Chen, Q.J. (2015) Egg cell-specific promoter-controlled CRISPR/Cas9 efficiently generates homozygous mutants for multiple target genes in Arabidopsis in a single generation. *Genome Biol.* **16**, 144–152.

Weeks, D.P., Spalding, M.H. and Yang, B. (2015) Use of designer nucleases for targeted gene and genome editing in plants. *Plant Biotechnol. J.* **14**, 483–495.

Wells, R., Trick, M., Soumpourou, E., Clissold, L., Morgan, C., Werner, P., Gibbard, C. *et al.* (2014) The control of seed oil polyunsaturate content in the polyploid crop species *Brassica napus*. *Mol. Breed.* **33**, 349–362.

Yan, L., Wei, S., Wu, Y., Hu, R., Li, H., Yang, W. and Xie, Q. (2015) High efficiency genome editing in Arabidopsis using Yao promoter-driven CRISPR/Cas9 system. *Mol. Plant*. **8**, 1820–1823.

Zubr, J. (1997) Oil seed crop: *Camelina sativa*. *Ind. Crop Prod.* **6**, 113–119.

Genome editing of the disease susceptibility gene *CsLOB1* in citrus confers resistance to citrus canker

Hongge Jia[1], Yunzeng Zhang[1], Vladimir Orbović[2], Jin Xu[1], Frank F. White[3], Jeffrey B. Jones[3] and Nian Wang[1,*]

[1]*Citrus Research and Education Center, Department of Microbiology and Cell Science, Institute of Food and Agricultural Sciences (IFAS), University of Florida, Lake Alfred, FL, USA*

[2]*Citrus Research and Education Center, IFAS, University of Florida, Lake Alfred, FL, USA*

[3]*Department of Plant Pathology, IFAS, University of Florida, Gainesville, FL, USA*

Correspondence
email nianwang@ufl.edu

Keywords: *Xanthomonas citri*, Cas9, sgRNA, *Citrus paradisi*.

Summary

Citrus is a highly valued tree crop worldwide, while, at the same time, citrus production faces many biotic challenges, including bacterial canker and Huanglongbing (HLB). Breeding for disease-resistant varieties is the most efficient and sustainable approach to control plant diseases. Traditional breeding of citrus varieties is challenging due to multiple limitations, including polyploidy, polyembryony, extended juvenility and long crossing cycles. Targeted genome editing technology has the potential to shorten varietal development for some traits, including disease resistance. Here, we used CRISPR/Cas9/sgRNA technology to modify the canker susceptibility gene *CsLOB1* in Duncan grapefruit. Six independent lines, $D_{LOB}2$, $D_{LOB}3$, $D_{LOB}9$, $D_{LOB}10$, $D_{LOB}11$ and $D_{LOB}12$, were generated. Targeted next-generation sequencing of the six lines showed the mutation rate was 31.58%, 23.80%, 89.36%, 88.79%, 46.91% and 51.12% for $D_{LOB}2$, $D_{LOB}3$, $D_{LOB}9$, $D_{LOB}10$, $D_{LOB}11$ and $D_{LOB}12$, respectively, of the cells in each line. $D_{LOB}2$ and $D_{LOB}3$ showed canker symptoms similar to wild-type grapefruit, when inoculated with the pathogen *Xanthomonas citri* subsp. citri (Xcc). No canker symptoms were observed on $D_{LOB}9$, $D_{LOB}10$, $D_{LOB}11$ and $D_{LOB}12$ at 4 days postinoculation (DPI) with Xcc. Pustules caused by Xcc were observed on $D_{LOB}9$, $D_{LOB}10$, $D_{LOB}11$ and $D_{LOB}12$ in later stages, which were much reduced compared to that on wild-type grapefruit. The pustules on $D_{LOB}9$ and $D_{LOB}10$ did not develop into typical canker symptoms. No side effects and off-target mutations were detected in the mutated plants. This study indicates that genome editing using CRISPR technology will provide a promising pathway to generate disease-resistant citrus varieties.

Introduction

Citrus varieties are high value tree crops with plantings in over one hundred countries. The fruit provides numerous benefits to human, including providing vitamins, fibre, calcium, potassium, folate and lowering health risks. Citrus production faces many biotic and abiotic challenges. Among them, the bacterial pathogens *Xanthomonas citri* ssp. citri (Xcc) and *Candidatus* Liberibacter asiaticus are the causal agents for citrus canker and HLB disease, respectively. Breeding disease-resistant varieties is the most efficient and sustainable approach to control plant diseases. However, traditional citrus breeding has often been hindered by polyembryony, pollen-ovule sterility, sexual and graft incompatibilities, and extended juvenility (Davey *et al.*, 2005). Various biotechnology methods have been used to develop modified and novel citrus varieties (Chen *et al.*, 2013; Dutt *et al.*, 2015; Fu *et al.*, 2011). However, no genetically modified varieties have been commercialized. The lack of commercial releases has been attributed to the lack of consumer acceptance of transgene technology. Recent developments in targeted genome editing technologies, however, have facilitated the process to establish genetically modified cultivars that lack transgenes in the final line (Doudna and Charpentier, 2014).

We previously identified *CsLOB1* as a critical citrus disease susceptibility gene for citrus canker (Hu *et al.*, 2014). *CsLOB1* is a member of the Lateral Organ Boundaries Domain (LBD) gene family of plant transcription factors. All strains of Xcc and a related pathogen *X. fuscans* subsp. aurantifolii (Xfa) encode transcription activator-like (TAL) effectors that recognize an effector binding element (EBE) in the promoter of *CsLOB1* and induce expression of the disease susceptibility gene (Hu *et al.*, 2014). Furthermore, the EBEs of individual critical TAL effectors in various canker causing strains overlap (Hu *et al.*, 2014). Thus, the EBE region of *CsLOB1* may be the Achilles' heel of citrus canker and presents an attractive target for genomic engineering of broad resistance to citrus canker. Previously, genome modifications of EBE regions of susceptibility genes *Os11N3*, *Os14N3* and *Os12N3* (also called *OsSWEET14*, *OsSWEET11* and *OsSWEET13*, respectively) have generated resistance in rice to bacterial blight, which is incited by *X. oryzae pv. oryzae* using a set of TAL effector genes related to the critical TAL effectors of Xcc and Xga (Blanvillain-Baufumé *et al.*, 2016; Li *et al.*, 2012; Zhou *et al.*, 2015). In our recent study, genome modification of the EBE of one single allele of *CsLOB1* gene in grapefruit Duncan (*Citrus paradisi* Macf.) alleviated the canker symptoms due to a specific TAL effector (Jia *et al.*, 2016). However, the modified grapefruit line, which is a hybrid, is still susceptible to wild-type Xcc as only one *CsLOB1* allele was altered, and mutation of the EBEs of both alleles of *CsLOB1* is required to generate reduced symptom plants (Jia *et al.*, 2016). Here, we reported our progress to generate canker-resistant citrus by disrupting the coding region of both alleles of CsLOB1 using Cas9/sgRNA.

Results

We first targeted the CsLOB1 coding region using Cas9/sgRNA in a transient assay on Duncan grapefruit (Citrus × paradisi), as grapefruit is one of the most canker susceptible citrus varieties. Grapefruit contains two alleles of CsLOB1, Type I and Type II (Jia et al., 2016) resulting from grapefruit being a hybrid of maternal donor pummelo (C. maxima) and paternal donor sweet orange (C. sinensis) (Velasco and Licciardello, 2014) (Figure 1). The two alleles of CsLOB1 showed polymorphisms at both nucleotide and protein levels. The sgRNA was selected to target a conserved region of the 1st exon in both alleles

```
Type I   ATTGTCATTCTTGCCTTTTCCTTTCTCTATATAAACCCCTTTTGCCTTGAACTTTGTTTC  60
         |||||||||||||||||||||||||||||||||||||||||||||||||| |||||||||||
Type II  ATTGTCATTCTTGCCTTTTCCTTTCTCTATATAAACCCCTTTTGCCTT-AACTTTGTTTC  59

Type I   AACTAAAGCAGCTCCTCCTCATCCCTTACTGTCTTTGCTTTCTCACTAACTACTACAACC  120
         |||||||||||||||||||||||||||||||||| ||||||||||||||||||||||||
Type II  AACTAAAGCAGCTCCTCCTCATCCCTTACTGTCTTCGCTTTCTCACTAACTACTACAACC  119

Type I   CAACAGTTTTCTTCTCTCAAAAATGGAATGCAAACACAAAATTAATGTAGCAATCCCAAT  180
         ||||||||||||||||||||||||||||||||| |||||||||||||||||||||||||||
Type II  CAACAGTTTTCTTCTCTCAAAAATGGAATGCAGACACAAAATTAATGTAGCAATCCCAAT  179

Type I   CACTAATATGAAGAACACTCAATTCTCATCTCCATCTACTTTCTCTACTTCTCCTCCTTC  240
         || |||||||||||| |||||||||||||||||||||||||||||||||||||||||||
Type II  CATTAATATGAAGAATACTCAATTCTCATCTCCATCTACTTTCTCTACTTCTCCTCCTTC  239

Type I   TCAATCTTCTCCACGCTTCCCTTCTCCTAATCATCAACAATTGTCTTCTCCAGAATCTTC  300
         |||||||||||| ||||| |||||||||||||||||||||||||||||||| |||||||
Type II  TCAATCTTCTCCATGCTTCCATTCTCCTAATCATCAACAATTGTCTTCTCCACAATCTTC  299

Type I   TCCAAGCTTTAAAGCTTCTCCTTCACAATCCTCTCCAAATCTTGCAGCTCCCCTCTCTCC  360
         ||||||||||||||||||||||||||||||||||||||||||||||| |||||||||||
Type II  TCCAAGCTTTAAAGCTTCTCCTTCACAATCCTCTCCAAATCTTGCAGATCCCCTCTCTCC  359

Type I   GCCGCCTATAGTTCTTAGCCCTTGTGCTGCTTGCAAAATCCTCCGCCGCAGATGCGTCGA  420
         ||||||||||||||||||||||||||||||| ||||||||||||||||||||||||||||
Type II  GCCGCCTATAGTTCTTAGCCCTTGTGCTGCCTGCAAAATCCTCCGCCGCAGATGCGTCGA  419

Type I   GAAATGTGTTTTAGCTCCATATTTTCCACCAACCGAACCATACAAGTTCACCATTGCTCA  480
         ||||||||||||||||||||||||||||||||||||||||||||||||||||||||||||
Type II  GAAATGTGTTTTAGCTCCATATTTTCCACCAACCGAACCATACAAGTTCACCATTGCTCA  479

Type I   TAGAGTCTTCGGTGCTAGCAATATCATCAAGTTCTTGCAGGTATGCACTTCTTTTGTATG  540
         ||| |||||||||||| |||||||||||||||||||||||||||||||||||||||||||
Type II  TAGGGTCTTCGGTGCAAGCAATATCATCAAGTTCTTGCAGGTATGCACTTCTTTTGTATG  539

Type I   TGATAAATTCAAACTAATTAAATGTCCAACCA-TTTTTTTTCTAATTGGGAGAAAAAAAA  599
         |||||||||||||||||||||||||||||||| |||||||||||||||||||||||||||
Type II  TGATAAATTCAAACTAATTAAATGTCCAACCATTTTTTTTTTCTAATTGGGAGAAAAAAAA  599

Type I   AACTTGTTAATTGTTTTATTTTCATCAATTAGTTGTGTGATTAGACTTTGGAGTGGTTGA  659
         ||||||||||||||||||||||||||||||||||||||||||||||||||||||||||||
Type II  AACTTGTTAATTGTTTTATTTTCATCAATTAGTTGTGTGATTAGACTTTGGAGTGGTTGA  659

Type I   TTGTTCCACTCTTTTTTTGGAAACTTACGGACTTCTCTAATCAAAAGAAAAGAGAGTGTGA  719
         ||||||||||||||||||||||||||||||||||||||||||||||||||||||||||||
Type II  TTGTTCCACTCTTTTTTTGGAAACTTACGGACTTCTCTAATCAAAAGAAAAGAGAGTGTGA  719

Type I   CATTTCAACTGA  731
         ||||||||||||
Type II  CATTTCAACTGA  731
```

Figure 1 Alignment of Type I CsLOB1 and Type II CsLOB1 in Duncan grapefruit. Two alleles of CsLOB1, Type I and Type II, are present in Duncan grapefruit. Part of the promoter regions and coding sequences are shown, in which the difference was indicated by purple, and the PthA4 effector binding elements were highlighted by blue. The intron was highlighted in grey. The translation start site was highlighted in green. The sgRNA-targeting region, which is conservative on both alleles, was highlighted in red. The primers were underlined, which were used to analyse indel mutation in genome-modified Duncan by targeted next-generation sequencing.

(Figures 1 and S1). To facilitate the screen process, a binary vector GFP-p1380N-Cas9/sgRNA:cslob1, which contains a GFP reporter gene, was constructed (Figure S1). Citrus plants transformed with GFP-p1380N-Cas9/sgRNA:cslob could be readily monitored with GFP fluorescence. First, Xcc-facilitated *Agrobacterium*-mediated infiltration (Jia and Wang, 2014) and transient expression in citrus leaves were used to test GFP-p1380N-Cas9/sgRNA:cslob1 function. Four days after infiltration, GFP fluorescence was observed at the inoculation site, whereas no GFP signal was observed at the site of infiltration with the control vector p1380-AtHSP70BP-GUSin (Jia and Wang, 2014) (Figure 2a). PCR amplification and sequencing confirmed the targeted modification of *CsLOB1* (Figure 2b and c). Therefore, GFP-p1380N-Cas9/sgRNA:cslob1 is functional for *CsLOB1* coding region targeting.

Via *Agrobacterium*-mediated transformation, Duncan grapefruit epicotyls were used as explants to create transgenic citrus plants (Orbović and Grosser, 2015). Six independent transgenic lines, $D_{LOB}2$, $D_{LOB}3$, $D_{LOB}9$, $D_{LOB}10$, $D_{LOB}11$ and $D_{LOB}12$, were selected based on GFP fluorescence and verified by PCR analyses (Figure 3a and b). To calculate the mutation frequency and determine the genotype for *CsLOB1* locus, targeted next-generation sequencing of the six transgenic lines was performed on amplified fragments using primers targeting a 380-bp region cover the targeted site. In total, more than 50 000 paired-end reads were generated for each sample, and, after filtering and quality trimming, the reads were grouped to clusters with a threshold of 100% pairwise identity using UCLUST (Edgar, 2010) (Table S1). Based on the sequencing results, the mutation rate was 31.58%, 23.80%, 89.36%, 88.79%, 46.91% and 51.12%

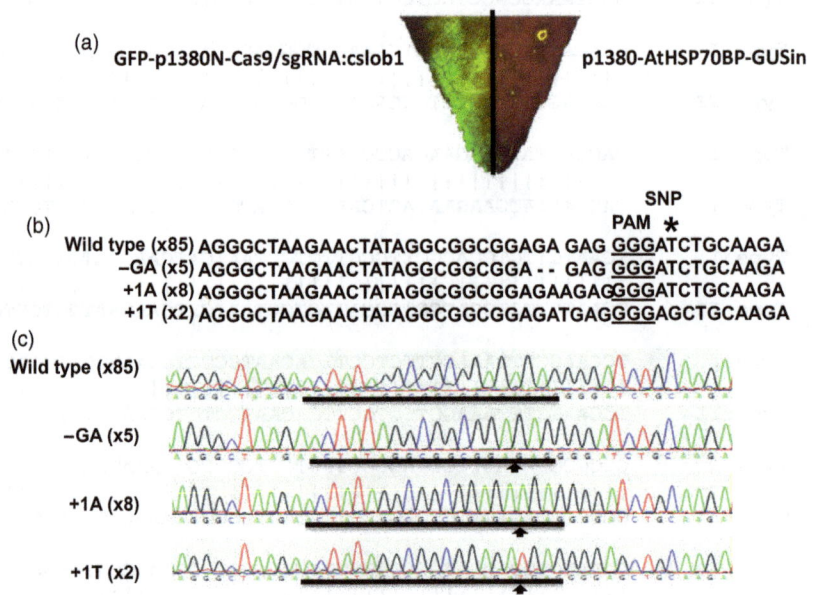

Figure 2 Function analysis of GFP-p1380N-Cas9/sgRNA:cslob1 in Duncan leaves with the aid of GFP. (a) Four days after agroinfiltration with *Agrobacterium* cells harbouring GFP-p1380N-Cas9/sgRNA:cslob1, GFP fluorescence was readily observed in Duncan grapefruit leaf. *Agrobacterium* cells harbouring p1380-AtHSP70BP-GUSin were used as a negative control. (b) GFP-p1380N-Cas9/sgRNA:cslob1-directed modification to *CsLOB1* coding region. The GFP-p1380N-Cas9/sgRNA:cslob1-targeted sequence in *CsLOB1* was shown in red, and the mutations were shown in purple. (c) The representative chromatograms of *CsLOB1* and its mutations. The targeted sequence within *CsLOB1* was underlined by black lines, and the mutant site was indicated with an arrow. Single nucleotide polymorphism (SNP) was indicated by an asterisk (*).

Figure 3 Analysis of GFP-p1380N-Cas9/sgRNA:cslob1-transformed Duncan grapefruit. (a) six GFP-p1380N-Cas9/sgRNA:cslob1-transformed Duncan grapefruit plants ($D_{LOB}2$, $D_{LOB}3$, $D_{LOB}9$, $D_{LOB}10$, $D_{LOB}11$ and $D_{LOB}12$) were GFP positive. The wild-type grapefruit plant did not show GFP. (b) The six transgenic lines contain Cas9/sgRNA as indicated by PCR amplification using primers 35SP-5-P1 and NosP-3-P2. Plasmid GFP-p1380N-Cas9/sgRNA:cslob1 was used as a positive control. M, 1 kb DNA ladder; WT, wild type. C. The six *CsLOB1*-modified lines showed differential resistance to Xcc. At 4 days postinoculation with Xcc (5×10^8 CFU/mL) using needleless syringe, canker symptoms were observed on normal grapefruit, $D_{LOB}2$ and $D_{LOB}3$, but absent or reduced on $D_{LOB}9$, $D_{LOB}10$, $D_{LOB}11$ and $D_{LOB}12$.

for $D_{LOB}2$, $D_{LOB}3$, $D_{LOB}9$, $D_{LOB}10$, $D_{LOB}11$ and $D_{LOB}12$, respectively (Figure 4a, Table S2). More than half of the mutations were 1-bp insertions of A or T, resulting in frame shift (Figure 4b, Table S2). The majority of deletions were short, ranging from 2 bp to 22 bps. Most of the 2-bp deletions were GA deletions (Figure 4b and Table S2). The 1-bp insertions took place at the 4th bp upstream of the PAM site (Figure 4b). The GA and GAGA deletions also occurred three base pairs upstream of the PAM site (Figure 4b), which is consistent with the previous report that Cas9 nuclease cleaves target DNA at a position three base pairs upstream of the PAM sequence (Jinek et al., 2012). Lines $D_{LOB}9$ and $D_{LOB}10$ were further confirmed using Sanger sequencing analysis (Figure S2).

The susceptibility of the six CsLOB1-modified Duncan grapefruit plants was tested by challenging with Xcc at the concentration of 5×10^8 CFU/mL. $D_{LOB}2$ and $D_{LOB}3$ showed canker symptoms similar to wild-type Duncan grapefruit. No canker symptoms were observed on $D_{LOB}9$, $D_{LOB}10$, $D_{LOB}11$ and $D_{LOB}12$ at 4 DPI (Figure 3c). Pustules caused by Xcc were observed on $D_{LOB}9$, $D_{LOB}10$, $D_{LOB}11$ and $D_{LOB}12$ in later stages, which were much reduced compared to that on wild-type grapefruit. The pustules on $D_{LOB}9$ and $D_{LOB}10$ did not develop into typical canker symptoms, whereas reduced canker symptoms were observed on $D_{LOB}11$ and $D_{LOB}12$ (Figure S3). The appearance of the pustules

on $D_{LOB}9$ and $D_{LOB}10$ might result from wild-type cells or mutants that could not abolish the CsLOB1 function. We did not observe any visible phenotype change for CsLOB1-edited grapefruit lines (Figure S4).

We analysed potential off-target mutagenesis. Seven potential off-target sequences were identified from genomic data (Table S3). No off-target mutations in amplified fragments encompassing the sites were identified in the six CsLOB1-modified plants (Table S3). Due to the somatic nature of the plants and only ten random colonies per putative off-target site were sequenced, the possibility of off-target changes in a fraction of the cells cannot be ruled out.

Discussion

This study tested the proof of concept that we can generate canker-resistant plants by modifying susceptibility gene CsLOB1. In this study, the mutation rate was 89.36% and 88.79% for $D_{LOB}9$ and $D_{LOB}10$, respectively, and both lines showed canker resistance (Figure 3c), even though pustules can be observed at later stage (Figure 3). The mutation rate for $D_{LOB}11$, and $D_{LOB}12$ was 46.91% and 51.12%, respectively. Both $D_{LOB}11$ and $D_{LOB}12$ showed enhanced resistance against Xcc with more pustules than $D_{LOB}9$ and $D_{LOB}10$ (Figures 3c, S3). On the other hand, the

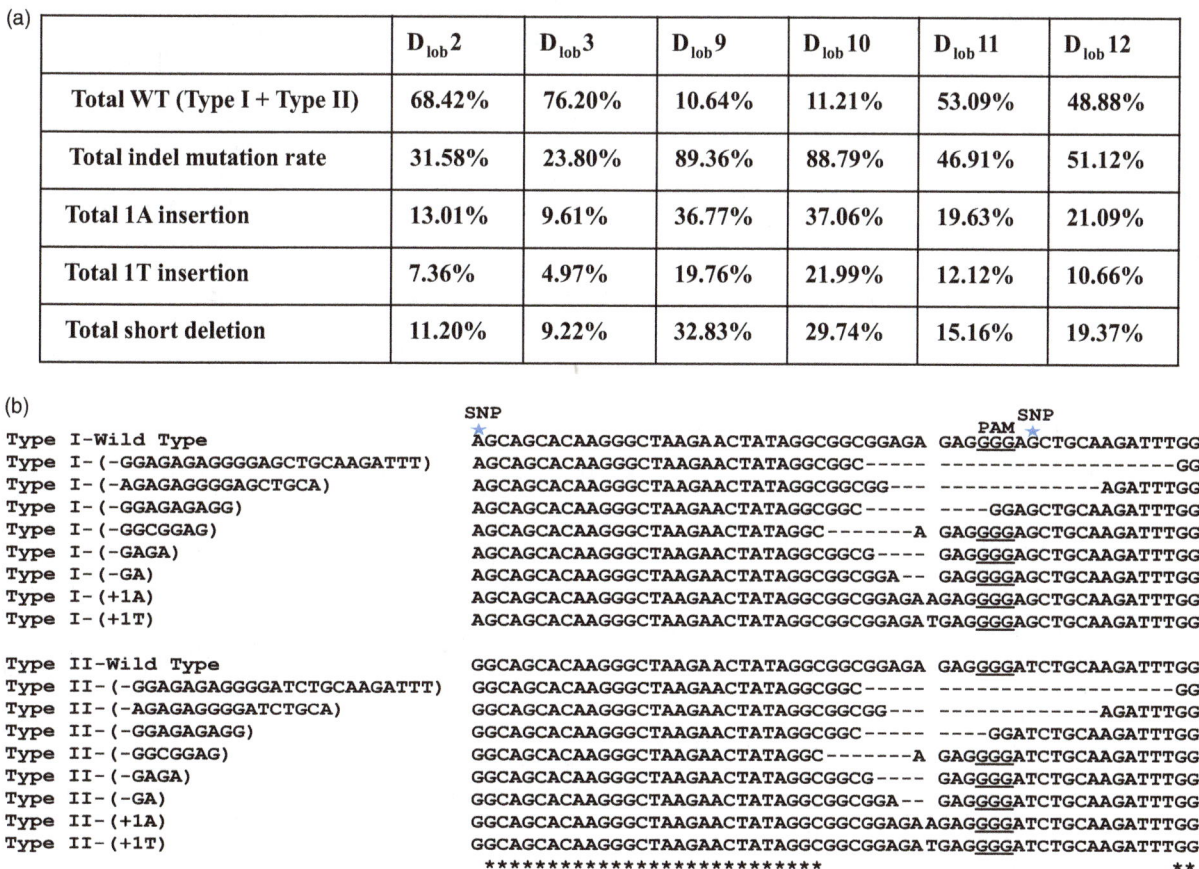

(a)

	$D_{lob}2$	$D_{lob}3$	$D_{lob}9$	$D_{lob}10$	$D_{lob}11$	$D_{lob}12$
Total WT (Type I + Type II)	68.42%	76.20%	10.64%	11.21%	53.09%	48.88%
Total indel mutation rate	31.58%	23.80%	89.36%	88.79%	46.91%	51.12%
Total 1A insertion	13.01%	9.61%	36.77%	37.06%	19.63%	21.09%
Total 1T insertion	7.36%	4.97%	19.76%	21.99%	12.12%	10.66%
Total short deletion	11.20%	9.22%	32.83%	29.74%	15.16%	19.37%

(b)

```
                                              SNP                                        SNP
                                                                                    PAM
Type I-Wild Type                     AGCAGCACAAGGGCTAAGAACTATAGGCGGCGGAGA GAGGGGAGCTGCAAGATTTGG
Type I-(-GGAGAGAGGGGAGCTGCAAGATTT)   AGCAGCACAAGGGCTAAGAACTATAGGCGGC----- ------------------GG
Type I-(-AGAGAGGGGAGCTGCA)           AGCAGCACAAGGGCTAAGAACTATAGGCGGCGG--- -------------AGATTTGG
Type I-(-GGAGAGAGG)                   AGCAGCACAAGGGCTAAGAACTATAGGCGGC----- ----GGAGCTGCAAGATTTGG
Type I-(-GGCGGAG)                     AGCAGCACAAGGGCTAAGAACTATAGGC-------A GAGGGGAGCTGCAAGATTTGG
Type I-(-GAGA)                        AGCAGCACAAGGGCTAAGAACTATAGGCGGCG---- GAGGGGAGCTGCAAGATTTGG
Type I-(-GA)                          AGCAGCACAAGGGCTAAGAACTATAGGCGGCGGA-- GAGGGGAGCTGCAAGATTTGG
Type I-(+1A)                          AGCAGCACAAGGGCTAAGAACTATAGGCGGCGGAGA AGAGGGGAGCTGCAAGATTTGG
Type I-(+1T)                          AGCAGCACAAGGGCTAAGAACTATAGGCGGCGGAGA TGAGGGGAGCTGCAAGATTTGG

Type II-Wild Type                    GGCAGCACAAGGGCTAAGAACTATAGGCGGCGGAGA GAGGGGATCTGCAAGATTTGG
Type II-(-GGAGAGAGGGGATCTGCAAGATTT)  GGCAGCACAAGGGCTAAGAACTATAGGCGGC----- ------------------GG
Type II-(-AGAGAGGGGATCTGCA)          GGCAGCACAAGGGCTAAGAACTATAGGCGGCGG--- -------------AGATTTGG
Type II-(-GGAGAGAGG)                  GGCAGCACAAGGGCTAAGAACTATAGGCGGC----- ----GGATCTGCAAGATTTGG
Type II-(-GGCGGAG)                    GGCAGCACAAGGGCTAAGAACTATAGGC-------A GAGGGGATCTGCAAGATTTGG
Type II-(-GAGA)                       GGCAGCACAAGGGCTAAGAACTATAGGCGGCG---- GAGGGGATCTGCAAGATTTGG
Type II-(-GA)                         GGCAGCACAAGGGCTAAGAACTATAGGCGGCGGA-- GAGGGGATCTGCAAGATTTGG
Type II-(+1A)                         GGCAGCACAAGGGCTAAGAACTATAGGCGGCGGAGA AGAGGGGATCTGCAAGATTTGG
Type II-(+1T)                         GGCAGCACAAGGGCTAAGAACTATAGGCGGCGGAGA TGAGGGGATCTGCAAGATTTGG
                                     *************************           **
```

Figure 4 Indel mutation rates and mutation genotypes in six CsLOB1-edited grapefruit lines. (a) Mutation rate for each CsLOB1-edited grapefruit line. Targeted next-generation sequencing was conducted for each line, and more than 50 000 paired-end reads were generated for each sample. (b) Representative indel mutation genotypes in Type I CsLOB1 plus Type II CsLOB1. The mutations included 1-bp insertion and short deletions. It should be noted that the AGAGAGGGGA(G/T)CTGCA deletion and GGAGAGAGGGGA(G/T)CTGCAAGATT deletion removed the PAM and the SNP nucleotide. Star indicates SNP (single nucleotide polymorphism) used for differentiating type I and type II alleles of CsLOB1.

mutation rate for $D_{LOB}2$ and $D_{LOB}3$ was 31.58% and 23.80%, respectively. Neither $D_{LOB}2$ nor $D_{LOB}3$ showed resistance to citrus canker (Figures 3c, S3). The appearance of the pustules on $D_{LOB}9$, $D_{LOB}10$, $D_{LOB}11$ and $D_{LOB}12$ at 7 DPI might result from wild-type cells or mutants that could not abolish the *CsLOB1* function. This recessive resistance due to mutation of *CsLOB1* is expected to be durable and efficient against all Xanthomonas pathotypes causing citrus canker because they all rely on induction of the susceptibility gene *CsLOB1* to induce canker symptoms. This is consistent with previous studies that mutation of the coding region of susceptibility gene will lead to disease resistance. Mutation of the susceptibility gene *OsSWEET13* corresponding to PthXo2 of *X. oryzae* pv. oryzae has generated disease-resistant rice (Zhou et al., 2015). We need to point out that the genome-modified citrus lines are not suitable for application to control citrus canker at this moment as they still contain Cas9 and sgRNA in the genome, thus are considered as transgenic, and require rigorous registration process before allowed for commercialization. Recently, USDA has granted nonregulatory status for the genome modified common white button mushroom (*Agaricus bisporus*) resisting against browning (Waltz, 2016b) and high amylopectin corn generated by knocking out the endogenous waxy gene Wx1 (Waltz, 2016a) because both do not contain foreign DNAs and are without off-target mutations. Consequently, we need to generate nontransgenic canker-resistant citrus varieties to facilitate the de-regulation.

No phenotypic changes were observed for the *CsLOB1*-modified plants. CsLOB1 belongs to the LBD proteins which are transcription factors in the regulation of plant growth and development (Husbands et al., 2007). The biological function of CsLOB1 remains to be determined. RNA-Seq analysis of the expression profiles associated with CsLOB1 identified many downstream genes, for example cell organization, cell division, cell cycle, cell wall degradation and cell wall modification (Zhang et al., 2016). *PtaLBD1* derived from populus is a homolog of *CsLOB1*. It was reported that *PtaLBD1* was involved in secondary woody growth in poplar (Yordanov et al., 2010). The expression of *PtaLBD1-SRDX*, harnessed for dominant-negative suppression of PtaLBD1, suppressed stem diameter growth. No phenotypic changes observed for the *CsLOB1*-modified plants are probably due to the fact that citrus contains multiple LOB genes with similar functions, for example *CsLOB1*, *CsLOB2* and *CsLOB3*. Induction of *CsLOB2* and *CsLOB3* using custom-designed TAL effectors leads to similar canker symptoms due to induction of *CsLOB1* by PthA4 (Zhang et al., 2016), indicating similar functions of *CsLOB1*, *CsLOB2* and *CsLOB3*. Thus, redundancy of *CsLOB1* might be the main reason for the lack of phenotypic effect of the mutation. No off-target mutations were observed in the genome-modified plants, which might also contribute to the lack of phenotypic effect of the mutation. We could not totally rule out other phenotypic effects, for example flowering, as the genome-modified lines will take 2 to 3 more years to flower.

In summary, we have shown that mutation of the coding region of both alleles of the susceptibility gene *CsLOB1* can generate citrus canker-resistant plants. Future work needs to focus on generating *CsLOB1*-modified citrus varieties which do not contain foreign DNA and comprise no off-target mutations for application purpose. Importantly, this study showed that we can generate disease-resistant citrus varieties using CRISPR technology and provide a long-term and efficient control measurement for other citrus diseases including HLB.

Materials and methods

Plasmid construction

The CaMV 35S promoter was amplified using primers CaMV35-5-*Xho*I (5′-A<u>CTCGAG</u>ACTAGTACCATGGTGGACTCCTCTTAA-3′) and sgRNA-cslob1-P1 (5′-phosphorylated-TATAGTCCTCTCCAA ATGAAATGAACTTC-3′), and the sgRNA-NosT fragment was amplified using primers sgRNA-cslob1-P2 (5′-phosphorylated-GGCGGCGGAGAGAGGTTTTAGAGCTAGAAATAGCAA-3′) and NosT-3-*Asc*I (5′-ACCTGGGCCC<u>GGCGCGCC</u>GATCTAGTAACATA GATGA-3′). Through three-way ligation, *Xho*I-digested CaMV35S and *Asc*I-cut sgRNA-NosT were inserted into *Xho*I-*Asc*I-treated p1380N-Cas9 to form p1380N-Cas9/sgRNA:cslob1. The p1380N-Cas9 was described previously (Jia and Wang, 2014).

Using a pair of primers 35T-P1 (5′-AGGT<u>GGATCC</u>GAGCTC GAAAATTTCTCCATAATAAT.

GTGTGAGT -3′) and 35T-P2 (5′-AGGT<u>ATTAAT</u>AAGCTTCGGG GGATCTGGATTTTAGTA CT-3′), the CaMV 35S terminator was amplified and cloned into *Bam*HI-*Ase*I-digested p1380N-Cas9 to produce p1380-35S-35T. The cassava vein mosaic virus promoter (CsVMV) was amplified using primers CsVMV-5-*Spe*I (5′-AGGT <u>ACTAGT</u>AAGCTTGCATGCCCGCGCC AGAAGGTAATTATCCAA G-3′) and CsVMV-3-*Sal*I (5′-AGGT<u>GTCGAC</u>AAACTTACAAA TTTC TCTGAAG-3′) from plasmid AtSUC2-NPR1 (Dutt *et al.*, 2015), and the GFP fragment was amplified using primers GFP-5-*Xho*I (5′-AGGT<u>CTCGAG</u>ATGAAGACTAATCTTT TTCTCT-3′) and GFP2 (5′-TC<u>GAGCTC</u>TTAAAGCTCATCATGTTTGTAT-3′) from p1380-35S-GFP (Jia and Wang, 2014). Through three-way ligation, *Spe*I-CsVMV-*Sal*I fragment and *Xho*I-GFP-*Sac*I were inserted into *Spe*I-*Sac*I-treated p1380-35S-35T to form p1380-CsVMV-GFP-35T. After digestion with *Hind*III, the *Hind*III-CsVMV-GFP-35T-*Hind*III fragment from p1380-CsVMV-GFP-35T was cloned into p1380N-Cas9/sgRNA:cslob1 to obtain GFP-p1380N-Cas9/sgRNA:cslob1.

The binary vector GFP-p1380N-Cas9/sgRNA:cslob1 was introduced into *A. tumefaciens* strain EHA105 competent cells by the freeze–thaw method. Recombinant *Agrobacterium* cells were employed for citrus transformation or Xcc-facilitated agroinfiltration.

Duncan *CsLOB1* sequencing and analysis

Using a Wizard Genomic DNA Purification Kit (Promega), genomic DNA was extracted from wild-type Duncan, or transgenic plants, or the GFP-positive Duncan leaves treated by *Xanthomonas citri* ssp. citri (Xcc)-facilitated agroinfiltration of GFP-p1380N-Cas9/sgRNA:cslob1 (Figure 3a). To analyse *CsLOB1* gene in detail, PCR was performed with the Phusion DNA polymerase (New England Biolabs) and a pair of primers, CsLBDP-5-P1 (5′-ATTGTCATTCTTGCCTTTTCCTTTCT-3′) and CsLOB1-3-P2 (5′-TCAGTTGAAATGTCACACTCTCTT-3′), flanking part of *CsLOB1* promoter and its coding region. By blunt end cloning, the PCR products were inserted into the PCR-BluntII-TOPO vector (Life Technologies). The colonies were randomly selected for DNA sequencing, and the results were visualized by Chromas Lite program.

For PCR product direct sequencing, CsLBDP-5-P1 and CsLOB1-3-P2 were used to amplify the DNA fragments from genomic DNA. The PCR products were purified and subjected to direct sequencing using primer CsLOB1-P2 (5′-TGAGCAATGGTGAACT TGTATGGTTC-3′). The results were analysed by Chromas Lite program.

Xcc-facilitated agroinfiltration in Duncan grapefruit

Duncan grapefruit (Citrus paradisi) was grown in a glasshouse at temperatures ranging from 25 to 30 °C. Before Xcc-facilitated agroinfiltration was carried out, the plants were pruned for uniform shoot establishment.

The detailed protocol for Xcc-facilitated agroinfiltration in citrus leaves was described previously (Jia and Wang, 2014), with minor modification. Briefly, Duncan leaves were inoculated with a culture of actively growing XccΔgumC re-suspended in sterile tap water (5 × 10^8 CFU/mL). Twenty-four hours later, the XccΔgumC-treated leaf areas were agroinfiltrated with recombinant Agrobacterium cells harbouring GFP-p1380N-Cas9/sgRNA: cslob1 or p1380-AtHSP70BP-GUSin (Jia and Wang, 2014). Four days after agroinfiltration, leaves were subjected to GFP observation or genomic DNA extraction.

GFP detection

Four days after Xcc-facilitated agroinfiltration with GFP-p1380N-Cas9/sgRNA:cslob1 or p1380-AtHSP70BP-GUSin, GFP fluorescence in the treated leaves was visualized under illumination of an EBQ 100 isolated light source using a Zeiss Stemi SV11 dissecting microscope equipped with an Omax camera. The leaf was photographed using the Omax Toupview software.

Agrobacterium-mediated Duncan grapefruit transformation

Citrus transformation was performed as reported before (Orbović and Grosser, 2015). In detail, about 2923 Duncan epicotyl explants were co-incubated with recombinant Agrobacterium cells harbouring binary vector GFP-p1380N-Cas9/sgRNA:cslob1. Five weeks later, about 839 shoots sprouted from these explants after co-incubation. All explants were inspected for the presence of GFP fluorescence. In the initial screen, 15 shoots were designated as positive and micro-grafted on 'Carrizo' citrange rootstock plants [Citrus sinensis (L.) Osbeck × Poncirus trifoliata (L.) Raf.]. Out of these shoots, seven died upon grafting in in vitro conditions before they were transferred to pots. Additional two plants were discarded based on unsatisfactory level of GFP fluorescence detected in their tissue during secondary inspection. The six remaining GFP-positive plants were used for further analysis.

The GFP-p1380N-Cas9/sgRNA:cslob1-transformed plants were subjected to PCR analysis with a pair of primers, 35SP-5-P1 (5′-ATCAAAGGCCATGGAGTCAAA-3′) and NosP-3-P2 (5′-TTGTCGTTTCCCGCCTTCAGT-3′).

Next-generation sequencing analysis

Genomic DNA from six transgenic plants was used as template for PCR amplification using a pair of primers, CsLOB1-P1 (5′-TCTCACTAACTACTACAACCCAACAG-3′) and CsLOB1-P2 (Figure 1). All PCR products were pooled to construct the DNA library for sequencing using an Illumina HiSeq 2500 platform at Novogene (Beijing, China). For each sample, more than 50 000 paired-end reads were generated. After de-multiplex, barcode and primer deletion using custom Perl script, the raw reads were quality trimmed using sickle software with parameters average quality 30 and reads length threshold 200 bp (Fass et al., 2011). The remaining high-quality reads were clustered with a threshold of 100% pairwise identity using UCLUST (Edgar, 2010). The representative sequences from abundant clusters with relative abundance >1% were aligned using MEGA 6 (Tamura et al., 2013) and further analysed for mutation genotype.

Xcc infection assay

Wild-type Duncan grapefruit and CsLOB1-modified grapefruit lines were grown in a glasshouse. The same age leaves were inoculated with Xcc (5 × 10^8 CFU/mL) using a needleless syringe. After inoculation, citrus canker formation was observed and photographed at different time points.

Analysis of potential off-targets

To analyse potential off-targets of GFP-p1380N-Cas9/sgRNA: cslob1 in CsLOB1-modified grapefruit lines, we analysed the putative off-targets using a web-based software (http://cbi.hza u.edu.cn/cgi-bin/CRISPR). Genomic DNA from CsLOB1-modified grapefruit lines was used as template, and the primers listed in Table S3 were used to amplify the fragment containing the off-targets. Finally, the PCR products were ligated with PCR-BluntII-TOPO vector for sequencing analysis.

Acknowledgements

This study has been supported by Florida Citrus Research and Development Foundation. We thank Dr. Manjul Dutt for plasmid AtSUC2-NPR1.

References

Blanvillain-Baufumé, S., Reschke, M., Solé, M., Auguy, F., Doucoure, H., Szurek, B., Meynard, D. et al. (2016) Targeted promoter editing for rice resistance to Xanthomonas oryzae pv. oryzae reveals differential activities for SWEET14-inducing TAL effectors. Plant Biotechnol. doi:10.1111/pbi.12613.

Chen, X., Barnaby, J., Sreedharan, A., Huang, X., Orbovic, V., Grosser, J., Wang, N. et al. (2013) Over-expression of the citrus gene CtNH1 confers resistance to bacterial canker disease. Physiol. Mol. Plant Pathol. 84, 115–122.

Davey, M.R., Anthony, P., Power, J.B. and Lowe, K.C. (2005) Plant protoplasts: status and biotechnological perspectives. Biotechnol. Adv. 23, 131–171.

Doudna, J.A. and Charpentier, E. (2014) Genome editing. The new frontier of genome engineering with CRISPR-Cas9. Science, 346, 1258096.

Dutt, M., Barthe, G., Irey, M. and Grosser, J. (2015) Transgenic citrus expressing an Arabidopsis NPR1 gene exhibit enhanced resistance against Huanglongbing (HLB; Citrus Greening). PLoS ONE, 10, e0137134.

Edgar, R.C. (2010) Search and clustering orders of magnitude faster than BLAST. Bioinformatics, 26, 2460–2461.

Fass, J.N., Joshi, N.A., Couvillion, M.T., Bowen, J., Gorovsky, M.A., Hamilton, E.P., Orias, E. et al. (2011) Genome-scale analysis of programmed DNA elimination sites in Tetrahymena thermophila. G3: Genes – Genomes – Genet. 1, 515–522.

Fu, X.Z., Chen, C.W., Wang, Y., Liu, J.H. and Moriguchi, T. (2011) Ectopic expression of MdSPDS1 in sweet orange (Citrus sinensis Osbeck) reduces canker susceptibility: involvement of H$_2$O$_2$ production and transcriptional alteration. BMC Plant Biol. 11, 55.

Hu, Y., Zhang, J., Jia, H., Sosso, D., Li, T., Frommer, W.B., Yang, B. et al. (2014) Lateral organ boundaries 1 is a disease susceptibility gene for citrus bacterial canker disease. Proc. Natl. Acad. Sci. USA, 111, E521–E529.

Husbands, A., Bell, E.M., Shuai, B., Smith, H.M. and Springer, P.S. (2007) LATERAL ORGAN BOUNDARIES defines a new family of DNA-binding transcription factors and can interact with specific bHLH proteins. Nucleic Acids Res. 35, 6663–6671.

Jia, H. and Wang, N. (2014) Xcc-facilitated agroinfiltration of citrus leaves: a tool for rapid functional analysis of transgenes in citrus leaves. Plant Cell Rep. 33, 1993–2001.

Jia, H., Orbovic, V., Jones, J.B. and Wang, N. (2016) Modification of the PthA4 effector binding elements in Type I CsLOB1 promoter using Cas9/sgRNA to produce transgenic Duncan grapefruit alleviating XccΔpthA4:dCsLOB1.3 infection. Plant Biotechnol. J. 14, 1291–1301.

Jinek, M., Chylinski, K., Fonfara, I., Hauer, M., Doudna, J.A. and Charpentier, E. (2012) A programmable dual-RNA-guided DNA endonuclease in adaptive bacterial immunity. Science, 337, 816–821.

Li, T., Liu, B., Spalding, M.H., Weeks, D.P. and Yang, B. (2012) High-efficiency TALEN-based gene editing produces disease-resistant rice. *Nat. Biotechnol.* **30**, 390–392.

Orbović, V. and Grosser, J.W. (2015) Citrus transformation using juvenile tissue explants. *Methods Mol. Biol.* **1224**, 245–257.

Tamura, K., Stecher, G., Peterson, D., Filipski, A. and Kumar, S. (2013) MEGA6: molecular evolutionary genetics analysis version 6.0. *Mol. Biol. Evol.* **30**, 2725–2729.

Velasco, R. and Licciardello, C. (2014) A genealogy of the citrus family. *Nat. Biotechnol.* **32**, 640–642.

Waltz, E. (2016a) CRISPR-edited crops free to enter market, skip regulation. *Nat. Biotechnol.* **34**, 582.

Waltz, E. (2016b) Gene-edited CRISPR mushroom escapes US regulation. *Nature*, **532**, 293.

Yordanov, Y.S., Regan, S. and Busov, V. (2010) Members of the LATERAL ORGAN BOUNDARIES DOMAIN transcription factor family are involved in the regulation of secondary growth in Populus. *Plant Cell*, **22**, 3662–3677.

Zhang, J., Carlos Huguet Tapia, J., Hu, Y., Jones, J., Wang, N., Liu, S. and White, F.F. (2016) Homologs of *CsLOB1* in citrus function as disease susceptibility genes in citrus canker. *Mol. Plant Pathol.* doi: 10.1111/mpp.12441.

Zhou, J., Peng, Z., Long, J., Sosso, D., Liu, B., Eom, J.S., Huang, S. *et al.* (2015) Gene targeting by the TAL effector PthXo2 reveals cryptic resistance gene for bacterial blight of rice. *Plant J.* **82**, 632–643.

16

Overexpression of a tomato miR171 target gene *SlGRAS24* impacts multiple agronomical traits via regulating gibberellin and auxin homeostasis

Wei Huang, Shiyuan Peng, Zhiqiang Xian, Dongbo Lin, Guojian Hu, Lu Yang, Maozhi Ren and Zhengguo Li*

Genetic Engineering Research Center, School of Life Sciences, Chongqing University, Chongqing, China

*Correspondence
email zhengguoli@cqu.edu.cn

Summary

In *Arabidopsis*, the miR171-GRAS module has been clarified as key player in meristem maintenance. However, the knowledge about its role in fruit crops like tomato (*Solanum lycopersicum*) remains scarce. We previously identified tomato *SlGRAS24* as a target gene of *Sly-miR171*. To study the role of this probable transcription factor, we generated transgenic tomato plants underexpressing *SlGRAS24*, overexpressing *SlGRAS24*, overexpressing *Sly-miR171* and expressing β-glucuronidase (GUS) under the *SlGRAS24* promoter (pro*SlGRAS24*-GUS). Plants overexpressing *SlGRAS24* (*SlGRAS24*-OE) had pleiotropic phenotypes associated with multiple agronomical traits including plant height, flowering time, leaf architecture, lateral branch number, root length, fruit set and development. Many GA/auxin-related genes were down-regulated and altered responsiveness to exogenous IAA/NAA or GA_3 application was observed in *SlGRAS24*-OE seedlings. Moreover, compromised fruit set and development in *SlGRAS24*-OE was also observed. These newly identified phenotypes for *SlGRAS24* homologs in tomato were later proved to be caused by impaired pollen sacs and fewer viable pollen grains. At anthesis, the comparative transcriptome results showed altered expression of genes involved in pollen development and hormone signalling. Taken together, our data demonstrate that *SlGRAS24* participates in a series of developmental processes through modulating gibberellin and auxin signalling, which sheds new light on the involvement of hormone crosstalk in tomato development.

Keywords: Auxin, gibberellin, miR171, SlGRAS24, tomato (*Solanum lycopersicum*).

Introduction

The GRAS family of plant proteins is responsible for regulating many aspects of growth, development and responses to the biotic and abiotic environment. GRAS family members are diverse proteins that typically have five conserved motifs in the C-terminus (Bolle, 2004). GRAS proteins usually act as transcription factors, but not all GRAS protein functions have been described. We previously identified a tomato (*Solanum lycopersicum*) GRAS transcription factor gene *SlGRAS24* as target of tomato miR171 (Huang *et al.*, 2015) and aimed to discover more about its function in tomato development here.

SlGRAS24 is phylogenetically clustered into the HAIRY MERISTEM (HAM) subfamily of GRAS genes and shows the highest sequence identity with *Arabidopsis thaliana* AtSCL6 (Huang *et al.*, 2015). In 2002, a GRAS transcription factor named HAM was identified as a component of a novel non-cell-autonomous signalling pathway maintaining shoot indeterminacy in *Petunia hybrida*, *ham* mutants displayed arrest in lateral organ and stem production (Stuurman *et al.*, 2002). In the same year, two *Arabidopsis* orthologs of *Petunia HAM* were proved to be endogenous targets of post-transcriptional degradation by *miR171*, a member of a miRNA family conserved in different plant species (Llave *et al.*, 2002). Actually, a total of three *GRAS* genes are regulated by *miR171* in *Arabidopsis*, *SCL6/SCL6-IV*, *SCL22/SCL6-III* and *SCL27/SCL6-II* (also known as the *HAM* or *LOM* (*LOST MERISTEMS*) genes because of their mutant

phenotypes) (Reinhart *et al.*, 2002). *LOM1* and *LOM2* genes promote incorporation of peripheral zone cells into leaf primordia and help to maintain a polar organization of the shoot meristem (Schulze *et al.*, 2010). Further research showed that these miR171 target genes were not only required for shoot apical meristem maintenance, but for maintenance of root indeterminacy (Engstrom *et al.*, 2011). More recently, it is found that HAM proteins act as conserved interacting cofactors with WUS/WOX proteins. They share common targets and their physical interaction is important in driving downstream transcriptional programmes and in promoting shoot stem cell proliferation (Zhou *et al.*, 2015). *Arabidopsis* overexpressing *miR171* and the triple *scl6* mutants have similar pleiotropic phenotypes, where shoot branching, plant height, chlorophyll accumulation, primary root elongation, flower structure, and leaf shape and patterning were all altered (Wang *et al.*, 2010). In barley and rice, overexpression of *miR171* affects phase transitions and floral meristem determinacy (Curaba *et al.*, 2013; Fan *et al.*, 2015). miR171-GRAS module controls flowering time (phase transition) and trichome distribution via inhibiting the activity of miR156-targeted SPL proteins (Xue *et al.*, 2014). This module is also critical for mediating GA-DELLA signalling in the coordinate regulation of chlorophyll biosynthesis and leaf growth in light (Ma *et al.*, 2014). Moreover, it has been extensively studied about the role of miR171 upon various stresses in different species, including *Arabidopsis*, barley, maize and *Solanum tuberosum* (Hwang *et al.*, 2011; Kantar *et al.*, 2010; Kong *et al.*, 2010; Liu *et al.*, 2008).

HAM gene function may be conserved but the dramatic expansion in HAM homologs diversity in flowering plants strongly suggests the evolution of novel functions or functional subspecialization in angiosperms (Wu *et al.*, 2014). Indeed, elevated rates of evolution in flowering plant HAM homologs indicate a refinement of HAM function in response to selective pressures (Engstrom *et al.*, 2011). In tomato, six *SlGRAS* genes are clustered into the HAM subfamily, including *SlGRAS24* and *SlGRAS40*, which are confirmed to be targeted for mRNA cleavage by *miR171*, and *SlGRAS8*, a suspected target gene whose translation is repressed by *miR171* (Huang *et al.*, 2015). *SlGRAS24* contains a conserved MIR-binding sequence which is perfectly matched with *Sly-miR171* (Huang *et al.*, 2015). Despite the close evolutionary relationship between *SlGRAS24* and *SlGRAS40*, their transcripts in different tissues/organs and in response to hormone and abiotic stress differ greatly (Huang *et al.*, 2015), indicating that they might have different functions and participate in distinct physiological processes.

In this study, the tomato GRAS transcription factor gene *SlGRAS24*, an ortholog of *Arabidopsis AtSCL6* (Huang *et al.*, 2015), was characterized. First, the expression pattern of the gene was studied in wild-type (WT) and transgenic plants expressing β-glucuronidase (GUS) under the *SlGRAS24* promoter (pro*SlGRAS24-GUS*). Transgenic tomato plants overexpressing *SlGRAS24* exhibited phenotypes similar to those observed in other species suggesting functional conservation among these homologs. Some new characteristics such as abnormal axillary bud emergence, reduced fruit set ratio, arrested fruit and seed development were also observed, indicating that *SlGRAS24* has additional specific functions in tomato. Moreover, the promoter of *SlGRAS24* was analysed and found to be associated with both gibberellin and auxin signalling pathways. Hormone-related transcripts were thus quantified and the floral transcriptome was analysed in plants overexpressing *SlGRAS24*. Collectively, our study of SlGRAS24 protein unravels its role in vegetative growth and reproductive tissues and advances our understanding about hormone crosstalk in tomato.

Results

Sly-miR171 and SlGRAS24 are ubiquitously but differentially expressed in tomato

We hypothesized that the *SlGRAS24* gene and its regulator *Sly-miR171* would have similar functions in tomato as their respective orthologs in *Arabidopsis* (Wang *et al.*, 2010). We investigated this first by quantifying the expression of *Sly-miR171* and *SlGRAS24* in the 'Micro-Tom' cultivar (WT) by qRT-PCR. Both genes were detectable in all tissues with the highest expression levels found in flowers (Fig. 1A). The levels of both transcripts dropped during fruit development and the lowest level of expression was at the ripening stage. Overall, *Sly-miR171* and *SlGRAS24* mRNAs had similar transcription patterns, which were consistent with research on their *Arabidopsis* orthologs (Wang *et al.*, 2010). However, floral organs were an exception because *SlGRAS24* mRNA was most abundant in stamens where *Sly-miR171* mRNA expression was at its lowest level. The expression data here suggest that miR171-SlGRAS24 regulatory networks are needed throughout vegetative and reproductive development in tomato.

Spatial expression of *SlGRAS24* was monitored by histochemical staining of transgenic tomato in which GUS reporter's expression was driven by the upstream promoter sequence of *SlGRAS24* gene (Fig. 1B). In pro*GRAS24-GUS* homozygous seedlings, GUS reporter activity was strong in leaf primordia and root tips, but quite weak in cotyledons and hypocotyls. GUS activity was also expressed in flower primordia, in the young leaves' margins of shoot apices, in internodes and in axillary buds. In reproductive tissues, *SlGRAS24* was highly expressed in stamens and stigmas, and predominantly expressed in seeds of young fruits. These results suggest that *SlGRAS24* expression is spatiotemporally regulated and *SlGRAS24* may have specific functions in developmental processes.

SlGRAS24 is a transcription factor targeted to the nucleus

To determine the subcellular localization of SlGRAS24 protein, the vector 35S-*SlGRAS24-GFP* was transiently expressed in tobacco protoplasts. Confocal imaging of protein fluorescence showed that the green fluorescence signal of 35S-*SlGRAS24-GFP* was exclusively detected in the nucleus, whereas the cells transformed with the vector containing *GFP* alone displayed fluorescence throughout the cells (Fig. 1C). A yeast two-hybrid experiment was used to examine the transcriptional activity of SlGRAS24. A GAL4 DNA-binding domain SlGRAS24 fusion protein was expressed in yeast cells, which were then assayed for their ability to activate transcription from the GAL4 sequence. SlGRAS24 promoted yeast growth in the absence of histidine and adenine, and showed X-α-gal activity, whereas the control vector pGBKT7 did not (Fig. 1D). These results suggest that SlGRAS24 has transcriptional activity and is targeted to the nucleus in plant cells.

Phenotypic characterization of SlGRAS24-OE lines

To assess the physiological importance of the *SlGRAS24*-encoded protein, the tomato 'Micro-Tom' genotype was transformed with sense or antisense constructs of the *SlGRAS24* gene to produce several independent overexpressing (*SlGRAS24*-OE) or underexpressing (*SlGRAS24*-AS) homozygous lines. qRT-PCR was performed to evaluate the expression of *SlGRAS24* in transgenic plants, and the results showed that the *SlGRAS24* was successfully up-regulated or down-regulated in all transgenic lines tested (Table 1). Thereby, two most up- or down-regulated lines of each genotype were then chosen for the following characterization. Interestingly, *SlGAR24*-AS lines did not differ from WT tomato (Fig. 2A and B). This was not unexpected as no phenotype had been observed in *Arabidopsis* plants in which only one of the HAM genes was mutated (Wang *et al.*, 2010). We hypothesized that there is functional redundancy among GRAS family members or multiple *miR171* target genes in tomato. We tested this by generating transgenic plants overexpressing the precursor of tomato *miR171* to silence *miR171* target genes including *SlGRAS24* (Table 1). *SlGRAS24* expression was inhibited to a similar degree in *SlGAR24*-AS and *SlymiR171*-OE plants comparing to WT, but only *SlymiR171*-OE demonstrated taller plants with earlier phase transition time (Fig. 2C and D). The *SlGRAS24*-OE lines showed pleiotropic phenotypes, some of which were in line with the phenotypes of transgenic *Arabidopsis* plants overexpressing *SlGRAS24* orthologs AtHAMs (Wang *et al.*, 2010). Flower opening was significantly delayed in *SlGRAS24*-OE, which was consistent with the delayed phase transition time from vegetative to reproductive development. The tissue sections of apical meristems showed no formed floral primordium in *SlGRAS24*-OE lines, 25 days postgermination (dpg) (Fig. 2E). In line with that, transcript levels of *SlFT* and *SlCO1*, two key regulators controlling flowering time, undergone absolutely

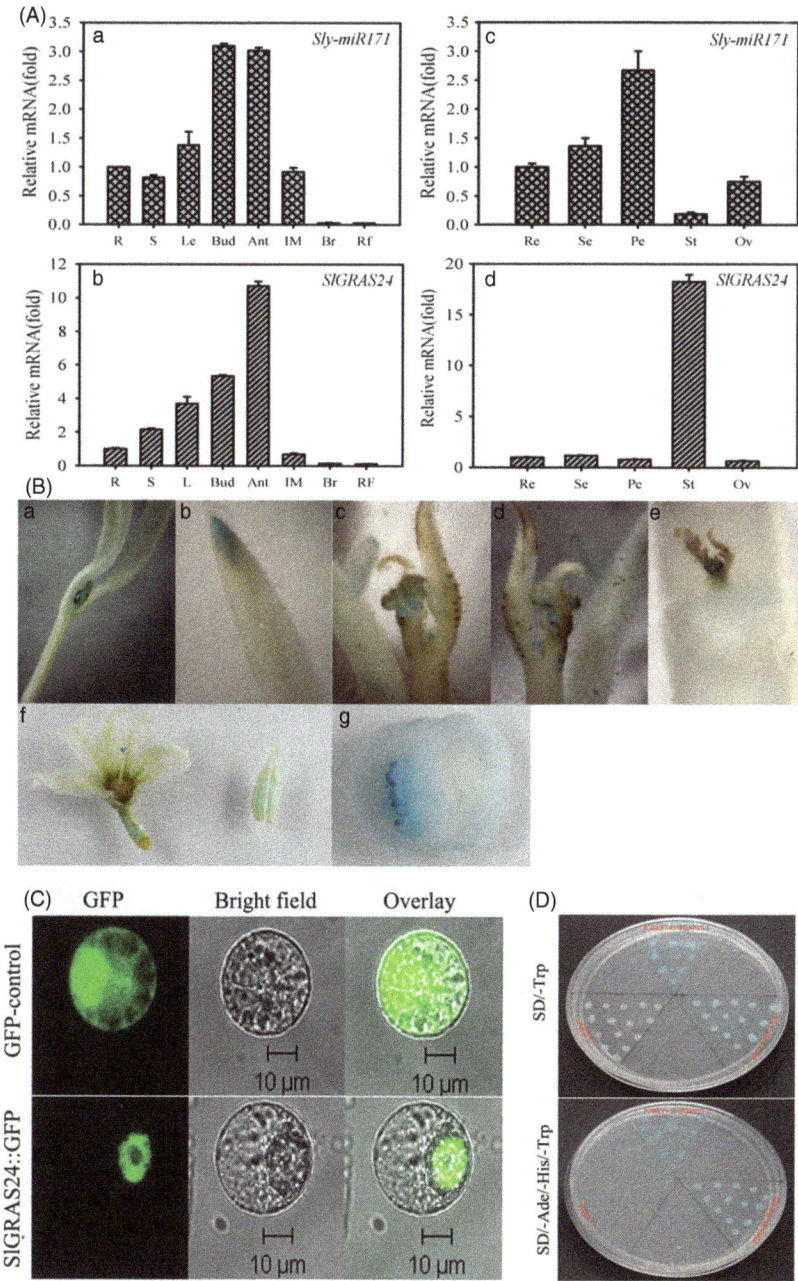

Figure 1 Expression patterns, subcellular localization and transcriptional activity of SlGRAS24. A, Tissue profiling analysis of Sly-miR171 (a, c) and SlGRAS24 (b, d) in different organs of wild-type tomato. R, root; S, stem; L, leaf; Bud, bud flower; Ant, anthesis flower; IM, immature green fruit; Br, colour breaker fruit; RF, ripening fruit; Re, receptacle; Se, sepal; Pe, petal; St, stamen; Ov, ovary. The expression data of root and receptacle were normalized to 1, respectively. Error bars show the standard error between three biological replicates performed (n = 3). B, Expression patterns of SlGRAS24 via GUS staining: (a, b) young seedlings; (c, d) shoot apices of phase transition stage plants; (e) nodal stem and axillary buds; (f) anthesis flowers; (g) immature green fruits. C, Subcellular localization of SlGRAS24. The photographs were taken under bright light, in the dark field for the GFP-derived green fluorescence and merged, respectively. D, Analysis of the transactivation activity of SlGRAS24. Up, SD/-Trp medium; below, SD/-Ade/-His/-Trp medium; both contained X-α-gal for assaying another yeast reporter (MEL1) gene.

Table 1 Summary of phenotypes in transgenic tomato plants overexpressing SlGRAS24, underexpressing SlGRAS24 and overexpressing Sly-miR171

Genotype	Line number	SlGRAS24 expression (relative to WT)	Phenotype
35S-SlGRAS24	L1–L16	8.4–88.8 folds	Dwarf plants, delayed phase transition time, abnormal leaves, inhibited root growth, increased lateral shoots, decreased flower number, impaired fruit set, compromised fruit and seed development
35S-asSlGRAS24	L1–L7	0.48–0.65 folds	No apparent phenotype
35S-SlymiR71	L1–L5	0.46–0.74 folds	Higher plants, earlier phase transition time and flowering time

opposite expression tendency in samples harvested at three different stages (20 dpg, 30 dpg, and 40 dpg) (Fig. 2E). The leaves of transgenic plants were both shorter and narrower than

WT leaves and leaf margins were not serrated (Fig. 2F). Microscopic analysis showed that leaves were thicker in SlGRAS24-OE lines, which might be attributed to having much larger lower epidermal cells (Fig. 2F). WT plants had axillary buds at the internodes, but abnormal axillary bud emergence was observed on stems of *SlGRAS24*-OE plants (Fig. 2G). Besides, *SlGRAS24*-OE plants had more lateral branches and abnormal flower bud emergence (Fig. S1). Primary and lateral root growth of *SlGRAS24*-OE lines was strongly suppressed compared to WT

Figure 2 Phenotypic characterization of wild-type and transgenic plants. A, Image of 30-day-old plants of different genotypes. OE L10 and OE L15, two independent *SlGRAS24*-overexpressing lines; AS L3 and AS L5, two independent *SlGRAS24*-downexpressing lines; miR171 OE L4 and miR171 OE L15, two independent Sly-miR171-overexpressing lines. B, Expression levels of *SlGRAS24* in plants shown in A. Expression of *SlGRAS24* in WT was normalized to 1. Error bars show the standard error between three biological replicates performed (n = 3). C, Height of plants shown in A. D, Days to floral transition (phase transition time) of plants shown in A. For C and D, error bars show the standard error between three biological replicates (n = 3) with more than ten plants for each replicate performed. E, Longitudinal sections of shoot apices from 25-day-old WT (a) and *SlGRAS24*-OE (b, c) plants; arrows indicate the emerged flower bud (a) and flower primordium (b, c); expression analysis of two flowering time regulator *FT* (d) and *CO1* (e) in leaves from WT and *SlGRAS24*-OE plants; dpg, days postgermination. F, Leaves from different positions (at nodes 1-6 from cotyledons) in WT and *SlGRAS24*-OE plants (a), transverse sections of leaves from node 5 in the WT (b) and *SlGRAS24*-OE (c) plants. G, Axillary buds of WT (a) and *SlGRAS24*-OE (b, c) plants; arrows indicate the positions of new formed axillary buds; transverse sections of stems in WT (d) and *SlGRAS24*-OE (e) plants. H, Roots in WT and *SlGRAS24*-OE plants. Different letters above bars indicate significant differences among different genotypes (Student's *t*-test, *P* < 0.05).

(Fig. 2H). More detailed information about the phenotypes of SlGRAS24-OE transgenic tomato plants is shown in Table 2. Noticeably, the fruit set ratio of SlGRAS24-OE tomato plants severely decreased, and fruit development was defective (Table 2).

Overexpression of SlGRAS24 disrupts fertilization

SlGRAS24-OE plants showed reduced fruit set and fruits were smaller with fewer seeds (Table 2, Fig. 3A, B and C). As tomato is the most important model for fruit development, it is meaningful to investigate the molecular mechanism underlying these defects. We found that SlGRAS24-OE plants occasionally produced flowers with dehiscent stamens, but the ovary and ovules in SlGRAS24-OE flowers were as in WT (Fig. 3D). However, SlGRAS24-OE flowers had smaller pollen sacs and collapsed anthers (Fig. 3D). TTC staining for pollen viability showed that fewer pollen viable grains in SlGRAS24-OE flowers undergoing anthesis (first day of flower opening) than that in WT flowers (Fig. 3D). Considering that SlGRAS24 transcripts were most abundant in anthesis flowers (Fig. 1A and B), expression pattern was studied in more detailed by examining GUS staining in flowers from -2 days postanthesis (dpa) to 4 dpa (Fig. 3E) and by qPCR analysing in flowers and fruits at different stages (Fig. 3F). For GUS staining, in stamens and stigmas, the strongest staining was at 0 dpa, and it became much weaker at 2 dpa, till there was almost no expression at 4 dpa. By contrast, almost no staining was observed in −2 dpa ovaries, but the expression increased since 0 dpa ovaries but was limited to ovules (Fig. 3E). Consistently, qPCR results also showed most abundant SlGRAS24 transcript in flowers at the anthesis stage (Fig. 3F).

Cross-fertilization assay was carried out to examine the fertility of transgenic flowers (Table 3). Fruit set ratios were 90%, 22% and 15% in self-pollinated WT flowers, OE L10 flowers and OE L15 flowers, respectively. When WT flowers were used as the female recipient, the fruit set ratio increased slightly to 33% and 25% with OE L10 and OE L15 pollen, respectively. There was 100% fruit set when WT pollen was used to pollinate OE pistils. The results also showed that WT × WT fruits and WT × OE fruit (pollen × pistil crosses) contained more seeds than OE × WT and OE × OE fruit. These experiments demonstrate that SlGRAS24 is necessary for normal stamen development and overexpression of SlGRAS24 has a negative impact on fertilization in tomato.

Overexpression of SlGRAS24 inhibits cell division and expansion in fruit

As SlGRAS24-OE plants have smaller fruits, the histology of WT and SlGRAS24-OE ovaries was analysed from day 3 to day 15 (Fig. 4A and Fig. S2). The pericarp of SlGRAS24-OE ovaries contains smaller cells and fewer cell layers than WT pericarp, indicating that overexpression of SlGRAS24 led to an inhibition of cell division and expansion in early fruit development. We analysed transcript levels of four genes involved in cell division (CYCLIN DEPENDENT KINASE (SlCDKB2.1), CYCLIN (SlCycB2.1) and SlCycD3.1, which encodes a G1 cyclin) and cell expansion (XYLOGLUCAN ENDOTRANSGLUCOSYLASE/HYDROLASE 1SlXTH1) (Fig. 4B). Relative to WT ovaries, the expressions of the three cell division genes were low in 3 dpa ovaries and increased at 7 dpa and 15 dpa. By contrast, expression of the SlXTH1 cell expansion gene was always lower in SlGRAS24-OE ovaries than WT ovaries.

SlGRAS24 is involved in GA and auxin signalling

The 2.2-kb SlGRAS24 promoter sequence was analysed in silico using the PLACE program (http://www.dna.affrc.go.jp/PLACE/signalup.html). Several cis-acting elements were identified including the canonical auxin response element (AuxRE) at position -599, two GA-responsive elements (GARE) at positions −523 and −826, and other elements related to GA and auxin (Fig. 5A). This strongly suggests that SlGRAS24 is regulated by both two hormones. SlGRAS24 transcript levels were compared in leaves treated or untreated by GA_3 or IAA, respectively. SlGRAS24 expression increased significantly within 1 h in response to either

Table 2 Phenotypes of wild-type (WT) and SlGRAS24-overexpressing (SlGRAS24-OE) transgenic tomato plants (L10 and L15)

Parameter	WT	OE L10	OE L15
Plant height (one month old, cm)	9.5 ± 0.3	3.9 ± 0.4**	3.7 ± 0.3**
Plant height (two months old, cm)	19.4 ± 1.1	11.1 ± 0.5**	9.6 ± 1.2**
Plant height (three months old, cm)	20.3 ± 1.9	13.8 ± 1.2**	13.1 ± 1.4**
Plant height (four months old, cm)	20.8 ± 2.5	18.1 ± 2.3	21.4 ± 2.7
Stem diameter of sixth internode (two months old, cm)	5.4 ± 0.5	5.3 ± 0.3	5.3 ± 0.3
Leave length/width of sixth node (two months old, cm)	5.8 ± 0.5/2.5 ± 0.2	4.3 ± 0.6/1.6 ± 0.4*	3.7 ± 0.5/1.3 ± 0.3**
Primary root length (two months old, cm)	36.2 ± 3.3	15.4 ± 2.8**	12.6 ± 1.9**
Lateral root number (two months old, n)	47.6 ± 4.1	26.6 ± 3.3**	14.8 ± 3.5**
Lateral branch number (two months old, n)	0.5 ± 0.2	2.2 ± 0.3**	2.9 ± 0.4**
Leaves to first inflorescence (n)	6.8 ± 0.3	6.5 ± 0.4	6.6 ± 0.2
Days to first visible flower bud	27.6 ± 1.3	35.4 ± 2.4**	39.5 ± 3.5**
Days to anthesis of first flower	42.1 ± 0.7	52.7 ± 2.6**	57.3 ± 2.8**
Days to colour breaker of first mature fruit	80.3 ± 1.9	90.9 ± 2.7**	97.2 ± 3.6**
Flowers in the two first inflorescences (n)	21.7 ± 2.1	16.0 ± 1.6**	13.2 ± 1.3**
Fruit set ratio	90 ± 1.4%	22 ± 2.3%**	15 ± 1.8%**
Fruits per plant (n)	26.3 ± 2.8	11.2 ± 2.1**	7.5 ± 1.4**
Fruit production (g per plant)	85.8 ± 10.3	24.6 ± 4.7**	11.2 ± 2.8**

Values are means of 10–12 plants, ±SE. The statistical significance of mean differences was analysed using a t-test: *P < 0.05, **P < 0.01.

(A)

(B)

(C)

(D)

(E)

(F)

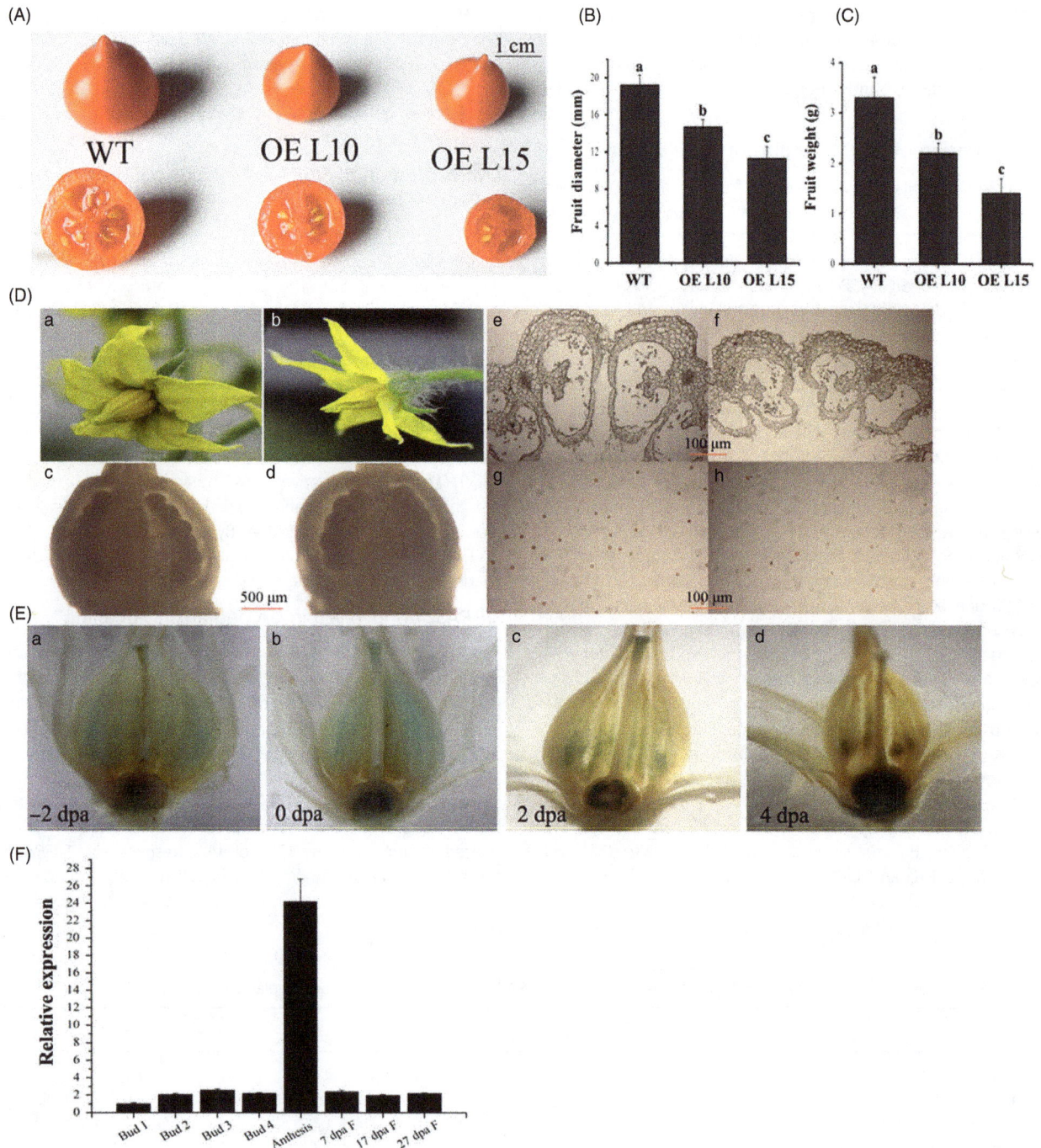

Figure 3 Overexpression of *SlGRAS24* causes smaller fruits with less seeds due to impaired fertilization. A, Fruits of WT and *SlGRAS24*-OE plants. B and C, Diameter and weight of WT and *SlGRAS24*-OE fruits. Error bars show the standard error between three biological replicates (n = 3) with more than ten fruits for each replicate performed. Different letters above bars indicate significant differences among different genotypes (Student's *t*-test, *P* < 0.05). D, Anthesis flowers in WT (a) and *SlGRAS24*-OE showing dehiscent stamen (b). Ovary and ovule of WT (c) and *SlGRAS24*-OE (d) anthesis flowers. Transverse sections of anthers of WT (e) and *SlGRAS24*-OE (f) anthesis flowers. Comparison of the pollen energy of WT (g) and *SlGRAS24*-OE (h) with TTC staining. E, GUS staining analysis of -2 dpa (a), 0 dpa (b), 2 dpa (c) and 4 dpa flowers from pro*SlGRAS24-GUS* transgenic plants. Dpa, days postanthesis. F, Expression level of *SlGRAS24* in flowers and fruits at different developmental stages. Buds were divided into 4 developmental stages from 1 to 4 according to the length of bud. Bud 1 stands for no more than 1 mm, bud 2 stands for between 2 and 3 mm, bud 3 stands for between 4 and 5 mm, and bud 4 stands for between 6 and 7 mm. Error bars show the standard error between three biological replicates performed (n = 3).

GA$_3$ or IAA (Fig. 5B). GA and auxin responsiveness of the promoters were tested using pro*GRAS24-GUS* transgenic seedlings incubated in solutions containing GA$_3$ or IAA for 3 h. Compared with untreated seedlings (mock), GUS staining revealed that GA$_3$ or IAA treatment led to ectopic expression of the *GUS* gene (Fig. 5C). qPCR analysis showed that both *GUS* and *SlGRAS24* transcripts significantly increased in response to GA$_3$ or IAA treatment in pro*GRAS24-GUS* seedlings (Fig. 5D).

Table 3 Cross-fertilization assay. Emasculated wild-type flowers were fertilized with SlGRAS24-OE pollen and the number of fruit and number of seed in each fruit were assessed at the ripe stage. Conversely, tomato pollen from wild-type flowers was used to fertilize emasculated SlGRAS24-OE flowers. Spontaneous self-pollinated flowers from each genotype were used as control. For each cross-fertilization assay, the capacity of the T1 seeds to grow on kanamycin-containing medium was assessed. Results are representative of data from two independent lines (OE L10 and OE L15)

Cross	Fruit set (Fruits Developed/ No. of Attempts)	Fruit set ratio (%)	Seeds (No. per fruit)	F1 kanamycin resistance (%)
♀ WT × ♂ WT	18/20	90	22.8 ± 3.5	0
♀ OE L10 × ♂ OE L10	4/18	22	6.6 ± 2.3	100
♀ OE L15 × ♂ OE L15	3/20	15	5.5 ± 1.8	100
♀ WT × ♂ OE L10	5/14	36	5.4 ± 2.4	100
♀ WT × ♂ OE L15	4/16	25	4.6 ± 0.8	100
♀ OE L10 × ♂ WT	12/12	100	14.2 ± 2.1	100
♀ OE L15 × ♂ WT	10/10	100	11.3 ± 1.6	100

Expression of GA- and auxin-related genes is differently regulated in SlGRAS24-OE seedlings

To further study the role of SlGRAS24 in GA and auxin pathways, the expression levels of a panel of 21 tomato genes were monitored in WT and SlGRAS24-OE seedlings in response to GA_3 or IAA treatment (Fig. 6A). The panel was made up of 5 GA biosynthetic enzymes (SlGA20ox1, SlGA20ox2, SlGA20ox4, SlGA3ox1 and SlGA3ox2), 3 GA deactivating enzymes (SlGA2ox1, SlGA2ox2 and SlGA2ox4), a key regulator of GA signalling pathway (SlDELLA), 4 auxin/indole-3-acetic acid (Aux/IAA) transcription factors (SlIAA2, SlIAA4, SlIAA4 and SlIAA9), 4 auxin response gene (ARF) transcription factors (SlARF5, SlARF6, SlARF7 and SlARF8) and 4 PIN-FORMED (PIN) auxin efflux transport proteins (SlPIN1, SlPIN3, SlPIN5 and SlPIN6). Without hormone treatment, 17 genes were down-regulated and 2 genes were up-regulated in SlGRAS24-OE seedlings compared to WT, which suggested that overexpression of SlGRAS24 disrupts GA and auxin homeostasis in transgenic plants. Furthermore, some of these genes responded differently to GA_3 and/or IAA in SlGRAS24-OE seedlings as compared to WT. For instance, upon GA_3 treatment, SlDELLA expression decreased in WT, but significantly increased in SlGRAS24-OE seedlings. SlGA2ox4 was induced by both GA_3 and IAA treatment in WT, while in SlGRAS24-OE it was not induced in response to GA_3 and was inhibited in response to IAA. SlIAA2 was up-regulated 2.45-fold by IAA treatment in WT, but about 12-fold in OE seedlings. SlARF8 was down-regulated under both hormone treatments in WT, while in SlGRAS24-OE it increased in response to GA_3 treatment and did not respond to IAA treatment. Comparing WT and SlGRAS24-OE seedlings, the different responsiveness of the GA-related genes during IAA treatment, and conversely the auxin-related genes during GA_3 treatment, might indicate that SlGRAS24 acts as an integrator between GA and auxin pathways.

It is thus possible that SlGRAS24 plays a role in regulating the expression of hormone-related genes in tomato, particularly genes associated with GA or auxin biosynthesis, transport and signal transduction. To assess whether changes in hormone levels accompanied changes in gene expression, endogenous IAA and GA_3 were quantified using HPLC-MS/MS (Fig. 6B). SlGRAS24-OE seedlings contained more IAA and GA_3 than in WT seedlings. This was somewhat unexpected as most GA/auxin-related genes were down-regulated.

Altered responsiveness to GA_3 and IAA application for SlGRAS24-OE plants

Exogenous IAA and/or GA3 were applied to WT and SlGRAS24-OE seedlings to investigate whether overexpression of SlGRAS24 altered other aspects of GA and auxin responsiveness (Fig. 7A). Without hormone treatment, the primary roots of SlGRAS24-OE seedlings were distinctly shorter than those of WT. In the presence of 1.0 µM IAA treatment, primary root growth was inhibited in both WT and SlGRAS24-OE seedlings (Fig. 7A and B), while SlGRAS24-OE seedlings had longer primary roots and fewer lateral roots than WT seedlings (Fig. 7A, B, and C), indicating that auxin responsiveness was reduced when SlGRAS24 was overexpressed. Synthetic auxin NAA stimulated more and longer adventitious roots to form from excised WT cotyledons than from SlGRAS24-OE cotyledons (Fig. 7E), another indication of a reduced auxin response in SlGRAS24-OE seedlings. Similarly, 50 µM GA3 inhibited the primary root growth of both WT and SlGRAS24-OE tomato seedlings, reducing the initial difference in length (Fig. 7A and B). Besides, outgrowth of the first true leaves from the shoot apex was severely suppressed in SlGRAS24-OE seedlings compared with WT (Fig. 7A and D).

Under normal condition, germination was inhibited in seeds from SlGRAS24-OE compared to that from WT (Fig. 7F). More sensitive phenotypes were observed in SlGRAS24-OE seeds when 10 µM GA3 or 5 µM paclobutrazol, a GA biosynthesis inhibitor, was applied since no observation of marked difference when they were applied to WT seeds (Fig. 7F), implying that SlGRAS24 is likely involved in seed germination through modulating GA signalling. However, the germination rates were not fully recovered to WT level when GA_3 was applied (Fig. 7F). Similarly, the dwarf phenotype and delayed flowering time of SlGRAS24-OE plants were only partially rescued by spraying with 20 µM GA3 (Fig. 7G, H and I), suggesting that some GA-independent pathways are involved in SlGRAS24-mediated regulation of plant growth and seed germination.

Overexpression of SlGRAS24 causes transcriptome changes in flowers at anthesis

To detect the changes in transcript levels that may be involved in the flower–fruit transition, a comparative transcriptome analysis was conducted using flowers at the onset of anthesis (0 dpa) of two SlGRAS24-OE transgenic lines (L10 and L15) and WT controls. Under the criteria of false discovery rate <0.05 and log2 fold change ≥1, a total of 1671 and 1436 unigenes were differentially expressed in L10 and L15, respectively, compared with WT controls (File S1 and S2). Functional annotation of putative gene products indicated that overexpression of SlGRAS24 affected multiple processes including transcription, signal transduction, primary and secondary metabolite biosynthesis, phytohormone biosynthesis, photosynthesis and stress responses, to name a few. Based on the properties of SlGRAS24 and the phenotypes of SlGRAS24-OE transgenic plants, we focused on genes involved in pollen development, hormonal biosynthesis/signalling and genes encoding transcription factors (Table 4). A total of 11 genes were selected for supplementary qRT-PCR analysis, including 3 pollen-related genes, 5 hormone-

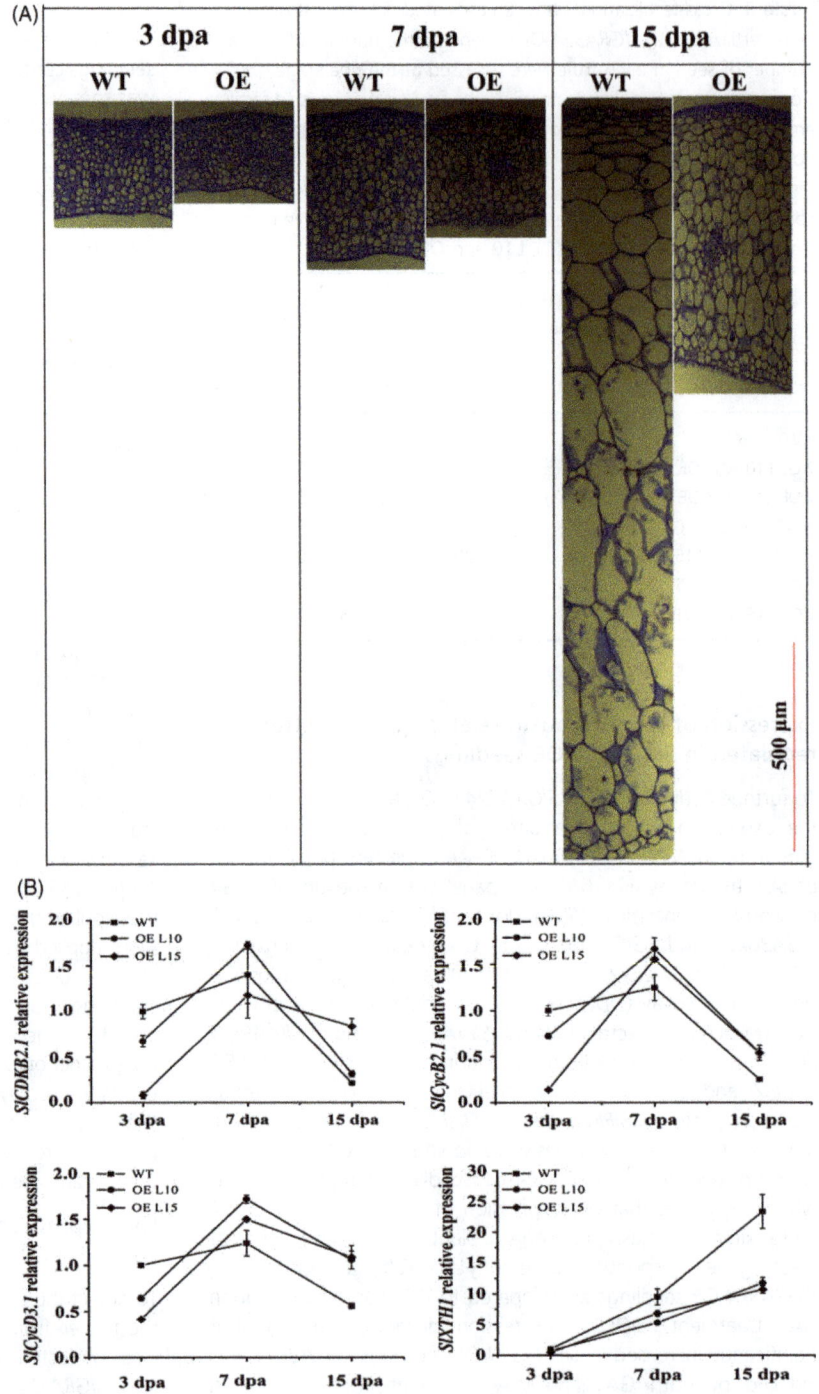

Figure 4 Histology and qPCR analysis of *SlGRAS24*-OE fruits. A, Transverse sections of 3 dpa, 7 dpa and 15 dpa WT and *SlGRAS24*-OE fruit pericarps. B, qRT-PCR analysis of cell division and expansion genes in 3 dpa, 7 dpa and 15 dpa WT and *SlGRAS24*-OE fruits. Dpa, days postanthesis. Error bars show the standard error between three biological replicates performed (n = 3).

related genes and 3 transcription factor genes which were stamen development regulators (Fig. 8). For all the genes tested, qPCR analysis results validated the transcriptomic data.

Discussion

The miR171-GRAS regulatory network participates in various physiological processes, including shoot meristem maintenance, axillary bud formation, flowering time, chlorophyll biosynthesis and trichome distribution. Tomato plants underexpressing *SlGRAS24* did not differ from WT, while overexpression of *Sly-miR171* did cause plants to grow taller and flower earlier

(Fig. 2A), providing another clue about the functional redundancy among GRAS family members or *miR171* target genes. Indeed, it has been shown the existence of functional redundancy among GRAS proteins in *Arabidopsis*. GRAS protein SCL3 mutant line *scl3-1* does not show any phenotype when compared with WT, but it confers important role in promoting gibberellin signalling by antagonizing master growth repressor DELLA protein (Zhang et al., 2011). Loss function of only one of the three *miR171* target genes in *Arabidopsis* does not have visible effects on plant growth, while transgenic plants overexpressing MIR171c (35S-109mmpro*MIR171c*) and *scl6-II scl6-III scl6-IV* triple mutant plants exhibit a similar reduced shoot branching phenotype

(A)

(B)

(C)

Mock   +GA3   +IAA

(D)

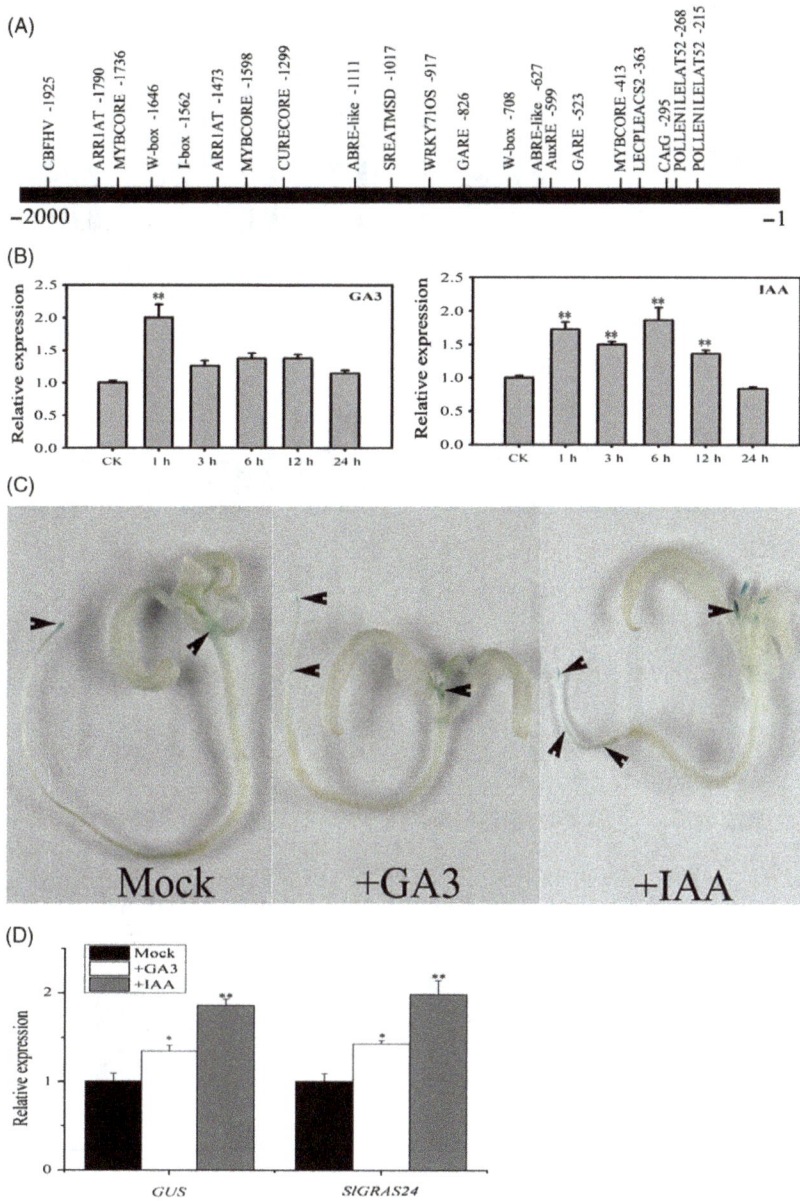

Figure 5 *SlGRAS24* is involved in GA and auxin signalling. A, Promoter region and the putative *cis*-acting elements. B, qRT-PCR analysis of *SlGRAS24* mRNA using leaves of 2-month-old WT plants after treatment with 100 μM GA$_3$ or 100 μm IAA. Expression of *SlGRAS24* in 0-h treated plants was normalized to 1. Different letters above bars indicate significant differences among different treatment time points (Student's *t*-test, *P* < 0.05). C, Expression pattern of pro*SlGRAS24-GUS* in 15-day-old seedlings and exogenous GA$_3$ or IAA treatment (20 μm for 3 h); arrows indicate places with clearly GUS staining. D, qRT-PCR analysis of *GUS* and *SlGRAS24* mRNA in 15-day-old pro*SlGRAS24-GUS* transgenic seedlings shown in C. Expression of *SlGRAS24* or *GUS* in untreated seedlings was normalized to 1. All samples were collected at the indicted time points for three biological replicates (n = 3). Different letters above bars indicate significant differences among different genotypes (Student's *t*-test, *P* < 0.05).

(Wang *et al.*, 2010). On the other hand, overexpression of *SlGRAS24* caused alteration of many important agronomical traits such as plant height, flowering time, leaf architecture, inflorescence architecture, lateral branch number, root length, fruit set and development (Table 2), making it a good target gene for generating new elite crop varieties with optimal flowering time and plant architectures, thus to meet the increasing demand for food, feed and biofuel production. Interestingly, we found that the plant height of *SlGRAS24*-OE lines was severely suppressed at the early development stage, but the suppression gradually disappeared due to the growth of later branches. Overexpression of *SlGRAS24* led to apical dominance inhibition and this inhibition accompanied the whole life cycle of the OE lines. At the mean time, because of apical dominance inhibition, the lateral branches showed vigorous growth, in number and length. As we know, it has long been a target of breeding selection to affect tiller and panicle/ear branching complexity because it significantly affects crop yield (Springer, 2010).

The relationship between GA and GRAS proteins has been extensively studied (Sun *et al.*, 2012), but some GRAS proteins have also been found to be involved in auxin signalling. For instance, an auxin-induced GRAS protein AtSCL15 plays crucial role during seed maturation programme (Gao *et al.*, 2015). The SCR/SHR complex comprised of two GRAS members has been proved to take part in root development by modulating both GA and auxin signalling (Heo *et al.*, 2011; Rovere *et al.*, 2015; Zhang *et al.*, 2011). Interactions between GA and auxin are established by intricate crosstalk and self-regulatory mechanisms involving the expression of auxin transport and GA metabolism genes. However, many important aspects of this relationship remain undiscovered. Plant hormones are implicated in the growth and development of shoot apical meristem (SAM) and root apical meristem (RAM) (Benková and Hejátko, 2009; Perilli *et al.*, 2012; Shani *et al.*, 2006). Important roles have been attributed to GA/auxin signalling in SAM and RAM (Weiss and Ori, 2007). It has been documented that *HAM* genes are required for the

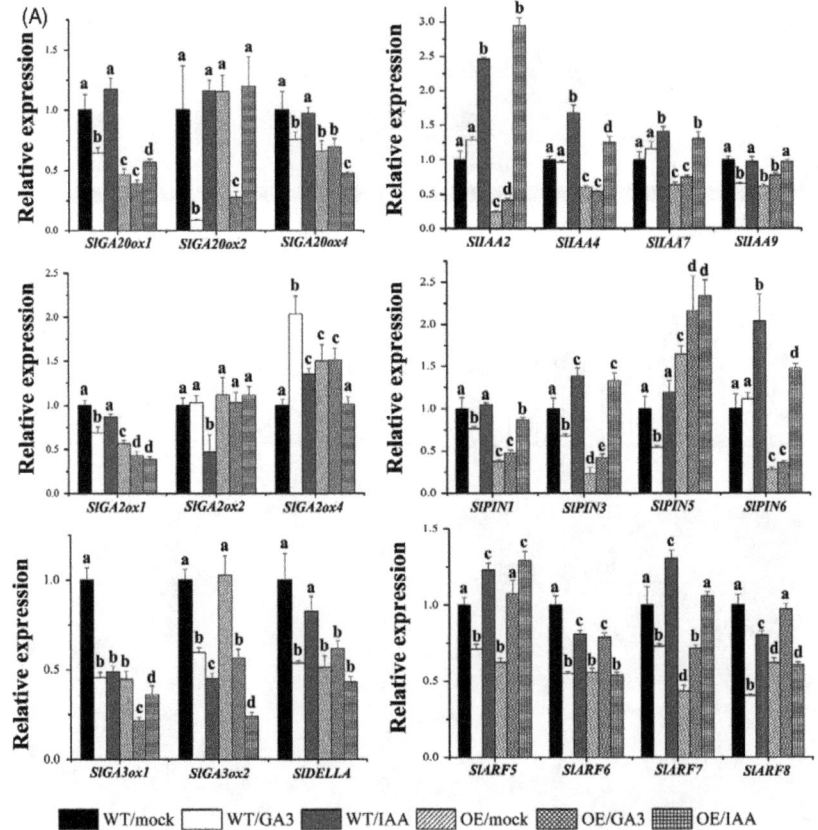

Figure 6 Expression analysis of GA/auxin-related genes and endogenous IAA/GA$_3$ content characterization. A, qRT-PCR analysis of GA- and auxin-related genes in 10-day-old WT and SlGRAS24-OE seedlings as well as in response to GA$_3$ or IAA treatment (20 μM for 3 h). B, Endogenous IAA and GA$_3$ content in 10-day-old WT and SlGRAS24-OE seedlings. Error bars show the standard error between three biological replicates performed (n = 3). Different letters above bars indicate significant differences among different treatments/genotypes (Student's t-test, P < 0.05).

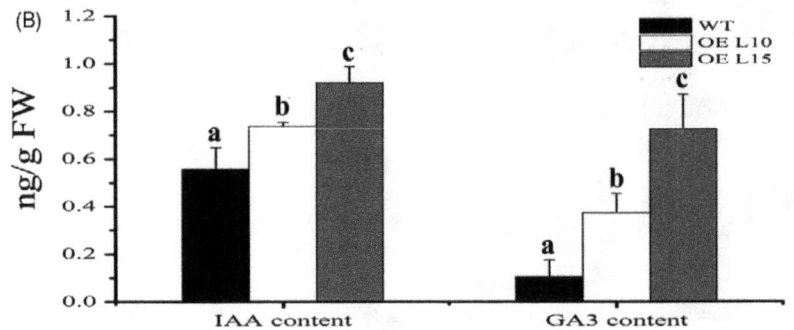

maintenance of both SAM and RAM (Engstrom et al., 2011), suggesting that HAM genes might exert functions in SAM and RAM by modulating GA and auxin signalling. In Arabidopsis, although it is found that Atham1, 2, 3 exhibits root apex auxin maxima that are comparable to WT in spatial expression and intensity (Engstrom et al., 2011), it did not link AtHAM gene action directly to auxin signalling. Here, we demonstrate that tomato SlGRAS24 is a key transcription factor for coordinated regulation of GA and auxin signalling pathways, which may provide clues to the mechanisms underlying the significantly increased height of Atham1, 2, 3 mutant plants and their shorter primary roots (Engstrom et al., 2011; Wang et al., 2010). Atham1, 2, 3 mutant plants also have altered lateral branch formation (Schulze et al., 2010; Wang et al., 2010), a process which is also regulated by auxin and GA (Martínez-Bello et al., 2015; Müller and Leyser, 2011; Ni et al., 2015). Recently, it was found that overexpression of a sunflower (Helianthus annuus L.) GRAS-like gene altered the GA content and axillary meristem outgrowth of transgenic Arabidopsis plants (Fambrini et al., 2015). Tomato plants overexpressing SlGRAS24 were dwarf with

short primary roots, fewer lateral roots and more lateral branches (Fig. 2 and Table 2). Thus, it is plausible that HAM genes are regulators of endogenous GA/auxin balance in SAM, RAM and axillary meristems to control meristem maintenance and organ production.

Although cis-acting elements including AuxRE and GARE were found in the promoter region of SlGRAS24 gene and SlGRAS24 transcripts were up-regulated under exogenous GA$_3$ and IAA treatment (Fig. 5), we observed that most GA/auxin-related genes were significantly down-regulated in SlGRAS24-OE tomato seedlings. This was despite these seedlings containing more endogenous IAA than WT (Fig. 6B) and raises the possibility that there is a negative feedback loop between SlGRAS24 expression and auxin metabolism. The reduced auxin responsiveness of SlGRAS24-OE seedlings under IAA or NAA treatment (Fig. 7) implies that SlGRAS24 might function in the downstream signalling response rather than in the upstream biosynthesis. There are more than ten kinds of active GAs in plant (GA9, GA4, GA34, GA7, GA51, GA19, GA20, GA1, GA8, GA5, GA3, GA29, etc.). Although the GA3 was elevated in OE seedlings (Fig. 6B),

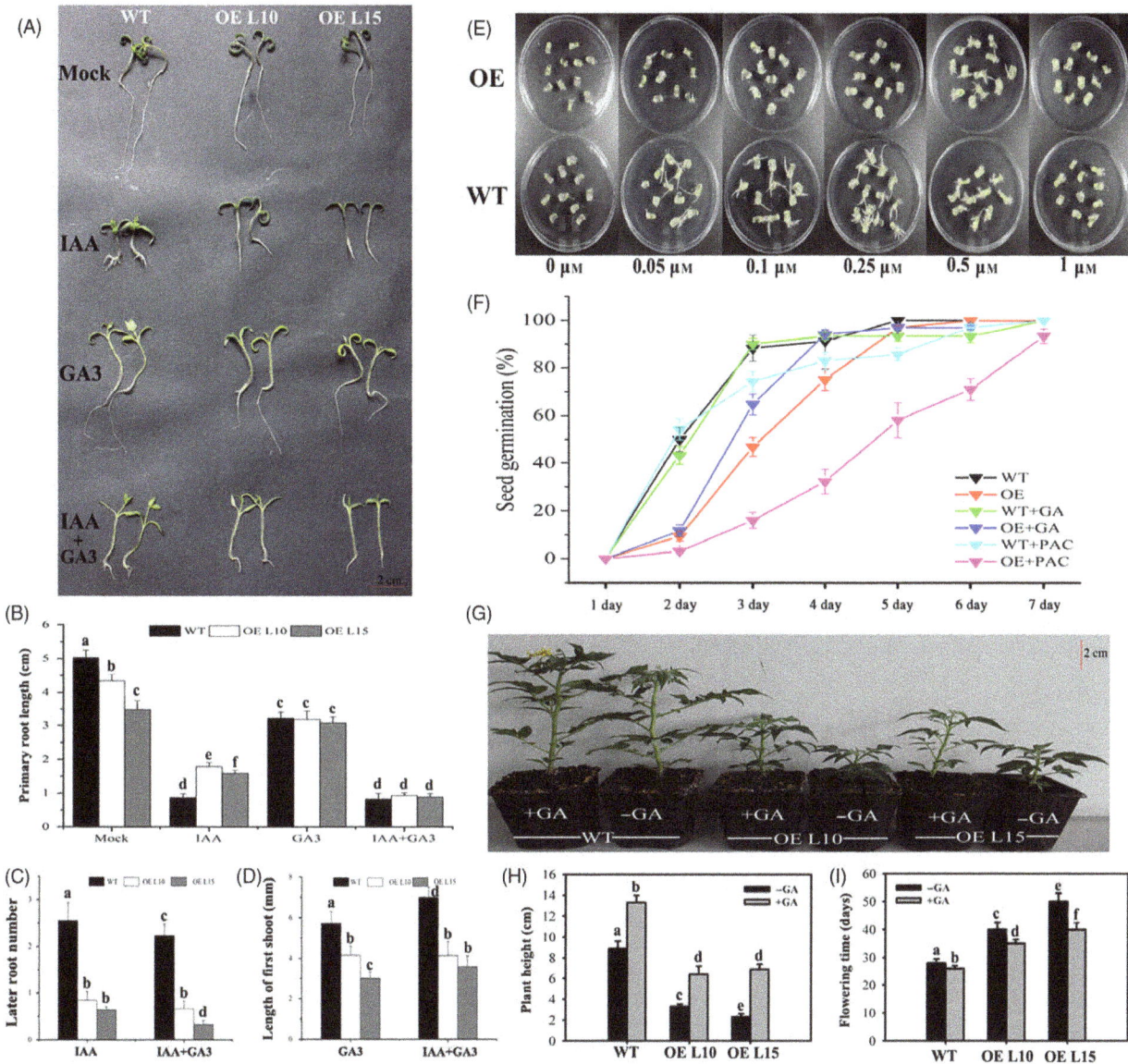

Figure 7 Altered responsiveness to exogenous IAA/GA$_3$ for *SlGRAS24*-OE lines. A, Phenotypes of two-week-old WT and *SlGRAS24*-OE seedlings grown on MS/2 medium containing 1 μM IAA and/or 50 μM GA$_3$. B, The length of primary root in WT and *SlGRAS24*-OE seedlings shown in A. C, The number of lateral roots in WT and *SlGRAS24*-OE seedlings under IAA or IAA+GA$_3$ treatment. D, Length from hypocotyls to true leaves in WT and *SlGRAS24*-OE seedlings under GA$_3$ or IAA+GA$_3$ treatment. E, Auxin dose–response assay of cotyledon explants. The explants were treated with increasing concentrations (0, 0.05, 0.1, 0.25, 0.5 and 1.0 μM) of NAA. F, Germination assay of WT and *SlGRAS24*-OE line as well as in response to GA$_3$ (10 μM) or PAC (5 μM) treatment. G, *SlGRAS24*-OE dwarfism partially rescued by exogenous GA$_3$ application. H and I, Plant height and phase transition time in GA$_3$ treated plants shown in G. Error bars show the standard error between three biological replicates (n = 3) with more than ten plants for each replicate performed. Different letters above bars indicate significant differences among different genotypes (Student's *t*-test, *P* < 0.05).

we do believe that the total GA content is reduced since most GA synthetic genes were down-regulated (Fig. 6A) and the observation of dwarfism phenotype and lower seed germination rate. Noticeably, the expression of *SlDELLA*, which acts as repressor of GA signalling, was down-regulated (Fig. 6). In contrast to the slender and earlier flowering time phenotype exhibited by *pro* mutant (a point mutation in the *DELLA* gene) tomato plants (Carrera *et al.*, 2012), dwarfism and late flowering time phenotypes were observed in *SlGRAS24*-OE plants, indicating the inhibition of DELLA-dependent GA responses. Moreover, application of GA$_3$ to *SlGRAS24*-OE plants only partially rescued the dwarf phenotype and the germination rate to the WT level

(Fig. 7), suggesting that these typically GA-related phenotypes are not merely due to the alteration of the GA signalling.

As transcription factors, GRAS proteins have been shown to participate in various biological processes in dozens of higher plant species. However, less is known about their roles during reproductive developmental stages. In lily (*Lilium longiflorum*), a GRAS protein named LiSCL was found to be involved in microsporogenesis (Morohashi *et al.*, 2003). In *Arabidopsis*, *Atham1, 2, 3* mutants occasionally produced flowers with three or five petals during the very early stages of flowering (Wang *et al.*, 2010), which indicated the potential active roles of miR171-targeted *GRAS* genes in flower organ formation. Here,

Table 4 Nonexhaustive list of genes differently regulated ($P < 0.05$) between wild-type and *SlGRAS24*-OE tomato anthesis flowers. Genes indicated with asterisks were validated by qPCR

	ITAG 2.40 Tomato	*Arabidopsis* orthologue	Functional annotation	Log2 fold (OE L10/WT)	Log2 fold (OE L15/WT)
Pollen development-related genes	Solyc01g008680.2		S2 self-incompatibility locus-linked pollen 3.2 protein	−1.39	−1.31
	Solyc01g008740.1	AT5G12180	CDPK17, involved in pollen tube growth	−1.43	−1.2
	Solyc01g059910.2*	AT3G51550	Mediates male–female gametophyte interactions during pollen tube reception	−2.12	−1.47
	Solyc01g067370.2*		Pollen-specific lysine-rich protein SBgLR	−1.26	−1.26
	Solyc02g076860.2	AT4G18596	Pollen allergen Phl p 11	−1.2	−1.37
	Solyc05g026360.2	AT5G56750	Pollen-specific protein SF21	−1.31	−1.16
	Solyc06g008240.2	AT4G04900	RIC10, involved in pollen tube growth	−1.44	−1.11
	Solyc06g008650.2	AT5G42232	Pollen allergen ole e 6	−1.21	−1.43
	Solyc09g065450.2		Pollen allergen ole e 6	−1.45	−1.2
	Solyc10g081700.1*	AT3G11690	Pollen preferential protein	−1.26	−1.31
	Solyc12g014240.1	AT1G29140	Pollen ole e 1 allergen and extensin	−1.16	−1.23
	Solyc12g014580.1		Pollen allergen Ole e 6	−1.14	−1.29
	Solyc10g081460.1	AT5G36940	CAT3, involved in amino acid transport	−1.1	−1.06
	Solyc09g010090.2	AT3G52600	Cell wall invertase 2, involved in sucrose catabolic process	−1.33	−1.23
	Solyc01g107830.2	AT3G53160.1	UDP-glycosyltransferase superfamily protein	−2.66	−3.4
Hormone signalling-related genes	Solyc05g025920.2	AT5G20810	SAUR-like auxin-responsive protein	−1.55	−1.59
	Solyc06g063060.2	AT1G54070	Auxin-repressed protein-like protein	−1.58	−1.28
	Solyc09g056360.2*	AT3G25290	Auxin-responsive family protein	−1.05	−1.21
	Solyc10g011660.2	AT2G46370	Auxin-responsive GH3 family protein	−1.29	−1.18
	Solyc12g009280.1*	AT5G20810	SAUR-like auxin-responsive protein	−1.22	−1.37
	Solyc10g009640.1*	AT2G46370	Jasmonic acid-amido synthetase JAR1	−1.52	−1.33
	Solyc03g044740.2	AT3G50440	Methyl jasmonate esterase	1.11	2.08
	Solyc12g009220.1*	AT1G19180	Jasmonate ZIM-domain protein 1 (JAZ1)	1.68	2.25
	Solyc00g095860.1	AT4G08040	ACS11, involved in ethylene biosynthetic process	1.92	2.2
	Solyc10g076450.1*	AT3G51770	Ethylene-overproduction protein 1	−1.32	−1.31
	Solyc10g076320.1	AT5G01700	PP2C, ABA response	−1.4	−1.28
	Solyc07g005680.2	AT4G30610	BRI1 SUPPRESSOR 1 (BRS1), involved in brassinosteroid signalling via BRI1	−1.28	−1.45
	Solyc12g008900.1	AT1G75450	Cytokinin oxidase/dehydrogenase 2	1.04	0.98
Transcription factors	Solyc01g079260.2	AT2G47260	WRKY transcription factor 4	1.15	1.28
	Solyc01g087990.2*	AT5G13790	MADS-box transcription factor 3	−1.17	−1.02
	Solyc01g094320.2	AT2G45800	GATA type zinc finger transcription factor family protein	−1.19	−1.28
	Solyc01g100460.2	AT1G75390	Basic-leucine zipper (bZIP) transcription factor	−1.4	−1.34
	Solyc02g073580.1	AT4G18650	Basic-leucine zipper (bZIP) transcription factor	2.29	2.1
	Solyc03g116890.2	AT1G80840	WRKY transcription factor 2	2.82	3.17
	Solyc04g009440.2	AT5G63790	NAC transcription factor	1.61	1.84
	Solyc04g056360.2*	AT4G26440	WRKY transcription factor 78	−1.45	−1.58
	Solyc05g051830.2	AT1G22130	MADS-box transcription factor 1	−1.44	−1.56
	Solyc06g034030.2	AT5G56840	MYB-like transcription factor family protein	−1.55	−1.24
	Solyc06g068570.2	AT1G16060	AP2-like ethylene-responsive transcription factor	1.39	1.15
	Solyc08g006470.2	AT2G24500	C2H2 zinc finger protein FZF	1.86	1.64
	Solyc08g008280.2	AT5G24110	WRKY transcription factor 30	1.77	1.83
	Solyc09g015770.2	AT3G56400	WRKY transcription factor 6	1.49	1.17
	Solyc09g014990.2	AT5G56270	WRKY-like transcription factor	2.4	2.67
	Solyc11g020950.1	AT1G58110	Basic-leucine zipper (bZIP) transcription factor	−1.41	−1.05
	Solyc11g044740.1	AT1G10200	GATA type zinc finger transcription factor family protein	−1.21	−1.265
	Solyc11g045310.1	AT2G03060	MADS-box transcription factor	−1.29	−1.23
	Solyc12g044610.1*	AT3G16350	MYB transcription factor	−1.33	−1.24

we found that *SlGRAS24* played pivotal roles in late stamen development. *SlGRAS24* transcripts accumulated most in flowers during anthesis, predominantly in stamens (Fig. 1A and B). Overexpression of *SlGRAS24* in tomato impaired pollen sac and pollen development and as a consequence the fruit set ratio was lower than in WT (Fig. 3 and Table 2). Pollen development in flowering plants is a highly programmed process requiring many genes. In the nonexhaustive list of genes differentially regulated between *SlGRAS24*-OE and WT flowers (Table 4), 15 are related to pollen development, 13 are associated with hormones and 19

Pollen development-related genes

Hormone signalling-related genes

Transcription factors

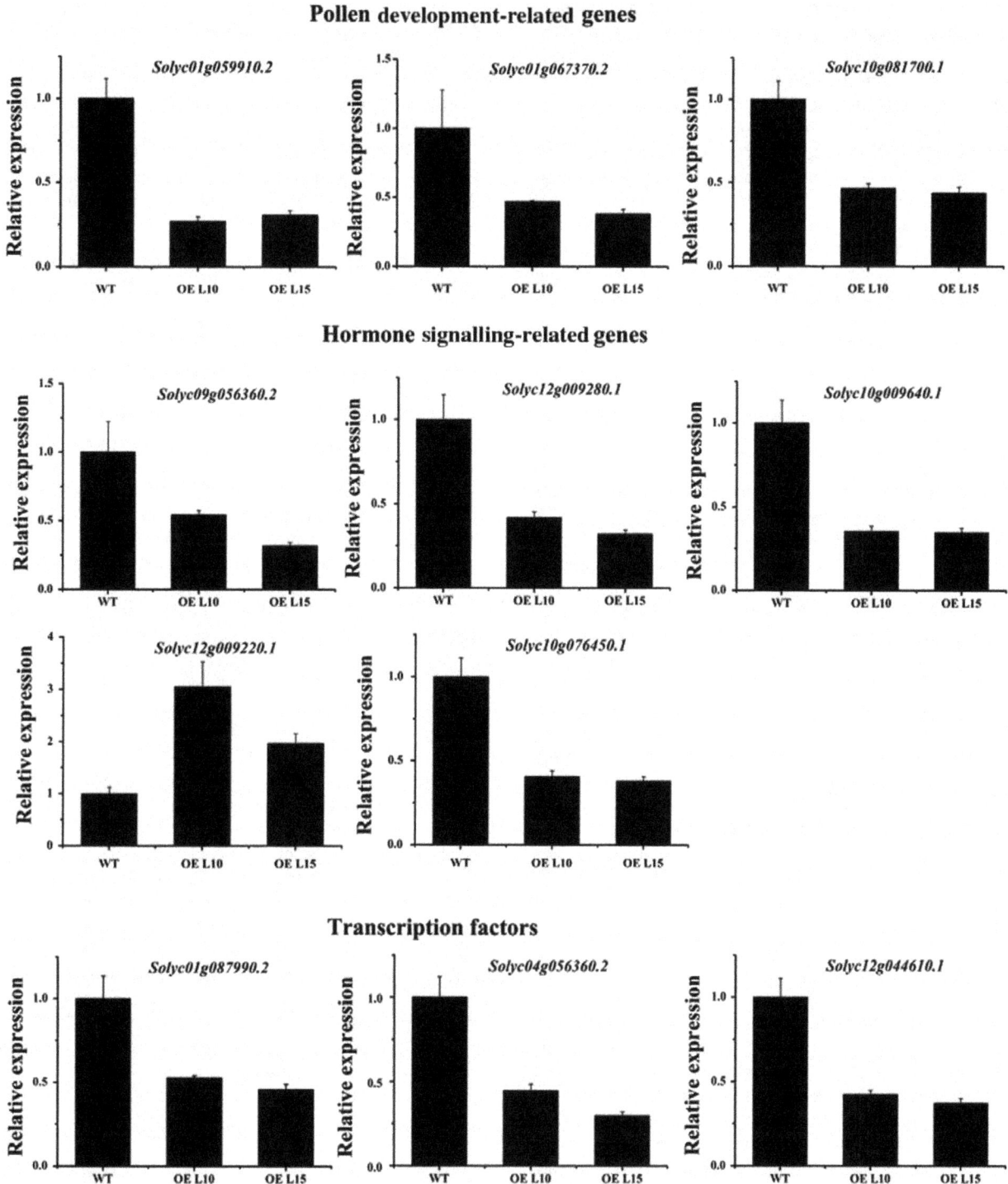

Figure 8 qPCR validation of transcriptomic data. 3 pollen development-related genes, 5 hormone signalling-related genes and 3 transcription factor genes were selected and validated by qRT-PCR. Error bars show the standard error between three biological replicates performed (n = 3).

are transcription factors. Most phytohormones play crucial roles in the regulation of pollen development either directly (Song et al., 2013) or indirectly (Dobritzsch et al., 2015; Ji et al., 2011). Our results demonstrated altered expression of genes related to all hormones except GA (Table 4). Young tomato flower buds have a high content in metabolites of GA pathways, which decrease progressively during ovary development. Active GA levels are very low in anthesis-stage flowers during pollination, after which the total GA content within the ovary increases again

(Fos et al., 2000). As we only compared the transcriptome of flowers undergoing anthesis, it is understandable why GA-related gene expression did not differ between WT and SlGRAS24-OE lines. Many transcription factors are involved in regulating pollen development in a dynamic regulatory network. In our work, the transcription factors differentially regulated by SlGRAS24 are mainly from MYB, WRKY, MADS, zinc finger and bZIP gene families. In Arabidopsis, a number of MYB proteins have been documented as important regulators of pollen development

(Cheng *et al.*, 2009; Higginson *et al.*, 2003; Mandaokar and Browse, 2009; Mandaokar *et al.*, 2006; Yang *et al.*, 2007). Two pollen-specific transcription factors, WRKY34 and its close homolog WRKY2, are required for male gametogenesis (Guan *et al.*, 2014). A subset of pollen-specific MIKC*MADS box proteins (AGL30/65/66/94/104) are expressed preferentially during pollen maturation and double mutant combinations reveal the important roles these genes play in pollen germination and pollen fitness (Verelst *et al.*, 2007). Interestingly, we found decreased expression of *Solyc04g056360.2* and *Solyc11g045310.1*, which are the homologous gene of *Arabidopsis WRKY34* (AT4G26440) and *AGL30* (AT2G03060, a member of AtMIKC*MADS complexes), respectively. It has been proved that *WRKY34* is one of the direct target genes of AtMIKC*MADS complexes in pollen. During pollen maturation, *WRKY34* is suppressed by several MIKC*MADS box transcription factors (Verelst *et al.*, 2007). Thus, we speculate that *SlGRAS24* might participate in the MIKC*MADS-WRKY34 regulatory network in tomato pollen development.

Upon flower fertilization, fruit and seed development occurs concomitantly, orchestrated by various phytohormones (McAtee *et al.*, 2013). In *SlGRAS24*-OE transgenic plants, fruits were smaller with fewer seeds (Fig. 3A). It is believed that pollen quantity and/or quality are closely associated with fruit and seed set (Aizen and Harder, 2007; Burd, 1994). In normal conditions, the developing seed continually sends signals to the surrounding tissue to expand and there is usually a positive correlation between seed number and fruit size (Nitsch, 1970). Therefore, we assumed that overexpression of *SlGRAS24* led to impaired pollen sac development and pollen viability, which resulted in less efficient pollination/fertilization and hence smaller fruits with fewer seeds. Both cell division and cell elongation programmes were significantly suppressed in the smaller transgenic fruits (Fig. 4), as would be predicted by the 'seed control' hypothesis that the seeds communicate through hormones to the surrounding tissue(s) to promote fruit growth firstly through cell division and then cell expansion (Ozga *et al.*, 2002).

Experimental procedures

Plant materials and growth conditions

Tomato plants (*Solanum lycopersicum* cv. Micro-Tom) were grown on soil in controlled glass house conditions with 14-h light: 10-h dark cycles, 25 °C day: 20 °C night temperature, 60% relative humidity and weekly irrigation with plant nutrient solution. For gene expression analysis, roots, stems and leaves were collected from 1-month-old plants, and flowers were harvested at the bud and anthesis stages and fruits at the immature, breaker and red stages. Receptacles, sepals, petals, stamens and ovaries were harvested from flowers at the anthesis stage. For each tissue/organ type, samples were collected from at least six healthy plants, mixed and then frozen in liquid nitrogen immediately. Sampling was done three independent times.

Plasmid construction and generation of transgenic plants

Four DNA fragments, the *SlGRAS24* promoter, the precursor of *miR171*, the full-length *SlGRAS24* coding sequence and a partial *SlGRAS24* coding sequence were amplified from tomato genomic DNA or cDNA. Primer sequences used for amplification are listed in Table S1. The *SlGRAS24* promoter sequence was fused with

GUS in an expression vector. The *miR171* precursor and *SlGRAS24* full-length coding sequence were cloned into the modified binary vector pLP100 in the sense orientation, while the partial *SlGRAS24* coding sequence was cloned in the antisense orientation, all under the CaMV 35S promoter. Transgenic plants were generated by *Agrobacterium tumefaciens*-mediated transformation according to Huang *et al.* (2016). For each construct, more than 6 independent lines with consistent phenotypes were obtained. Homozygous lines from T2 or later generations were used for experiments.

Subcellular localization and transactivation activity assay of SlGRAS24

The *SlGRAS24* open reading frame without the stop codon was amplified and cloned into the pGreen0029 vector. The recombinant plasmid containing the *SlGRAS24-GFP* fusion gene and the control plasmid with *GFP* alone were transformed into tobacco (*Nicotiana tabacum* L.) protoplasts according to Ren *et al.* (2011). For transactivation assays, the coding region of *SlGRAS24* was amplified and ligated into the yeast expression vector pGBKT7 (Clontech, Japan) to produce pBD-*SlGRAS24*. According to the manufacturer's instructions, pBD-*SlGRAS24*, pGBKT7 (plasmid for negative control) and pGBKT7-53+pGADT7-T (plasmid combination for positive control) were transformed separately into the yeast strain AH109. Transformants were selected on SD/-Trp or SD/-Ade/-His/-Trp medium (Clontech, USA). The transactivation activity of each protein was evaluated by comparing growth on permissive and selective medium and the activity of X-α-Gal (5-bromo-4-chloro-3-indoxyl-α-d-galacto-pyranoside).

Histochemical and histological analysis

GUS activity was assayed by submerging plant samples in 0.5 mg/mL X-Gluc solution (0.1 M sodium phosphate buffer pH 7.0, 10 mM EDTA, 0.1% Triton X-100, 0.5 mM potassium ferrocyanide, 0.5 mM potassium ferricyanide), infiltrating them under vacuum and incubating them at 37 °C. Samples were destained in 70% ethanol.

Histological preparations were performed according to Gabe (1968). Specific tissues/organs were embedded in FAA solution (50% (v/v) ethanol, 5% (v/v) acetic acid and 3.7% (v/v) paraformaldehyde). Samples were then placed under vacuum for 10 min, incubated at room temperature for 24 h, then dehydrated in an ethanol gradient and embedded in paraffin (Paraplast Plus, Sigma). Observations were carried out under a light microscope (OLYMPUS BX-URA2, Japan).

Hormone treatments for gene expression analysis

Two-month-old WT tomato plants were sprayed with 100 μM GA$_3$ or 100 μM IAA. Leaves were harvested 0, 1, 3, 6, 12 and 24 h after spraying, frozen in liquid nitrogen and stored at −80 °C until RNA extraction. Meanwhile, 15-d-old pro*SlGRAS24*-GUS transgenic seedlings were soaked in liquid MS/2 medium containing 20 μM GA$_3$ or 20 μM IAA for 3 h, and then the whole seedlings were used for GUS staining analysis or directly frozen in liquid nitrogen and stored at −80 °C. Similarly, 10-d-old WT and *SlGRAS24*-OE transgenic seedlings were soaked in liquid MS/2 medium containing 20 μM GA$_3$ or 20 μM IAA for 3 h, and whole seedlings were frozen in liquid nitrogen and stored at −80 °C. Seedlings soaked in liquid MS/2 medium without hormone were used as controls. All treatments were performed three independent times.

Hormone treatments for analysis of plant development

Surface-sterilized seeds of WT and each T2 transgenic tomato line were germinated in the dark. After the emergence of the radicle, seeds were transferred onto MS/2 medium containing 10^{-6} M IAA and/or 5×10^{-5} M GA$_3$. The seedlings were grown in a controlled growth chamber for 2 weeks with a 16-h light: 8-h dark photoperiod and 25 °C day: 20 °C night temperature. For auxin dose–response experiments, cotyledon explants from 10-d-old WT and SlGRAS24-OE seedlings were incubated on MS/2 medium containing the indicated auxin (α-naphthalene acetic acid, NAA) concentrations in the same growth chamber conditions for 10 d.

For seed germination assays, surface-sterilized seeds from T2 transgenic lines and WT were germinated on MS/2 medium containing 10 μM GA$_3$ or 5 μM paclobutrazol (PAC), a GA biosynthesis inhibitor, then placed in the same growth chamber conditions described above. Seeds germinated on MS/2 medium were used as controls. Germination rates based on radicle tip emergence were scored daily for 7 d after sowing. GA$_3$ was also applied by spraying the aerial part of 10-d-old plants grown in the glasshouse with 20 μM GA$_3$ solution containing 0.1% Tween-80 every other day for 3 weeks. Control plants were sprayed with the equivalent solvent solution.

All experiments were repeated three times with about 30 seeds or 10 plants for each genotype.

Extraction and quantification of endogenous IAA and GA$_3$

Pure IAA and GA3 were purchased from Sigma Chemical Co. (St Louis, MO). Isotopically labelled internal standards [^{2}H$_5$]IAA and [^{2}H$_2$]GA$_3$ were purchased from ICON Isotopes (Summit, NJ). Plant hormones were extracted and quantified as previously described (Pan et al., 2008) with some modifications. Approximately 0.2 g of 10-d-old WT or SlGRAS24-OE seedlings was frozen in liquid nitrogen and ground into powder. Two millilitres of isopropanol–HCl buffer solution (2:0.002, v/v) was added to the powder and shaken for 30 min at 4 °C. Dichloromethane (4 mL) was added, shaken for 30 min and then centrifuged at 13 000 g for 5 min. After centrifugation, the lower organic phase was transferred to a 10-mL tube and evaporated in a constant stream of nitrogen. Each sample was kept in the dark and resolubilized in 150 μL methanol containing 0.1% formic acid, and then the solution was filtered using a 0.45 μm microfilter for HPLC-MS/MS analysis. Samples were injected into a reversed-phase column (C18 ZORBAX 300SB 3 μm, 4.6 × 150 mm, Agilent, CA) using a binary solvent system composed of methanol (solvent A) and water with 0.1% formic acid (solvent B) as the mobile phase. The column thermostat was set at 30 °C. Separations were performed using a solvent gradient which started at 20% methanol for 2 min, increased linearly to 80% over 14 min and maintained for 5 min, and returned to 20% methanol for 0.1 min, when it was allowed to equilibrate for 5 min. A hybrid triple quadrupole/linear ion trap mass spectrometer (SCIEX 6500 QTrap, Applied Biosystems, Foster City, CA) was used with nebulizer gas pressure set at 75 psi, drying gas pressure at 65 psi, curtain gas pressure at 15 psi, source voltage at 4.5 kV and source temperature at 500 °C.

Pollen viability assay

Pollen activity was evaluated by soaking pollen grains in 0.1% 2, 3, 5-triphenyl-2 h-tetrazolium chloride (TTC) solution. Viable pollen stains red because the NADH/NADPH produced reduces TTC to 1,3,5-triphenylformazan (TPF), which is red. Stained pollen grains were observed under the microscope.

Digital gene expression profiling

Total RNA was extracted from anthesis-stage flowers of SlGRAS24-OE plants (lines 10 and 15) and WT controls using RNeasy® Plant Mini Kit (Qiagen) following the manufacturer's protocol for RNA-Seq. RNA quantity and quality were assayed in the Agilent 2100 Bioanalyzer (Agilent Technologies). Two independent RNA samples from transgenic or WT plants were sent to Illumina Cluster Station and Illumina HiSeq™ 2000 System (BGI Inc.) for RNA library construction and deep sequencing. RSeQC-2.3.2 program (http://code.google.com/p/rseqc/) was used to assess the quality of RNA-Seq data. Clean tags were obtained after quality filtering of sequences and mapped to the annotated genome sequence of S. lycopersicum in the Tomato Sol Genomic Network database (http://solgenomics.net/), and transcript abundance was also normalized by the fragments per kilobase of exon per million mapped reads (FRKM) method using Cuffdiff software (http://cufflinks.cbcb.umd.edu/) to identify differentially expressed genes (DEGs). The expression level of a gene from RNA-Seq was normalized by the tags per million method. The criteria for defining differentially expressed genes were a false discovery rate (FDR) <0.05 with a P value <0.05. The raw transcriptome reads reported here have been deposited in the NCBI Short Read Archive under accession no. SRA473616.

Real-time quantitative PCR

One microgram of total RNA (RNeasy® Plant Mini Kit, Qiagen) was used to synthesize first-strand cDNA (RevertAid™ First Strand cDNA Synthesis Kit, Fermentas). Transcript levels were determined by absolute qPCR according to the methodology described in Huang et al. (2015) using specific primers. Primer sequences used for qRT-PCR are listed in Table S2. Amounts of mRNA in samples were quantified using three biological replicates.

Statistical analysis

All data in this report were obtained from at least three independent experiments with three technical replicates each. For data analysed with Student's t-test, the differences between treatments were considered as significant when $P < 0.05$.

Acknowledgements

This work was supported by the National Basic Research Program of China (2013CB127101), the National High Technology Research and Development Program of China (2012AA101702) and the National Natural Science Foundation of China (31272166, 31401924). The authors declare no conflict of interests.

References

Aizen, M.A. and Harder, L.D. (2007) Expanding the limits of the pollen-limitation concept: effects of pollen quantity and quality. Ecology, **88**, 271–281.

Benková, E. and Hejátko, J. (2009) Hormone interactions at the root apical meristem. Plant Mol. Biol. **69**, 383–396.

Bolle, C. (2004) The role of GRAS proteins in plant signal transduction and development. Planta, **218**, 683–692.

Burd, M. (1994) Bateman's principle and plant reproduction: the role of pollen limitation in fruit and seed set. *Bot. Rev.* **60**, 83–139.

Carrera, E., Ruiz-Rivero, O., Peres, L.E.P., Atares, A. and Garcia-Martinez, J.L. (2012) Characterization of the *procera* tomato mutant shows novel functions of the SlDELLA protein in the control of flower morphology, cell division and expansion, and the auxin-signaling pathway during fruit-set and development. *Plant Physiol.* **160**, 1581–1596.

Cheng, H., Song, S.S., Xiao, L.T., Soo, H.M., Cheng, Z.W., Xie, D.X. and Peng, J.R. (2009) Gibberellin acts through jasmonate to control the expression of *MYB21*, *MYB24*, and *MYB57* to promote stamen filament growth in *Arabidopsis*. *PLoS Genet.* **5**, e1000440.

Curaba, J., Talbot, M., Li, Z. and Helliwell, C. (2013) Over-expression of microRNA171 affects phase transitions and floral meristem determinancy in barley. *BMC Plant Biol.* **13**, 6.

Dobritzsch, S., Weyhe, M., Schubert, R., Dindas, J., Hause, G., Kopka, J. and Hause, B. (2015) Dissection of jasmonate functions in tomato stamen development by transcriptome and metabolome analyses. *BMC Biol.* **13**, 1–18.

Engstrom, E.M., Andersen, C.M., Gumulak-Smith, J., Hu, J., Orlova, E., Sozzani, R. and Bowman, J.L. (2011) *Arabidopsis* homologs of the petunia *hairy meristem* gene are required for maintenance of shoot and root indeterminacy. *Plant Physiol.* **155**, 735–750.

Fambrini, M., Mariotti, L., Parlanti, S., Salvini, M. and Pugliesi, C. (2015) A *GRAS*-like gene of sunflower (*Helianthus annuus* L.) alters the gibberellin content and axillary meristem outgrowth in transgenic *Arabidopsis* plants. *Plant Biol.* **17**, 1123–1134.

Fan, T., Li, X., Yang, W., Xia, K., Ouyang, J. and Zhang, M. (2015) Rice *osa-miR171c* mediates phase change from vegetative to reproductive development and shoot apical meristem maintenance by repressing four OsHAM transcription factors. *PLoS ONE*, **10**, e0125833.

Fos, M., Nuez, F. and Garci'a-Marti'nez, J.L. (2000) The gene *pat-2*, which induces natural parthenocarpy, alters the gibberellin content in unpollinated tomato ovaries. *Plant Physiol.* **122**, 471–480.

Gabe, M. (1968) *Techniques Histologiques*. Paris: Masson.

Gao, M.J., Li, X., Huang, J., Groop, G.M., Gjetvaj, B., Lindsay, D.L., Wei, S. *et al.* (2015) SCARECROW-LIKE15 interacts with HISTONE DEACETYLASE19 and is essential for repressing the seed maturation programme. *Nat. Commun.* **6**, 7234.

Guan, Y., Meng, X., Khanna, R., LaMontagne, E., Liu, Y. and Zhang, S. (2014) Phosphorylation of a WRKY transcription factor by MAPKs is required for pollen development and function in *Arabidopsis*. *PLoS Genet.* **10**, e1004384.

Heo, J.O., Chang, K.S., Kim, I.A., Lee, M.H., Lee, S.A., Song, S.K., Lee, M.M. *et al.* (2011) Funneling of gibberellin signaling by the GRAS transcription regulator scarecrow-like 3 in the *Arabidopsis* root. *Proc. Natl Acad. Sci. USA*, **108**, 2166–2171.

Higginson, T., Li, S.F. and Parish, R.W. (2003) AtMYB103 regulates tapetum and trichome development in *Arabidopsis thaliana*. *Plant J.* **35**, 177–192.

Huang, W., Xian, Z.Q., Kang, X., Tang, N. and Li, Z.G. (2015) Genome-wide identification, phylogeny and expression analysis of GRAS gene family in tomato. *BMC Plant Biol.* **15**, 209.

Huang, W., Xian, Z.Q., Hu, G.J. and Li, Z.G. (2016) SlAGO4A, a core factor of RNA-directed DNA methylation (RdDM) pathway, plays an important role under salt and drought stress in tomato. *Mol. Breeding*, **36**, 1–13.

Hwang, E.W., Shin, S.J., Yu, B.K., Byun, M.O. and Kwon, H.B. (2011) miR171 family members are involved in drought response in *Solanum tuberosum*. *J Plant Biol.* **54**, 43–48.

Ji, X., Dong, B., Shiran, B., Talbot, M.J., Edlington, J.E., Hughes, T., White, R.G. *et al.* (2011) Control of abscisic acid catabolism and abscisic acid homeostasis is important for reproductive stage stress tolerance in cereals. *Plant Physiol.* **156**, 647–662.

Kantar, M., Unver, T. and Budak, H. (2010) Regulation of barley miRNAs upon dehydration stress correlated with target gene expression. *Funct. Integr. Genomic.* **10**, 493–507.

Kong, Y., Elling, A.A., Chen, B. and Deng, X. (2010) Differential expression of microRNAs in maize inbred and hybrid lines during salt and drought stress. *Am. J. Plant Sciences*, **1**, 69.

Liu, H.H., Tian, X., Li, Y.J., Wu, C.A. and Zheng, C.C. (2008) Microarray-based analysis of stress-regulated microRNAs in *Arabidopsis thaliana*. *RNA*, **14**, 836–843.

Llave, C., Xie, Z., Kasschau, K.D. and Carrington, J.C. (2002) Cleavage of Scarecrow-like mRNA targets directed by a class of *Arabidopsis* miRNA. *Science*, **297**, 2053–2056.

Ma, Z.X., Hu, X.P., Cai, W.J., Huang, W.H., Zhou, X., Luo, Q., Yang, H.Q. *et al.* (2014) *Arabidopsis* miR171-targeted scarecrow-like proteins bind to GT cis-elements and mediate gibberellin-regulated chlorophyll biosynthesis under light conditions. *PLoS Genet.* **10**, e1004519.

Mandaokar, A. and Browse, J. (2009) MYB108 acts together with MYB24 to regulate jasmonate-mediated stamen maturation in *Arabidopsis*. *Plant Physiol.* **149**, 851–862.

Mandaokar, A., Thines, B., Shin, B., Lange, B.M., Choi, G., Koo, Y.J., Yoo, Y.J. *et al.* (2006) Transcriptional regulators of stamen development in *Arabidopsis* identified by transcriptional profiling. *Plant J.* **46**, 984–1008.

Martínez-Bello, L., Moritz, T. and López-Díaz, I. (2015) Silencing C_{19}-GA2-oxidases induces parthenocarpic development and inhibits lateral branching in tomato plants. *J. Exp. Bot.* **66**, 5897–5910.

McAtee, P., Karim, S., Schaffer, R. and David, K. (2013) A dynamic interplay between phytohormones is required for fruit development, maturation, and ripening. *Front. Plant Sci.* **4**, 79.

Morohashi, K., Minami, M., Takase, H., Hotta, Y. and Hiratsuka, K. (2003) Isolation and characterization of a novel GRAS gene that regulates meiosis-associated gene expression. *J. Biol. Chem.* **278**, 20865–20873.

Müller, D. and Leyser, O. (2011) Auxin, cytokinin and the control of shoot branching. *Ann. Bot.* **107**, 1203–1212.

Ni, J., Gao, C., Chen, M.S., Pan, B.Z., Ye, K. and Xu, Z.F. (2015) Gibberellin promotes shoot branching in the perennial woody plant *Jatropha curcas*. *Plant Cell Physiol.* **56**, 1655–1666.

Nitsch, J.P. (1970) Hormonal factors in growth and development. In *The Biochemistry of Fruits and Their Products*, (Hulme, A. ed), pp. 427–472. London: Academic Press.

Ozga, J.A., van Huizen, R. and Reinecke, D.M. (2002) Hormone and seed-specific regulation of pea fruit growth. *Plant Physiol.* **128**, 1379–1389.

Pan, X.Q., Welti, R. and Wang, X.M. (2008) Simultaneous quantification of major phytohormones and related compounds in crude plant extracts by liquid chromatography-electrospray tandem mass spectrometry. *Phytochemistry*, **69**, 1773–1781.

Perilli, S., Di Mambro, R. and Sabatini, S. (2012) Growth and development of the root apical meristem. *Curr. Opin. Plant Biol.* **15**, 17–23.

Reinhart, B.J., Weinstein, E.G., Rhoades, M.W., Bartel, B. and Bartel, D.P. (2002) MicroRNAs in plants. *Gene. Dev.* **16**, 1616–1626.

Ren, Z.X., Li, Z.G., Miao, Q., Yang, Y.W. and Deng, W. (2011) The auxin receptor homologue in *Solanum lycopersicum* stimulates tomato fruit set and leaf morphogenesis. *J. Exp. Bot.* **62**, 2815–2826.

Rovere, F.D., Fattorini, L., D'Angeli, S., Veloccia, A., Del Duca, S., Cai, G., Falasca, G. *et al.* (2015) *Arabidopsis* SHR and SCR transcription factors and AUX1 auxin influx carrier control the switch between adventitious rooting and xylogenesis in planta and in *in vitro* cultured thin cell layers. *Ann. Bot.* **115**, 617–628.

Schulze, S., Schäfer, B.N., Parizotto, E.A., Voinnet, O. and Theres, K. (2010) *LOST MERISTEMS* genes regulate cell differentiation of central zone descendants in *Arabidopsis* shoot meristems. *Plant J.* **64**, 668–678.

Shani, E., Yanai, O. and Ori, N. (2006) The role of hormones in shoot apical meristem function. *Curr. Opin. Plant Biol.* **9**, 484–489.

Song, S.S., Qi, T.C., Huang, H. and Xie, D.X. (2013) Regulation of stamen development by coordinated actions of jasmonate, auxin, and gibberellin in *Arabidopsis*. *Mol. Plant*, **6**, 1065–1073.

Springer, N. (2010) Shaping a better rice plant. *Nat. Genet.* **42**, 475–476.

Stuurman, J., Jäggi, F. and Kuhlemeier, C. (2002) Shoot meristem maintenance is controlled by a GRAS gene mediated signal from differentiating cells. *Gene. Dev.* **16**, 2213–2218.

Sun, X.L., Jones, W.T. and Rikkerink, E. (2012) GRAS proteins: the versatile roles of intrinsically disordered proteins in plant signaling. *Biochem. J.* **442**, 1–12.

Verelst, W., Twell, D., de Folter, S., Immink, R., Saedler, H. and Munster, T. (2007) MADS-complexes regulate transcriptome dynamics during pollen maturation. *Genome Biol.* **8**, R249.

Wang, L., Mai, Y.X., Zhang, Y.C., Luo, Q. and Yang, H.Q. (2010) MicroRNA171c-targeted *SCL6-II*, *SCL6-III*, and *SCL6-IV* genes regulate shoot branching in *Arabidopsis*. *Mol. Plant*, **3**, 794–806.

Weiss, D. and Ori, N. (2007) Mechanisms of cross talk between gibberellin and other hormones. *Plant Physiol.* **144**, 1240–1246.

Wu, N.N., Zhu, Y., Song, W.L., Li, Y.X., Yan, Y.M. and Hu, Y.K. (2014) Unusual tandem expansion and positive selection in subgroups of the plant GRAS transcription factor superfamily. *BMC Plant Biol.* **14**, 373.

Xue, X.Y., Zhao, B., Chao, L.M., Chen, D.Y., Cui, W.R., Mao, Y.B., Wang, L.J. *et al.* (2014) Interaction between two timing microRNAs controls trichome distribution in *Arabidopsis*. *PLoS Genet.* **10**, e1004266.

Yang, C.Y., Xu, Z.B., Song, J., Conner, K., Vizcay Barrena, G. and Wilson, Z.A. (2007) *Arabidopsis* MYB26/MALE STERILE35 regulates secondary thickening in the endothecium and is essential for anther dehiscence. *Plant Cell*, **19**, 534–548.

Zhang, Z.L., Ogawa, M., Fleet, C.M., Zentella, R., Hu, J., Heo, J.O., Lim, J. *et al.* (2011) Scarecrow-like 3 promotes gibberellin signaling by antagonizing master growth repressor DELLA in *Arabidopsis*. *Proc. Natl Acad. Sci. USA*, **108**, 2160–2165.

Zhou, Y., Liu, X., Engstrom, E.M., Nimchuk, Z.L., Pruneda-Paz, J.L., Tarr, P.T., Yan, A. *et al.* (2015) Control of plant stem cell function by conserved interacting transcriptional regulators. *Nature*, **517**, 377–380.

17

RNAi-mediated endogene silencing in strawberry fruit: detection of primary and secondary siRNAs by deep sequencing

Katja Härtl, Gregor Kalinowski, Thomas Hoffmann, Anja Preuss and Wilfried Schwab*

Biotechnology of Natural Products, Technische Universität München, Freising, Germany

*Correspondence

email wilfried.schwab@tum.de
GenBank™ accession numbers: *FaCHS*
(AY997297), *FaOMT* (AF220491).

Keywords: RNA interference,
Fragaria × ananassa, transitive gene
silencing, chalcone synthase,
methyltransferase, RNAseq.

Summary

RNA interference (RNAi) has been exploited as a reverse genetic tool for functional genomics in the nonmodel species strawberry (*Fragaria × ananassa*) since 2006. Here, we analysed for the first time different but overlapping nucleotide sections (>200 nt) of two endogenous genes, *FaCHS* (chalcone synthase) and *FaOMT* (O-methyltransferase), as inducer sequences and a transitive vector system to compare their gene silencing efficiencies. In total, ten vectors were assembled each containing the nucleotide sequence of one fragment in sense and corresponding antisense orientation separated by an intron (inverted hairpin construct, ihp). All sequence fragments along the full lengths of both target genes resulted in a significant down-regulation of the respective gene expression and related metabolite levels. Quantitative PCR data and successful application of a transitive vector system coinciding with a phenotypic change suggested propagation of the silencing signal. The spreading of the signal in strawberry fruit in the 3′ direction was shown for the first time by the detection of secondary small interfering RNAs (siRNAs) outside of the primary targets by deep sequencing. Down-regulation of endogenes by the transitive method was less effective than silencing by ihp constructs probably because the numbers of primary siRNAs exceeded the quantity of secondary siRNAs by three orders of magnitude. Besides, we observed consistent hotspots of primary and secondary siRNA formation along the target sequence which fall within a distance of less than 200 nt. Thus, ihp vectors seem to be superior over the transitive vector system for functional genomics in strawberry fruit.

Introduction

Eukaryotic organisms, including plants, animals and fungi, have developed a double-stranded RNA (dsRNA)-induced gene silencing mechanism to protect their cells against invading nucleic acids (Brodersen and Voinnet, 2006). In this RNA interference (RNAi) pathway, dsRNA is specifically and rapidly degraded into small interfering RNAs (siRNAs) of 21 to 25 nucleotides (nt) by RNase III-like enzymes, known as Dicer-like proteins (DCL; Eamens et al., 2008; Mlotshwa et al., 2008). The siRNAs are incorporated into the RNA-induced silencing complex (RISC), which eventually leads to the degradation of any complementary single-stranded RNA in the cytoplasm. In addition, siRNAs are recruited into RNA-induced initiation of the transcriptional silencing complex (RITS) causing chromatin modifications (Noma et al., 2004). Foldback of self-complimentary intron-hairpin sequences (ihp), hybridization of sense and antisense sequences and the action of an RNA-dependent RNA polymerase (RdRP) can give rise to dsRNA from endogenous, viral and transgenic RNA molecules. Studies have shown that RNAi is implicated in the suppression of transposon activity, resistance to viral infection, post-transcriptional and post-translational regulation of gene expression, and the epigenetic regulation of chromatin structure (Kusaba, 2004; Wang and Metzlaff, 2005).

RNAi has attracted great attention as a reverse genetic tool for studies of gene function in numerous organisms (Gilchrist and Haughn, 2010; Hoffmann et al., 2006; Lipardi et al., 2003; Zhai

et al., 2009). The sequence specificity of RNAi-mediated gene inactivation allows silencing of individual genes as well as multiple members of a multigene family (Miki et al., 2005). Transitivity, the spread of RNA silencing along primary target sequences, is an aspect of RNAi that has not been well understood (Bleys et al., 2006). Endogene-derived mRNAs can become a production source of secondary siRNAs that correspond to regions of the target gene outside the original trigger (Sanders et al., 2002; Sijen et al., 2001). Secondary siRNA itself can trigger degradation of homologous transcripts through a process called transitive silencing. The production of secondary siRNAs from regions outside of the sequence initially targeted by primary siRNAs was first described in *Caenorhabditis elegans* where it proceeds over a distance of a few hundred nucleotides in the 3′–5′ direction (Sijen et al., 2001).

In plants, spreading of the RNAi signal has been shown to occur in both the 3′–5′ and the 5′–3′ directions along transgene RNAs in a primer-dependent and primer-independent manner (Bleys et al., 2006; Braunstein et al., 2002; Himber et al., 2003; Klahre et al., 2002; Kościańska et al., 2005; Miki et al., 2005; Petersen and Albrechtsen, 2005; Vaistij et al., 2002; Van Houdt et al., 2003). Silencing induced by 3′ fragments spread only for a limited distance of up to 332 nt, with a possible limit of 600 nt, while gene silencing in the 5′–3′ direction has been shown to spread over a distance of at least 1000 nt (Petersen and Albrechtsen, 2005; Vaistij et al., 2002). Although transitive RNA silencing of transgenes has been clearly demonstrated, many studies failed to prove transitivity along endogenous sequences. It was assumed that endogenous sequences are resistant to transitivity by some inherent properties (Vaistij *et al.*, 2002; Himber *et al.*, 2003; Kościańska *et al.*, 2005; Miki *et al.*, 2005; Petersen and Albrechtsen, 2005).

In plants, animals and fungi, RdRPs have been proposed to be a part of the RNAi silencing system (Schwach *et al.*, 2005). Their likely biochemical role is the production of dsRNA, which is consistent with the *in vitro* activity of RdRPs from *Lycopersicum esculentum* (Schiebel 1998), *Neurospora crassa* (Makeyev and Bamford, 2002) and *Schizosaccharomyces pombe* (Motamedi *et al.*, 2004). Models of transitive silencing suggest that RdRPs utilize siRNAs as primers and homologous transcripts as template for dsRNA synthesis (Alder *et al.*, 2003; Sijen *et al.*, 2001). The reaction amplifies the silencing response and mediates the production of secondary siRNAs. RdRPs add ribonucleotides to the 3′ end of a growing RNA chain, which would explain spreading of the silencing in the 3′–5′ direction along the target transcript. The 5′–3′ spreading of the silencing signal can be explained by RdRP-mediated unprimed dsRNA synthesis at the 3′ end of target mRNAs and/or by siRNA-primed dsRNA synthesis using an antisense transcript as template (Baulcombe, 2004; Vaistij *et al.*, 2002).

In this paper, we describe the down-regulation and the spread of silencing of two endogenous strawberry genes *Fragaria × ananassa* chalcone synthase (*FaCHS*; Lunkenbein *et al.*, 2006a) and *F. × ananassa* O-methyltransferase (*FaOMT*; Wein *et al.*, 2002) by ihp constructs and transitive RNAi vectors. To systematically investigate the sequence dependency of RNAi directed to *FaOMT* and *FaCHS*, we prepared ihp constructs harbouring sequence fragments (219–303 bp) along the full lengths of both target genes. All sequence fragments resulted in a significant down-regulation of the respective gene expression. The spreading of the silencing signal could be demonstrated by agroinfiltration of transitive RNAi vectors and was confirmed by deep sequencing of small RNAs. The data suggest that both direct ihp and transitive methods are useful tools in plant functional genomics and biotechnology, but homologous ihp sequences are superior due to their silencing efficacy, which was also demonstrated by the abundance of primary siRNAs in comparison with secondary siRNAs.

Results

Efficient silencing of *FaOMT* and *FaCHS* is independent of the sequence fragments used in the ihp construct

RNA silencing with an ihp construct (pBI-*FaCHSi*) containing a 303-bp-long sense and antisense sequence of *FaCHS* separated by a *Fragaria* intron sequence has been used to suppress chalcone synthase activity in strawberry (*Fragaria × ananassa*) fruit, resulting in a significantly reduced level of red fruit pigments (Hoffmann *et al.*, 2006). Here, we tested the general applicability of ihp constructs to silence endogenous genes in fruits of the nonmodel strawberry plant and systematically analysed the outcome of the silencing technique, to devise a method for the application of RNAi to high-throughput plant functional genomics in this species. In addition to *FaCHS*, the flavour-related *FaOMT* gene (AF220491; Wein *et al.*, 2002; Zorrilla-Fontanesi *et al.*, 2012) was chosen. Repression of *FaOMT* transcripts led to a near total loss of 2,5-dimethyl-4-methoxy-3(2H)-furanone (DMMF), the product formed by the encoded enzyme from the substrate 4-hydroxy-2,5-dimethyl-3(2H)-furanone (HDMF). In addition,

lower levels of feruloyl glucose in comparison with caffeoyl glucose were observed (Lunkenbein *et al.*, 2006a,b). Since it has been shown that the potency of siRNAs can vary drastically due to their thermodynamic properties, and that complex target structures may affect the activity of siRNAs (Kurreck, 2006; Overhoff *et al.*, 2005; Schubert *et al.*, 2005), we systematically investigated the efficiency of RNAi-mediated gene silencing by different ihp constructs covering the complete ORFs of *FaOMT* and *FaCHS*. Both gene sequences were divided into five (A to E) overlapping fragments (Figures 1-3). These fragments were used for the assembly of the ihp constructs pBI-*FaOMT_Ai* to *Ei* and pBI-*FaCHS_Ai* to *Ei*. Subsequently, they were agroinfiltrated into ripening strawberry fruits, together with the control vector pBI-Intron (Figure 1). Quantitative PCR was applied to assess the efficacy of target gene silencing (Figures 2 and 3). The relative expression levels of *FaOMT* and *FaCHS* in native versus pBI-Intron treated fruit were statistically not significant ($P = 7.2$ E-1 and $P = 5.7$ E-1), indicating that injection of the control construct did not alter the expression of the target genes. However, agroinfiltration of ihp constructs resulted in a significant ($P < 1.0$ E-2) reduction in the transcript levels of *FaOMT* (Figure 2) and *FaCHS* (Figure 3) by more than 80% (relative gene expression < 0.2) in comparison with the levels in untreated fruits, with the exception of pBI-*FaCHS_Ci*. Although the median for pBI-*FaCHS_Ci* is less than 0.2 (reduction by more than 80%), the *P*-value is slightly higher than 1.0 E-2 due to the high variance of the values. The uneven distribution of the *A. tumefaciens* strains in the tissue is an inherent problem of the agroinfiltration technique and causes high variation of the expression level. The data indicate that all five sequences comprising 240 up to 303 nt of both genes can act as efficient inducers as well as targets of post-transcriptional silencing.

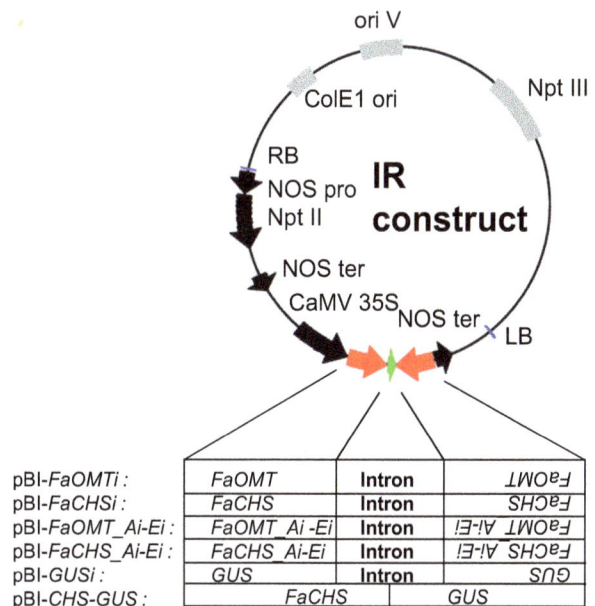

pBI-*FaOMTi* :	*FaOMT*	Intron	*FaOMT*
pBI-*FaCHSi* :	*FaCHS*	Intron	*FaCHS*
pBI-*FaOMT_Ai-Ei* :	*FaOMT_Ai -Ei*	Intron	*FaOMT Ai-Ei*
pBI-*FaCHS_Ai-Ei* :	*FaCHS_Ai-Ei*	Intron	*FaCHS Ai-Ei*
pBI-*GUSi* :	*GUS*	Intron	*GUS*
pBI-*CHS-GUS* :	*FaCHS*		*GUS*

Figure 1 Schematic diagram of binary vectors for the determination of RNAi-mediated *FaOMT* and *FaCHS* transcript degradation. The binary vectors shown are derivatives of pBI121. The expressed part of each plasmid relevant to this study is shown. Details about sizes and positions of the cloned fragments are summarized in Table S1.

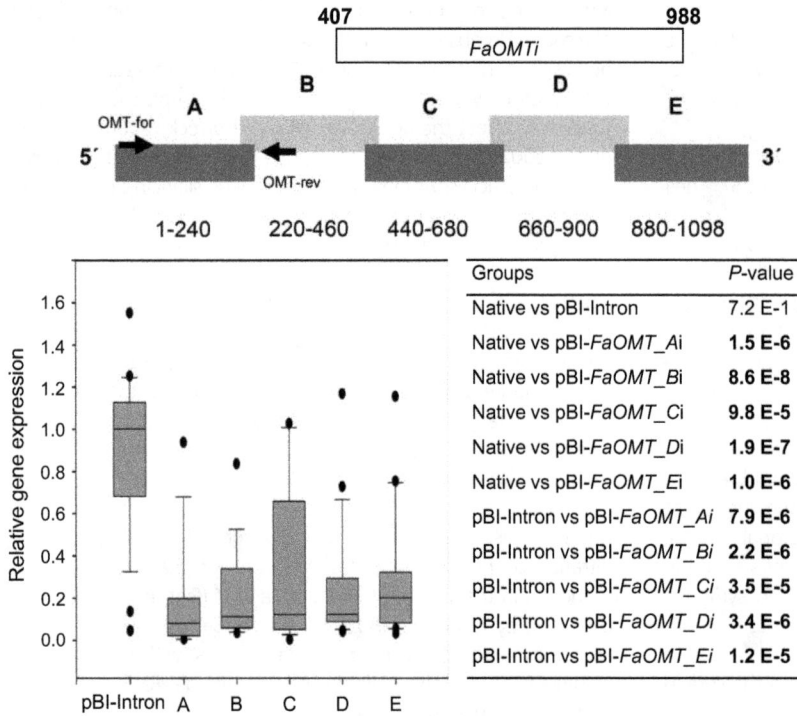

Groups	P-value
Native vs pBI-Intron	7.2 E-1
Native vs pBI-*FaOMT*_Ai	**1.5 E-6**
Native vs pBI-*FaOMT*_Bi	**8.6 E-8**
Native vs pBI-*FaOMT*_Ci	**9.8 E-5**
Native vs pBI-*FaOMT*_Di	**1.9 E-7**
Native vs pBI-*FaOMT*_Ei	**1.0 E-6**
pBI-Intron vs pBI-*FaOMT*_Ai	**7.9 E-6**
pBI-Intron vs pBI-*FaOMT*_Bi	**2.2 E-6**
pBI-Intron vs pBI-*FaOMT*_Ci	**3.5 E-5**
pBI-Intron vs pBI-*FaOMT*_Di	**3.4 E-6**
pBI-Intron vs pBI-*FaOMT*_Ei	**1.2 E-5**

Figure 2 RNAi-mediated gene silencing of *FaOMT*. Division of *FaOMT* into sections A to E with information about the nucleic acid positions, and the relative position of the PCR product obtained with primer pair OMT-for and OMT-rev (top). A longer fragment *FaOMTi* of 581 nt was also used (refer to Figure 7). Relative *FaOMT* gene expression in strawberry fruit infiltrated with pBI-Intron and ihp constructs *FaOMT*_Ai to *FaOMT*_Ei (A–E) in relation to the mean expression in native untreated fruit which was set to 1 (bottom left). Number of biological replicates: native untreated fruit ($N = 17$), pBI-Intron ($N = 27$), A ($N = 14$), B ($N = 18$), C ($N = 18$), D ($N = 21$), E ($N = 21$). The Wilcoxon–Mann–Whitney U-test was used to assess intergroup significance (bottom right). Statistically significant differences ($P < 1.0$ E-2) are shown in bold.

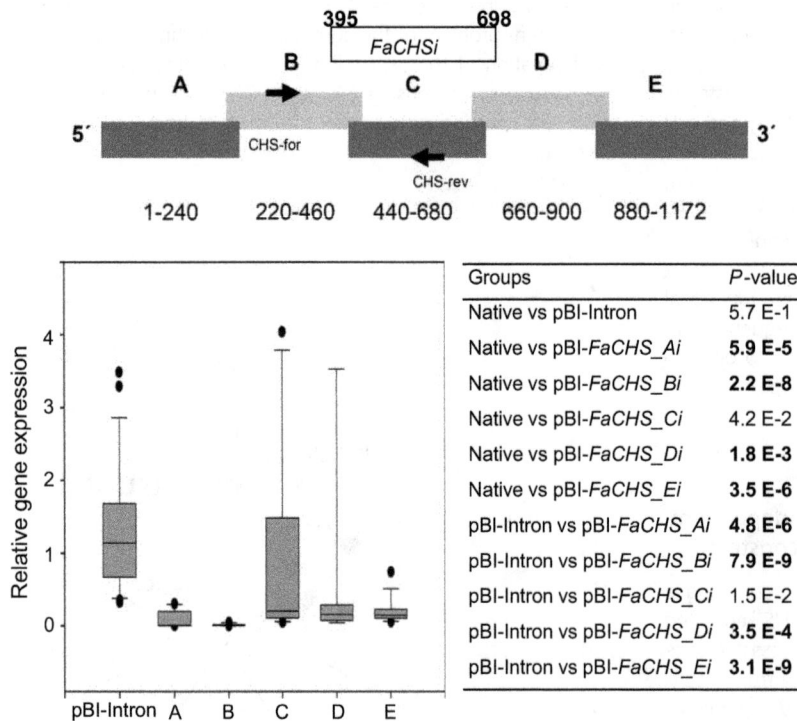

Groups	P-value
Native vs pBI-Intron	5.7 E-1
Native vs pBI-*FaCHS*_Ai	**5.9 E-5**
Native vs pBI-*FaCHS*_Bi	**2.2 E-8**
Native vs pBI-*FaCHS*_Ci	4.2 E-2
Native vs pBI-*FaCHS*_Di	**1.8 E-3**
Native vs pBI-*FaCHS*_Ei	**3.5 E-6**
pBI-Intron vs pBI-*FaCHS*_Ai	**4.8 E-6**
pBI-Intron vs pBI-*FaCHS*_Bi	**7.9 E-9**
pBI-Intron vs pBI-*FaCHS*_Ci	1.5 E-2
pBI-Intron vs pBI-*FaCHS*_Di	**3.5 E-4**
pBI-Intron vs pBI-*FaCHS*_Ei	**3.1 E-9**

Figure 3 RNAi-mediated gene silencing of *FaCHS*. Division of *FaCHS* into sections A to E with information about the nuclei acid positions and the relative position of the PCR product obtained with primer pair CHS-for and CHS-rev (top). Fragment *FaCHSi* was successfully used by Hoffmann *et al.* (2006). Relative *FaCHS* gene expression in strawberry fruit infiltrated with pBI-Intron and ihp constructs *FaCHS*_Ai to *FaCHS*_Ei (A–E) in relation to the mean expression in native untreated fruit which was set to 1 (bottom left). Number of biological replicates: native untreated fruit ($N = 23$), pBI-Intron ($N = 26$), A ($N = 10$), B ($N = 10$), C ($N = 10$), D ($N = 9$), E ($N = 15$). The Wilcoxon–Mann–Whitney U-test was used to assess intergroup significance (bottom right). Statistically significant differences ($P < 1.0$ E-2) are shown in bold.

Efficient RNA silencing manifests itself in altered metabolite levels

Because *FaCHS* encodes an essential enzyme in the anthocyanin biosynthetic pathway, the reduction in *FaCHS* transcript levels results in a loss of pigmentation in infected fruits when compared to control fruits. All five ihp constructs produced a similar chimeric phenotype, indicating a significant reduction in pelargonidin 3-O-glucoside (Figure 4). This was confirmed by LC-MS analyses (data

not shown). In contrast, down-regulation of *FaOMT* by pBI-*FaOMT*_Ai to Ei did not yield an altered visible phenotype. However, LC-MS analysis revealed differences in the ratio of HDMF to the total amount of furanones (DMMF and HDMF). Strawberries agroinfiltrated with ihp constructs targeted to *FaOMT* produced significantly lower levels ($P < 1.0$ E-2) of the methylated product DMMF as compared with the substrate HDMF, resulting in a higher proportion of HDMF relative to the total amount of furanones (Figure 5, top). All five constructs used

Untreated   pBI-*FaCHS_Ai*   pBI-*FaCHS_Bi*

pBI-*FaCHS_Ci*   pBI-*FaCHS_Di*   pBI-*FaCHS_Ei*

Figure 4 Phenotypes of untreated strawberry fruit and fruit agroinfiltrated with pBI-*FaCHS_Ai* to pBI-*FaCHS_Ei*.

were similarly effective. The normalized levels of HDMF in the fruit expressing pBI-Intron were not significantly different from the levels in untreated fruit ($P = 6.4$ E-1).

The normalized levels of feruloyl glucose another potential product formed by FaOMT in fruit were not affected by agroinfiltration of pBI-*FaOMT_Ai* to *Ei* when compared to the levels in untreated and pBI-Intron infiltrated fruit (Figure 5, bottom). However, statistical analysis revealed a significant difference ($P = 3.0$ E-3) between the normalized level in untreated and pBI-Intron control fruit, indicating an induction of the phenylpropanoid biosynthesis by agroinfiltration. The increased formation of phenylpropanoic acids probably counteracts the metabolic effects of *FaOMT* silencing on the levels of the phenolic acids.

Transitive silencing with transcriptional fusion of *GUS* ihp and a partial sequence of *FaCHS*

We assembled two constructs (Figure 1) to probe whether the RNA silencing activity of a silencing inducer RNA can be transmitted to the sequence of an endogenous strawberry gene (secondary target) by fusing part of this sequence to the silencing-inducing sequence in a single transcript (primary target; Bleys *et al.*, 2006). One nucleotide sequence represented a silencing inducer and the other a primary target. A construct carrying an ihp of a partial GUS sequence (pBI-*GUSi*) was used as source of primary siRNAs, while a fusion of the homologous GUS sequence with a *FaCHS* fragment (pBI-*CHS-GUS*) served as primary target (Van Houdt *et al.*, 2003). T-DNAs were introduced into ripening strawberry fruit by agroinfiltration 14 days after pollination. Strawberries harvested in the full ripe stage showed the typical chimeric phenotype of *FaCHS* deficient fruits but did not feature

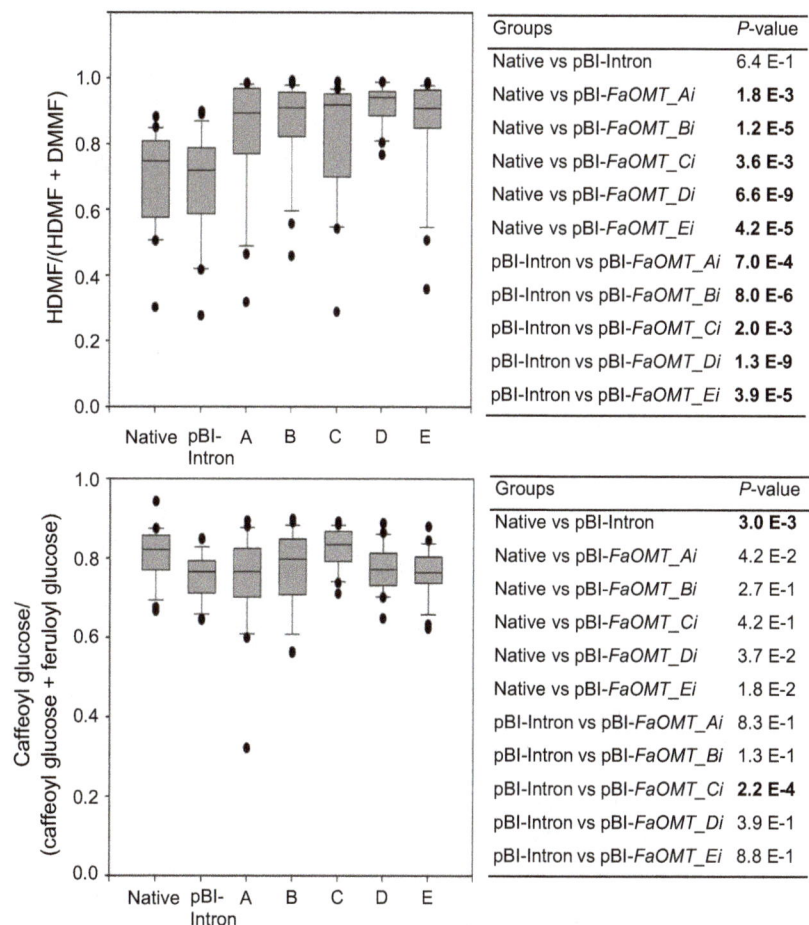

Groups	P-value
Native vs pBI-Intron	6.4 E-1
Native vs pBI-*FaOMT_Ai*	**1.8 E-3**
Native vs pBI-*FaOMT_Bi*	**1.2 E-5**
Native vs pBI-*FaOMT_Ci*	**3.6 E-3**
Native vs pBI-*FaOMT_Di*	**6.6 E-9**
Native vs pBI-*FaOMT_Ei*	**4.2 E-5**
pBI-Intron vs pBI-*FaOMT_Ai*	**7.0 E-4**
pBI-Intron vs pBI-*FaOMT_Bi*	**8.0 E-6**
pBI-Intron vs pBI-*FaOMT_Ci*	**2.0 E-3**
pBI-Intron vs pBI-*FaOMT_Di*	**1.3 E-9**
pBI-Intron vs pBI-*FaOMT_Ei*	**3.9 E-5**

Groups	P-value
Native vs pBI-Intron	**3.0 E-3**
Native vs pBI-*FaOMT_Ai*	4.2 E-2
Native vs pBI-*FaOMT_Bi*	2.7 E-1
Native vs pBI-*FaOMT_Ci*	4.2 E-1
Native vs pBI-*FaOMT_Di*	3.7 E-2
Native vs pBI-*FaOMT_Ei*	1.8 E-2
pBI-Intron vs pBI-*FaOMT_Ai*	8.3 E-1
pBI-Intron vs pBI-*FaOMT_Bi*	1.3 E-1
pBI-Intron vs pBI-*FaOMT_Ci*	**2.2 E-4**
pBI-Intron vs pBI-*FaOMT_Di*	3.9 E-1
pBI-Intron vs pBI-*FaOMT_Ei*	8.8 E-1

Figure 5 Metabolite analyses. Normalized levels of HDMF (top left) and caffeoyl glucose (bottom left) in native untreated strawberry fruit and fruit infiltrated with pBI-Intron and ihp constructs *FaOMT_Ai* to *FaOMT_Ei* (A–E) calculated as ratio of HDMF to the total concentration of HDMF and DMMF and as ratio of caffeoyl glucose to the total concentration of caffeoyl glucose and feruloyl glucose. Number of biological replicates: native untreated fruit ($N = 22$), pBI-Intron ($N = 26$), A ($N = 22$), B ($N = 23$), C ($N = 21$), D ($N = 21$), E ($N = 23$). The Wilcoxon–Mann–Whitney U-test was used to assess intergroup significance (top right and bottom right). Statistically significant differences ($P < 1.0$ E-2) are shown in bold.

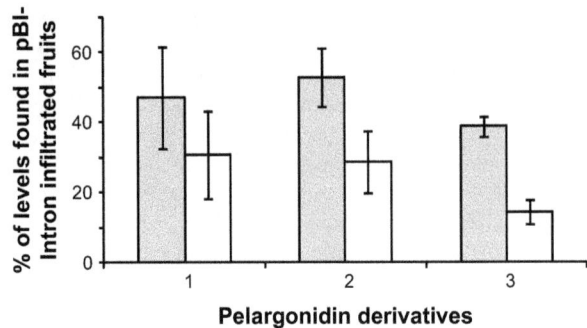

Figure 6 Efficacy of ihp-PTGS versus transitive RNA silencing. Comparison of the levels of pelargonidin derivatives (1, pelargonidin 3-O-glucoside, 2, pelargonidin 3-O-glucoside-6'-O-malonate, 3, (epi) afzelechin-pelargonidin 3-O-glucoside) in strawberry fruit agroinfiltrated with pBI-FaCHSi (white bars) and with a silencing inducer pBI-GUSi in combination with pBI-CHS-GUS (grey bars) carrying one region homologous to the silencing gene (GUS) and another homologous to the target (FaCHS). Infiltration of ihp construct of FaCHS results in a strong down-regulation of pelargonidin derivatives (white bars) while transitive silencing (grey bars) is less efficient.

the extensive white sections as seen with fruit agroinfiltrated with FaCHS ihp constructs (data not shown). LC-MS analysis confirmed that the levels of pelargonidin derivatives were not as strongly suppressed by transitive silencing as by ihp constructs of FaCHS sequences (Figure 6). In a control experiment, only the FaCHS-GUS construct was agroinfiltrated but without pBI-GUSi. The result clarified that the silencing effect is indeed caused by transitive silencing and not by FaCHS-GUS-induced sense cosuppression as infiltration of FaCHS-GUS does not affect the pigmentation of strawberry fruit (Figure S1). It appears that transitive silencing of the endogenous FaCHS gene(s) by chimeric recombinant GUS-CHS fusions, together with the inducer pBI-GUSi, is less efficient than the down-regulation by ihp constructs targeting directly FaCHS sequences.

Detection of small RNAs by next-generation sequencing

To test whether the reduction in endogene mRNA is caused by transitive silencing in addition to digestion with exonucleases such as XRN4, we assessed the accumulation of sequence-specific small RNAs by next-generation sequencing after agroinfiltration of pBI-FaOMT_Ci and pBI-FaOMTi into ripening strawberry fruits and focused on primary and secondary siRNAs targeting the inducer regions and the full-length FaOMT.

We analysed the abundance of small RNAs in FaOMT_Ci and FaOMTi fruits to quantify siRNAs derived from the target sequence (primary siRNAs) and regions outside of the target (secondary siRNAs). Mapping of the quality checked sequences to the reference genome resulted in 7 245 462 (62% of total reads, FaOMT_Ci_short) and 11 263 154 (66% of total, FaOMTi_long) transcript counts, of which 9.5% (FaOMT_ Ci short) and 9% (FaOMTi_long) arise from the FaOMT sequence (gene 12447; accession number AF220491; Figure 7). The size distribution of total genomewide short RNAs showed the predominance of siRNAs of 21 nt and to a lesser extent of 22 nt and 24 nt, for both constructs (Figure S2a). Almost all of the siRNAs derived from FaOMT_Ci and FaOMTi were mapped to their target regions and are primary siRNAs (Figure 7c). This conclusion was supported by the dominance of short RNAs of 21 nt, 22 nt and 24 nt that mapped in the FaOMT_Ci and FaOMTi target region (Figure S2b).

However, low-abundant short RNAs were also visible upstream (5′) of both, the FaOMT_Ci and FaOMTi target sequences, and in particular downstream (3′) of both inducer sequences (Figure 7c insets), confirming the formation of secondary siRNAs. They almost exclusively consisted of short RNAs of 21 nt and 22 nt (Figure S2c).

There is uneven, nonrandom distribution of siRNAs within the target sequences suggesting the presence of hotspots for siRNA formation. Interestingly, the positions of these hotspots differed between the sense and antisense strands but not between primary and secondary siRNAs (Figure 7d). The similarity of the abundance profiles but not the absolute counts of the secondary siRNAs produced by FaOMT_Ci and the primary siRNAs evoked by FaOMTi indicates that related enzymes are involved in the cleavage of the precursor RNAs. The size distribution of siRNAs mapped in the inducer regions FaOMT_Ci and FaOMTi revealed, almost exclusively, siRNAs of 21, 22 and 24 nt (Figure S2b).

Discussion

We have exploited RNAi as a reverse genetic tool to develop a rapid system to perform functional genomics in strawberry fruit (F. × ananassa) (Griesser et al., 2008; Hoffmann et al., 2006; Matthew, 2004; Song et al., 2015). To systematically investigate the silencing efficacy of ihp constructs that carry different gene sequence segments, we divided the ORFs of FaCHS and FaOMT into five fragments (240–293 nt). In total, ten vectors were assembled each containing the nucleotide sequence of one fragment in sense and corresponding antisense orientation separated by an intron obtained from a F. × ananassa quinone oxidoreductase gene. Bioinformatic analysis of the modelled mRNA structures suggested different silencing efficiencies of the sequences A to E due to strongly varying theoretical accessibilities (ranging from 0.48 to 1.90 for FaOMT fragments and 0.05 to 1.88 for FaCHS fragments) of local target sites as has been reported by Schubert et al., 2005;. Accessible local target sites are defined as mRNA regions with more than 10 unpaired nt (Overhoff et al., 2005). However, the efficiencies of the applied constructs to silence the endogenous FaCHS and FaOMT genes did not show significant differences. Thus, it appears that additional factors govern the efficiency of the post-transcriptional silencing process in plant cells. Until now, all characterized sRNAs get incorporated into ARGONAUTE (AGO) proteins to contribute to RISC formation (Fang and Qi, 2016). Recently, it was suggested that subcellular compartmentalization of AGO-miRNA-RISC complexes and sequestration of miRNAs and siRNAs could be additional factors determining the silencing efficiency (Fang and Qi, 2016; Wu et al., 2015).

Although our results did not show any significant difference in the silencing efficiency of the various constructs, the uneven distribution of primary siRNAs along the target sequence (Figure 7), that is the detection of siRNA hotspots and low-abundant siRNAs, suggests that the use of shorter inducer sequence would eventually lead to unequal silencing efficacies. As the distance between siRNA hotspots is less than 200 nt inducer, sequences of more than 200 nt in length are sufficient to initiate the stable down-regulation of endogenous genes.

Quantitative PCR analyses with primer pairs specific to the 5′ regions of the target RNAs (located in FaOMT fragment A to B and FaCHS fragment B to C) revealed substantial degradation of primer target sites after agroinfiltration of all ihp RNA sequences

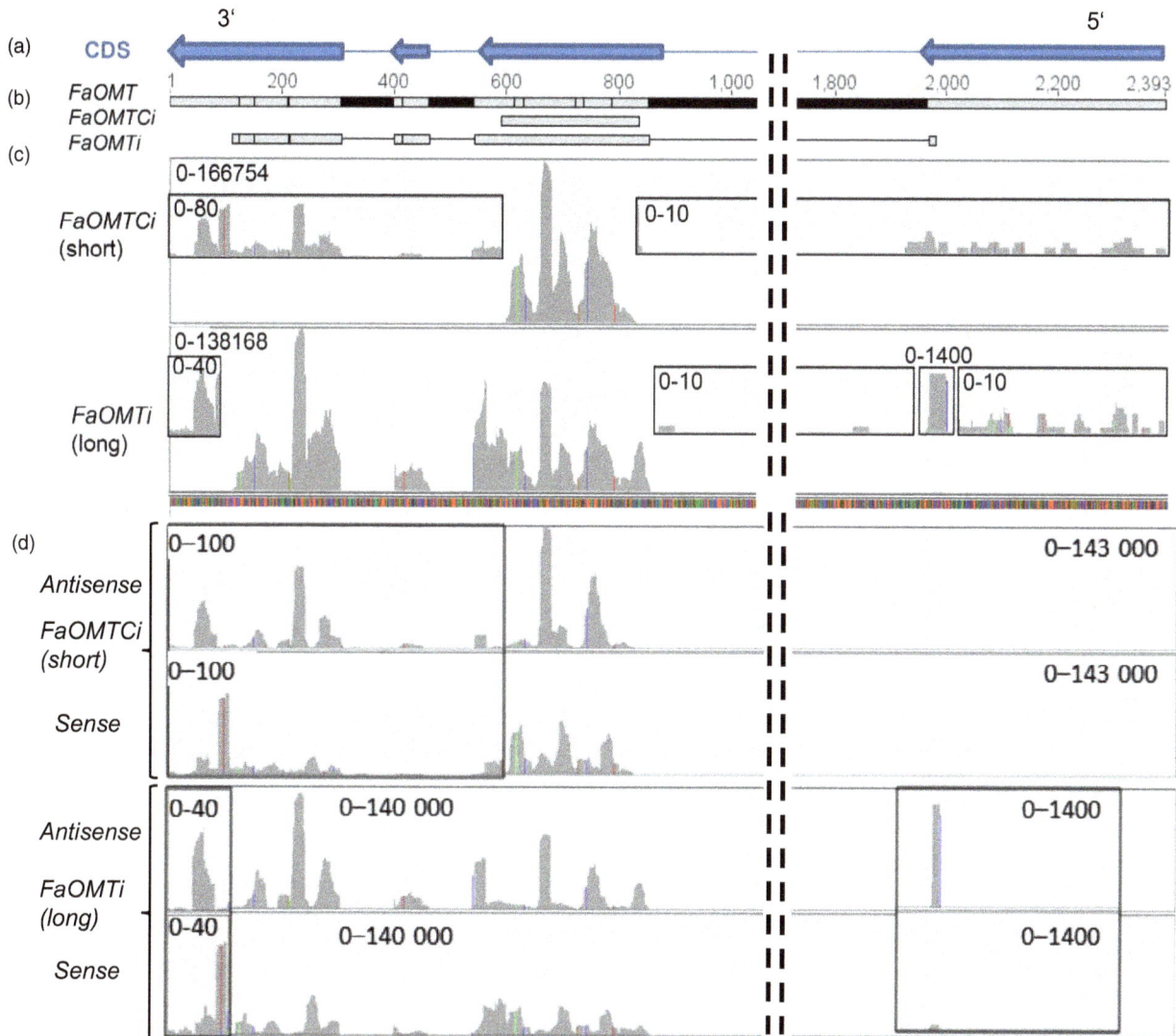

Figure 7 Detection of small RNAs derived from the *FaOMT* gene. Coding sequence (CDS, a). *FaOMT* gene with exons (grey)/introns (black) and number of nucleotides as well as locations of gene fragments *FaOMT_Ci* and *FaOMTi* used for the assembly of RNAi constructs (b). Abundance of siRNAs derived from different regions of *FaOMT* after agroinfiltration of the short fragment *FaOMT_Ci* and the long sequence *FaOMTi*. Insets show secondary siRNAs (c). Abundance of antisense and sense siRNAs derived from different regions of *FaOMT* after agroinfiltration of the short fragment *FaOMT_Ci* and the long sequence *FaOMTi* (d). Bar charts were computed with the Integrated Genomics Viewer (igv; Robinson *et al.*, 2011). Numbers denote the range of the abundances.

used (Figures 2 and 3). Degradation of mRNA outside the primary RNAi target sites might be explained by digestion of the aberrant mRNA with exonucleases such as XRN4 (Vazquez and Hohn, 2013). Alternatively, this observation might provide a hint of the spreading of the RNAi signal from the 3′ to the 5′ region over a distance of up to 1000 nt and in the reverse direction of up to 500 nt. Spreading over a distance of at least 1000 nt from the 5′ end to the 3′ end has been shown by virus-induced gene silencing (VIGS), while 3′–5′ spreading extended at least through 332 nt, with a possible limit of 600 nt (Petersen and Albrechtsen, 2005; Vaistij *et al.*, 2002). Thus, agroinfiltration of ihp constructs appears to be equally effective in spreading of the RNAi signal as VIGS.

It has been proposed that the interaction of sense transcripts with the antisense siRNAs might change the structure of the RNA or of a ribonucleoprotein complex and thereby allow RdRP to access the 3′ end of the target RNA. Synthesis of dsRNA from

the 3′ end would result in siRNA production corresponding to the entire transcript sequence and could explain RdRP-mediated spreading in a primer-independent way (Vaistij *et al.*, 2002).

There are only a few reports showing the spread of RNAi along endogenous genes as we have shown by agroinfiltration of transitive vectors into strawberry fruit (Figure 6). By contrast, transitivity has frequently been observed along reporter transgenes (Sanders *et al.*, 2002). Since most studies failed to demonstrate spreading along endogenous plant genes, it has been concluded that transgenes, and not endogenous genes, are good templates for generation of secondary siRNAs. Reasons for the plant RdRP's apparent preference for the transgene transcripts as a template in the transitivity remained unclear (Miki *et al.*, 2005; Petersen and Albrechtsen, 2005; Shimamura *et al.*, 2007; Vaistij *et al.*, 2002). Transgenes and RNAi probes are mostly expressed under the control of strong promoters (e.g. CaMV 35S). Thus, the primary silencing inducer targeting the transgene gives rise to a large number of primary siRNAs that eventually exceed a critical threshold level, which may lead to the activation of RdRP, and finally to efficient transitive silencing. Consistent with this hypothesis, it has been shown that a hairpin construct controlled by the strong 35S promoter induced a stronger silencing phenotype than the same construct controlled by the weak nopaline synthase promoter (Chuang and Meyerowitz, 2000). In addition, the efficiency of transitive silencing of endogens depends on the degree of sequence homology to the primary target (Bleys et al., 2006). However, strong silencing of the target RNA would mean that the template RNA for RdRP is scarce, which would also reduce the efficiency of spreading (Vaistij et al., 2002). Taken together, both the level of siRNAs and the target transcript might determine the onset of transitivity. Transcript levels of numerous endogenous genes are less abundant and therefore less prone to transitive silencing (Vaistij et al., 2002).

Attempts at explaining the absence of RNA silencing of the phytoene desaturase (PDS) endogenous gene have evoked the characteristics of endogenous RNAs that inhibit RdRP or prevent interactions with siRNAs (Vaistij et al., 2002). However, in the meantime, PDS alleles have been successfully down-regulated by a hairpin construct in sugarcane and tobacco, which invalidates the former hypothesis (Osabe et al., 2009; Wang et al., 2010). Thus, the assumption that transgenes are better templates for production of secondary siRNAs is no longer considered a likely explanation for the inability to detect transitivity in certain studies. Our results support the hypothesis that there is no principle distinction between transgenes and endogenous genes, except their promoters with respect to their active recruitment in RNA silencing including spreading of the signal (Sanders et al., 2002).

In Arabidopsis, it has been shown that both transgenes and endogenes can be silenced by secondary transitive signals (Bleys et al., 2006). Since insertion of additional nt sequences between the region targeted by the silencing inducer and the upstream region homologous to a transgenic target (GUS) led to a delay of the onset of transitive silencing, it has been suggested that transitivity requires time to accumulate a certain steady state level of secondary siRNAs that results in a corresponding maximum plateau level of silencing. Thus, ihp constructs driven by a strong promoter will reach the level earlier.

RNA silencing was discovered in plants as a mechanism whereby invading nucleic acids such as viruses are suppressed. In this context, it appears logical that transgenes that are expressed under the control of viral promoters (e.g. CaMV 35S) are efficiently silenced by transitive RNAi, while endogenous transcripts seem to be protected (Bleys et al., 2006). This observation supports the proposed biological function of this process as a natural antiviral response. Spreading of the RNAi signal by transitive silencing even intensifies the efficacy of viral gene suppression. However, it could also lead to uncontrollable degradation of plant genes that show partial homology to the inducer sequence. In strawberry, the introduction of the Vitis vinifera stilbene synthase gene under the control of the 35S promoter inadvertently caused the down-regulation of the endogenous chalcone synthase gene transcripts, probably due to transitive silencing (Hanhineva et al., 2009).

To detect primary and secondary siRNAs and to clearly prove the spreading of the RNAi signal, deep sequencing of small RNAs in FaOMT_Ci and FaOMTi fruits was performed for the first time in strawberry fruit and revealed 21, 22 and 24 nt small RNAs (Figure 7d) as the most dominant sizes, consistent with the siRNA size distribution observed in angiosperms (Chávez Montes et al.,

2014). The high percentage of 21nt RNAs indicated that they were predominantly produced by the activity of DCL4-like nucleases (Fusaro et al., 2006). DCL2 and DCL3 can substitute DCL4, producing 22nt and 24nt RNAs (Ghildiyal and Zamore, 2009; Small, 2007). In plants, AGO proteins sort miRNAs and siRNAs based on size and the identity of the 5' nucleotide. 21-mers typically associate with AGO1 and guide mRNA cleavage and subsequent secondary siRNA production, whereas 24-mers associate with AGO4 and 6 promoting the formation of repressive chromatin (Ghildiyal and Zamore, 2009). Furthermore, it has been shown that also 22nt miRNAs can trigger siRNA production, but apparently asymmetrically positioned bulged bases in the miRNA:miRNA* duplex within the AGO proteins are sufficient to trigger transitivity (Manavella et al., 2012). When only small RNAs were considered that mapped into their target sequence (FaOMT_Ci and FaOMTi), a marked reduction of the proportion of long RNAs (>25 nt) was observed but a substantial number of short RNAs (<21 nt) were still counted (Figure S2). Interestingly, 9.5% and 9% of the siRNAs isolated from FaOMT_Ci and FaOMTi fruits, respectively, targeted the FaOMT sequence. Thus, the RNAi machinery was efficiently recruited by the ihp vectors. The proportion of long RNAs (>21 nt) was increased in FaOMTi fruits in comparison with FaOMT_Ci samples, whereas the percentage of short RNAs (<21 nt) was higher in FaOMT_Ci fruit. The reason for this observation is unclear, but the very short RNAs might be secondary degradation products of siRNAs.

Clustering of siRNAs along the inducer sequences was observed, indicating that siRNAs are apparently nonrandomly distributed (Figure 7). Similar asymmetry in siRNA distribution has been observed in studies on the RNAi-mediated silencing of the GFP gene (Llave et al., 2002) and plant virus-induced RNA silencing (Molnár et al., 2005). One explanation is that loading of siRNAs into AGO proteins, which are part of the RISC complex, is specified by the 5' terminal nucleotide as most 21 nt siRNAs with 5' terminal U are predominantly associated with AGO (Mi et al., 2008). Secondly, the structure of the dsRNA might determine the specificity and efficiency of DCL activity (Vermeulen et al., 2005). Besides, it was shown that the position of transacting siRNA-generating loci (TAS genes) can restrict siRNA production to certain regions (Rajeswaran et al., 2012). Also the distinctive distribution pattern of mapped short sequences in FaOMT_Ci and FaOMTi infiltrated fruits indicates certain hotspots of siRNA production (Figure 7c and d).

Primary siRNAs were identified by mapping to the inducer sequence, whereas secondary siRNAs are produced from regions outside of the sequence initially targeted by primary siRNAs. Low-abundant secondary siRNAs were detected upstream and in particular downstream of the inducer sequences (Figure 7c insets) indicating that formation by primer-independent synthesis by RdRP strongly prevails. The low numbers of small RNAs upstream of the inducer sequence make transitive silencing by siRNA-primed cRNA synthesis very unlikely. Similarly, silencing induced by a central ß-glucuronidase (GUS) gene fragment in Nicotiana benthamiana carrying a GUS transgene spread only into downstream regions (Petersen and Albrechtsen, 2005), and secondary siRNAs originated preferentially from the 3' region of the inducer region in tobacco (Shimamura et al., 2007). However, in Caenorhabditis elegans, it has been shown that secondary siRNA has a 5' triphosphate, precluding cloning methods that rely on a single 5' phosphate (Miska and Ahringer, 2007). Thus, this species might be underrepresented in the large-scale sequencing studies. Di- and triphosphate groups at the 5' termini of secondary siRNAs also indicate that they are likely to be primary, unprimed RdRP products (Carthew and Sontheimer, 2009).

The strong but similar sense/antisense bias of primary and secondary siRNAs (Figure 7d) suggests a related formation pathway. It is conceivable that secondary siRNAs are also formed by DCL4 3' of the primary targeted *FaOMT*, as mRNA cleavage, rather than priming of RdRP by primary siRNAs, was proposed as the signal for siRNA amplification (Ghildiyal and Zamore, 2009).

Overall, in our study, primary siRNAs were three orders of magnitude more abundant than secondary siRNAs (Figure 7) confirming the observation that ihp constructs, compared with the transitive construct, generated higher frequencies (data not shown) and efficacies of loss-of-function phenotypes (Filichkin *et al.*, 2007; Figure 6).

Our results demonstrate that endogenous genes can be efficiently down-regulated in strawberry fruit by RNAi-mediated silencing after agroinfiltration of ihp constructs carrying sequences of 200 up 300 nt homologous to an endogene within 14 days. Transitive vectors are less effective probably due to lower levels of secondary siRNAs targeting the gene of interest. Since the silencing efficiency is independent of the calculated theoretical accessibilities of the nucleotide sequences used, and thus independent of the gene fragment, the method represents a versatile tool to perform functional genomics research in strawberry.

Materials and methods

Constructs

The pBI-Intron control construct was assembled as described by Hoffmann *et al.* (2006). pBI-Intron contained *GUS* separated by the first intron of the *F. × ananassa* quinone oxidoreductase gene (AY158836, nucleotides 4107–4561). The assay constructs were made by using pBI121 (Jefferson, 1987). The second intron of the *F. × ananassa* quinone oxidoreductase gene (AY158836, nucleotides 4886–4993) was PCR-amplified from strawberry *F. × ananassa* cv. Elsanta genomic DNA and cloned into BamHI–Ecl136II cut pBI121 to replace *GUS*. For cloning of the *FaCHS* and *FaOMT* silencing constructs, the genes were divided into five parts, respectively, that overlapped each other to ensure complete coverage of the full sequences (Figures 2 and 3). These parts were PCR-amplified by primers introducing restriction sites and overhanging nucleotides. After restriction digestion, the pieces were inserted into the 5' and 3' arms of the intron in sense and antisense direction allowing the formation of self-complementary ihp structures. The resulting plasmids were named pBI-*FaCHS_Ai* to pBI-*FaCHS_Ei* and pBI-*FaOMT_Ai* to pBI-*FaOMT_Ei*. To generate pBI-*GUSi*, a 414-bp fragment of *GUS* (AF485783) was inserted into the 5' and 3' arms of the intron. For pBI-*CHS-GUS* construction, a 244-bp *CHS* fragment was PCR-amplified and cloned into SnaBI-Ecl136II-cut pBI121 (Jefferson, 1987). All primer sequences including restriction sites, amplicon sizes and amplicon positions are summarized in Table S1.

Plant material

The octoploid strawberry *F. × ananassa* cv. Elsanta (Kraege Beerenobst, Telgte, Germany) was purchased as frigo plants (Kraege, Telgte, Germany) and used for transient gene experiments (Hoffmann *et al.*, 2006). Standard growing conditions were maintained at 25 °C and a 16-h photoperiod under 120 μmol/m^2/s irradiance provided by Osram Fluora lamps (München, Germany). For genetic and molecular analysis, fruits were injected 14 days after pollination and harvested 3–24 days

after injection. Fruits harvested 28 days after pollination were used as controls.

Agroinfiltration

Infiltration of Agrobacterium AGL0 (Lazo *et al.*, 1991) containing the different constructs was performed according to Hoffmann *et al.* (2006) and Spolaore *et al.* (2001).

qPCR

Total RNA was extracted from mature fruit using the cetyltrimethylammonium bromide (CTAB) extraction procedure (Asif *et al.*, 2000; Liao *et al.*, 2004). RNA samples were treated with RNase-free DNase I (Fermentas, St. Leon-Rot, Germany) for 1 h at 37 °C. First-strand cDNA synthesis was performed in duplicate in a 20 μL reaction volume, with 1 μg of total RNA as the template, random primer (random hexamer, 100 pmol) and M-MLV reverse transcriptase (200 U, Invitrogen, Karlsruhe, Germany) according to the manufacturer's instructions. Real-time PCR was performed with 2 μL of cDNA on a StepOnePlus System (Applied Biosystems, Foster City, CA) using SYBR Green PCR Master MIX (Applied Biosystems). To monitor dsDNA synthesis, data were analysed with ABI StepOne Software v2.0. Relative quantification of the *FaCHS* (AI795154) and *FaOMT* (AF220491) transcripts was performed using the *Actin* gene (AB116565) from *F. × ananassa* as a reference (Almeida *et al.*, 2007). Quantitative PCR primers can be found in Table S1, with supplemental information about amplicon lengths and positions. All reactions were run two times with two sets of cDNAs. The specificity of the PCR amplification was checked with a melting curve analysis following the final step of the PCR. For each sample, threshold cycles (Ct, cycle at which the increase of fluorescence exceeded the threshold setting) were determined. Relative gene expression data were normalized to the mean expression of *FaOMT* ($n = 17$) and *FaCHS* ($n = 23$) in native, untreated fruits (set to 1). The relative expression ratio was calculated according to Pfaffl (2001).

Metabolite analysis

Identification and quantification of strawberry fruit metabolites were performed by liquid chromatography-UV-electrospray ionization-mass spectrometry (LC-MS) as described by Hoffmann *et al.* (2006) and Griesser *et al.* (2008). Box plots of signal intensities were generated by Sigma Plot (SPSS, Chicago, IL), and statistical significance levels were calculated using the Wilcoxon–Mann–Whitney *U*-test (Hart, 2001).

Bioinformatic analysis and processing of transcriptomic data

PBi-*FaOMTi* and pBi-*FaOMT_Ci* were agroinfiltrated into ripening strawberry fruit according to Hoffmann *et al.* (2006). Total RNA was extracted from mature fruit using the CTAB extraction procedure (Asif *et al.*, 2000; Liao *et al.*, 2004). The RNA samples were treated with RNase-free DNase I (Fermentas, St. Leon-Rot, Germany) for 1 h at 37 °C and sent to Eurofins Genomics (Ebersberg, Germany), where library preparation and RNA sequencing were carried out. The library was prepared with the TruSeq(TM) SBS v5 kit (Illumina Inc.). Final cDNA libraries were purified and size-selected by capillary electrophoresis (<51 nt). Sequencing was performed on an Illumina HiSeq 2000 platform after a PhiX library (Illumina Inc.) was spiked in the channel to estimate the sequencing quality. The following software packages were used for base calling: HiSeq control software v. 1.4.8,

RTA 1.12.4.2, CASAVA 1.7.0 and OLB-1.9.0 (all Illumina Inc.). Prior to read sorting, the 3′ adaptor sequence was trimmed from the raw reads using cutadapt 0.9.3 (Martin, 2011). Sequencing (<50 nt) yielded 14 288 304 reads (OMT_short) and 21 722 563 reads (OMT_long), respectively.

Subsequent data processing was performed on the Galaxy server (Blankenberg et al., 2010b; Giardine et al., 2005; Goecks et al., 2010). The application programming interface (API) was used to set up pipeline analyses (Sloggett et al., 2013), and the Galaxy Data Manager was employed for handling built-in reference data (Blankenberg et al., 2014a). Tools were installed and maintained via Galaxy ToolShed (Blankenberg et al., 2014b). Before mapping, reads were filtered according to their quality score and length (15–50 nt; min. quality 20) with the fastq_filter v. 1.0.0 tool implemented in Galaxy (Blankenberg et al., 2010a). Overall read quality was tracked with the FastQC software by S. Andrews (http://www.bioinformatics.babraham.ac.uk/projects/fastqc/).

After quality clipping, 11 693 897 (81.84% of total reads, OMT_short) and 17 084 937 (78.65%, of total reads OMT_long) reads were subjected to the mapping against the *F. vesca* reference genome (version 2.0.a1: downloaded from Genome Database for Rosaceae GDR www.rosaceae.org; Tennessen et al., 2014). Mapping was done by TopHat read aligner v 2.0.14 (Kim et al., 2013) for single-end data in default settings. Aligned reads were counted from resulting bam files by HTSeq-count v. 0.6.0 (Anders et al., 2015) in "Union" mode for stranded reads with a minimum alignment quality of 10. The gene prediction input file was also downloaded from GDR (Tennessen et al., 2014). The coverage of small RNAs was visualized with the Integrated Genomics Viewer (igv, Robinson et al., 2011).

Modelling of mRNA structures

Modelling of the *FaCHS_Ai* to *Ei* and *FaOMT_Ai* to *Ei* mRNA structures and calculation of the theoretical accessibilities was carried out online via the *Mfold* web server for nucleic acid folding and hybridization prediction (http://unafold.rna.albany.edu/?q=mfold/RNA-Folding-Form) in default parameters as published by Zuker (2003).

Acknowledgements

We thank Robert Kurtzer and Hannelore Meckl for technical assistance.

Funding

This work was supported by the DFG (grant no. SCHW634/10-2, SCHW634/24-1) and BMBF (grant no. PLANT-KBBE, FraGE-NOMICS, 0315463).

References

Alder, M.N., Dames, S., Gaudet, J. and Mango, S.E. (2003) Gene silencing in *Caenorhabditis* by transitive RNA interference. *RNA*, **9**, 25–32.

Almeida, J.R.M., D'Amico, E., Preuss, A., Carbone, F., de Vos, C.H.R., Deiml, B., Mourgues, F. et al. (2007) Characterization of major enzymes in flavonoid and proanthocyanidin biosynthesis during fruit development in strawberry (*Fragaria* × *ananassa*). *Arch. Biochem. Biophys.* **465**, 61–71.

Anders, S., Pyl, P.T. and Huber, W. (2015) HTSeq–a Python framework to work with high-throughput sequencing data. *Bioinformatics*, **31**, 166–169.

Asif, M.H., Dhawan, P. and Nath, P. (2000) A simple procedure for the isolation of high quality RNA from ripening banana fruit. *Plant Mol. Biol. Rep.* **18**, 109–115.

Baulcombe, D. (2004) RNA silencing in plants. *Nature*, **431**, 356–363.

Blankenberg, D., Gordon, A., Von Kuster, G., Coraor, N., Taylor, J. and Nekrutenko, A. (2010a) Manipulation of FASTQ data with Galaxy. *Bioinformatics*, **26**, 1783–1785.

Blankenberg, D., Von Kuster, G., Coraor, N., Ananda, G., Lazarus, R., Mangan, M., Nekrutenko, A. et al. (2010b) Galaxy: a web-based genome analysis tool for experimentalists. *Curr. Protoc. Mol. Biol.* **89**, 19.10.1–19.10.21.

Blankenberg, D., Johnson, J.E., Taylor, J. and Nekrutenko, A. (2014a) Wrangling Galaxy's reference data. *Bioinformatics*, **30**, 1917–1919.

Blankenberg, D., Von Kuster, G., Bouvier, E., Baker, D., Afgan, E., Stoler, N., Taylor, J. et al. (2014b) Dissemination of scientific software with Galaxy ToolShed. *Genome Biol.* **15**, 403.

Bleys, A., Vermeersch, L., Van Houdt, H. and Depicker, A. (2006) The frequency and efficiency of endogene suppression by transitive silencing signals is influenced by the length of sequence homology. *Plant Physiol.* **142**, 788–796.

Braunstein, T.H., Moury, B., Johannessen, M. and Albrechtsen, M. (2002) Specific degradation of 3′ regions of GUS mRNA in posttranscriptionally silenced tobacco lines may be related to 5′–3′ spreading of silencing. *RNA*, **8**, 1034–1044.

Brodersen, P. and Voinnet, O. (2006) The diversity of RNA silencing pathways in plants. *Trends Genet.* **22**, 268–280.

Carthew, R.W. and Sontheimer, E.J. (2009) Origins and mechanisms of miRNAs and siRNAs. *Cell*, **136**, 642–655.

Chávez Montes, R.A., De Fátima Rosas-Cárdenas, F., De Paoli, E., Accerbi, M., Rymarquis, L.A., Mahalingam, G., Marsch-Martínez, N. et al. (2014) Sample sequencing of vascular plants demonstrates widespread conservation and divergence of microRNAs. *Nat. Commun.* **5**, 3722.

Chuang, C.F. and Meyerowitz, E.M. (2000) Specific and heritable genetic interference by double-stranded RNA in *Arabidopsis thaliana*. *Proc. Natl Acad. Sci. USA*, **97**, 4985–4990.

Eamens, A., Wang, M.-B., Smith, N.A. and Waterhouse, P.M. (2008) RNA silencing in plants: yesterday, and tomorrow. *Plant Physiol.* **147**, 356–468.

Fang, X. and Qi, Y. (2016) RNAi in plants: an argonaute-centered view. *Plant Cell*, **28**, 272–285.

Filichkin, S.A., DiFazio, S.P., Brunner, A.M., Davis, J.M., Yang, Z.K., Kalluri, U.C., Arias, R.S. et al. (2007) Efficiency of gene silencing in Arabidopsis: direct inverted repeats vs. transitive RNAi vectors. *Plant Biotech. J.* **5**, 615–626.

Fusaro, A.F., Matthew, L., Smith, N.A., Curtin, S.J., Dedic-Hagan, J., Ellacott, G.A., Watson, J.M. et al. (2006) RNA interference-inducing hairpin RNAs in plants act through the viral defence pathway. *EMBO Rep.* **7**, 1168–1175.

Ghildiyal, M. and Zamore, P.D. (2009) Small silencing RNAs: an expanding universe. *Nat. Rev. Genet.* **10**, 94–108.

Giardine, B., Riemer, C., Hardison, R.C., Burhans, R., Elnitski, L., Shah, P., Zhang, Y. et al. (2005) Galaxy: a platform for interactive large-scale genome analysis. *Genome Res.* **15**, 1451–1455.

Gilchrist, E. and Haughn, G. (2010) Reverse genetics techniques: engineering loss and gain of gene function in plants. *Brief. Funct. Genomic. Proteomic.* **9**, 103–110.

Goecks, J., Nekrutenko, A. and Taylor, J. (2010) Galaxy: a comprehensive approach for supporting accessible, reproducible, and transparent computational research in the life sciences. *Genome Biol.* **11**, R86.

Griesser, M., Hoffmann, T., Bellido, M.L., Rosati, C., Fink, B., Kurtzer, R., Aharoni, A. et al. (2008) Redirection of flavonoid biosynthesis through the downregulation of an anthocyanidin glucosyltransferase in ripening strawberry (*Fragaria* × *ananassa*) fruit. *Plant Physiol.* **146**, 1528–1539.

Hanhineva, K., Kokko, H., Siljanen, H., Rogachev, I., Aharoni, A. and Kärenlampi, S.O. (2009) Stilbene synthase gene transfer caused alterations in the phenylpropanoid metabolism of transgenic strawberry (*Fragaria* × *ananassa*). *J. Exp. Bot.* **60**, 2093–2106.

Hart, A. (2001) Mann-Whitney test is not just a test of medians: differences in spread can be important. *Br. Med. J.* **323**, 391–393.

Himber, C., Dunoyer, P., Moissiard, G., Ritzenthaler, C. and Voinnet, O. (2003) Transitivity-dependent and –independent cell-to-cell movement of RNA silencing. *EMBO J.* **22**, 4523–4533.

Hoffmann, T., Kalinowski, G. and Schwab, W. (2006) RNAi-induced silencing of gene expression in strawberry fruit (*Fragaria × ananassa*) by agroinfiltration: a rapid assay for gene function analysis. *Plant J.* **48**, 818–826.

Jefferson, A.R. (1987) Assaying chimeric genes in plants: the GUS gene fusion system. *Plant Mol. Biol. Rep.* **5**, 387–405.

Kim, D., Pertea, G., Trapnell, C., Pimentel, H., Kelley, R. and Salzberg, S.L. (2013) TopHat2: accurate alignment of transcriptomes in the presence of insertions, deletions and gene fusions. *Genome Biol.* **14**, R36.

Klahre, U., Crété, P., Leuenberger, S.A., Iglesias, V.A. and Meins, F. (2002) High molecular weight RNAs and small interfering RNAs induce systemic posttranscriptional gene silencing in plants. *Proc. Natl Acad. Sci. USA*, **99**, 11981–11986.

Kościańska, E., Kalantidis, K., Wypijewski, K., Sadowski, J. and Tabler, M. (2005) Analysis of RNA silencing in agroinfiltrated leaves of *Nicotiana benthamiana* and *Nicotiana tabacum*. *Plant Mol. Biol.* **59**, 647–661.

Kurreck, J. (2006) siRNA efficiency: structure or sequence – that is the question. *J. Biomed. Biotechnol.* **4**, 1–7 Article ID 83757.

Kusaba, M. (2004) RNA interference in crop plants. *Curr. Opin. Biotechnol.* **15**, 139–143.

Lazo, G.R., Pascal, A.S. and Ludwig, R.A. (1991) A DNA transformation-competent *Arabidopsis* genomic library in *Agrobacterium*. *Biotechnology*, **9**, 963–967.

Liao, Z., Chen, M., Guo, L., Gong, Y., Tang, F., Sun, X. and Tang, K. (2004) Rapid isolation of high-quality total RNA from *Taxus* and *Ginkgo*. *Prep. Biochem. Biotechnol.* **34**, 209–214.

Lipardi, C., Wei, Q. and Paterson, B.M. (2003) RNA silencing in Drosophila. *Acta Histochem. Cytochem.* **36**, 123–134.

Llave, C., Kasschau, K.D., Rector, M.A. and Carrington, J.C. (2002) Endogenous and silencing-associated small RNAs in plants. *Plant Cell*, **14**, 1605–1619.

Lunkenbein, S., Coiner, H., De Vos, C.H., Schaart, J.G., Boone, M.J., Krens, F.A., Schwab, W. *et al.* (2006a) Molecular characterization of a stable antisense chalcone synthase phenotype in strawberry (*Fragaria × ananassa*). *J. Agric. Food Chem.* **54**, 2145–2153.

Lunkenbein, S., Salentijn, E.M.J., Coiner, H.A., Boone, M.J., Krens, F.A. and Schwab, W. (2006b) Up- and down-regulation of *Fragaria × ananassa* O-methyltransferase: impacts on furanone and phenylpropanoid metabolism. *J. Exp. Bot.* **57**, 2445–2453.

Makeyev, E.V. and Bamford, D.H. (2002) Cellular RNA-dependent RNA polymerase involved in posttranscriptional gene silencing has two distinct activity modes. *Mol. Cell*, **10**, 1417–1427.

Manavella, P.A., Koenig, D. and Weigel, D. (2012) Plant secondary siRNA production determined by microRNA-duplex structure. *Proc. Natl Acad. Sci. USA*, **109**, 2461–2466.

Martin, M. (2011) Cutadapt removes adapter sequences from high-throughput sequencing reads. *EMBnet. J.* **17**, 10–12.

Matthew, L. (2004) RNAi for plant functional genomics. *Comp. Funct. Genomics*, **5**, 240–244.

Mi, S., Cai, T., Hu, Y., Chen, Y., Hodges, E., Ni, F., Wu, L. *et al.* (2008) Sorting of small RNAs into *Arabidopsis* argonaute complexes is directed by the 5′ terminal nucleotide. *Cell*, **133**, 116–127.

Miki, D., Itoh, R. and Shimamoto, K. (2005) RNA silencing of single and multiple members in a gene family of rice. *Plant Physiol.* **138**, 1903–1913.

Miska, E.A. and Ahringer, J. (2007) RNA interference has second helpings. *Nature Biotechnol.* **25**, 302–303.

Mlotshwa, S., Pruss, G.J., Peragine, A., Endres, M.W., Li, J., Chen, X., Poethig, R.S. *et al.* (2008) Dicer-LIKE2 plays a primary role in transitive silencing of transgenes in Arabidopsis. *PLoS ONE*, **3**, e1755.

Molnár, A., Csorba, T., Lakatos, L., Várallyay, E., Lacomme, C. and Burgyán, J. (2005) Plant virus-derived small interfering RNAs originate predominantly from highly structured single-stranded viral RNAs. *J. Virol.* **79**, 7812–7818.

Motamedi, M.R., Verdel, A., Colmenares, S.U., Gerber, S.A., Gygi, S.P. and Moazed, D. (2004) Two RNAi complexes, RITS and RDRC, physically interact and localize to noncoding centromeric RNAs. *Cell*, **119**, 789–802.

Noma, K., Sugiyama, T., Cam, H., Verdel, A., Zofall, M., Jia, S., Moazed, D. *et al.* (2004) RITS acts in *cis* to promote RNA interference–mediated transcriptional and post-transcriptional silencing. *Nature Genet.* **36**, 1174–1180.

Osabe, K., Mudge, S.R., Graham, M.W. and Birch, R.G. (2009) RNAi mediated down-regulation of PDS gene expression in sugarcane (*Saccharum*), a highly polyploidy crop. *Trop. Plant Biol.* **2**, 143–148.

Overhoff, M., Alken, M., Far, R.K.-K., Lemaitre, M., Lebleu, B., Sczakiel, G. and Robbins, I. (2005) Local RNA target structure influences siRNA efficacy: a systematic global analysis. *J. Mol. Biol.* **348**, 871–881.

Petersen, B.O. and Albrechtsen, M. (2005) Evidence implying only unprimed RdRP activity during transitive gene silencing in plants. *Plant Mol. Biol.* **58**, 575–583.

Pfaffl, M.W. (2001) A new mathematical model for relative quantification in real-time RT-PCR. *Nucleic Acids Res.* **29**, 2002–2007.

Rajeswaran, R., Aregger, M., Zvereva, A.S., Borah, B.K., Gubaeva, E.G. and Pooggin, M.M. (2012) Sequencing of RDR6-dependent double-stranded RNAs reveals novel features of plant siRNA biogenesis. *Nucleic Acids Res.* **40**, 6241–6254.

Robinson, J.T., Thorvaldsdóttir, H., Winckler, W., Guttman, M., Lander, E.S., Getz, G. and Mesirov, J.P. (2011) Integrative genomics viewer. *Nat. Biotechnol.* **29**, 24–26.

Sanders, M., Maddelein, W., Depicker, A., Van Montagu, M., Cornelissen, M. and Jacobs, J. (2002) An active role for endogenous ß-1,3-glucanase genes in transgene-mediated co-suppression in tobacco. *EMBO J.* **21**, 5824–5832.

Schubert, S., Grünweller, A., Erdmann, V.A. and Kurreck, J. (2005) Local RNA target structure influences siRNA efficacy: systematic analysis of intentionally designed binding regions. *J. Mol. Biol.* **348**, 883–893.

Schwach, F., Vaistij, F.E., Jones, L. and Baulcombe, D.C. (2005) An RNA-dependent RNA polymerase prevents meristem invasion by potato virus X and is required for the activity but not the production of a systemic silencing signal. *Plant Physiol.* **138**, 1842–1852.

Shimamura, K., Oka, S.-I., Shimotori, Y., Ohmori, T. and Kodama, H. (2007) Generation of secondary small interfering RNA in cell-autonomous and non-cell autonomous RNA silencing in tobacco. *Plant Mol. Biol.* **63**, 803–813.

Sijen, T., Fleenor, J., Simmer, F., Thijssen, K.L., Parrish, S., Timmons, L., Plasterk, R.H. *et al.* (2001) On the role of RNA amplification in dsRNA-triggered gene silencing. *Cell*, **107**, 465–476.

Sloggett, C., Goonasekera, N. and Afgan, E. (2013) BioBlend: automating pipeline analyses within Galaxy and CloudMan. *Bioinformatics*, **29**, 1685–1686.

Small, I. (2007) RNAi for revealing and engineering plant gene functions. *Curr. Opin. Biotechnol.* **18**, 148–153.

Song, C., Ring, L., Hoffmann, T., Huang, F.-C., Slovin, J. and Schwab, W. (2015) Acylphloroglucinol biosynthesis in strawberry fruit. *Plant Physiol.* **169**, 1656–1670.

Spolaore, S., Trainotti, L. and Casadoro, G. (2001) A simple protocol for transient gene expression in ripe fleshy fruit mediated by Agrobacterium. *J. Exp. Bot.* **52**, 845–850.

Tennessen, J.A., Govindarajulu, R., Ashman, T. and Liston, A. (2014) Evolutionary origins and dynamics of octoploid strawberry subgenomes revealed by dense targeted capture linkage maps. *Genome Biol. Evol.* **6**, 3295–3313.

Vaistij, F.E., Jones, L. and Baulcombe, D.C. (2002) Spreading of RNA targeting and DNA methylation in RNA silencing requires transcription of the target gene and a putative RNA-dependent RNA polymerase. *Plant Cell*, **14**, 857–867.

Van Houdt, H., Bleys, A. and Depicker, A. (2003) RNA target sequences promote spreading of RNA silencing. *Plant Physiol.* **131**, 245–253.

Vazquez, F. and Hohn, T. (2013) Biogenesis and biological activity of secondary siRNAs in plants. *Scientifica*, Article ID783253 http://dx.doi.org/10.1155/2013/783253.

Vermeulen, A., Behlen, L., Reynolds, A., Wolfson, A., Marshall, W.S., Karpilow, J. and Khvorova, A. (2005) The contributions of dsRNA structure to Dicer specificity and efficiency. *RNA*, **11**, 674–682.

Wang, M.-B. and Metzlaff, M. (2005) RNA silencing and antiviral defense in plants. *Curr. Opin. Plant Biol.* **8**, 216–222.

Wang, M., Wang, G. and Ji, J. (2010) Suppression of the phytoene desaturase gene influence on the organization and function of phytosystem II (PSII) and antioxidant enzyme activities in tobacco. *Environ. Exp. Bot.* **67**, 460–466.

Wein, M., Lavid, N., Lunkenbein, S., Lewinsohn, E., Schwab, W. and Kaldenhoff, R. (2002) Isolation, cloning and expression of a multifunctional O-methyltransferase capable of forming 2,5-dimethyl-4-methoxy-3(2H)-

furanone, one of the key aroma compounds in strawberry fruits. *Plant J.* **31**, 755–765.

Wu, J., Yang, Z., Wang, Y., Zheng, L., Ye, R., Ji, Y., Zhao, S. *et al.* (2015) Viral-inducible Argonaute18 confers broad-spectrum virus resistance in rice by sequestering a host microRNA. *eLife*, **4**, e05733.

Zhai, Z., Sooksa-nguan, T. and Vatamaniuk, O.K. (2009) Establishing RNA interference as a reverse-genetic approach for gene functional analysis in protoplasts. *Plant Physiol.* **149**, 642–652.

Zorrilla-Fontanesi, Y., Rambla, J.L., Cabeza, A., Medina, J.J., Sánchez-Sevilla, J.F., Valpuesta, V., Botella, M.A. *et al.* (2012) Genetic analysis of strawberry fruit aroma and identification of O-methyltransferase FaOMT as the locus controlling natural variation in mesifurane content. *Plant Physiol.* **159**, 851–870.

Zuker, M. (2003) Mfold web server for nucleic acid folding and hybridization prediction. *Nucleic Acids Res.* **31**, 3406–3415.

PERMISSIONS

LIST OF CONTRIBUTORS

Weijuan Fan, Hongxia Wang, Yinliang Wu, Nan Yang and Peng Zhang
National Key Laboratory of Plant Molecular Genetics, CAS Center for Excellence in Molecular Plant Sciences, Institute of Plant Physiology and Ecology, Shanghai Institutes for Biological Sciences, Chinese Academy of Sciences, Shanghai, China

Jun Yang
Shanghai Key Laboratory of Plant Functional Genomics and Resources, Shanghai Chenshan Plant Science Research Center, Chinese Academy of Sciences, Shanghai Chenshan Botanical Garden, Shanghai, China

Chen Klap, Ester Yeshayahou, Tzahi Arazi, Suresh K. Gupta, Sara Shabtai, Yehiam Salts and Rivka Barg
The Institute of Plant Sciences, The Volcani Center, Agricultural Research Organization, Rishon LeZion, Israel

Anthony M. Bolger
Institut für Biologie I, RWTH Aachen, Aachen, Germany

Björn Usadel
Institut fur Biologie I, RWTH Aachen, Aachen, Germany
Institut für Bio-und Geowissenschaften 2 (IBG-2) Plant Sciences, Forschungszentrum Jülich, Jülich, Germany

Dongfa Sun
College of Plant Science and Technology, Huazhong Agricultural University, Wuhan, Hubei, China

Zhigang Li
College of Plant Science and Technology, Huazhong Agricultural University, Wuhan, Hubei, China
Department of Genetics and Biochemistry, Clemson University, Clemson, SC, USA

Man Zhou, Ning Yuan, Peipei Wu, Qian Hu, Shuangrong Yuan and Hong Luo
Department of Genetics and Biochemistry, Clemson University, Clemson, SC, USA

Haiyan Jia
Department of Genetics and Biochemistry, Clemson University, Clemson, SC, USA

The Applied Plant Genomics Laboratory of Crop Genomics and Bioinformatics Centre, and National Key Laboratory of Crop Genetics and Germplasm Enhancement, Nanjing Agricultural University, Nanjing, Jiangsu, China

Fangyuan Gao
Department of Genetics and Biochemistry, Clemson University, Clemson, SC, USA
Crop Research Institute, Sichuan Academy of Agricultural Sciences, Chengdu, Sichuan, China

Jean-Yves Paul, Harjeet Khanna, Jennifer Kleidon, Phuong Hoang, Jason Geijskes, Ella Zaplin, Anthony James, Bulukani Mlalazi, Pradeep Deo, Douglas Becker, James Tindamanyire, Robert Harding and James Dale
Centre for Tropical Crops and Biocommodities, Queensl and University of Technology, Brisbane, Qld, Australia

Jeff Daniells
Agri-Science Queensl and Department of Agriculture and Fisheries, South Johnstone, Qld, Australia

Yvonne Rosenberg
PlantVax Inc, Rockville, MD, USA

Geofrey Arinaitwe and Wilberforce Tushemereirwe
National Agricultural Research Laboratories, National Agricultural Research Organization, Kampala, Uganda

Priver Namanya
Centre for Tropical Crops and Biocommodities, Queensl and University of Technology, Brisbane, Qld, Australia
National Agricultural Research Laboratories, National Agricultural Research Organization, Kampala, Uganda

Fernando Pérez-Martín, Fernando J. Yuste-Lisbona, Estela Giménez, Antonia Fernández-Lozano, Ana Ortíz-Atienza, Manuel García-Alcázar, Laura Castañeda, Rocío Fonseca, Carmen Capel, Jorge L. Quispe, Juan Capel, Trinidad Angosto and Rafael Lozano
Centro de Investigación en Biotecnología Agroalimentaria (BITAL), Universidad de Almería, Almería, Spain

Benito Pineda, María Pilar Angarita-Díaz, Begoña García-Sogo, Teresa Antón, Sibilla Sánchez, Alejandro Atarés, Geraldine Goergen, Jorge Sánchez and Vicente Moreno
Instituto de Biología Molecular y Celular de Plantas (UPV-CSIC), Universidad Politécnica de Valencia, Valencia, Spain

Michael Gomez Selvaraj, Milton Valencia, Beata Dedicova and Manabu Ishitani
International Center for Tropical Agriculture (CIAT), Cali, Colombia

Satoshi Ogawa
International Center for Tropical Agriculture (CIAT), Cali, Colombia
Japan Society for the Promotion of Science, The University of Tokyo, Bunkyo-ku, Tokyo, Japan

Takuma Ishizaki
Tropical Agriculture Research Front (TARF), Japan International Research Center for Agricultural Sciences (JIRCAS), Ishigaki, Okinawa, Japan

Takuya Ogata, Kyouko Yoshiwara, Kyonoshin Maruyama and Kazuo Nakashima
Biological Resources and Post-harvest Division, Japan International Research Center for Agricultural Sciences (JIRCAS), Tsukuba, Ibaraki, Japan

Fuminori Takahashi and Kazuo Shinozaki
RIKEN Center for Sustainable Resource Science, Yokohama, Kanagawa, Japan
RIKEN Center for Sustainable Resource Science, Tsukuba, Ibaraki, Japan

Miyako Kusano
RIKEN Center for Sustainable Resource Science, Yokohama, Kanagawa, Japan
RIKEN Center for Sustainable Resource Science, Tsukuba, Ibaraki, Japan
Graduate School of Life and Environmental Sciences, University of Tsukuba, Tsukuba, Ibaraki, Japan

Kazuki Saito
RIKEN Center for Sustainable Resource Science, Yokohama, Kanagawa, Japan
RIKEN Center for Sustainable Resource Science, Tsukuba, Ibaraki, Japan
Department of Molecular Biology and Biotechnology, Graduate School of Pharmaceutical Sciences, Chiba University, Chiba, Japan

Vikas K. Singh, Aamir W. Khan, Rachit K. Saxena, Pallavi Sinha, Sandip M. Kale, Swathi Parupalli, Vinay, Sameer Kumar and Mamta Sharma
International Crops Research Institute for the Semi-Arid Tropics, Patancheru, Telangana State, India

Anuradha Ghanta and Kalinati Narasimhan Yamini
Agricultural Research Station (ARS)-Tandur, Professor Jayashankar Telangana State Agricultural University (PJTSAU), Hyderabad, Telangana State, India

Sonnappa Muniswamy
Agricultural Research Station (ARS)-Gulbarga, University of Agricultural Sciences (UAS), Raichur, Karnataka, India

Rajeev K. Varshney
International Crops Research Institute for the Semi-Arid Tropics, Patancheru, Telangana State, India
School of Plant Biology and Institute of Agriculture, The University of Western Australia, Crawley, WA, Australia

Anna Stein, Birgit Samans, Sarah V. Schiessl, Christian Obermeier and Rod J. Snowdon
Department of Plant Breeding, IFZ Research Centre for Biosystems, Land Use and Nutrition, Justus Liebig University, Giessen, Germany

Olivier Coriton, Mathieu Rousseau-Gueutin, Anne-Marie Chèvre
IGEPP, INRA, Agrocampus Ouest, Université de Rennes 1, Le Rheu, France

Isobel A.P. Parkin
Agriculture and Agri-Food Canada, Saskatoon, Canada

Laura Stoffels, Henry N. Taunt and Saul Purton
Algal Biotechnology Group, Institute of Structural and Molecular Biology, University College London, London, UK

Bambos Charalambous
Research Department of Infection, University College London Medical School, London, UK

Jie-Li Wang, Sheng Chen, Xiang-Feng Zheng, Zheng Wang, Ke-Ming Zhu, Li-Na Ding and Xiao-Li Tan
Institute of Life Sciences, Jiangsu University, Zhenjiang, China

Min-Qiang Tang and Sheng-Yi Liu
The Oil Crops Research Institute (OCRI) of the Chinese Academy of Agricultural Sciences (CAAS), Wuhan, China

Hui-Xian Mo, Sheng-Jun Li and Yun-Hai Li
State Key Laboratory of Plant Cell and Chromosome Engineering, Institute of Genetics and Developmental Biology (IGDB), Chinese Academy of Sciences (CAS), Beijing, China

Qing-gang Zhu and Chu-li Deng
Zhejiang Provincial Key Laboratory of Horticultural Plant Integrative Biology, Zhejiang University, Hangzhou, China

Xue-ren Yin, Kun-song Chen and Miao-miao Wang
Zhejiang Provincial Key Laboratory of Horticultural Plant Integrative Biology, Zhejiang University, Hangzhou, China
The State Agriculture Ministry Laboratory of Horticultural Plant Growth, Development and Quality Improvement, Zhejiang University, Hangzhou, China

Zheng-rong Luo
Key Laboratory of Horticultural Plant Biology, Ministry of Education, Huazhong Agricultural University, Wuhan, China

Ning-jing Sun
Department of Horticultural Sciences, College of Agriculture, Guangxi University, Nanning, China

Donald Grierson
Zhejiang Provincial Key Laboratory of Horticultural Plant Integrative Biology, Zhejiang University, Hangzhou, China
Plant & Crop Sciences Division, School of Biosciences, University of Nottingham, Loughborough, UK

Baogen Wang, Xiaohua Wu, Yaowen Hu, Wen Zhou, Zhongfu Lu, Xinyi Wu and Xinyi Wu
Institute of Vegetables, Zhejiang Academy of Agricultural Sciences, Hangzhou, China

Guojing Li and Pei Xu
Institute of Vegetables, Zhejiang Academy of Agricultural Sciences, Hangzhou, China
State Key Lab Breeding Base for Sustainable Control of Plant Pest and Disease, Zhejiang Academy of Agricultural Sciences, Hangzhou, China

Timothy J. Close and María Muñoz-Amatriaín
Department of Botany and Plant Sciences, University of California-Riverside, Riverside, CA, USA

Bao-Lam Huynh and Philip A. Roberts
Department of Nematology, University of California-Riverside, Riverside, CA, USA

Olga Yurchenko and John M. Dyer
USDA-ARS, US Arid-Land Agricultural Research Center, Maricopa, AZ, USA

Jay M. Shockey
USDA-ARS, Southern Regional Research Center, New Orleans, LA, USA

Satinder K. Gidda, Maxwell I. Silver and Robert T. Mullen
Department of Molecular and Cellular Biology, University of Guelph, Guelph, ON, Canada

Kent D. Chapman
Department of Biological Sciences, University of North Texas, Denton, TX, USA

Wen Zhi Jiang, Edgar B. Cahoon and Donald P. Weeks
Department of Biochemistry and Center for Plant Science Innovation, University of Nebraska, Lincoln, NE, USA

Isabelle M. Henry, Peter G. Lynagh and Luca Comai
Department of Plant Biology and UC Davis Genome Center, University of California, Davis, CA, USA

Hongge Jia, Yunzeng Zhang, Jin Xu and Nian Wang
Citrus Research and Education Center, Department of Microbiology and Cell Science, Institute of Food and Agricultural Sciences (IFAS), University of Florida, Lake Alfred, FL, USA

Vladimir Orbović
Citrus Research and Education Center, IFAS, University of Florida, Lake Alfred, FL, USA

Frank F. White and Jeffrey B. Jones
Department of Plant Pathology, IFAS, University of Florida, Gainesville, FL, USA

Wei Huang, Shiyuan Peng, Zhiqiang Xian, Dongbo Lin, Guojian Hu, Lu Yang, Maozhi Ren and Zhengguo Li
Genetic Engineering Research Center, School of Life Sciences, Chongqing University, Chongqing, China

Katja Härtl, Gregor Kalinowski, Thomas Hoffmann, Anja Preuss and Wilfried Schwab
Biotechnology of Natural Products, Technische Universität München, Freising, Germany

Index

www.ingramcontent.com/pod-product-compliance
Lightning Source LLC
Chambersburg PA
CBHW070155240326

41458CB00126B/4950